Food Biotechnology

Food Biotechnology

Editor: Donald Nash

Callisto Reference
www.callistoreference.com

Callisto Reference,
118-35 Queens Blvd., Suite 400,
Forest Hills, NY 11375, USA

Visit us on the World Wide Web at:
www.callistoreference.com

ISBN: 978-1-63239-945-8 (Hardback)

Printed in China.

Cataloging-in-Publication Data

Food biotechnology / edited by Donald Nash.
p. cm.
Includes bibliographical references and index.
ISBN 978-1-63239-945-8
1. Food--Biotechnology. I. Nash, Donald.
TP248.65.F66 F66 2018
664.024--dc23

Table of Contents

Preface ... **IX**

Chapter 1 **Using of Combined Treatment between Edible Coatings Containing Ethanolic Extract of Papaya (*Carica Papaya* L.) Leaves and Gamma Irradiation for Extending Shelf-Life of Minced Chicken Meat** ... **1**
Abdeldaiem M. H.

Chapter 2 **Evaluation of the Effect of Packaging Materials and Storage Temperatures on Quality Degradation of Extra Virgin Olive Oil from Olives Grown in Palestine** ... **12**
Jehad Abbadi, Ibrahim Afaneh, Ziad Ayyad, Fuad Al-Rimawi, Wadie Sultan, Khalid Kanaan

Chapter 3 **Combining Different Types of Prebiotic Plant Isolates Toward, Enhancing the Growth of Probiotic Organisms** ... **25**
Fathima Hajara Aslam, Jagath Wansapala, Manel Perera

Chapter 4 **Influence of Chemical Properties of Wheat-Lupine Flour Blends on Cake Quality** ... **32**
Abdelrahman R. Ahmed

Chapter 5 **Vinification By-products and their Phenolic Compounds** ... **41**
Milene Teixeira Barcia, Paula Becker Pertuzatti, Vivian Caetano Bochi, Isidro Hermosín-Gutiérrez, Helena Teixeira Godoy

Chapter 6 **Preparation and Characterization of Microwave-made Oat and Fruit Cupcakes** ... **47**
Paula Becker Pertuzatti, Denise Soares Dorneles de Oliveira, Kelly Cristina Campagnolo Port, Larissa Alves Vieira, Mariana Bento da Cruz, Gabrielle Lázara Ribeiro Rodrigues de Barros Vinhal

Chapter 7 **Effects of Pre-treatments and Drying Methods on Chemical Composition, Microbial and Sensory Quality of Orange-Fleshed Sweet Potato Flour and Porridge** ... **51**
Fana Haile, Shimelis Admassu, Abrehet Fisseha

Chapter 8 **Constraints and Compliances of Traceability in Low Grown Orthodox Black Tea Manufacturing Process** ... **58**
ChandanaVindika Kumara Lokunarangodage, Indira Wickramasinghe, Kamburawala Kankanamge Don Somathilaka Ranaweera

Chapter 9 **Nutritional and Functional Characterization of Undecorticated Groundnut (*Arachis hypogaea* L.) Seeds from Bosso Market, Minna, Nigeria** ... **66**
Mustapha S., Mohammed U. M., Adeosun N. O., Mathew T. J., Muhammed S. S., Ibn-Aliyu A

Chapter 10 **Biotransformation in Temperate Climate Fruit: A Focus on Berries** ... **72**
Paula Becker Pertuzatti, Milene Teixeira Barcia

Chapter 11 **Effect of Honey and Sugar Solution on the Shelf Life and Quality of Dried Banana (*Musa paradisiaca*) Slices** 78
Md. Sultan Mahomud, Md. Keramot Ali, Md. Mostafijur Rahman, Md. Hafijur Rahman, Tajnuba Sharmin, Md. Jiaur Rahman

Chapter 12 **Amino Acid and Fatty Acid Profile of Twenty Wild Plants used as Spices in Cameroon** 85
Armand Abdou Bouba, Roger Ponka, Goudoum Augustin, Nicolas Njintang Yanou, Mehanni Abul-Hamd El-Sayed, Didier Montet, Joël Scher, Carl Moses Mbofung

Chapter 13 **Effect of Some Thermal Processing Techniques on the Anti- nutritional Factors of *Canavalia plagiosperma* Piper Seeds** 94
Serah O. Alagbaoso, Justina N. Nwosu, Ijeoma .M. Agunwa, Njideka E. Njoku, Linda O. Akajiaku, Chioma N. Eluchie

Chapter 14 **Effect of Phytol on Dehydrogenase Activity of Bacterial Isolates from Grilled Meat** 101
Eluchie, Oranusi, Akujobi, Alagbaoso

Chapter 15 **Effect of Wrapping Materials on Physico-chemical and Microbiological Qualities of Fermented Melon Seed (*Citrullus colocynthis* L.) used as Condiment** 107
Anne I. Peter-Ikechukwu, Ngozi O. Kabuo, Serah O. Alagbaoso, Njideka E. Njoku, Chioma N. Eluchie., Williams O. Momoh

Chapter 16 **Development and Validation of Multi-Residue Analysis of 82 Pesticides in Grapes and Pomegranate as per the Requirements of the European Union (EU) and Codex Alimentarius using GC-MS/MS with Compound based Screening** 113
Gouri Satpathy, Yogesh Kumar Tyagi, Rajinder Kumar Gupta

Chapter 17 **Utilization of Pulse Processing Waste (*Cajanus cajan* Husk) for Developing Metal Adsorbent: A Value-added Exploitation of Food Industry Waste** 122
Vishal. R. Parate, Mohammed. I. Talib

Chapter 18 **Extraction and Preservation of Cashew Juice using Sorbic and Benzoic Acids** 131
Ngozi O .Kabuo, Moses Ojukwu, Gloria C Omeire, J. C Ibeabuchi

Chapter 19 **Some Physicochemical Characteristics and Storage Stability of Crude Palm Oils (*Elaeis guineensis* Jacq)** 138
Goudoum Augustin, Makambeu Ndomche Anne, Abdou Bouba Armand, Mbofung Carl Moses

Chapter 20 **Modeling and Evaluation of Thermal Diffusivity and Activation Energy of Potato slices in Forced Convection Multi Tray Solar Dryer** 144
Srivastava A. K, Shukla S. K, Singh U. K

Chapter 21 **Some Quality and Micro-structural Characteristics of Soup Enriched with Debittered *Moringa Oleifera* Seeds Flour** 150
Radha C., Ogunsina B. S., Hebina Babu K. T.

Chapter 22 **Diffusion Coefficient Estimation in Shrinking Solids. A Case Study: Tomato** 155
C. Martínez-Vera, I. Anaya-Sosa, M. G. Vizcarra-Mendoza

Chapter 23 **Physicochemical, Structural and Rheological Properties of Chestnut (*Castanea sativa*) Starch** 160
Marcio Schmiele, Georgia Ane Raquel Sehn, Valéria da Silva Santos, Thaís de Souza Rocha, Eveline Lopes Almeida, Elizabeth Harumi Nabeshima, Yoon Kil Chang, Caroline Joy Steel

Chapter 24 **Optimisation of Kernels Preparation Conditions Involved in the Press Extraction of Shea (*Vitellaria paradoxa* Gaertner F.) Butter** 167
Mohagir A. M., Bup N. D, Abi C. F., Kamga R., Kapseu C.

Chapter 25 **Effect of Wheat Bran Supplement with Sugarcane Bagasse on Growth, Yield and Proximate Composition of Pink Oyster Mushroom (*Pleurotus djamor*)** 175
M. T. Hasan, M. H. A. Khatun, M. A. M. Sajib, M. M. Rahman, M. S. Rahman, M. Roy, M. N. Miah, K. U. Ahmed

Chapter 26 **Effect of Preservation Methods of Oil Palm (*Elaeis guineensis*) Sap on Health Status of Male Wistar Rats** 183
Theophilus Maduabuchukwu Ikegwu, Gabriel Ifeanyi Okafor, Izuchukwu Shedrach Ochiogu

Chapter 27 **Assessment of the Effects of Microbial Fermentation on Selected Anti-Nutrients in the Products of Four Local Cassava Varieties from Niger State, Nigeria** 191
Etsuyankpa M. B., Gimba C. E., Agbaji E. B., Omoniyi K. I., M.M. Ndamitso, Mathew J. T.

Chapter 28 **Proximate, Anti-Nutritional and Phytochemical Composition of the Yellow Variety of the *Synsepalum Dulcificum* (Miracle Fruit) Berry** 199
Njoku N. E., Ubbaonu C. N., Alagbaoso S. O., Agunwa I. M, Eluchie C. N.

Chapter 29 **Adoptive Trials on the Yam Pounding Machine for Cowpea Dehulling** 206
Sanni L. A., Ogunsina B. S., Adegbenjo A. O., Ajayi M. O.

Chapter 30 **Evaluation of Some Chemical Composition, Minerals Fatty Acid Profiles, Antioxidant and Antimicrobial Activities of Tulsi (*Ocimum sanctum*)** 212
Sandip I. Vidhani, Vijay G. Vyas, Heena J. Parmar, Viren M. Bhalani, Mohammad M. Hassan, Ahmed Gaber, Baljibhai A. Golakiya

Chapter 31 **Proximate Composition, Fatty Acid and Mineral Contents of Four Freshwater Fish from Maga Lake (Far North Region of Cameroon)** 218
Noël Tenyang, Roger Ponka, Bernard Tiencheu, Fabrice Tonfack Djikeng, Hilaire Macaire Womeni

Chapter 32 **Kithul Flour (*Caryota urens*) as a Potential Flour Source for Food Industry** 224
J A A C Wijesinghe, I. Wicramasinghe, K. H Saranandha

Chapter 33 **Effect of Starch Type on the Physicochemical and Textural Properties of Beef Patties Formulated with Local Spices** 233
P. D. Mbougueng, D. Tenin, C. Tchiégang, J. Scher

Permissions

List of Contributors

Index

Preface

Food science is the study of food and it deals with the principles underlying food processing and the ways in which this field can be improved. Food engineering is the industrial process that is used in manufacturing food. It is the application of science and engineering to the biochemical and physical nature of food items and the methods of food processing to make it more effective and efficient. Different approaches, evaluations, methodologies and advanced studies on food science and engineering have been included in this book. It aims to serve as a resource guide for students and experts alike and contribute to the growth of discipline.

Significant researches are present in this book. Intensive efforts have been employed by authors to make this book an outstanding discourse. This book contains the enlightening chapters which have been written on the basis of significant researches done by the experts.

Finally, I would also like to thank all the members involved in this book for being a team and meeting all the deadlines for the submission of their respective works. I would also like to thank my friends and family for being supportive in my efforts.

Editor

1

Using of Combined Treatment between Edible Coatings Containing Ethanolic Extract of Papaya (*Carica Papaya* L.) Leaves and Gamma Irradiation for Extending Shelf-Life of Minced Chicken Meat

Abdeldaiem M. H.*

Atomic Energy Authority, Nuclear Research Center, Egypt
*Corresponding author: abdeldaiem2015@yahoo.com

Abstract The objective of this work was to study the combined effect of edible coating containing ethanolic extract of papaya (*Carica papaya* L.) leaves at ratio 2% and gamma irradiation at dose levels of 0, 2, 4 and 6 kGy on the chemical, microbiological and sensorial qualities of minced chicken thighs meat during cold storage (4 ± 1°C). Samples of minced chicken thighs meat were divided into three groups; uncoated (control), coated with edible coating (without any additives), and irradiated coated with edible coating samples at 0, 2, 4 and 6 kGy. The obtained results showed that gamma irradiation at 2, 4 and 6 kGy with edible coating reduced the initial total bacterial count, psychrophilic bacteria and lactic acid bacteria and prolonged shelf-life of the samples under investigation. Coated samples and irradiated at 2 kGy reduced the counts of Enterobacteriaceae, *Staphylococcus aureus* and *Bacillus cereus* as well as eliminating *Salmonella* spp, while coated samples irradiated at 4 and 6 kGy completely eliminated these bacteria. On the other hand, combination treatments of minced chicken thighs meat samples showed slight increase in thiobarbituric acid reactive substances (TBARS) post irradiation and during cold storage, but had no effects on their total volatile nitrogen (TVN) contents, while a gradual increase in these chemical quality indexes was observed during cold storage. Finally, combination treatment had no adverse effects on the sensory properties of minced chicken thighs meat samples. Therefore the incorporation of ethanolic extract of papaya leaves in edible coating materials increases the bacterial inhibitory effect of gamma irradiation and is suitable for preservation of minced chicken thighs meat or meat and chicken products.

Keywords: *edible coating, ethanolic extract, papaya leaves, gamma irradiation, chicken minced*

1. Introduction

In recent years there has been an increased consumer demand for convenience or "case ready" meat and meat products requiring minimal home preparation [1]. Accordingly, less poultry is marketed today as whole carcasses and more as a variety of further prepared or fully cooked products [2]. Processed chicken based products such as burgers have been distributed through wholesalers and restaurants, and also widely consumed by the people. Furthermore, local industries have grown up to accomplish the demands from these products [3]. The food poisoning microorganisms causing outbreaks were mainly Salmonellae and *S. aureus* with an incidence of 8.99 and 11.54, respectively, while incidence of other food poisoning microorganisms causing 4.07% of the cases [4].

Processed raw poultry meat naturally harbors bacteria, most of which are responsible for the spoilage of poultry meat. However, poultry products can harbor food-borne pathogens, like *Salmonella stereotypes*, *Campylobacter jejuni*, *Listeria monocytogens*, *C. perfringens* and *S. aureus* [5]. Poultry and poultry products rank first or second in foods associated with disease in most of the countries all over the world which in the USA ranked third of the reported foodborne disease outbreaks [6]. The total bacterial count of chicken products as sausage, burger, luncheon and frankfurter was 10^7, 10^7, 10^6 and 10^6 cfu/g, respectively, while *Staphylococcus aureus* was isolated from the same products at incidence of 40%, 70%, 20% and 40%, respectively [7].

Edible coatings or films (biodegradable) have been used as moisture barriers, oxygen barriers, mechanical property modifiers, and food additive carriers in various food products. Antimicrobial edible coatings or films acting as a protective barrier can be used to retard food spoilage, thus extending food shelf life [8,9]. Edible films that can be used as coatings for foods can be include proteins (gelatin, wheat protein, maize protein, casein, etc.), polysaccharides (cellulose and its derivatives, starch and its derivatives, etc.), and lipids [10]. A number of coating materials have been tested in attempt to maintain quality and prolong shelf life of meat products. The chitosan-coated Atlantic cod and herring had reduced moisture loss and lipid oxidation. Immobilizing antimicrobials into film forming solutions is a very advantageous technology for food preservation [11]. The resulting bio-active films or coatings provide more inhibitory effects against spoilage and pathogenic bacteria by lowering the diffusion processes and maintaining high concentrations of the active molecules on the food surfaces [12]. Applications of natural products with both antioxidants and antimicrobial additives in meat products may be useful to prolong the shelf life for pork. Therefore, minimizing lipid oxidation and delaying or inhibiting product contamination, growth of spoilage, and pathogenic organisms in the product are major keys for improving fresh meat shelf life and increasing consumer safety [13].

Plants have the major advantage of still being the most effective and cheaper alternative sources of. The local use of natural plants as primary health remedies, due to their pharmacological properties, is quite common in Asia, Latin America and Africa [14]. Papaya (*Carica papaya* L.) belongs to the family Caricaceae. It has the following common names; pawpaw tree, papaya, papayer, tinti, pepol, chich put, fan kua, wan shou kuo, kavunagaci, kepaya etc. The parts that are usually used include the leaves, fruit, seed, latex, and root. The plant is also described in a documented property forms and it act as analgesic, amebicide, antibacterial, cardiotonic, cholagogue, digestive, emenagogue, febrifuge, hypotensive, laxative, pectoral, stomachic and vermifuge [15]. Papaya leaves are known to contain proteolytic enzymes (papain, chymopapain), alkaloids (carpain, carpasemine), sulfurous compounds (benzyl isothiocyanate), flavonoids, triterpenes, organic acids and oils [16]. The papaya leaves had antimicrobial activity against both gram-negative and gram-positive bacteria is an indication that the plant is a potential source for production of drugs with a broad spectrum of activity and its extracts possess compounds with antibacterial properties that can be used as antibacterial agents in novel drugs for the treatment of gastroenteritis, uretritis, otitis media, and wound infections [17,18].

On the other hand, irradiation as a method of meat preservation has excellent potential to improve meat safety and extend shelf-life [19]. The approval of meat irradiation by the FDA [20] has made consumers more confident and attracted the interest of industries concerned with food quality, and irradiation technology is rapidly entering into commercial reality throughout the world. The current trends consist to develop combination of mild treatments with low dose gamma irradiation to improve food safety and shelf-life extension [21]. The reduced the irradiation dose required for complete elimination of *Salmonella* on fresh poultry by combining gamma irradiation with marinating in natural plant extract [22].

Thus the present work was undertaken to study the effect of gamma irradiation at doses of 0, 2, 4 and 6 kGy in combination treatments with edible coatings containing ethanolic extract of papaya (*Carica papaya* L.) leaves as natural antimicrobial and antioxidant on quality of minced chicken thighs meat during cold storage (4 ± 1°C).

2. Materials & Methods

2.1. Preparation of Ethanolic Extract of Papaya (*Carica Papaya* L.) Leaves

Dried powdered of papaya (Carica papaya L.) leaves were macerated in ethanol (70%) by shaking for 7 days in the dark at room temperature. The ethanolic extract solution was then filtered and concentrated in a rotary evaporator under reduced pressure at 40°C.

2.2. Preparation of Coating Solution

The thirty gram pectin was dissolved in 990 ml distilled water. After the addition of 10 ml polyethylene glycol, the solution was homogenized with a homogenizer for 1 h and homogenized again for 30min. Finally, ethanolic extract of papaya (*Carica papaya* L.) leaves at ratio 2% (v / v) was added and the mixture was further stirred until the ethanolic extract of papaya leaves dissolved. Also, the final solution was sonicated about 1 h in order to remove air bubble or dissolved air.

2.3. Preparation of Minced Chicken Thigh Meat and Coating Application

Fresh chicken thighs samples were obtained from a local market and transported to the laboratory under chilled condition in ice boxes. Skin, bone and visible fat accompanying each type of meat were removed. The chicken thighs were minced in an electric mincer (either treated with 2% ethanolic extract of papaya or untreated) and homogenized to obtain minced chicken thigh meat which shaped handily to pieces of 50 ± 3 g. The fillet pieces were dipped in each freshly prepared coating solution for 1-2 min, drained for 2 min, and dried with mild flow of air in an air dryer at room temperature for 30 min. Then, the untreated and treated samples chicken minced chicken thigh meat were divided into three groups, the first used was control, second was minced chicken thighs meat treated with 2% ethanolic extract of papaya then irradiated at dose levels of 2, 4 and 6 kGy, then every group was packaged in low density polyethylene plastic bags.

2.4. Irradiation Treatments

For irradiation treatments, samples of coated minced chicken thighs meat containing ethanolic extract of papaya leaves at ratio 2% were exposed to gamma irradiation at doses of 0, 2, 4 and 6 kGy using an experimental ^{60}Co Russian gamma chamber, in Cyclotron Project, Nuclear Research Center, Atomic Energy Authority, Abou Zaabal, Egypt. After irradiation, all samples were stored at 4 ± 1°C.

2.5. Microbiological Analysis:

Colony forming units for total bacterial count were counted by plating on plate count agar medium and incubation at 30°C for 3-5 days [23]. Total psychrophilic bacteria were enumerated on plate count agar medium after incubation at 5°C for 7 days [24]. Lactic acid bacteria were counted by the pour plate over layer method as in [25]. Enterobacteriaceae were counted on violet red bile glucose agar medium after incubation for 20-24 h at 37°C [26]. *Staphylococcus aureus* was counted using Baird–Parker medium after incubated at 35°C for 24-48 h [27]. *Bacillus cereus* was counted using Mannitol-egg Yolk-Polymyxin (MYP) agar and incubated at 37°C for 16-24 hours as in [26]. The detection of *Salmonella* was carried out using the most probable number technique. After enrichment at 37°C for 24 h in selenite broth, the cultures were streaked on Brilliant green agar and incubated at 37°C for 24 h, then colonies were biochemically examined in triple sugar iron agar [28].

2.6. Chemical Analysis

Total volatile nitrogen (TVN) was determined as in [29]. Measurement of lipid peroxidation:

2.6.1. Thiobarbaturic Acid Value (TBARs)

Lipid oxidation was determined as a 2-thiobarbituric acid reactive substances (TBARS) value by using the method described as in [30] with some modifications. Minced chicken thighs meat (5 g) and 15ml of deionized distilled water were homogenized with 50 µl butylated hydroxyanisol (7.2%) for 15 s. One milliliter of the meat homogenate was transferred to a disposable test tube and then 1ml of thiobarbituric acid/trichloroacetic acid (20mM TBA in 15% trichloroacetic acid) solution was added. The mixture was vortexed and incubated in a boiling water bath for 15 min, then cooled in cold water for 10min, and centrifuged for 15min at 2500 *g* at 4°C. Absorbance was measured at 532 nm and lipid oxidation was reported as mg malondialdehyde / kg meat.

2.7. Sensory Evaluation

Irradiated and non-irradiated coated minced chicken thighs meat containing ethanolic extract of papaya leaves at ratio 2% samples were periodically examined (every 3 days) for their appearance, texture and odor post treatments and during cold storage at 4 ± 1°C to determine the shelf-life of the samples. The panel consisted of ten members from our laboratory and scores were obtained as [31] by rating the above quality characteristics using the following rating scale: 9 = excellent, 8 = very good, 7 = good, 6 = below good-above fair, 5 = fair, 4 = below fair-above poor, 3 = Poor, 2 = very poor and 1 = extremely poor.

2.8. Statistical Analysis

The obtained data were exposed to analysis of variance. Duncan's multiple range tests at 5% level was used to compare between means. The analysis was carried out using the PROC ANOVA procedure of Statistical Analysis System [32].

3. Resultes & Discussion

The composition of minced chicken meat makes it favorable for microbial growth, therefore minced chicken meat during storage spoils as a result mainly from microorganisms. The data in Table 1 exhibit that the effects of combination treatment between gamma irradiation and edible coating containing 2% ethanolic extract of papaya leaves during cold storage (4 ± 1°C) on the microbial load in minced chicken thighs meat. Samples of control had an initial counts of 7.8 x 10^5, 4.1.2 x 10^4, 3.6 x 10^3, 8.5 x 10^2 and 4.8 x 10^3 cfu/g for total bacterial count, Psychrophilic bacteria, lactic acid bacteria, Enterobacteriaceae and the total molds and yeasts, respectively. On the other hand, treated samples of minced chicken thighs meat with edible coating (without any additives) and edible coating containing 2% ethanolic extract of papaya leaves reduced the counts of total bacterial count, Psychrophilic bacteria, lactic acid bacteria, Enterobacteriaceae and the total molds and yeasts to 6.9 x 10^5, 3.6 x 10^4, 2.5 x 10^3, 7.4 x 10^2 and 4.3 x 10^3 & 3.4 x 10^5, 1.9 x 10^4, 8.7 x 10^2, 1.2 x 10^2 and 2.3 x 10^3 cfu/g, respectively.

Also, irradiating edible coating containing 2% ethanolic extract of papaya leaves minced chicken thighs meat samples at dose levels of 2, 4 and 6 kGy markedly decrease the counts of total bacterial count, Psychrophilic bacteria and lactic acid bacteria by 93.21, 96.34 and 96.67; 99.22, 98.66 and 99.56 & 99.95, 99.94 and 99.17%, respectively. These results agreed with those of Ouattara et al. [33] found that combined effect of edible coating containing thyme oil and *trans*-cinnamaldehyde and gamma irradiation at dose of 3 kGy had synergistic effects in reducing the aerobic plate counts and *Pseudomonas putida* of pre-cooked shrimp. While, the irradiation at dose levels of 2, 4 and 6 kGy of samples under investigation completely eliminated Enterobacteriaceae bacteria from these samples as it remained undetectable upon cold storage periods of the treated samples [34,35]. Meanwhile, the irradiated samples under investigation at dose levels of 0, 2, 4 and 6 kGy inhibition the growth of lactic acid bacteria till 9, 18, 24 and 30 days of storage, respectively. Also, the Table 1 shows that the initial count of total molds and yeasts were 4.8 x 10^3 cfu/g in control samples of minced chicken thigh meat. The initial counts of molds and yeasts reached 4.5 x 10^3, 2.3 x 10^3 and 9.8 x 10^1 cfu/g in samples of coated, edible coating containing ethanolic extract of 2% papaya leaves & irradiated edible coating containing ethanolic extract of papaya leaves at dose level of 2 kGy, respectively. While, irradiated samples at dose levels 4 and 6 kGy induced inhibition growth till 21 and 30 days of storage, respectively. Cha´vez-Quintal et al. [36] found that the ethanolic extract of papaya leaves exhibited the broadest action spectrum against three phytopathogenic fungi, *Rhizopus stolonifer*, *Fusarium* spp. and *Colletotrichum gloeosporioides*. During cold storage, gradual increase in the total bacterial count and psychrophilic bacteria was observed in all samples, but the rate of increase was higher in control samples than irradiated samples. Furthermore, samples were rejected when the count total bacterial count reached 1 x 10^7 cfu/g. Thus, the shelf-life samples of uncoated control, coated with edible coating (without any additives), and irradiated coated with edible coating containing 2% ethanolic extract of papaya leaves at doses of 0, 2, 4 and 6 kGy by 6, 9, 15, 30, 39 and 54 days, respectively. These results are in agreement with the results of Yusha et al. [37], who reported that the ethanolic extract of powdered papaya leaves were active against *Escherichia coli*, *Klebsiella pneumoniae* and Proteus *mirabilis* all concentrations used in disc diffusion and microbroth dilution technique (In vitro). Romasi et al. [38] found that extracts of papaya leaves could inhibit the growth of *Rhizopus stolonifer*, *Bacillus stearothermophilus*, *Listeria monocytogenes*, *Pseudomonas* sp., and *Escherichia coli* by agar diffusion method. On the other hand, the growth of Enterobacteriaceae were not appeared in samples of combination treatment between gamma irradiation at doses 0 and 2 kGy and edible coating containing 2% ethanolic extract of papaya leaves till 9 and 15 day of cold storage, respectively. This is in agreement with the results of Alabi et al. [39] found that The aqueous, ethanol and acetone extract of both the dried and fresh papaya leaves had broad spectrum antimicrobial activity against Gram-negative and Gram-positive bacteria and fungi of medical importance.

Table 1. Effects of combination treatments between gamma irradiation and edible coating containing ethanolic extract of papaya leaves during cold storage (4 ± 1°C) on the microbial load in minced chicken thighs meat

Microbial determinations (cfu/g)	Storage (days)	Control	Edible coating	Edible coating with ethanolic extract of papaya leaves			
				0 kGy	2 kGy	4 kGy	6 kGy
Total aerobic count	0	$7.8x10^{5}$	$6.9x10^{5}$	$3.4x10^{5}$	$5.3x10^{4}$	$6.1x10^{3}$	$3.7x10^{2}$
	3	$5.2x10^{6}$	$9.2x10^{5}$	$6.7x10^{5}$	$6.4x10^{4}$	$8.3x10^{3}$	$5.2x10^{2}$
	6	$9.3x10^{6}$	$3.2x10^{6}$	$8.3x10^{5}$	$7.6x10^{4}$	$9.7x10^{3}$	$7.2x10^{2}$
	9	$8.4x10^{7}$R	$9.6x10^{6}$	$9.5x10^{5}$	$8.8x10^{4}$	$2.7x10^{4}$	$8.3x10^{2}$
	12		$7.1x10^{7}$R	$4.2x10^{6}$	$9.5x10^{4}$	$5.2x10^{4}$	$9.8x10^{2}$
	15			$9.4x10^{6}$	$3.6x10^{5}$	$6.7x10^{4}$	$2.9x10^{3}$
	18			$6.5x10^{7}$R	$6.9x10^{5}$	$8.2x10^{4}$	$5.7x10^{3}$
	21				$8.5x10^{5}$	$1.3x10^{5}$	$6.6x10^{3}$
	24				$1.6x10^{6}$	$3.8x10^{5}$	$8.9x10^{3}$
	27				$4.1\ x10^{6}$	$6.7x10^{5}$	$3.8x10^{4}$
	30				$7.9\ x10^{6}$	$8.3x10^{5}$	$7.5x10^{4}$
	33				$3.9x10^{7}$R	$4.2x10^{6}$	$1.9x10^{5}$
	36					$7.5\ x10^{6}$	$4.7x10^{5}$
	39					$9.1x10^{6}$	$8.5x10^{5}$
	42					$4.3x10^{7}$ R	$9.7x10^{5}$
	45						$2.9\ x10^{6}$
	48						$6.1x10^{6}$
	51						$7.6\ x10^{6}$
	54						$8.9x10^{6}$
	57						$5.8\ x10^{7}$R
Psychrophilic bacteria	0	$4.1x10^{4}$	$3.5x10^{4}$	$1.9x10^{4}$	$1.5x10^{3}$	$5.5x10^{2}$	$2.3x10^{1}$
	3	$5.9x10^{4}$	$6.5x10^{4}$	$4.4x10^{4}$	$6.6x10^{3}$	$9.2x10^{2}$	$4.6x10^{1}$
	6	$9.6x10^{4}$	$9.2x10^{4}$	$7.9x10^{4}$	$7.9x10^{3}$	$2.6x10^{3}$	$6.1x10^{1}$
	9	$5.2x10^{5}$R	$3.8x10^{5}$	$1.3x10^{5}$	$2.8x10^{4}$	$5.5x10^{3}$	$8.5x10^{1}$
	12		$6.9x10^{5}$R	$5.6x10^{5}$	$4.9x10^{4}$	$7.2x10^{3}$	$3.1x10^{2}$
	15			$9.6x10^{5}$	$7.8x10^{4}$	$9.8x10^{3}$	$5.8x10^{2}$
	18			$3.5x10^{6}$R	$9.6x10^{4}$	$3.1x10^{4}$	$7.5x10^{2}$
	21				$2.4x10^{5}$	$6.6x10^{4}$	$8.9x10^{2}$
	24				$5.1x10^{5}$	$8.8x10^{4}$	$9.7x10^{2}$
	27				$7.2x10^{5}$	$1.8x10^{5}$	$3.9x10^{3}$
	30				$9.5\ x10^{5}$	$4.6x10^{5}$	$6.4x10^{3}$
	33				$4.6\ x10^{6}$R	$6.3x10^{5}$	$8.4x10^{3}$
	36					$8.9\ x10^{5}$	$1.7x10^{5}$
	39					$1.2\ x10^{6}$	$4.3x10^{5}$
	42					$3.3\ x10^{6}$ R	$6.4x10^{5}$
	45						$7.8\ x10^{5}$
	48						$9.8x10^{5}$
	51						$3.6\ x10^{6}$
	54						$5.2x10^{6}$
	57						$8.7x10^{6}$R
Lactic acid bacteria	0	$3.6x10^{3}$	$2.5x10^{3}$	$8.7x10^{2}$	$1.2x10^{2}$	$1.6x10^{1}$	$3.0x10^{1}$
	3	$4.6x10^{3}$	$2.7x10^{3}$	<10	<10	<10	<10
	6	$8.7x10^{3}$	$5.4x10^{3}$	<10	<10	<10	<10
	9	$5.2x10^{4}$ R	$7.8x10^{3}$	<10	<10	<10	<10
	12		$2.6x10^{4}$ R	$6.2x10^{2}$	<10	<10	<10
	15			$8.7x10^{2}$	<10	<10	<10
	18			$3.9x10^{3}$ R	<10	<10	<10
	21				$1.6x10^{1}$	<10	<10
	24				$3.3x10^{1}$	<10	<10
	27				$6.5\ x10^{1}$	$1.2x10^{1}$	<10
	30				$9.8\ x10^{1}$	$3.8x10^{1}$	<10
	33				$7.9\ x10^{2}$R	$6.7x10^{1}$	$4.0x10^{1}$
	36					$8.2\ x10^{1}$	$2.1x10^{1}$
	39					$2.7\ x10^{2}$	$5.7x10^{1}$
	42					$6.8\ x10^{2}$R	$8.3x10^{1}$
	45						$9.7\ x10^{1}$
	48						$2.7\ x10^{2}$
	51						$6.9x10^{2}$
	54						$8.5x10^{2}$
	57						$4.6\ x10^{3}$R
Enterobacteriaceae	0	$8.5x10^{2}$	$6.4x10^{2}$	$1.2x10^{2}$	Nil	Nil	Nil
	3	$4.2x10^{3}$	$9.2x10^{2}$	<10	Nil	Nil	Nil
	6	$8.8x10^{3}$	$4.9x10^{3}$	<10	Nil	Nil	Nil
	9	$5.9x10^{4}$ R	$7.9x10^{3}$	<10	Nil	Nil	Nil
	12		$1.6x10^{4}$ R	$1.2x10^{1}$	Nil	Nil	Nil
	15			$4.6x10^{1}$	Nil	Nil	Nil

	18			7.1x10^{1}R	Nil	Nil	Nil
	21				Nil	Nil	Nil
	24				Nil	Nil	Nil
	27				Nil	Nil	Nil
	30				Nil	Nil	Nil
	33				Nil R	Nil	Nil
	36					Nil	Nil
	39					Nil	Nil
	42					Nil R	Nil
	45						Nil
	48						Nil
	51						Nil
	54						Nil
	57						Nil R
	0	4.8 x10^{3}	4.3 x10^{3}	2.3 x10^{3}	2.3 x10^{1}	<10	<10
	3	8.5 x10^{3}	7.2 x10^{3}	5.7x10^{3}	3.9 x10^{1}	<10	<10
	6	2.7 x10^{4}	9.8 x10^{3}	8.9x10^{3}	6.7x10^{1}	<10	<10
	9	7.6 x10^{4}R	3.7 x10^{4}	2.1x10^{4}	9.4x10^{1}	<10	<10
	12		8.7 x10^{4}R	6.8x10^{4}	3.4x10^{2}	<10	<10
	15			1.9x10^{5}	5.6x10^{2}	<10	<10
	18			5.8x10^{5}R	7.2x10^{2}	<10	<10
	21				9.5x10^{2}	<10	<10
	24				4.7x10^{3}	7.0 x10^{1}	<10
Total molds and yeasts	27				7.9x10^{3}	1.2 x10^{2}	<10
	30				2.6x10^{4}	2.3 x10^{2}	<10
	33				6.9x10^{4}R	4.1 x10^{2}	<10
	36					6.2 x10^{2}	4.0 x10^{1}
	39					8.7 x10^{2}	9.0x10^{1}
	42					3.6 x10^{3}R	1.8x10^{2}
	45						2.8x10^{2}
	48						5.3x10^{2}
	51						6.9x10^{2}
	54						8.4x10^{2}
	57						1.8x10^{3}R

R= Rejected

3.1. Food Borne Pathogens

Table 2 illustrates the effects of combination treatment gamma irradiation and edible coating during cold storage (4 ± 1°C) on food borne pathogens in minced chicken thighs meat. The results show that the initial counts of control samples were 7.8 x 10^{2} and 8.9 x 10^{2} cfu/g for *Staphylococcus aureus* and *B. cereus*, respectively. Moreover, bacteria of *Salmonella* spp were detected in these samples. AL-Dughaym and Altabari [40] revealed that revealed that the mean total bacterial count was ranged from 2.7 × 10^{4} cfu/g for nuggets A to 3.3 × 10^{7} cfu/g for burger B and the other products in the range of 10^{5}-10^{6} cfu/g. While *Staphylococcus aureus* mean count ranged from less than 10^{2} cfu/g for all samples, accept 10^{4} and 10^{6} cfu/g for mince and frankfurter samples, respectively. *Escherichia coli* isolated from 70% of the samples and *Salmonella arizona* was isolated at once from thigh samples from poultry meat products were collected from local markets.

Subjecting samples of minced chicken thighs meat to edible coating (without any additives) and edible coating containing 2% ethanolic extract of papaya leaves reduced the counts of *Staphylococcus aureus* to 4.1 x 10^{2} and 2.7 x 10^{2} cfu/g and counts of *B. cereus* to 5.3 x 10^{2} and 1.6 x 10^{2} cfu/g, respectively. While, subjecting coated samples under investigation to gamma irradiation at dose levels of 2, 4 and 6 kGy completely eliminated *Staphylococcus aureus* bacteria from these samples as it remained undetectable upon cold storage periods [34,35]. Meanwhile, bacteria of *Salmonella* spp was detected in samples edible coating (without any additives) while, these bacteria were not detected in samples of minced chicken thighs meat treated by edible coating containing 2% ethanolic extract of papaya leaves & irradiated edible coating containing 2% ethanolic extract of papaya leaves at dose of 2 kGy. These results are in agreement with previous reports Abdeldaiem [41], found that edible coatings containing 1% oregano essential oil samples of bolti fish fillets and irradiated at 1 kGy reduced the counts of Enterobacteriaceae, *Staphylococcus aureus* and *Bacillus cereus* as well as eliminating *Vibrio* spp and *Salmonella* spp, while coated samples irradiated at 3 and 5 kGy completely eliminated these bacteria. Also, Anibijuwon and Udeze [17] and Okunola et al. [18] concluded that papaya leaves had antimicrobial activity against both gram-negative and gram-positive bacteria is an indication that the plant is a potential source for production of drugs with a broad spectrum of activity and its extracts possess compounds with antibacterial properties that can be used as antibacterial agents in novel drugs for the treatment of gastroenteritis, uretritis, otitis media, and wound infections. Also, Romasi et al. [38] reported that the extracts of papaya leaves showed could inhibit *B. stearothermophilus, L. monocytogenes, Pseudomonas* sp., and *E. coli* by using agar diffusion method. Moreover, Rahman et al. [42] revealed that the ethanolic extract of papaya leaves inhibit the growth of selected microorganisms. Various Gram negative (*Escherichia coli, Proteus mirabilis, Pseudomonas aeruginosa, Salmonella typhi, Salmonella paratyphi A, Shigella flexenert)* and Gram positive bacteria (*Staphylococcus aureus, Staphylococcus epidermidis, Bacillus subtilis, Micrococcus luteus*) in vitro. During cold storage, gradual increases of *Staphylococcus aureus* and *B. cereus* were observed during cold storage in samples of control, edible coating containing 2% ethanolic extract of papaya leaves & irradiated edible coating containing 2% ethanolic extract of papaya leaves at dose of 2 kGy. While, irradiation at doses of 4 and 6 kGy combined with edible coating containing 2% ethanolic extract of papaya leaves completely eliminated these pathogen from samples minced chicken thighs meat. Moreover, Alabi et al. [39] mentioned that the plant kingdom synthesizes diverse active compounds, some of the active compounds do occur singly or in combination with other inactive substances which inhibit greatly the life processes of microbes, especially the pathogenic microbes. Medicinal plants are cheap and renewable source of pharmacologically active substances. Also, Azmi et al. [43] mentioned that the mainly component in papaya leaves were flavonoids, alkaloids, saponins, xanthine alkaloids, terpenoids and anthranol glycosides which could partially explain the pharmacological properties of this plant and demonstrates its importance in alimentation and daily intake especially for gout patient.

Table 2. Effects of combination treatments between gamma irradiation and edible coating containing ethanolic extract of papaya leaves during cold storage (4 ± 1°C) on food borne pathogens in minced chicken thighs meat

Microbial determinations (cfu/g)	Storage (days)	Control	Edible coating	Edible coating with ethanolic extract of papaya leaves			
				0 kGy	2 kGy	4 kGy	6 kGy
Staphylococcus aureus	0	$7.8x10^2$	$4.1x10^2$	$2.7x10^2$	Nil	Nil	Nil
	3	$1.9x10^3$	$6.2x10^2$	$3.6x10^2$	Nil	Nil	Nil
	6	$4.7x10^3$	$7.7x10^2$	$6.2x10^2$	Nil	Nil	Nil
	9	$8.5x10^3$R	$9.3x10^2$	$7.4x10^2$	Nil	Nil	Nil
	12		$4.2x10^3$R	$8.7x10^2$	Nil	Nil	Nil
	15			$3.3x10^3$	Nil	Nil	Nil
	18			$6.8x10^3$R	Nil	Nil	Nil
	21				Nil	Nil	Nil
	24				Nil	Nil	Nil
	27				Nil	Nil	Nil
	30				Nil	Nil	Nil
	33				Nil R	Nil	Nil
	36					Nil	Nil
	39					Nil	Nil
	42					Nil R	Nil
	45						Nil
	48						Nil
	51						Nil
	54						Nil
	57						Nil R
Bacillus cereus	0	$8.9x10^2$	$5.3x10^2$	$1.6x10^2$	$4.5x10^1$	N.D	N.D
	3	$9.6x10^2$	$7.5x10^2$	$2.9x10^2$	$6.9x10^1$	N.D	N.D
	6	$3.5x10^3$	$8.9x10^2$	$4.1x10^2$	$7.7x10^1$	N.D	N.D
	9	$8.1x10^3$R	$2.6x10^3$	$5.8x10^2$	$8.1x10^1$	N.D	N.D
	12		$6.7x10^3$R	$7.8x10^2$	$9.3x10^1$	N.D	N.D
	15			$9.8x10^3$	$2.8x10^2$	N.D	N.D
	18			$4.2x10^3$R	$4.3x10^2$	N.D	N.D
	21				$6.5x10^2$	N.D	N.D
	24				$7.2x10^3$	N.D	N.D
	27				$8.9x10^2$	N.D	N.D
	30				$2.1x10^3$	N.D	N.D
	33				$5.5x10^3$R	N.D	N.D
	36					N.D	N.D
	39					N.D	N.D
	42					N.D R	N.D
	45						N.D
	48						N.D
	51						N.D
	54						N.D
	57						N.D R
***Salmonella* spp**	0	+	+	N.D	N.D	N.D	N.D
	3	+	+	N.D	N.D	N.D	N.D
	6	+	+	N.D	N.D	N.D	N.D
	9	+ R	+	N.D	N.D	N.D	N.D
	12		+ R	N.D	N.D	N.D	N.D
	15			N.D	N.D	N.D	N.D
	18			N.D R	N.D	N.D	N.D
	21				N.D	N.D	N.D
	24				N.D	N.D	N.D
	27				N.D	N.D	N.D
	30				N.D	N.D	N.D
	33				N.D R	N.D	N.D
	36					N.D	N.D
	39					N.D	N.D
	42					N.D R	N.D
	45						N.D
	48						N.D
	51						N.D
	54						N.D
	57						N.D R

R = Rejected + = Positive N.D = not detected

3.2. Chemical Evaluation

Moreover, TVN can be considered as a reliable indicative measure for the quality of various food articles especially chicken meat and chicken cut-up meat. In general, TVN in chicken cut-up meat may be increased as the days of storage increased [44]. As the results in Table 3 show the changes in the total volatile basis nitrogen (TVBN) and thiobarbituric acid reactive substances (TBARs) as a chemical indexes due to combination treatment and cold storage, respectively. The TVBN maybe considered as a quality index for minced chicken meat because its increase is related to the activity of spoilage bacteria and endogenous enzymes [45]. Moreover, Mexis et al. [46] mentioned that As autolysis of muscle proteins proceeds, compounds of alkaline reaction such as amines and ammonia build up producing objectionable odors. Thus, TVB-N has been used as a spoilage indicator in several animal food substrates. Furthermore, samples treated by edible coating (without any additive) or by combination treatment between gamma irradiation at doses of 0, 2, 4 and 6 kGy and edible coating containing 2% ethanolic extract of papaya leaves induced no changes in the contents of TVBN. During cold storage, an increase in these compounds was observed, but the rate of increase was lower in samples of combination treatment especially at higher doses. Mexis et al. [46] reported that the samples of ground chicken were rejected when the TVB-N ranged between 42.5 and 57.5 mg/kg. Balamatsia et al. [47] who reported TVB-N values of 45.8, 40.0 and 36.0 mg N/100 g for chicken breast fillets packaging in air, under vacuum and modified atmosphere packaging (MAP) respectively, at the limit of sensory acceptability (6 and 9-10 day of storage respectively at 4°C). Patsias et al. [48] reported initial TVB-N values of 12 mg N/100 g which increased to 13.0 and 49.0 mg N/100 g for raw chicken fillets stored in air and under MAP after 9 day of storage.

Table 3. Effects of combination treatments between gamma irradiation and edible coating containing ethanolic extract of papaya leaves during cold storage (4 ± 1°C) on chemical characteristics in minced chicken thighs meat

Chemical quality index	Storage period (days)	Control	Edible coating	Edible coating with ethanolic extract of papaya leaves			
				0	2	4	6
The total volatile basic nitrogen (TVBN mg N/100g wet matter)	0	11.28	11.79	11.35	11.74	11.92	12.05
	3	18.07	18.42	17.14	13.43	12.58	12.44
	6	26.15	22.57	19.62	15.34	13.93	12.98
	9	35.73 R	27.29	24.08	18.17	14.77	13.02
	12		31.14	27.13	21.48	16.61	13.66
	15		35.06R	29.26	22.93	18.44	13.12
	18			32.41	24.60	19.96	14.35
	21			35.18R	27.83	22.38	14.94
	24				30.07	24.84	15.62
	27				31.68	25.99	16.77
	30				33.54	27.56	18.48
	33				36.12R	29.16	20.63
	36					30.09	21.87
	39					32.74	22.42
	42					35.66R	22.91
	45						24.46
	48						26.55
	51						28.78
	54						32.12
	57						36.09R
Thiobarbituric acid-reactive substances (mg malonaldehyde/kg)	0	0.41	0.43	0.42	0.48	0.55	0.59
	3	0.54	0.58	0.47	0.51	0.56	0.60
	6	0.62	0.62	0.52	0.53	0.58	0.61
	9	0.79 R	0.69	0.59	0.56	0.59	0.62
	12		0.73	0.68	0.58	0.61	0.63
	15		0.89R	0.74	0.61	0.62	0.65
	18			0.79	0.67	0.64	0.66
	21			0.89R	0.71	0.65	0.67
	24				0.75	0.67	0.69
	27				0.77	0.69	0.72
	30				0.79	0.71	0.73
	33				0.91R	0.73	0.74
	36					0.78	0.76
	39					0.85	0.77
	42					0.96 R	0.79
	45						0.81
	48						0.83
	51						0.86
	54						0.91
	57						0.98 R

R = Rejected

Table 4. Changes in sensory attributes of minced chicken thighs meat as affected by combination treatment between gamma irradiation and edible coating containing ethanolic extract of papaya leaves during cold storage (4 ± 1°C)

Storage period (days)	Appearance						Flavor						Color					
	Control	Edible coating	Edible coating with ethanolic extract of papaya leaves				Control	Edible coating	Edible coating with ethanolic extract of papaya leaves				Control	Edible coating	Edible coating with ethanolic extract of papaya leaves			
			0 kGy	2 kGy	4 kGy	6 kGy			0 kGy	2 kGy	4 kGy	6 kGy			0 kGy	2 kGy	4 kGy	6 kGy
0	$8.9^{Ba}\pm0.07$	$9.5^{Aa}\pm0.28$	$9.5^{Aa}\pm0.15$	$9.5^{Aa}\pm0.36$	$9.4^{Aa}\pm0.13$	$9.4^{A}{}^{a}\pm0.37$	$9.5^{Aa}\pm0.17$	$9.4^{Aa}\pm0.36$	$9.4^{Aa}\pm0.21$	$9.3^{Aa}\pm0.47$	$9.4^{Aa}\pm0.026$	$9.3^{A}{}^{a}\pm0.18$	$9.7^{Aa}\pm0.63$	$9.6^{Aa}\pm0.19$	$9.7^{Aa}\pm0.08$	$9.7^{Aa}\pm0.54$	$9.7^{Aa}\pm0.72$	$9.7^{A}{}^{a}\pm0.35$
3	$8.3^{Bb}\pm0.12$	$9.4^{Aa}\pm0.12$	$9.4^{Aa}\pm0.24$	$9.5^{Aa}\pm0.42$	$9.4^{Aa}\pm0.07$	$9.4^{A}{}^{a}\pm0.26$	$9.4^{Aa}\pm0.25$	$9.4^{Aa}\pm0.42$	$9.4^{Aa}\pm0.25$	$9.3^{Aa}\pm0.25$	$9.4^{Aa}\pm0.07$	$9.3^{A}{}^{a}\pm0.18$	$8.7^{Bb}\pm0.42$	$9.6^{Aa}\pm0.11$	$9.7^{Aa}\pm0.08$	$9.7^{Aa}\pm0.51$	$9.7^{Aa}\pm0.64$	$9.7^{A}{}^{a}\pm0.58$
6	$7.8^{Cc}\pm0.08$	$9.1^{Aa}\pm0.15$	$9.2^{Aa}\pm0.33$	$9.5^{Aa}\pm0.51$	$9.4^{Aa}\pm0.25$	$9.4^{A}{}^{a}\pm0.15$	$9.3^{Aa}\pm0.29$	$9.4^{Aa}\pm0.33$	$9.4^{Aa}\pm0.36$	$9.3^{Aa}\pm0.36$	$9.4^{Ba}\pm0.23$	$9.3^{A}{}^{a}\pm0.25$	$8.6^{Bb}\pm0.13$	$9.5^{Aa}\pm0.28$	$9.6^{Aa}\pm0.15$	$9.6^{Aa}\pm0.42$	$9.7^{Aa}\pm0.68$	$9.7^{A}{}^{a}\pm0.76$
9	$3.4^{Cd}\pm0.24$R	$8.9^{Bb}\pm0.07$	$9.2^{Aa}\pm0.12$	$9.5^{Aa}\pm0.38$	$9.3^{Aa}\pm0.29$	$9.4^{A}{}^{a}\pm0.18$	$3.1^{Cb}\pm0.14$R	$8.9^{Bb}\pm0.27$	$9.4^{Aa}\pm0.57$	$9.3^{Aa}\pm0.34$	$9.4^{Aa}\pm0.43$	$9.3^{A}{}^{a}\pm0.29$	$3.1^{Bc}\pm0.07$R	$9.5^{Aa}\pm0.26$	$9.5^{Aa}\pm0.16$	$9.6^{Aa}\pm0.12$	$9.7^{Aa}\pm0.73$	$9.6^{A}{}^{a}\pm0.64$
12		$3.5^{Cc}\pm0.35$R	$8.7^{Bb}\pm0.17$	$9.3^{Aa}\pm0.23$	$9.3^{Aa}\pm0.53$	$9.4^{A}{}^{a}\pm0.06$		$4.3^{Bc}\pm0.34$R	$9.3^{Aa}\pm0.56$	$9.2^{Aa}\pm0.16$	$9.4^{Aa}\pm0.38$	$9.2^{A}{}^{a}\pm0.22$		4.1^{Bb} 0.28 ±R	$9.5^{Aa}\pm0.28$	$9.5^{Aa}\pm0.19$	$9.6^{Aa}\pm0.18$	$9.6^{A}{}^{a}\pm0.12$
15			$8.7^{Bb}\pm0.18$	$9.5^{Aa}\pm0.12$	$9.2^{Aa}\pm0.58$	$9.3^{A}{}^{a}\pm0.51$			$9.3^{Aa}\pm0.12$	$9.2^{Aa}\pm0.58$	$9.4^{Aa}\pm0.29$	$9.2^{A}{}^{a}\pm0.41$			$8.9^{Bb}\pm0.27$	$9.4^{Aa}\pm0.25$	$9.6^{Ab}\pm0.47$	$9.5^{A}{}^{a}\pm0.11$
18			$4.3^{Bc}\pm0.09$R	$9.2^{Aa}\pm0.28$	$9.2^{Aa}\pm0.34$	$9.3^{A}{}^{a}\pm0.42$			$3.2^{Ba}\pm0.09$R	$9.1^{Aa}\pm0.12$	$9.3^{Aa}\pm0.16$	$9.2^{A}{}^{a}\pm0.46$			$4.3^{Bc}\pm0.64$R	$9.4^{Aa}\pm0.29$	$9.6^{Aa}\pm0.38$	$9.4^{A}{}^{a}\pm0.09$
21				$9.2^{Aa}\pm0.19$	$9.2^{Aa}\pm0.12$	$9.3^{A}{}^{a}\pm0.46$				$9.1^{Aa}\pm0.08$	$9.3^{Aa}\pm0.44$	$9.2^{A}{}^{a}\pm0.52$				$9.3^{Aa}\pm0.35$	$9.6^{Aa}\pm0.69$	$9.4^{A}{}^{a}\pm0.25$
24				$8.9^{Bb}\pm0.06$	$8.9^{Bb}\pm0.18$	$9.3^{A}{}^{a}\pm0.55$				$8.9^{Bb}\pm0.06$	$9.3^{Aa}\pm0.31$	$9.2^{A}{}^{a}\pm0.16$				$9.1^{Aa}\pm0.48$	$9.4^{Aa}\pm0.07$	$9.3^{A}{}^{a}\pm0.48$
27				$8.9^{Bb}\pm0.11$	$8.9^{Bb}\pm0.05$	$9.2^{A}{}^{a}\pm0.32$				$8.9^{Bb}\pm0.29$	$9.3^{Aa}\pm0.22$	$9.2^{A}{}^{a}\pm0.19$				$8.9^{Bb}\pm0.27$	$9.4^{Aa}\pm0.15$	$9.3^{A}{}^{a}\pm0.39$
30				$8.7^{Bb}\pm0.32$	$8.9^{Bb}\pm0.26$	$9.2^{A}{}^{a}\pm0.24$				$8.8^{Bb}\pm0.45$	$9.3^{Aa}\pm0.57$	$9.2^{A}{}^{a}\pm0.25$				$8.7^{Bb}\pm0.71$	$9.1^{Aa}\pm0.19$	$9.2^{A}{}^{a}\pm0.59$
33				$4.3^{Cc}\pm0.42$R	$8.8^{Bb}\pm0.46$	$9.1^{A}{}^{a}\pm0.19$				$6.4^{Bb}\pm0.52$R	$9.1^{Aa}\pm0.42$	$9.2^{A}{}^{a}\pm0.36$				$5.8^{Cc}\pm0.52$R	$9.1^{Ba}\pm0.27$	$9.2^{A}{}^{c}\pm0.67$
36					$8.8^{Bb}\pm0.42$	$9.1^{A}{}^{a}\pm0.13$					$8.9^{Bb}\pm0.08$	$9.2^{A}{}^{a}\pm0.28$					$8.9^{Bb}\pm0.54$	$9.1^{A}{}^{a}\pm0.74$
39					$8.7^{Bb}\pm0.19$	$9.1^{A}{}^{a}\pm0.15$					$8.9^{Bb}\pm0.16$	$9.1^{a}\pm0.41$					$8.8^{Bb}\pm0.65$	$9.1^{A}{}^{a}\pm0.53$
42					$4.2^{Cc}\pm0.27$R	$8.9^{B}{}^{b}\pm0.12$					$7.5^{Cc}\pm0.15$R	$9.1^{a}\pm0.49$					$6.2^{Bc}\pm0.39$R	$9.1^{A}{}^{a}\pm0.41$
45						$8.9^{b}\pm0.09$						$9.1^{a}\pm0.65$						$8.9^{b}\pm0.26$
48						$8.9^{b}\pm0.24$						$9.1^{a}\pm0.43$						$8.9^{b}\pm0.32$
51						$8.8^{b}\pm0.29$						$8.9^{b}\pm0.09$						$8.8^{b}\pm0.18$
54						$8.7^{b}\pm0.45$						$8.9^{b}\pm0.17$						$8.8^{b}\pm0.14$
57						$4.6^{c}\pm0.53$R						$8.6^{b}\pm0.47$R						$7.5^{c}\pm0.27$R

Capital and small letters were used for comparing between means in the rows and columns, respectively.
Means with the same letters are not significantly different ($p>0.05$).
R = Rejected

The TBARs value is an index of lipid oxidation measuring malondialdehyde (MDA) content. MDA formed through hydroperoxides, which are the initial reaction product of polyunsaturated fatty acids with oxygen [49]. The oxidative rancidity in fresh, frozen and cooked chicken breast and leg meat was evaluated by measuring malonaldehyde in fat with an improved thiobarbituric acid (TBARs) assay with antioxidant protection [50]. Exposing samples of coated minced chicken thighs meat containing 2% ethanolic extract of papaya leaves to gamma irradiation at doses levels of 2, 4 and 6 kGy induced slight increase in their TBARs values as compared to values of control, edible coating (without any additive) and edible coating containing 2% ethanolic extract of papaya leaves samples of minced chicken thighs meat. The slight increase in values of TBARs in irradiated samples may be mainly attributed to the strong antioxidant effect of ethanolic extract of papaya leaves which acts as a radical scavenger [51]. Imaga et al. [52] found that the extracts of papaya leaves contained flavonoids and the antioxidant vitamins A and C these compounds may be due to their inherent antioxidant nutrient composition. Ayoola et al. [53] reported that the ethanolic extract of papaya leaves showed high antioxidant (free radical scavenging) activities when compared to vitamin C and these extract showed the presence of flavonoids, terpenoids, saponins, tannins and reducing sugars. Abdeldaiem [41] showed slight increase in thiobarbituric acid reactive substances (TBARS) post irradiation and during cold storage of edible coatings containing 1% oregano essential oil of bolti fish fillets samples, but had no effects on their total volatile nitrogen (TVN) contents, while a gradual increase in these chemical quality indexes was observed during cold storage. The acceptability limit of TBARS value in this study was 1.0. Earlier workers reported that meat sample containing TBARS value less than 1 possesses no off odor [54]. Meanwhile, Jayasingh et al. [55] reported that for secondary oxidation products, such as, TBA, no legal threshold exists, but a limit of 1 mg malonaldehyde/kg meat has been suggested for sensory perceived rancidity. Shon et al. [56] illustrated that the soy protein isolate coating reduced thiobarbituric acid-reactive substances (TBARS) and peroxide value (PV) of cut pork stored at 4°C for 5 days compared with controls.

3.3. Sensory Evaluation

Table 4 showed that from these data all samples under investigation had an excellent scores for flavor and color, but appearance of coated samples with edible coating (without any additive) and coated with edible coating containing 2% ethanolic extract of papaya leaves treated by gamma irradiation at doses of 2, 4 and 6 kGy had higher scores than control samples. These may be due to brightness of coatings that coat minced chicken thighs meat samples. Similar results were observed as in [57] reported that irradiated pork patty coated with pectin-based material containing green tea leaf extract powder had high scores for appearance compared with control samples. Upon cold storage, samples of control and coated showed similar scores till the detection off odor and their total bacterial count more than 1 x 10^7 cfu/g and their rejection on day 9 and 12 of storage, respectively. While coated samples of minced chicken thighs meat with edible coating containing ethanolic extract of papaya leaves and treated by gamma irradiation at doses of 0, 2, 4 and 6 kGy were scored as good samples their rejection due to increasing their total bacterial count to more than 1 x 10^7 cfu/g on day 21, 33, 42 and 57 of storage, respectively. These results are in agreement with those of Ouattara et al. [58] they found that the shelf-life extension periods ranged from 3 to 10 days for uncoated shrimps and pizzas and from 7 to 20 days for irradiated edible coated shrimps and edible coated pizzas at dose of 4 kGy, respectively, compared to uncoated / unirradiated products. Furthermore, Lacroix et al. [59] mentioned that shelf-life extension periods for irradiated edible coated ground beef samples at doses 1, 2 and 3 kGy were 4, 7 and 10 days, respectively. Javanmard et al. [60] reported that the irradiation dose of 5 kGy can be effective to control bacterial pathogens in chicken meat, through its effectiveness in extending their frozen shelf-life without any significant effects on the sensory quality. Meanwhile, Abdeldaiem [41] concluded that the coated samples of bolti fish fillets with edible coating containing oregano EO and treated by gamma irradiation at doses of 0, 1, 3 and 5 kGy were scored as good samples their rejection due to increasing their total bacterial count to more than 1 x 10^7 cfu/g on day 18, 24, 33 and 39 of storage, respectively

4. Conclusion

In conclusion, The results obtained from this study showed that combination treatments between gamma irradiation at doses of 2, 4 and 6 kGy and edible coating films containing 2% of ethanolic extract of papaya leaves for improving the quality and safety of minced chicken thighs meat through its effectiveness in eliminating bacteria of public health and extending the refrigerated shelf-life to 30, 39 and 54 days for irradiated coated samples under investigation, respectively, compared to 6 days for uncoated control samples without any adverse changes in their chemical and sensory properties. Therefore the incorporation of ethanolic extract of papaya leaves in edible coating materials increases the bacterial inhibitory effect of gamma irradiation and is suitable for preservation of minced chicken meat.

References

[1] Stubbs, R.L.; Morgan, J.B; Ray, F.K. and Dolezal, H.G, "Effect of supplemental vitamin E on the color and case-life of top loin steaks and ground chuck patties in modified atmosphere case-ready retail packaging systems". Meat Sci., 61: 1-5. 2002.

[2] Ang, C.Y.W. and Lyon, B.G, "Evaluations of warmed-over flavor during chill storage of cooked broiler breast, thigh and skin by chemical, instrumental and sensory methods". J. Food Sci., 55: 644-648. 1990.

[3] Guerrero-Legarreta, I. and Hui, Y. H, "Handbook of Poultry Science and Technology: Secondary Processing". New Jersey: John Wiley & Sons, Inc. 2010.

[4] Altabari, G., Al-Dughaym, A.M, "The role of sanitary inspection of meat in relation of food poisoning". In: The Second Annual Scientific Meeting for Environment Hygiene (Meat Hygiene), Riyadh, pp. 180-203. 2002.

[5] Waldroup, A.L, "Contamination of raw poultry with pathogen". World's Poult. Sci., 52, 7. 1996.

[6] Bean, N.N. and Griffin, P.M, "Foodborne disease outbreaks in the United States", 1973 1987; pathogens, vehicles and trends, J. Food Prot., 53, 804. 1990.

[7] EL-khateib, T.; Abd El-Rahman, H; Hamdy, M. and Lotfi, A, "Poultry meat products in Egypt "Proximal chemical composition and microbiological quality" Fleischwirtsch. 68, 6, 756-757. 1988.

[8] Cagri, A.; Ustunol, Z. and Ryser, E, "Antimicrobial edible films and coatings". J Food Prot 67 (4): 833-8. 2004.

[9] Cha, D.S. and Chinnan, M.S, "Biopolymer-based antimicrobial packaging: a review". Crit Rev. Food Sci. Nutr., 44 (4), 223-37. 2004.

[10] Baldwin, E.A., Nisperos, M.O., Hagenmaier, R.D. and Baker, R.A, "Use of lipids in coatings for food products". Food Technol. 51 (6), 56-64. 1997.

[11] Jeon, Y.; Kamil, J. and Shahidi, F, "Chitosan as edible invisible film for quality preservation of herring and Atlantic cod". J Agric Food Chem., 50, 5167-78. 2002.

[12] Ouattara, B.; Simard, R. E.; Piette, J.G.; Begin, A. and Holley R.A, "Inhibition of surface spoilage bacteria in processed meats by application of antimicrobial films prepared with chitosan". Int. J. Food Microbiol., 62, 139-148. 2000.

[13] McCarthy, T.L.; Kerry, J.P.; Kerry, J.F.; Lynch, P.B. and Buckley, D.J, "Assessment of the antioxidant potential of natural food and plant extracts in fresh and previously frozen pork patties". Meat Sci., 57, 177-184. 2001.

[14] Bibitha, B.; Jisha, V.K.; Salitha, C.V.; Mohan, S. and Valsa, A.K. (2002). "Antibacterial activity of different plant extracts". Short Communiaction. Indian J. Microbiol. 42: 361-363.

[15] Afolayan, A.J, "Extracts from the shoots of Arctotis artotoides inhibit the growth of bacteria and fungi". Pharm. Biol. 41: 22-25. 2003.

[16] Osuna-Torres L, Tapia-Pe´rez ME, Aguilar-Contreras, A, "Plantas medicinales de la medicina tradicional mexicana para tratar afecciones gastrointestinales: Estudio etnobota´nico fitoquı ´mico y farmacolo´gico". Universidat de Barcelona, Barcelona. 2005.

[17] Anibijuwon, I.I. and Udeze, A.O, "Antimicrobial Activity of Carica Papaya (Pawpaw Leaf) on Som Pathogenic Organisms of Clinical Origin from South-Western Nigeria". Ethnobotanical Leaflets 13: 850-64. 2009.

[18] Okunola, A.; A.; Muyideen, T. H.; Chinedu, P. A.; Tomisin, J.; Harrison, A.; Victor, U. O. and Babatunde, E. E, "Comparative studies on antimicrobial properties of extracts of fresh and dried leaves of *Carica papaya* (L) on clinical bacterial and fungal isolates". Advances in Applied Science Research, 3 (5): 3107-3114. 2012.

[19] Dogbevi, M.; Vachon, C. and Lacroix, M, "Physicochemical and microbiological changes in irradiated fresh pork loins". Meat Sci., 51, 349-354. 1999.

[20] Food and Drug Administration (FDA), "Irradiation in the production, processing and handling of food". Fed. Regist. 62, 64107-64121. 1997.

[21] Lacroix, M. and Ouattara, B, "Combined industrial processes with irradiation to assure innocuity and preservation of food products: a review". Food Res. Inst., 33, 719-724. 2000.

[22] Mahrour, A.; Caillet, S.; Nketsa-Tabiri, J.; Lacroix, M, "Microbial and sensory quality of marinated and irradiated chicken". J. Food Prot. 66 (11), 2156-2159. 2003.

[23] APHA, "Compendium of Methods for the Microbiological Examination of Foods"; (2nd ed.), American Public Health Association, Washington D.C. 1992.

[24] APHA, "Standard methods for the examination of dairy products". 14th Ed. American Public Association, Washington D.C. 1985.

[25] Oxoid, "The Oxoid Manual of Culture Media Ingredients and other Laboratories Services" (5th ed.), Oxoid Limited, Hampshire, England. 1982.

[26] Roberts, D.; Hooper, W. and Greenwood, M, "Practical Food Microbiology; Methods for the Examination of Food for Microorganisms of Public Health Significance", Public Health Laboratory Service, London. 1995.

[27] Oxoid, "The Oxoid Manual", (8th ed.). Oxoid Ltd., Wade Road, Basingstoke, Hampshire, R G24 8PW, England. 1988.

[28] International Standards Organization, "ISO 7922 Geneva, Switzerland". 1978.

[29] Mwansyemela, N. A.; "Report on Studies of Routine Analysis for Food Chemistry at the Institute for Fisher Products TNO at Ijmuiden" – Holland from 2nd April to 15th September (1992).

[30] Ahn, D. U., Olson, D. G., Jo, C., Love, J., & Jin, S. K, "Volatiles production and lipid oxidation on irradiated cooked sausage as related to packaging and storage". Journal of Food Science, 64, 226-229. 1999.

[31] Wierbicki, E, "Food Irradiation Processing" IAEA Publications, Vienna, Austria, 79. 1985.

[32] SAS Program, "SAS/STAT User's Guide Release 6.12 edition" Cary, NC, USA: SAS Inst. Inc. 1996.

[33] Ouattara, B., Sabato, S.F. and Lacroix, M, "Combined effect of antimicrobial coating and gamma irradiation on shelf life extension of pre-cooked shrimp (*Penaeus* spp.)". International Journal of Food Microbiology, 68, 1-9. 2001.

[34] Hammad, A.A.I., El-Mongy, T.M .and Mabrouk, A.K, "Shelf-life extension and improvement of the microbiological quality of fresh sausage by irradiation". Egypt, J. Rad. Sci., Appl., 13 (1), 57. 2000.

[35] Rady, A.H., Khalaf, H.H., Badr, H.M. and Abdeldaiem, M.H, "Improving the quality of ready-to-eat meals by gamma irradiation. 1-Cooked meat and mashed potatoes". Egypt, J. Rad. Sci., Appl., 37 (1), 35-54. 2005.

[36] Cha´vez-Quintal, P., Gonza´lez-Flores, T. Rodrı´guez-Buenfil, I. and Gallegos-Tintore, S, "Antifungal Activity in Ethanolic Extracts of Carica papaya L. cv. Maradol Leaves and Seeds". Indian J Microbiol., 51 (1): 54-60. 2011.

[37] Yushau, M., Onuorah, F. C. and Murtala, Y, "In-vitro sensitivity pattern of some urinary tract isolates to *carica papaya* extracts". Bayero Journal of Pure and Applied Sciences, 2 (2): 75-78. 2009.

[38] Romasi, E.F., Karina, J. and Parhusip, A.J.N, "Antibacterial activity of papaya leaf extracts against pathogenic bacteria". Makara, Teknologi, 15 (2): 173-177. 2011.

[39] Alabi, O.A.; Haruna, M.T.; Anokwuru, C.P.; Jegede, T.; Abia, H.; Okegbe, V.U. and Esan, B.E, "Comparative studies on antimicrobial properties of extracts of fresh and dried leaves of *Carica papaya* (L) on clinical bacterial and fungal isolates". Advances in Applied Science Research, 3 (5): 3107-3114. 2012.

[40] AL-Dughaym, A.M. and Altabari. G.F, "Safety and quality of some chicken meat products in Al-Ahsa markets-Saudi Arabia". Saudi Journal of Biological Sciences, 17, 37-42. 2010.

[41] Abdeldaiem, "Effect of gamma irradiation and edible coatings containing oregano essential oil on quality of bolti fish fillets". Isotope and Rad. Res., 41 (1): 75-91. 2009.

[42] Rahman, S., Imran M., Muhammad N., Hassan N., Chisthi A. K., Khan A. F., Sadozai K. S. and Khan S. M, "Antibacetial screening of leaves and stem of *Carica papaya*". Journal of Medicinal Plants Research Vol. 5 (20), pp. 5167-5171. 2011.

[43] Azmi, S.M.N., Jamal, P. and Amid, A, "Purification of Xanthine Oxidase Inhibitor from *Carica papaya* Leaves using Reversed Phase Flash Column Chromatography (RPFCC)-High Performance Thin Layer Chromatography (HPTLC)". Australian Journal of Basic and Applied Sciences, 6 (1): 117-122. 2012.

[44] Reddy, S., Ilenrickson, R. l., Olson, H. C, "The influence of lactic acid culryres on ground beef quality Oklahoma Agri. Experiment station, Stillwater, Oklahoma". J. Food Sci. 35-787. 1970.

[45] Ruiz-Capillas, C. and Moral, A, "Sensory and biochemical aspects of quality of whole bigeye tuna (Thunnus obesus) during bulk storage in controlled atmospheres". Food Chemistry, 89 (3): 347-354. 2005.

[46] Mexis, S.F., E. Chouliara, E. and Kontominas, M.G, "Shelf life extension of ground chicken meat using an oxygen absorber and a citrus extract". LWT-Food Science and Technology 49, 21e27. 2012.

[47] Balamatsia, C. C., Rogga, K., Badeka, A., Kontominas, M. G. and Savvaidis, I. N, "Possible role of volatile amines as quality indicating metabolites in MAP packaged chicken fillets: correlation with microbiological and sensory attributes". Food Chemistry, 104, 1622e1628. 2007.

[48] Patsias, A., Badeka, A. V., Savvaidis, I. N. and Kontominas, M. G, "Combined effect of freeze chilling and MAP on quality parameters of raw chicken fillets". Food Microbiology, 25 (4), 575e581. 2008.

[49] Fernandez, J., Perez-Alvarez, J. and Fernandez-Lopez, J, "Thiobarbituric acid test for monitoring lipid oxidation in meat". Food Chemistry, 59 (3): 345-353. 1997.

[50] Abd El-Kader, Z. M, "Lipid oxidation in chicken as affected by cooking and frozen storage". Nahrung. 40: 21-42. 1996.

[51] Skerget, M., Kotnik, P., Hadolin, M., Hras, A. R., Simonic, M. and Knez, Z, "Phenols, proanthocyanidins, flavones and flavonols in some plant materials and their antioxidant activities". Food Chemistry, 89 (2): 191-198. 2005.

[52] Imaga, N.A., Gbenle,G.O., Okochi, V.I., Adenekan S., Duro-Emmanuel, T., Oyeniyi, B., Dokai, P.N. Oyenuga, M., Otumara, A. and Ekeh, F.C, "Phytochemical and antioxidant nutrient constituents of *Carica papaya* and *Parquetina nigrescens* extracts". Scientific Research and Essays, Vol. 5 (16): 2201-2205. 2010.

[53] Ayoola1, G.A., Coker1, H.A., Adesegun, S.A. Adepoju-Bello1, A.A., Obaweya1, K., Ezennia1, E.C. and Atangbayila, T.O, "Phytochemical Screening and Antioxidant Activities of Some Selected Medicinal Plants Used for Malaria Therapy in

Southwestern Nigeria". Tropical Journal of Pharmaceutical Research, September, 7 (3), 1019-1024. 2008.

[54] Tarladgis, B.G., Watts, B.M., Younathan, M.T. and Dugan, L.R.J, "A distillation method for quantitative determination of malonaldehyde in rancid foods". Journal of American Oil Chemist Society 37: 44-48. 1960.

[55] Jayasingh, P., Cornforth, D. P., Brennand, C. P., Carpenter, C. E. and Whittier, D. R. (2002). "Sensory evaluation of ground beef stored in high-oxygen modified atmosphere packaging". Journal of Food Science, 67 (9): 3493-3496. 2002.

[56] Shon, J., Kim, J., Eo, J. and Choi, Y, "Effect of Soy Protein Isolate Coating on Meat Quality of Pork Fresh Cut during Refrigerated Storage". J. Appl. Biol. Chem., 55 (1), 27-34. 2012.

[57] Kang, H.J., Jo, C., Kwon, J.H., Kim, J.H. Chung, H.J. and Byun, M.W, "Effect of a pectin-based edible coating containing green tea powder on the quality of irradiated pork patty". Food Control, 18, 430-435. 2007.

[58] Ouattara, B., Sabato, S.F. and Lacroix, M, "Use of gamma-irradiation technology in combination with edible coating to produce shelf-stable foods". Radiation Physics and Chemistry, 63, 305-310. 2002.

[59] Lacroix, M., Ouattara, B., Saucier, L., Giroux, M. and Smoragiewicz, W, "Effect of gamma irradiation in presence of ascorbic acid on microbial composition and TBARS concentration of ground beef coated with an edible active coating". Radiation Physics and Chemistry, 71, 71-75. 2004.

[60] Javanmard, M., Rokni, N., Bokaie, S. and Shahhosseini, G, "Effects of gamma irradiation and frozen storage on microbial, chemical and sensory quality of chicken meat in Iran". Food Control, 17, 469-473. 2006.

2

Evaluation of the Effect of Packaging Materials and Storage Temperatures on Quality Degradation of Extra Virgin Olive Oil from Olives Grown in Palestine

Jehad Abbadi[1,*], Ibrahim Afaneh[2], Ziad Ayyad[2], Fuad Al-Rimawi[3], Wadie Sultan[3], Khalid Kanaan[3]

[1]Department of Biology, College of Science and Technology, Al-Quds University, Abu Dies, Jerusalem, Palestine
[2]Department of Food Technology, College of Science and Technology, Al-Quds University, Abu Dies, Jerusalem, Palestine
[3]Department of Chemistry and Chemical Technology, College of Science and Technology, Al-Quds University, Abu Dies, Jerusalem, Palestine
*Corresponding author: jihadabbadi@yahoo.com

Abstract The quality of extra virgin olive oil (EVOO) is intimately affected by packaging material and storage temperature. In this study, the influence of packaging materials and elevated temperature on EVOO quality was investigated during six months. At ambient temperatures, oil maintained EVOO when stored in glass, polyethylene terephthalate (PET), high density polyethylene (HDPE), cans and Pottery in terms of chemical tests (acidity, peroxide value, K_{232}, and K_{270}). Loss of phenols was the highest in pottery-stored oil and the lowest was found in glass-stored oil. Only PET-stored oil maintained the EVOO grade in terms of sensory evaluation when stored at room temperature. At elevated temperature, oil stored in all packaging materials lost extra virgin quality in terms of chemical tests. The loss of phenols was the largest in HDPE and smallest in cans-stored oil. Sensory evaluation, maintained glass-stored oil and PET-stored oil as EVOO. This study has reaffirmed that at both storage temperatures, the best container in maintaining the EVOO quality was glass and the worst was pottery. Grading of stored olive oil under investigation using sensory evaluation solely was not sufficient. Also it was clear that the absorption coefficient K_{270} was the most sensitive determinant chemical test that determines the quality of stored olive oil and could be used as a rapid indicator test.

Keywords: *Olea europaea L., olive oil, oil oxidation, stability indicators, storage conditions, packaging materials*

1. Introduction

Olive trees (*Olea europaea* L.) is an important trees internationally, produce high nutritional and health quality edible oil. The global production of olive oil in 2012 was around 2,903,680 tons, from which around 22,950 tons are produced in Palestine. As olive oil production fluctuates from year to year, the mean annual production of olive oil globally during the recent ten years (2003-2012) was 2,946,288 tons and the average annual contribution in Palestine was 17,045 tons [1]. The European Union (EU) is the leading producer of olive oil and within the EU, the Mediterranean members are the biggest producers, accounting for 95% of world production and 85% of world consumption of olive oil [2].

Virgin and extra virgin olive oil is a genuine fruit juice obtained from olive drupes, using exclusively mechanical procedures, without further treatments or chemical additions. Several clinical data have shown that consumption of olive oil can provide heart health benefits, such as favorable effects on cholesterol regulation and LDL cholesterol oxidation, exerting anti-inflammatory, antithrombotic and antihypertensive effects [3]. Quality of olive oil is defined as the combination of its attributes that have significance in determining the degree of its acceptability by the consumer, and may be also defined from commercial, nutritional or organoleptic perspectives. The nutritional value of extra virgin olive oil (EVOO) originates from its high levels of oleic acid content and minor components, such as phenolic compounds that donate the oil its aroma [4]. Therefore, these quality parameters promote the consumption demands and price of olive oil in comparison with other edible oils ranking it superior among vegetable oils [5].

There is a need to develop reliable analytical methods to ensure compliance of olive oil quality with labeling, and to determine the genuineness of the product by the detection of eventual defects during adulterations, processing and storage conditions. Therefore, the International Olive Oil Council (IOOC) and European Communities Legislation (EC) define the identity characteristics of olive oil by specifying analytical methods and standard limit values of the quality parameters such as peroxide value (PV), acidity, Ultra violet (UV) absorbance values (K_{232} and K_{270}) and organoleptic characteristics (odor, taste and color) for olive oils in order to improve product quality, expand international trade, and raise its consumption. The chemical tests and the organoleptic properties categorize olive oil into extra virgin, virgin, and lampant oil indicating its edible quality and marketable values. The extra virgin olive oil is the highest grade and must contains zero defects and greater than zero positive attributes as evaluated by a certified taste panel, and must have a free acidity of less than 0.8%, peroxide value doesn't exceed 20 milliequivalent O_2 kg^{-1} oil and should have clear flavor that reflect the fruit from which it is produced [6,7].

Quality of olive oil is potentially affected by different factors including genetic (tree variety), agronomic (ripening stage, fertilization, irrigation, and harvesting practices), health of the drupe [8], environmental (temperature, day length, and sunlight duration), geographical [9] factors, and finally the postharvest processing including packaging materials and storage conditions [6]. Furthermore, an important European regulation allows the Protected Denomination of Origin (PDO) labeling of some EU EVOOs and this designation guarantees that the geographical origin of the product is closely in conjunction with the quality of the product [11]. The complex interference of these factors make only 50% of the world's olive oil production is classified as extra virgin grade [12].

In order to fulfill the expectations of consumers, good quality control of olive oil should be assured in the course of production and storage line. The quality of olive oil decreases during storage, and is attributable to oxidation that lead to rancidity [5], and to hydrolytic degradations causing partial loss of healthy minor constituents [13]. Preserving the positive attributes of oil is a matter of great concern for the olive oil industry during the time elapsing from production to bottling, and up to purchasing and consumption [14,15], because the variation of storage conditions during olive oil storage and transportation affect its quality [8,16]. During shelf life of bottled extra virgin olive oil, the bottle must be adequately protective against autoxidation that cause rancidity [7]. Several types of plastic films or metal containers can be used, but glass bottles of different shape and color are the most common [14,17]. Although, extra-virgin olive oil is usually packaged in glass, or plastic bottles, these packages have some disadvantages because their bottled contents may be subjected to oxidation [16]. Accordingly, oil producers need to pay a great deal of attention to the type of containers they place the oils in, after production and to the storage conditions they are kept in, before sale [14]. The influence of glass and high density polyethylene on oil quality during storage was frequently studied [17], while little information is known about the effect of high density polyethylene (HDPE), cans, and pottery jars. The effect of different packaging materials on the quality of olive oil is previously reported [7,14,17]. In the other hand, the non-optimal storage conditions, such as those occurring on a store shelf, may alter the qualitative characteristics of the product to the extent that they may eventually illegally differ from those indicated on the label. Thus, an investigation of the type and magnitude of the alterations in oil undergoes during its shelf life at elevated temperature may provide useful information about optimum practical storage or transport conditions that sustain high quality of olive oil for maximum storage period [7].

Although the effect of storage conditions, time and their consequences were studied for olive oils produced in many countries [9,18], there is no published studies - to our knowledge- corresponding to the effect of packaging materials and storage conditions on the quality of Palestinian olive oils except a recent investigation done by the research group of this investigation under different situations [15]. Therefore, the aim of this study is to evaluate different packaging materials (Glass bottles, PET plastic bottles, HDPE plastic bottles, tin plates, and pottery jars) in terms of their protective ability for quality indices of Palestinian extra virgin olive oil (acidity, peroxide value, K_{232}, K_{270}, phenolic compounds, sensory score 6.5) stored under different storage temperatures (18°C and 37°C) in a six months stability study. Additionally it is aimed to find the potential correlations between chemical quality indices with sensory evaluation test to optimize olive oil evaluation.

2. Materials and Methods

2.1. Experimental Design

Olive fruits of the cultivar 'Nabali Baladi' were handpicked in late October 2008 from an olives orchard located in Salfeet district of a Mediterranean climatic region of Palestine. The fruits were selected with no defects and at an optimal stage of ripening (5.5 N detachment force, 3.8 pigmentation index, and 57.5% water content). Washed olives were processed using stone mill and hydraulic press. The initial whole oil sample was filled temporarily in two 20-liter HDPE containers and directly transported to the laboratories of Al-Quds University. Extra virgin quality of the extracted oil was proved (peroxide value < 20, acidity < 0.8%, K_{232} < 2.5, and K_{270} < 0.25, iodine value 75-94, refractive index 1.4677-1.4700, Table 1). The 40 liters extra virgin olive oil was distributed into subsamples (300-ml each) that were bottled in different packaging materials (amber glass bottles, polyethylene terephthalate (PET), high density polyethylene (HDPE), tin plate cans hermetically sealed, and pottery jars with covers), maintaining 2% head space in each bottle. Bottled oil was stored under different storage temperatures (18 ± 1°C and 37 ± 1°C); in thermostatic and ventilated incubators (with 100 Lux normal white light inside for around 10 hours daily simulating the condition on shelves). The samples were rearranged weakly to insure uniform spacial distribution of the bottles. The bottles (in four replicates for each treatment) of different packaging materials were randomized in a complete randomized design (CRD) in each storage condition. The effect of each of these factors (packaging materials and temperature storage conditions) on the stability of the extra virgin olive oil was studied in a non orthogonal design by monitoring oil quality indicators that include: acidity (percent as oleic acid), peroxide value, ultraviolet extinction coefficients (K_{232} and K_{270}), total phenolic contents (expressed as mg of gallic acid kg^{-1} oil), and sensory attributes (Panel test) during six months of the experimental period (0, 30, 60, 90, and 180 days of storage).

Table 1. Quality of olive oil sample initially used in the study

Quality parameter	Value	Unit
Acidity	0.38	g oleic acid per 100 g oil (%)
Peroxide value	10.49	equivalent O_2 per kg oil
Iodine value	82.63	ml I_2 per 100g oil
Saponification value	188	mg KOH per g oil
K_{232}	1.68	absorbance
K_{270}	0.158	absorbance
Density	0.919	g per ml oil
Refractive index	1.46675	-
Sensorial evaluation	0 defect, 4.7 fruity, 5 pungency, 4.5 bitterness	

2.2. Determination of Oil Quality Indicators

Acidity (g oleic acid 100 g^{-1} oil) and peroxide value (milliequivalent O_2 kg^{-1} oil) were determined according to the AOAC [19]. Ultraviolet light absorption indexes (K_{232} and K_{270} extinction coefficients) were determined using the methods described in IOOC [20]. Total Phenol compounds were extracted according to Georgios et al, 2006 [21] and analyzed according to AOAC [19], and their content (mg gallic acid kg^{-1} oil) was determined spectrophotometrically at 765 nm. Sensory evaluation was run by taster team for sensory analyses in the Palestinian Standard Institution laboratory, Ramallah, Palestine. The test was performed by the analytical panel done by 13 trained technicians, working according to the method defined by the Standard IOOC [20]. The results obtained based on the ranking according to the median of notes from the tasters. Each bottle in each treatment was analyzed monthly for each mentioned chemical quality indicator up to six months. The sensory evaluation was inspected in three periods (0, 3, and 6 months).

2.3. Statistical Analysis

Four bottles of each treatment were independently analyzed in each sampling time. The results are expressed as mean ± standard deviation. All statistical analyses were carried out using SAS (SAS Institute Inc., Cary, USA, Release 8.02, 2001). Comparisons of means with respect to the influence of different storage conditions and different packaging materials were carried out using the GLM procedure considering a fully randomized design, treating main factors (packaging materials and storage conditions) separately using one-way analysis of variance. The Bonferroni procedure was employed with multiple t-tests in order to maintain an experiment wise of 5%.

Initially Pearson correlations were calculated to test the relation among quality indicators of stored olive oil at each storage condition separately and when data were pooled. The NOMISS option was used in order to obtain results consistent with subsequent multiple regression studies.

3. Results

3.1. Acidity

Our findings reveal that, acidity of EVOO increased dramatically with increasing storage time in all studied storage containers stored at elevated temperatures, except for that stored in pottery jars, where the highest acidity value was reached after 90 days, then was significantly reduced after 135 and 180 days of storage (Table 2). After 30 days of storage, glass containers retain the highest acidity values followed by pottery followed by cans followed by PET and the least was found in HDPE but the values were statistically not significant in both types of plastic containers. At the end of storage time, only glass and cans exceeded the limit for the extra virgin grade (0.8 % oleic acid), where they shared the highest acidity values in stored oil (0.81 and 0.82 for glass and cans respectively). At the end of storage period, the least acidity value was found in oil stored in pottery, while both types of plastic containers retained the same intermediate acidity values.

At room temperature, acidity of stored EVOO increased slightly but significantly with increasing time of storage. At the end of storage period, the least acidity value was reported in oil stored in pottery, while the other containers maintained similar values significantly. All storage containers protected stored EVOO in terms of acidity and maintained its extra virgin grade throughout storage period. Comparing the acidity in the same container type at the same storage time but different temperature treatments, acidity was higher under elevated temperature in all packaging materials.

Table 2. Acidity (% as oleic acid) at elevated temperatures compared to room temperature (between brackets)

days	Glass	PET	HDPE	Cans	Pottery
0	0.38 E, a (0.38 C), a	0.38 D, a (0.38 C), a	0.38 D, a (0.38 E), a	0.38 E, a (0.38 D), a	0.38 C, a (0.38 D), a
30	0.58* CD, a (0.42 CB), a	0.44 C, c (0.42 B), a	0.41* C, c (0.54 B), a	0.46* D, cb (0.40 C), a	0.50* B, b (0.39 C), a
45	0.57* D, a (0.42 CB), c	0.47* C, cb (0.42 B), c	0.50 B, b (0.50 C), a	0.59* C, a (0.47 CB), b	0.41 C, c (0.42 BC), c
90	0.63* C, b (0.43 B), c	0.53* B, c (0.49 A), ba	0.50 B, c (0.50 C), a	0.71* B, a (0.48 B), cb	0.53* A, c (0.45 CBA), ba
135	0.71* B, a (0.52 A), ba	0.54* B, b (0.50 A), b	0.57* A, b (0.43 D), a	0.73* B, a (0.53 BA), a	0.49 B, c (0.50 BA), b
180	0.81* A, a (0.53 A), a	0.58* A, b (0.51 A), a	0.57* A, b (0.58 A), a	0.82* A, a (0.56 A), a	0.49* B, c (0.51 A), b

Different capital letters within each column or small letters within each line indicate significant difference ($p < 0.05$, $n = 4$). * Indicates significance between different temperature treatments in the same cell of the table at a given P level ($p < 0.05$).

3.2. Peroxide Value

Our results highlighted that, PV of stored EVOO at elevated temperature showed different responses in different packaging materials as a function of storage time (Table 3). It fluctuated in glass containers; where it decreased significantly after 30 days of storage in dramatic manner tell 90 days of storage, then it increased and out-yielded the initial value but without significant difference. PV decreased drastically with time of storage in oil stored in both types of plastic containers, while increased in oil stored in pottery continuously with time of storage and overcame the limit of EVOO grade (20 milliequivalent O_2 kg^{-1} oil) before 135 days of storage. All storage containers except pottery retained the EVOO quality in terms of peroxide value during the experiment when stored at elevated temperature.

At ambient temperature, PV decreased significantly with time in oil stored in glass, PET and cans, while it didn't change significantly in oil stored in HDPE, and it was significantly elevated in oil stored in pottery. At the end of storage time, PV of oil stored at elevated temperature was found significantly higher than that stored at room temperature in glass and pottery, while the opposite was recorded for oil stored in cans. In the other hand, both types of plastic containers maintained peroxide values similar at both storage conditions.

Table 3. Peroxide value (as milliequivalent O_2 kg^{-1} oil) at elevated temperatures compared to room temperature (between brackets)

days	Glass	PET	HDPE	Cans	Pottery
0	10.50 A, a (10.50 A), a	10.50 A, a (10.50 A), a	10.50 A, a (10.50 A), a	10.50 A, a (10.50 A), a	10.50 C, a (10.50 C), a
30	10.50 A, (9.90 A), a	8.10 A, a (8.27 B), a	10.87 A, a (8.70 A), a	10.53* A, a (8.20 B), a	10.8 7* C, a (8.37 D), a
45	7.87* B, cb (7.37 B), b	9.03* BA, cb (8.50 B), a	9.50* A, ba (8.63 A), a	7.30 B, c (8.37 B), a	11.10* C, a (8.13 D), a
90	6.23* C, d (8.23 B), b	8.13* B, c (8.63 B), b	8.83 A, b (9.37 A), b	5.90* B, d (8.17 B), b	16.53* B, a (11.63 CB), a
135	6.73* B, b (8.37 B), c	8.3 B, b (8.67 B), c	9.77* A, b (9.23 A), b	6.20* B, b (8.87 B), cb	22.73* A, a (12.57 B), a
180	10.87* A, b (8.43 B), c	8.43 B, c (8.87 B), c	9.77 A, b (10.03 A), b	6.17* B, d (7.70 B), d	21.83* A, a (14.10 A), a

Different capital letters within each column or small letters within each line indicate significant difference ($p < 0.05$, $n = 4$). * Indicates significance between different temperature treatments in the same cell of the table at a given P level ($p < 0.05$).

3.3. Ultraviolet Extinction Coefficients

3.3.1. Extinction Coefficient at 232 nm (K_{232})

The extinction coefficient K_{232} of olive oil stored in HDPE at elevated temperature under study, increased continuously and significantly with extending time of storage (Table 4). The same response was recorded for oil stored in pottery in the first 135 days but this extinction coefficient was slightly and significantly decreased after 180 days compared to the previous measurement. In glass bottles, K_{232} fluctuated during storage, where it decreased significantly after 45 days, then reached its peak after 135 days, where it was significantly higher than the initial measurement, and at the end of the experiment went back to a value similar to the initial one. In PET, K_{232} showed a trend of increment during the experiment with a higher significant value at the end of the experiment compared with the baseline measurement. This quality index of oil stored in cans fluctuated during storage period; where the initial and final measurements were statistically similar. The extra virgin grade in terms of K_{232} (<2.5) was maintained in oil stored in glass, PET and cans even though they were stored for six months at elevated temperature. But oil stored in HDPE quitted this grade in terms of this quality index at the end of the experiment and that stored in pottery, exceeded 2.5 after 135 days and was marginal to the critical limit at the end of the storage period.

At ambient storage temperature, the extinction coefficient K_{232}, decreased slightly but significantly within the respective testing dates in oil samples stored in all packaging materials under study except for pottery jars, where a significant increase was reported after 90 days in pottery and the rate of increase was maintained tell the end of the experiment. None of the samples stored in either packaging material at ambient temperature exceeded the higher limit of K_{232} determining extra virgin quality of olive oil. Values of K_{232} measured at each testing time for each packaging material was found significantly higher in oil stored at elevated temperature compared to oil stored at ambient temperature, and this was true for all packages under study.

Table 4. K_{232} at elevated temperatures compared to room temperature (between brackets)

days	Glass	PET	HDPE	Cans	Pottery
0	2.02 B, a (2.02 A), a	2.02 C, a (2.02 A), a	2.02 D, a (2.02 BA), a	2.02 BC, a (2.02 A), a	2.02 E, a (2.02 D), a
30	2.04* B, b (1.77 D), b	2.03 C, b (2.03 A), a	2.09* D, ba (2.04 BA), a	2.15* A, a (2.02 A), a	2.09* D, ba (2.03 D), a
45	1.90* C, c (1.74 D), c	2.31* A, ba (2.01 BA), ba	2.26* C, b (2.08 A), a	1.91 D, c (1.84 B), cb	2.40* C, a (2.05 D), a
90	1.92* C, e (1.85 C), c	2.16* B, c (2.02 A), b	2.33* C, b (1.75 C), d	1.94* DC, d (1.88 B), c	2.37* C, a (2.23 B), a
135	2.34* A, c (1.90 CB), c	2.17* B, d (1.98 CB), b	2.49* B, b (1.95 BA), b	1.92 D, e (1.96 A), b	2.62* A, a (2.10 C), a
180	2.04* B, e (1.96 BA), b	2.31* A, c (1.96 C), b	2.60* A, a (1.87 CB), c	2.09* BA, d (1.83 B), c	2.48* B, b (2.36 A), a

Different capital letters within each column or small letters within each line indicate significant difference ($p < 0.05$, $n = 4$). * Indicates significance between different temperature treatments in the same cell of the table at a given P level ($p < 0.05$).

3.3.2. Extinction Coefficients at 270 nm (K_{270})

Extinction coefficient measured at 270 nm (K_{270}) of stored olive oil at elevated temperature increased progressively in significant values with increasing time of storage in all studied packaging materials under study (Table 5). At the end of the experiment, the highest K_{270} value was found in oil stored in HDPE, followed by pottery without significant difference, followed by PET, followed by cans, and the least value was recorded in oil stored in glass bottles. All storage containers deteriorate stored olive oil and quitted from extra virgin grade in terms of K_{270} (< 0.2) when oil was stored at elevated temperature but at different storage periods. PET bottles retained stored oil as extra virgin in terms of K_{270} for less than 135 days, and that stored in glass and cans for less than 90 days, and for that stored in HDPE and pottery for less than 45 days.

At ambient temperature, K_{270} was slightly increased in oil stored in glass, PET, and pottery, while it was not affected in oil stored in cans, but was significantly decreased in oil stored in HDPE. None of packaging materials under investigation elevated K_{270} of stored olive oil to the critical limit of extra virgin grade when oil stored at ambient temperature for six months. K_{270} values of oil stored at elevated temperature was higher than that stored at room temperature in all packaging materials under study in most storage periods.

Table 5. K_{270} at elevated temperatures compared to room temperature (between brackets)

days	Glass	PET	HDPE	Cans	Pottery
0	0.160 D, a (0.160 B), a	0.160 E, a (0.160 B), a	0.160 F, a (0.160 CB), a	0.160 D, a (0.160 A), a	0.160 D, a (0.160 C), a
30	0.187 CB, a (0.180 A), ba	0.193 DC, a (0.197 A), a	0.200* E, a (0.180 BA), ba	0.187* C, a (0.160 A), c	0.160* D, b (0.173 CB), cb
45	0.180* C, c (0.163 B), a	0.230* B, a (0.190 A), a	0.217* D, b (0.203 A), a	0.183* CB, c (0.163 A), a	0.213* C, b (0.197 A), a
90	0.203* BA, dc (0.160 B), bc	0.190* D, d (0.187 A), a	0.230* C, a (0.147 C), c	0.210* A, cb (0.167 A), ba	0.220* C, ba (0.187 BA), a
135	0.210* A, b (0.160 B), b	0.207* C, b (0.180 B), a	0.263* B, a (0.153 C), b	0.210* A, b (0.170 A), ba	0.247* B, a (0.167 C), ba
180	0.197 BCA, c (0.180 A), a	0.260* A, b (0.187 A), a	0.290* A, a (0.137 C), b	0.200* BA, c (0.170 A), a	0.280* A, a (0.190 BA), a

Different capital letters within each column or small letters within each line indicate significant difference ($p < 0.05$, $n = 4$). * Indicates significance between different temperature treatments in the same cell of the table at a given P level ($p < 0.05$).

3.4. Total Phenolic Compounds

Storage at elevated temperature significantly reduced total phenolic compounds of EVOO stored in all packaging materials under study (Table 6). Total phenols were significantly and highly reduced at all consequent storage periods in oil stored in PET, HDPE, and pottery, while in glass and cans, the successive reduction of phenolic compounds were reported until 135 days of storage but were significantly elevated at the end of storage period. Comparing phenolic compounds contents of stored olive oil at the end of storage period related to their initial contents in the same packaging material, the most reduced contents of phenolic compounds was found in HDPE followed by pottery followed by PET followed by glass and the least was recorded in oil stored in cans.

At room temperature storage condition, phenolic compounds were dramatically and significantly reduced with consecutive increase of storage period. At the end of storage period, the largest loss of phenolic compounds was found in pottery followed by HDPE, followed by cans and PET, and the least reduction of phenolic contents was recorded in glass.

Table 6. Total phenols at elevated temperatures compared to room temperature (between brackets)

days	Glass	PET	HDPE	Cans	Pottery
0	213.3 A, a (213.3 A), a	213.3 A, a (213.3 A), a	213.3 A, a (213.3 A), a	213.3 A, a (213.3 A), a	213.3 A, a (213.3 A), a
30	194.3* B, b (203.0 B), b	213.7 A, a (206.3 BA), b	196.7* CB, b (207.7 B), ba	190.0* CB, bc (214.3 A), a	185.3 B, c (185.3 B), c
45	188.7* C, c (201.3 B), b	212.7* A, a (202.33 CB), b	197.7* B, b (201.7 C), b	184.7* DC, cd (213.7 A), a	182.7* B, d (178.6 CB), c
90	182.7* D, c (200.3 CB), a	202.7* B, a (199.7 CB), a	188.0* DC, b (183.3 E), bc	181.7 DC, c (195.3 B), ba	143.7* C, d (170.0 DC), c
135	180.3* D, b (196.3 DC), a	197.0* B, a (195.0 C), ba	183.0* D, b (191.7 D), b	179.3* D, b (194.7 B), ba	140.7* C, c (161.3 D), c
180	191.7* CB, ba (196.0 D), a	182.7* C, c (184.6 D), b	123.3* E, b (167.3 F), c	196.7* B, a (184.3 B), b	133.0* D, c (161.0 D), d
Reduction‡ (%)	10.1 (8.1)	14.3 (13.5)	42.2 (21.6)	7.8 (13.6)	37.6 (24.5)

Different capital letters within each column or small letters within each line indicate significant difference ($p < 0.05$, $n = 4$). * Indicates significance between different temperature treatments in the same cell of the table at a given P level ($p < 0.05$).

‡Total reduction of phenolic compounds at the end of storage period based on the initial contents of total phenols.

Total phenols contents in oil stored in all packaging materials under study was found less in oil stored at elevated temperature than that stored at ambient temperature when compared in the same packaging material and the same testing date after all consecutive storage periods except for oil stored in cans at the end of the storage period where the opposite was recorded.

3.5. Sensory Evaluations

Sensory evaluation was done for all samples subjected to storage conditions in three periods (before storage, after three months, and after six months of storage). Sensory evaluation (Table 7) reveals that, oil stored in glass sustained the extra virgin grade under elevated temperature throughout the experiment, and also at room temperature till 90 days and then turned to virgin grade because of the appearance of sensory defects (Figure 1). The fruity of the glass-stored oil at both storage conditions decreased consequently with increasing time of storage. Caned oil responded the same at both storage conditions, and lost the extra virgin grade before 90 days of storage then remained in the virgin grade throughout the experiment, because fruity of oil was lost and sensory defects appeared during storage. Oil stored in PET sustained the extra virgin grade for six months without sensory defects but with marginal loss in fruity. In HDPE, oil at both storage conditions became virgin after 90 days. Oil fruity decreased largely before 90 days of storage but the sensory defects appeared at the end of the experiment. Because of the complete loss of oil fruity, and the appearance of high level of sensory defects (>2.5), oil stored in pottery quitted from the virgin grade at both storage conditions before 90 days of storage.

Table 7. Olive oil grading according to the sensory evaluation for oil samples stored at elevated temperatures compared to room temperature bottled in different packaging materials

Container	days	Sensory Defects		Sensory Fruity		Olive oil grade	
		Elevated T	Room T	Elevated T	Room T	Elevated T	Room T
	0	0.0	0.0	4.9	4.9	EVOO	EVOO
Glass	90	0.0	0.0	1.9	2.0	EVOO	EVOO
	180	0.0	0.8	1.0	1.0	EVOO	VOO
	0	0.0	0.0	4.9	4.9	EVOO	EVOO
Can	90	1.5	1.5	0.0	0.0	VOO	VOO
	180	2.3	2.3	0.5	0.5	VOO	VOO
	0	0.0	0.0	4.9	4.9	EVOO	EVOO
PET	90	0.0	0.0	3.0	3.0	EVOO	EVOO
	180	0.0	0.0	2.6	2.6	EVOO	EVOO
	0	0.0	0.0	4.9	4.9	EVOO	EVOO
HDPE	90	0.0	0.0	1.2	1.2	EVOO	EVOO
	180	1.9	1.9	1.3	1.3	VOO	VOO
	0	0.0	0.0	4.9	4.9	EVOO	EVOO
Pottery	90	3.0	3.0	0.0	0.0	OVOO	OVOO
	180	3.5	3.5	0.0	0.0	OVOO	OVOO

EVOO is extra virgin olive oil, VOO is virgin olive oil, OVOO is ordinary virgin olive oil.

Table 8. Olive oil grading according to the sensory evaluation and other stability indices for oil samples stored at different temperatures bottled in different packaging materials

Container	days	Acidity	PV	K_{232}	K_{270}	Sensory evaluation		Olive oil grade
						Defect	Fruity	
Elevated temperature								
	0	0.38	10.49	2.02	0.16	0.0	4.9	EVOO
Glass	90	0.63	6.23	1.92	0.20	0.0	1.9	EVOO
	180	**0.81**	10.88	2.04	0.19	0.0	1.0	VOO
	0	0.38	10.49	2.02	0.16	0.0	4.9	EVOO
Can	90	0.71	7.30	1.94	0.21	**1.5**	**0.0**	VOO
	180	**0.82**	6.21	2.09	0.20	**2.3**	0.5	VOO
	0	0.38	10.49	2.02	0.16	0.0	4.9	EVOO
PET	90	0.53	8.13	2.16	0.19	0.0	3.0	EVOO
	180	0.58	8.44	2.31	**0.26**	0.0	2.6	OVOO
	0	0.38	10.49	2.02	0.16	0.0	4.9	EVOO
HDPE	90	0.50	8.82	2.33	**0.23**	0.0	1.2	VOO
	180	0.57	9.77	**2.60**	**0.29**	**1.9**	1.3	OVOO
	0	0.38	10.49	2.02	0.16	0.0	4.9	EVOO
Pottery	90	0.53	16.53	2.37	**0.22**	**3.0**	**0.0**	OVOO
	180	0.50	**21.81**	2.48	**0.28**	**3.5**	**0.0**	OVOO
Room temperature								
	0	0.38	10.49	2.02	0.16	0.0	4.9	EVOO
Glass	90	0.51	8.42	1.82	0.22	0.0	2.0	EVOO
	180	0.66	8.23	2.07	**0.27**	**0.8**	1.0	VOO
	0	0.38	10.49	2.02	0.16	0.0	4.9	EVOO
Can	90	0.47	8.17	1.88	0.17	**1.5**	**0.0**	VOO
	180	0.56	7.72	1.83	0.17	**2.3**	0.5	VOO
	0	0.38	10.49	2.02	0.16	0.0	4.9	EVOO
PET	90	0.43	7.99	2.03	**0.24**	0.0	3.0	VOO
	180	0.52	8.55	1.85	**0.23**	0.0	2.6	VOO
	0	0.38	10.49	2.02	0.16	0.0	4.9	EVOO
HDPE	90	0.49	9.42	1.73	0.18	0.0	1.2	EVOO
	180	0.56	10.84	1.80	0.21	**1.9**	1.3	VOO
	0	0.38	10.49	2.02	0.16	0.0	4.9	EVOO
Pottery	90	0.51	11.63	2.23	0.19	**3.0**	**0.0**	O VOO
	180	0.42	14.12	2.36	0.19	**3.5**	**0.0**	O VOO

EVOO is extra virgin olive oil, VOO is virgin olive oil, OVOO is ordinary virgin olive oil.

3.6. Pearson Correlation with Oil Quality Parameters as Affected with Temperature Treatments

Pearson correlations between quality parameters of olive oil stored at room temperature (Table 9) show that peroxide value was positively and significantly correlated with K_{232} extinction coefficient but the correlation with phenolic contents was significantly negative, and also insignificantly correlated with K_{270}. K_{232} was significantly and positively correlated with K_{270} but was significantly and negatively correlated with phenolic contents. There was no significant correlation found between K_{270} and phenolic contents. At elevated temperatures, peroxide value was significantly and positively correlated with both extinction coefficients, and was also significantly and negatively correlated with phenolic contents. K_{232} was highly correlated with both K_{270} and phenolic compounds content but the correlation with the formers was positive while the correlation with the later was negative. K_{270} and phenolic contents was highly negatively correlated with each other. Pearson correlation between quality parameters of olive oil stored at both room and elevated temperature when all data was pooled (Table 10) shows that peroxide value was highly positively and significantly correlated with K_{232}. K_{270} was significantly and positively correlated with peroxide value and K_{232}. Phenolic contents showed highly negative and significant correlation with all quality parameters under study (peroxide value and K_{232}, and K_{270}).

Table 9. Pearson coefficients between quality parameters of oil stored at room temperature (above the diagonal) and at elevated temperature (below the diagonal)

	Peroxide	K_{232}	K_{270}	Phenols
Peroxide	-	0.608***	-0.055	-0.372***
K_{232}	0.559***	-	0.458***	-0.251*
K_{270}	0.311**	0.789***	-	-0.103
Phenols	-0.530***	-0.643***	-0.734***	-

Table 10. Pearson coefficients between quality parameters of olive oil stored at both room temperature and elevated temperature (pooled data)

	K_{232}	K_{270}	Phenols
Peroxide	0.550***	0.285***	-0.499***
K_{232}		0.779***	-0.546***
K_{270}			-0.584***

4. Discussion

The value of EVOO that determines it's commercial and health quality originates from its high oleic acid contents and the presence phenolic compounds that donates it the special aroma and antioxidant activity [12,22]. Olive oil quality and stability are principally affected by lipid oxidation, generating off-flavor (rancidity) and reduction in oil nutritional value causing health risks and even toxicity for consumers. Lipid peroxidation produces toxic compounds which causes lung damage. In addition to this effect, reactions between peroxidized lipids and proteins have been shown to cause loss of enzyme activities, polymerization, accelerated formation of brown pigments and the destruction of essential amino acids such as histidine, lysine, tryptophan and methionine. Aldehydes, ketones, hydrocarbons and furans, are known as the cleavage products of hydroperoxides, cause reduction in protein solubility, and reduction in nutritional value of proteins. As well, lipid oxidation provokes a decrease in nutritional values of some vitamins such as A, D, E and K. From the health point of view, lipid radicals and oxidation products contribute in aging, DNA damage, Parkinsonism, carcinogenesis, and coronary heart diseases [23].

As lipids oxidize, they form hydroperoxides, which are susceptible to further oxidation or decomposition to secondary reaction products, such as aldehydes, ketones, acids, and alcohols. In many cases, these compounds adversely affect flavor, aroma, taste, nutritional value, and overall quality. The oxidation process of triglycerides is complex because it always takes place by chain reactions either in dark involving free radicals, called autoxidation, or light-dependent reactions known as photooxidation [22]. Many catalytic systems such as light, temperature, enzymes (lipase), metals, and microorganisms, can accelerate lipids oxidation [24]. Variation during olive oil storage and transportation that enhance lipid oxidation is common, and may be attributed to natural or climatic condition and to extreme storage conditions [8,18]. In addition to storage conditions, the retention of oil quality for an extended period of time that allows its worldwide distribution is also highly affected by the type of packaging material [25]. Knowledge about packaging materials, and their interactions with the bottled oil, along with a deeper understanding of the oxidation pathways under various storage conditions provide necessary information for improving the quality of packaged olive oil during shelf life and transportation [17].

Therefore, in order to fulfill the consumer's requirements, good quality control of olive oil should be assured in the course of production and storage processes. The quality of olive oils is interpreted in terms of measurements of analytical parameters for which certain limit values are set. The most important quality requirements of olive oil in commercial transactions are: acidity, peroxide value, K_{232}, K_{270}, and total phenolic content in addition to the sensory evaluation. These parameters have been evaluated for the Palestinian olive oil samples under investigation as stability-indicators in terms of storage time in response to different packaging materials and storage temperatures.

4.1. Acidity

Acidity is mainly determined by titration using potassium hydroxide that measure the amount of free fatty acids (FFA's) present in the oil as oleic acid which is the major component in the triglycerides present in the olive oil, and should be less than 0.8% if the oil is extra virgin [26]. Although, acidity values are used as a basic criterion for classifying different categories of olive oil, it was not considered as the best criterion for evaluating olive oil quality by some investigators [27]. Acidity reflects oil stability and susceptibility to rancidity. The hydrolytic rancidity of oil due to presence of water and the catalytic action of the lipase (often derived by microorganisms) in oil as mentioned above, partially degrade triglycerides giving glycerol and free fatty acids, which increase acidity.

In agreement with our findings, acidity of EVOO stored in glass increased with increasing storage time but didn't exceed the limits during storage at room temperature [28,29], but exceeded the limit at elevated temperatures [28,30]. While in contrary to our results, other reporters [30,31] found that acidity of oil stored in glass didn't change significantly during storage at room temperature. Several studies conducted on olive oil shelf life attested the glass as the best material for the storage [32], in terms of its acidity, especially when oil was stored in the dark with respect to other packages [33]. As acidity values of oil stored at room temperature in glass, PET, HDPE, and cans didn't differ significantly in our experiment, other investigators clearly indicated the glass as the best (less value) in terms of acidity in the following ranking Glass > HDPE > PET [34]. Metal containers have the same water resistant properties as glass and may protect the product from oxygen, light, and microorganisms that could increase the acidity of oil through increasing the rate of hydrolysis of triglycerides. But when oil was stored at elevated temperature, our results reported both glass and cans as the worst packaging materials in terms of acidity of stored oil which exceeded the extra virgin grade limit, while plastic materials (PET and HDPE) where found better and pottery was reported as the best. This can be explained by the high thermal conductivity of glass and cans compared to plastic ones, and for the cooling effect of pottery on stored oil.

4.2. Peroxide Value

Peroxide value (PV), a measure of total peroxides in olive oil (meq. O_2 kg^{-1} oil) is a major guide of oil quality. The official determination method is based on the titration of iodine liberated from potassium iodide by peroxides present in the oil. In other words, the peroxide value is a measure of the active oxygen bound by the oil which reflects the hydroxyperoxide value, and measures the degree of lipid peroxidation. The higher the number means the greater degradation due to oxidation with an upper limit of 20 meq. O_2 kg^{-1} oil, but levels higher than 10 may mean less stable oil with a shorter shelf life [35]. In lipid oxidation reactions, many free radicals and oxygen species, such as singlet oxygen are involved. The main substrates for these reactions are unsaturated fatty acids and oxygen. The free radical mechanism of lipid oxidation is usually described in a three stages chain reaction including initiation, propagation, and termination steps. Initiation starts with the abstraction of a hydrogen atom adjacent to a double bond in a fatty acid molecule, by the catalytic effect of light, heat, or metal ions to form a free radical, where direct reaction of fatty acid molecule with oxygen does not take place frequently, because of the high activation energy. The resultant free radical reacts with atmospheric oxygen to form an unstable peroxy free radical may in turn abstract a hydrogen atom from another unsaturated fatty acid to form a hydroperoxide. A new alkyl free radical initiates further oxidation and contributes to the chain reaction, and this chain reaction is called propagation stage of autoxidation. The chain reaction may be terminated by formation of nonradical products resulting from combination of two radical species. The propagation stage in autoxidation process includes an induction period when hydroperoxides formation is minimal. The rate of oxidation of fatty acids increases in relation to their degree of unsaturation, therefore, oils that contain high proportions of polyunsaturated fatty acids may experience instability problems. The breakdown products of hydroperoxides, such as alcohols, aldehydes, ketones, furans, esters, lactones and hydrocarbons, generally cause off-flavors, and may also interact with other food components and change their functional and nutritional properties [36].

In accordance with our results, other investigators [5,30,37] found that PV of oils bottled in glass and PET stored at room temperature fluctuated during storage time and did not exceed the official limit during six months of storage. In the other hand, a linear increase in PV with storage time at room temperature in oil stored in glass bottles [9,10] and in tin plates [38] was reported. In the same line with our results in oil stored in glass, PET, and cans at room temperature and elevated temperature, PV decreased significantly with increasing storage time [7]. But in contrast with our findings except for oil stored in pottery, other scientists [22] reported an increase in the PV of oil samples stored under elevated temperature. In accordance with our results, fluctuation in the PV of oil samples stored at elevated temperature [10], and at shelf [15] was reported. The decrease in the PV with increasing time in many testing dates observed in our results in different packaging materials and at both storage temperatures, can be explained by the degradation of primary oxidation products (peroxides) to form secondary oxidation products which can be detected by K_{232} values. The results of PV was correlated with that obtained by K_{232} (Table 9, Table 10) and agreed with other reporters [7,28,39]. Generally, during the beginning of storage, PV in different packaging materials increased as a consequence of the action of both diluted and headspace oxygen in the containers and additionally, the temperature which induce a rapid deterioration of oil in terms of PV. After a period of storage, the PV progressively decrease because of the degradation of primary products into secondary products, which is more obvious in the samples packed in cans and glass containers and less in those packed in plastic and pottery. The oil samples packed in pottery and stored at both room and elevated temperatures have higher peroxide values compared to those stored in other containers. These results may point to the probable intrusion of oxygen and water through pottery, although it is impermeable to light with low thermal conductivity that retain primary oxidation products for longer time and delay their destruction to produce secondary oxidation products.

4.3 Ultraviolet Extinction Coefficients

Determination of the absorption coefficients in the ultraviolet region (232 nm and 270 nm) reflects the stage of oxidation for olive oil during storage [40], in which the shelf-life of virgin olive oil is determined by the increase in the K_{232} absorption coefficient [41], or by means of the time required to reach the upper legal limit of K_{270} absorption coefficient [7, 42]. Primary oxidation products in olive oil (fatty acid hydroperoxides and oxidized triacylglycerols) are measured as peroxide value and K_{232} absorption coefficient (measure the conjugated dienes), while secondary oxidation products (aldehydes, alcohols, ketones and hydrocarbons) are detected by K_{270} absorption coefficient [40,42]. Hydroperoxides are the initial products of oxidation -very sensitive and comparatively unstable- and used as indicator of the early stages of oxidative deterioration in the oxidation process [17,43], while the K_{270} index is used to study the behavior of the secondary oxidation products by the formation of dimers and polymers of triacylglycerides [42].

For instance, an increase in K_{232} and K_{270} values is very common between extraction of olive oil and its consumption as affected by storage time and conditions [22]. It is documented that heat affects olive oil quality by increasing the trienes formation, measured by K_{270} [9], more than the dienes measured as K_{232} [5]. In agreement with our findings, K_{270} values were affected by the heat exposure conditions more than that of K_{232}, with higher values reported in the samples stored at elevated temperature than in those kept at ambient temperature [44]. Such a response is due to the degradation of primary oxidation products (peroxides) to form secondary oxidation products, as K_{232} representing the amount of conjugated dienes of the primary oxidation products [7,28,39] and are transferred to trienes measured by K_{270} [45].

Our findings are in agreement with previously reported results [37] which found that K_{270} of oil stored in glass bottles and PET containers at elevated temperatures, exceeded the limit of extra virgin grade after two and three months of storage for glass and PET respectively. Also in the same line with our results, other investigators [30] reported an increase in K_{270} of oil samples stored in glass and PET at room temperature throughout the storage, but in contrary with our findings, they found that K_{270} values exceeded the limit (0.2) after two months of storage. The increase in K_{232} with increasing time of oil -in contrary with our findings- was reported [30,37] when oil bottled in glass and PET container stored at room temperature but the values did not exceed the official limit, and values in glass overcame that in PET. Because of the significant variation of K_{270} values during olive oil storage as a response to oil oxidation, this parameter may be of capital importance to control the quality of stored extra virgin olive oils in terms of determining the time at which they will lose their "extra" category [7].

4.4. Total Phenolic Compounds

Extra virgin olive oil, is one of the few oils being consumed without any chemical treatment. It has high resistance to oxidative deterioration mainly due to its fatty acid composition -high monounsaturated to polyunsaturated ratio- and to the presence of natural antioxidants, especially phenolic compounds, carotenoids, and tocopherols, therefore delay the oxidation of lipids and the production of the undesirable volatile compounds [8,22]. During oil storage, the hydrolysis, esterification and oxidation deplete the minor constituents, because of the action of phenolic compounds as antioxidants mainly at the initial stage of autoxidation [46] by scavenging free radicals and chelating metals. Accordingly, the determination of the minor constituents in olive oil is essential for the analytical assessment of its quality and self protection potential.

In agreement with previous reports [7,28,47], our findings showed that total polyphenolic contents of extra virgin olive oil under investigation decreased during storage in all means of packaging materials and storage conditions (Table 6); due to degradation of these compounds that was well fitted to first order kinetics. At the end of storage period, the phenolic compounds of samples stored at elevated temperature showed significantly higher reduction than those stored at ambient temperature [5,17,30,32] in all types of packaging materials except those were stored in cans. Some reporters [30] found that total phenols of oils bottled in glass and PET container didn't show significant decrease during storage, while others reported an increase in phenolic compounds contents with increasing time of storage [10,28], a situation found in our findings when oil was stored in glass and cans after 180 days of storage compared to the previous sampling date (135 days), which could be due to hydrolysis of secoiridoid derivatives in oil. As phenolic compounds act as natural antioxidants in oil and inhibit autoxidation of lipids (RH) by trapping intermediate peroxyl radicals [48], their reduction during storage is a result of oil oxidation [38].

The stability of virgin olive oil also depends on the presence of pro-oxidant substances as well as on factors linked to the storage conditions, namely the presence of oxygen, temperature and above all light exposure, therefore, the level of degradation of an oil results from a balance of all these factors [14]. The phenolic compounds act by giving an electron so that they can interrupt the radical reaction occurring with oxidation. The carotenoids act as electron acceptors, quenching the singlet oxygen. Finally, tocopherols act both as electron donors, slowing down the oxidative reaction, and as electron acceptors, determining the singlet oxygen quenching or scavenging, with consequent inhibition of the oxidation of lipids [49]. At the beginning of storage time, olive oil under this study contained 214 $\pm$ 1.5 mg kg^{-1} oil of total phenolic compounds, and this value was in consistent with the data (121-410 mg kg^{-1}) reported previously [15]. Afterwards, the total content of phenols decreased as a function of time, with various degree of reduction among the storage containers, and the decrease was more pronounced under elevated temperature storage condition. Table 6 showed that the lowest difference between the initial and final antiradical activity (percentage loss of total phenols) at ambient temperature was in glass bottles (8.1%), followed by PET (13.5%) and cans (13.6%), followed by HDPE (21.6%), and the highest reduction was found in pottery (24.5%) stored at room temperature. But concerning the reduction of phenolic compounds in glass and cans was more pronounced after 135 day of storage in a reduction percentage similar to each other and to PET (15.5% and 15.9% for glass and cans respectively). It was previously reported that glass bottles kept more phenolic compounds than that stored in PET containers [15]. The reduction of antioxidants in plastic containers could be due to their permeability to oxygen and the migration of active compounds between oil and packaging material. The large reduction found in oil stored in pottery could be due to the penetration of both oxygen and moisture which both accelerate the hydrolysis of fatty acids, formation of radicals and the depletion of antioxidants.

At elevated temperature, the highest reduction in phenol compounds was found in HDPE (42.2%), followed by pottery (37.6%), followed by PET (14.3%), followed by glass bottles (10.1%), and the least reduction was found in cans (7.8%). Although cans and glass bottles have the highest thermal conductivity, they showed the least reduction in phenolic compounds. This can be discussed by the effect of oxygen penetration on the oxidation of oil and the consequent reduction of antioxidant compounds including total phenols and the more oxygen penetration through PET, HDPE, and pottery stated clearly that phenolic compound loss intensity during storage is directly proportional to the attitude and degree of oxidation occurred in the presence of oxygen.

4.5. Correlation among PV, K_{232}, K_{270}, and Phenolic Compounds

As the oxidation process of olive oil triglycerides occur as a consecutive chain reaction, and each stage in this oxidation pathway could be monitored by quality indicator(s), olive oil quality indices are correlated to each others. PV is correlated with the K_{232} value not only at time zero but also during storage. The significant correlation between K_{232} and peroxide value is expected as both parameters reflect primary oxidation products of the oil and therefore positive correlation was observed and was previously reported [44]. Therefore, for safety issues, PV determination could be excluded from the routine control of olive oil and replaced by K_{232} determination and the use of unwanted chemicals used in PV analysis could be avoided. No significant correlation was found between K_{270} and peroxide value as K_{270} reflects the secondary oxidation products of the oil. Regarding the negative correlation between peroxide value and phenolic content, this correlation is expected because when phenolic content decreases (by oxidation), the peroxide value increases and this explain why a negative correlation was observed. K_{232} and K_{270} are positively correlated which implies that there is a direct relationship between primary oxidation products and secondary oxidation products i.e. as primary oxidation products increases, secondary oxidation products increases too [44].

A close look at Pearson coefficients of quality indicators of oil stored at elevated temperature as compared to ambient temperature (Table 9) reveals that the correlation was stronger at the former storage condition as compared to the later, indicating that, the deterioration rate at elevated temperature is higher. Moreover the correlation between PV and K_{232} -which both indicate the primary oxidation products- are similar at both temperature treatments while the correlation between both mentioned indicators and K_{270} which indicates secondary oxidation products was higher at elevated temperature (there was no correlation between PV and K_{270} at room temperature). This highlight that the rate of transfer from primary oxidation products to secondary oxidation products is higher at elevated temperature as compared to that at room temperature [9]. This was also clearly observed in the presence of high negative Pearson coefficients at elevated temperature between total phenols and K_{270} (secondary oxidation products) compared to insignificant correlation between both indicators at room temperature. Also the correlation between the phenols in one hand and both PV an K_{232} (primary oxidation products) in the other hand was more negative at elevated temperature as compared to that at room temperature. This indicate that the formation of both primary and secondary oxidation products contribute to the depletion of phenolic compounds at higher temperature while the main contributor in the depletion of phenolic compounds under room temperature was the presence of primary oxidation products proving the importance of phenolic compounds as antioxidants in early stages of autoxidation [46].

4.6. Sensory Evaluation

The consumer expresses his judgment on olive oil quality considering some sensory characteristics, such as the pungent taste, fruity and mild flavor. A wide range of preferences within this context can be found, because the sensory quality may match cultural aspects or simple dietary habits. Characteristic aroma and in particular green and fruity features of olive oil originates from many volatile compounds derived from the degradation of polyunsaturated fatty acids through a chain of enzymatic reactions known as the lipoxygenase pathway which takes place during the oil extraction process [50,51]. Beside volatile compounds, non-volatile compounds such as phenolic compounds also stimulate the tasting perception of bitterness and pungency. The concentrations of volatile compounds depend on the enzymatic activity [52], and though, the external parameters (e.g. climate, soil, harvesting and extraction conditions) may alter the inherent olive oil sensory profile [53]. The aroma of olive oil is attributed to aldehydes (hexanal, trans-2-hexenal, acetaldehyde), alcohols (methanol, hexan-1-ol, 3-methylbutan-1-ol), esters (methyl acetate, ethyl acetate, hexyl acetate), hydrocarbons (2-methylbutane, hexane, nonane), ketones (2-butanone, 3-methyl-2-butanone, 3-pentanone), furans and other undefined volatile compounds. The major volatiles in virgin olive oils are C6 and C5 volatile compounds [50,54].

Evaluating the quality of stored olive oil in terms of its grade of virginity as influenced by different packaging materials using both chemical and sensory tests is shown in Table 8. At room temperature, the best type of container was shared by glass and HDPE (sustained EVOO grade for more that 90 days and was found VOO after six months of storage), followed by cans and PET (was found VOO after 90 days and 180 days), and the worst container was pottery which was found ordinary virgin olive oil (OVOO) after 90 days of storage. At elevated temperature, glass containers were superior and pottery was inferior while the other types of containers were intermediate.

Considering both chemical and sensory tests (Table 8), results reveal that, the quality of olive oil stored at room temperature deviated from the extra virgin grade because of the absorption coefficient K_{270} (which was the only determinant chemical test) along with the sensory evaluation parameters (presence of sensory defect and/or absence of sensory fruity, Figure 1). At elevated temperature (Table 8), the most relevant chemical test contributed in the loss of oil quality was K_{270} followed by sensory evaluation parameters, followed by acidity and both PV and K_{232} were the least contributors. Table 8 revealed that, grading of stored olive oil under investigation using sensory evaluation without chemical analysis is not sufficient. Also it is clear that the absorption coefficient K_{270} was the most sensitive determinant chemical test that determines the quality of stored olive oil.

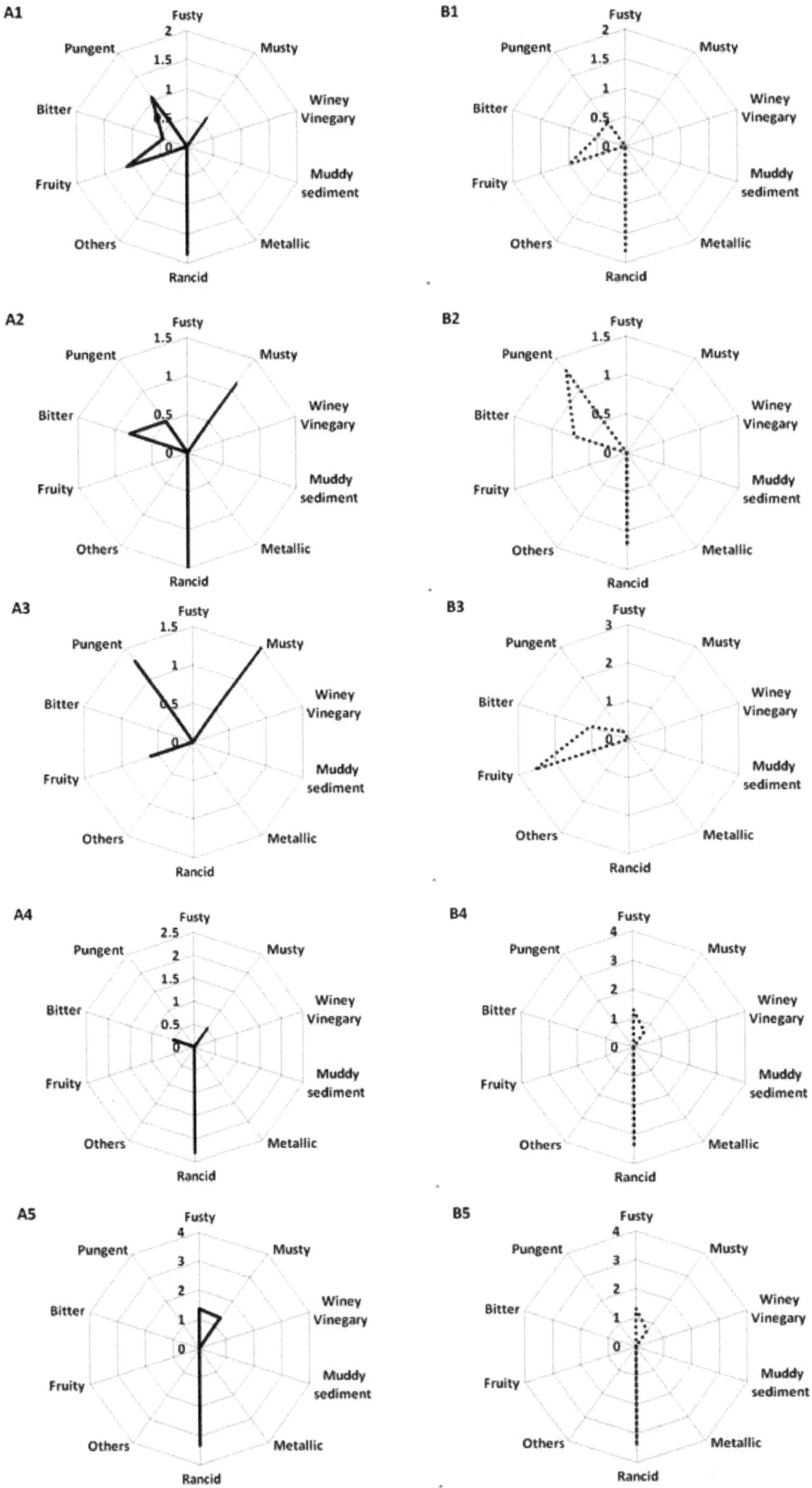

Figure 1. Evaluation of sensory attributes for EVOO stored at elevated temperature (A), and at room temperature (B) in glass bottles (1), PET (2), HDPE (3), cans (4), and pottery (5)

It was found that EVOO stored in glass bottles at low temperature maintained the extra virgin quality, whereas for that stored at elevated temperature (30°C) presented a sharp decrease in sensory score and lost its extra quality after less than two months of storage and become lambent due to loss of the positive attributes (fruity apple, green) and appearance of the negative ones (winy, muddy, rancid) [22]. A group of researchers [28] found a decrease in fruitiness during one year of oil storage and the rancid defect appear after 10-12 months at room temperature. Other investigators [55] found that the bitterness and pungency of virgin olive oil stored in glass bottles at increasing temperatures for 12-18 months decreased during storage time and the intensity of depletion was positively correlated with the increase in temperature of storage. Another research team [47] found that storage of olive oil in amber glass at low temperature results in lower amount of hexanal (off-flavor), but at ambient temperatures, positive attributes decrease throughout storage time.

5. Conclusions

As final statements and as a consequence of the results reported herein, olive oil storage and packaging are final steps of the production process and are as important as the other steps. The packaging material should ensure protection from storage conditions in order to maintain the olive oil quality. This study has reaffirmed that at ambient storage temperature, the best container in maintain the quality of stored oil is glass followed by HDPE, followed by both cans and PET, and the worst was pottery. At elevated temperature, glass was found the best primary packaging material, followed by PET, followed cans, followed by HDPE, and the worst container was pottery.

Deterioration agents can decrease the quality of olive oil during storage, so a correct control and monitoring of some quality indicators can be useful to predict the olive oil shelf life. The quality of olive oils is interpreted in terms of measurements of analytical parameters for which certain limit values are set. It was concluded that, grading of stored olive oil under investigation using sensory evaluation without chemical analysis is not sufficient. Also it is clear that the absorption coefficient K_{270} was the most sensitive determinant chemical test that determines the quality of stored olive oil.

Acknowledgments

The authors wish to express their gratitude to the staff in the Palestinian Standard Institution (PSI) for their help in running the sensorial analysis for the samples namely Miss Tagreed Shhadeh.

Abbreviations

EVOO: Extra virgin olive oil; **PET:** Polyethylene terephthalate; **HDPE:** High density polyethylene; **VOO:** Virgin olive oil; **EU:** European union; **LDL:** Low density lipoprotein; **IOOC:** International olive oil council; **PV:** Peroxide value; **EC:** European communities; **UV:** Ultra violet; **CRD:** Complete randomized design; **AOAC:** Association of official analytical communities; **FFA:** Free fatty acid.

References

[1] FAO statistics [Online]. http://faostat.fao.org, accessed on 20th June 2014.

[2] International Olive Oil Council. Sensory analysis of olive oil – Method- Organoleptic assessment of virgin olive oil., COI/T.20/Doc. No. 15/2nd Review. Madrid, Spain, 2007.

[3] Lairon, D. Intervention studies on Mediterranean diet and cardiovascular risk. *Mol. Nutr. Food Res.*., 51, 1209-1214, 2007.

[4] Hill, M. J.; Giacosa, A. The Mediterranean diet, *Eur. J. Cancer Prev.*, 1, 339-340, 1991.

[5] Vacca, V.; Caro, A.; Poiana, M. Effect of storage period and exposure conditions on the quality of bosana extra-virgin olive oil. *J. Food Qual.*, 29, 139-150, 2006.

[6] International Olive Oil Council (IOOC), Trade standard applying to olive oil and olive pomace oil, RES. COI/T.15/NC no. 3/Revision 1, 2003.

[7] Gutiérrez, F.; Fernández, J. L. Determinant parameters and components in the storage of virgin olive oil, prediction of storage time beyond which the oil is no longer of "extra" quality. *J. Agric. Food Chem.*, 50, 571-577, 2002.

[8] Velasco, J.; Dobarganes, C. Oxidative stability of virgin olive oil. *Eur. J. Lipid Sci. Technol.*, 104, 661-676, 2002.

[9] Gomez-Alonso, S.; Mancebo-Campos, V.; Salvador, M. D.; Fregapane, G. Evolution of major and minor components and oxidation indices of virgin olive oil during 21 months storage at room temperature. *Food Chem.*, 100, 36-42, 2007.

[10] Campos, V. M.; Salvador, M. D.; Fregapane, G. Comparative Study of Virgin Olive Oil Behavior under Rancimat Accelerated Oxidation Conditions and Long-TermRoom Temperature Storage. J. of Agric. Food Chem., 55, 8231-8236, 2007.

[11] Cosio, M. S.; Ballabio, D.; Benedetti, S.; Gigliotti, C. Geographical origin and authentication of extra virgin olive oils by an electronic nose in combination with artificial neural networks. *Anal. Chim. Acta.*, 567:202-210, 2006.

[12] Bongartz, A.; Oberg, D.G. Sensory evaluation of extra virgin olive oil (EVOO) extended to include the quality factor "harmony". *J. Agric. Sci. Tech. A.*, 1, 422-435, 2011.

[13] Dabbou, S; Gharbi, I; Dabbou, S.; Brahmi, F.; Nakbi, A.; Hammami, M. Impact of packaging material and storage time on olive oil quality. *Afr. J. Biotechnol.*, 10, 16937-16947, 2011.

[14] Caponio, F.; Bilancia, M. T.; Pasqualone, A.; Sikorska, E.; Gomes, T. Influence of the exposure to light on extra virgin olive oil quality during storage. *Eur. Food Res. Technol.*, 221, 92-98, 2005.

[15] Afaneh, I. A.; Abbadi, J.; Ayyad, Z.; Sultan, W.; Kanan, K. Evaluation of Selected Quality Degradation Indices for Palestinian Extra Virgin Olive Oil Bottled in Different Packaging Materials upon Storage under Different Lighting Conditions. *J. Food Sci. and Eng.*, 3, 267-283, 2013.

[16] Kalua, C. M.; Allen, M. S.; Bedgood, D. R.; Bishop, A.G.; Prenzler, P.D.; Robards, K. Olive oil volatile compounds, flavor development and quality: A critical review. *Food Chem.*, 100, 273-286, 2007.

[17] Kanavouras, A.; Munoz, P. H.; Coutelieris, F. A. Packaging of Olive Oil: Quality Issues and Shelf Life Predictions. Food Rev. Int., 22, 381-404, 2006.

[18] Kalua, C. M.; Bedgood, D. R.; Bishop A. G.; Prenzler P. D. Discrimination of storage conditions and freshness in Virgin Olive Oil. *J. Agric Food Chem.*, 54, 7144-7151, 2006.

[19] Official Methods of Analysis of the AOAC, 17th ed. AOAC, Arlington, Virginia USA, 2000.

[20] International Olive Oil Council (IOOC). Method of analysis: spectrophotometric investigation in the ultraviolet COI/T20/Doc. no. 19/Rev.1. 2001.

[21] Georgios, K.; Georgios, B.; Kiriaki, B. Stability and radical-scavenging activity of heated olive oil and other vegetable oils. *Eur. J. Lipid Sci. Tech.*, 108, 329-335, 2006.

[22] Gutierrez, F.; Fernandez, J. L. Determinant parameters and components in the storage of virgin olive oil. Prediction of storage time beyond which the oil is no longer of "Extra" Quality". J. Agric. Food Chem., 50, 571-577, 2002.

[23] Muik, B.; Lendl, B.; Molina-Diaz, A.; Ayora-Canada, M. J. Direct monitoring of lipid oxidation inedible oils by Fourier transform Raman spectroscopy. *Chem. Phys. Lipids.*, 134, 173-182, 2005.

[24] Garcia-Gonzalez, D. L.; Aparicio-Ruiz, R.; Aparicio, R. Virgin olive oil-chemical implications on quality and health. *Eur. J. Lipid Sci. Technol.*, 110, 1-6, 2003.

[25] Linares J.; Palma, M. G.; Iñigo M. Olive and olive pomace oil packing and marketing. Grasas Aceites., 57, 68-85, 2006.

[26] European Union Commission. Regulation EEC 2568/91 on the characteristics of olive oil and olive-residue oil and on the relevant methods of analysis. *Official J. of Eur. Commun.*, 248, 1991.

[27] Kiritsakis, A. K. Flavor components of olive oil-A review. *J. Am. Oil Chem. Soc.*, 75, 673-681, 1998.

[28] Fregapane, G. D.; Lavelli, V. R.; Salvador, M. Effect of filtration on virgin olive oil stability during storage. *Eur. J. Lipid Sci. Technol.*, 108: 134-142, 2006.

[29] Romani G. D. ; Labicci C. R. ; Cantini C. Evolution of minor polar compounds and antioxidant capacity during storage of bottled extra virgin olive oil. *J. Agric. Food Chem.*, 55, 1315-1320, 2007.

[30] Giovacchino L. D.; Mucciarella M. R.; Costantini N. Use of Nitrogen to Improve Stability of Virgin Olive Oil During Storage. *J. Am. Oil Chem. Soc.*, 79, 339-344, 2002.

[31] Gallardo-Guerrero, L.; Gandul-rojas, B.; Gandul-Rojas, B.; Mínguez-mosquera, M. I. Effect of Storage on the Original Pigment Profile of Spanish Virgin Olive Oil. J. Am. Oil Chem. Soc., 82, 33-39, 2005.

[32] Pristouri, G.; Badeka, A.; Kontominas, M. G. Effect of packaging material headspace, oxygen and light transmission, temperature and storage time on quality characteristics of extra virgin olive oil, *Food Control.*, 21, 412-418, 2010.

[33] Rababah, T. M.; Feng, H.; Yang, W.; Eriefejl, K.; Al-Omoush, M. Effects of type of packaging material on physicochemical and sensory properties of olive oil. *Int. J. Agric. Biol. Eng.*, 66-72, 2011.

[34] Ben Tekaya, I.; Ben Tekaya Ben Amor, I.; Belgaied, S.; El Atrache, A.; Hassouna M. Étude du conditionnement de l'huile d'olive dans les emballages en plastique. *Science des Aliments.*, 27, 214-233, 2007.

[35] Nouros, P. G.; Georgiou, C. A.; Polissiou, M. G. Direct parallel flow injection multichannel spectrophotometric determination of olive oil peroxide value. *Anal. Chim. Acta.*, 389, 239-245, 1999.

[36] Akoh, C. C.; Min, D. B. Lipid Oxidation of Edible Oils. *Food Lipids Chem. Nutr. Biotechnol.*, 54, 283–296, 2002.

[37] Sacchi, R.; Savarese, M.; Del Regno, A. Shelf Life of Vegetable Oils Bottled in Different Scavenging Polyethyleneterephthalate (PET) Containers. Pack. Technol. Sci., 21, 269-277, 2008.

[38] Anniva, C.; Grigoriadou, D.; Psomiadou, E.; Tsimidou, M. Z. Pheophytin degradation products as useful indices in the quality control of virgin olive oil. *J. Am. Oil Chem. Soci.*, 83, 371-375, 2006.

[39] Sharma, P.C.; Sharma, R. Storage behavior of olive (*Olea europaea* L.) oil in different packages. *J. Sci. Ind. Res.*, 65, 244-247, 2006.

[40] Kiritsakis, A.; Kanavouras, A.; Kiritsakis, K. Chemical analysis, quality control and packaging issues of olive oil. *Eur. J. Lipid Sci. Technol.*, 104, 628-638, 2002.

[41] Kiritsakis, A. K.; Dugan, L. R. Effect of selected storage conditions and packaging materials on olive oil quality. *J. Am. Oil Chem. Soc.*, 61: 1868-1870, 1984.

[42] Allien, J. C. Measurement of rancidity. In *Rancidity in Foods;* Allien, J. C.; Hamilton, R. J. Eds.; Elsevier Applied Science, London and New York, 23-51, 1989.

[43] Pagliarini, E.; Zanoni, B.; Giovanelli, G. Predictive study on Tuscan extra virgin olive oil stability under several commercial conditions. *J. Agric. Food Chem.*, 48, 1345-1351, 2000.

[44] Rodney J. M.; Graham, K. The Effect of Storage in Collapsible containers on Olive Oil Quality, Australian Government, rural industries research and development cooperation., RIRDC Publication No. 12/008, 2012.

[45] Bilancia, M.T., Caponio, F., Sikorskab, E., Pasqualonea, A. and Summo, C. Correlation of triacylglycerol oligopolymers and oxidised triacylglycerols to quality parameters in extra virgin olive oil during storage. *Food Res. Int.*, 40, 855-861, 2007.

[46] Deiana, M.; Rosa, A.; Cao, C. F.; Pirisi, F. M.; Bandino, G.; Dessi, M. A. Novel approach to study oxidative stability of extra virgin olive oils: importance of α- tocopherol concentration. *J. Agric. Food Chem.*, 50, 4342-4346, 2002.

[47] Morello, J. R.; Motilva, M. J.; Tovar, M. J.; Romero, M. P. Changes in commercial virgin olive oil (cv Arbequina) during storage, with special emphasis on the phenolic fraction. Food chem., 85, pp. 357-364, 2003.

[48] Psomiadou, E.; Tsimidou, M.; Simultaneous HPLC determination of tocopherols, carotenoids, and chlorophylls for monitoring their effect on virgin olive oil oxidation. *J. Agric. Food Chem.*, 46, 5132-5138, 1998.

[49] Morello, J. R.; Motilva, M. J.; Tovar, M. J.; Romero, M. P. Changes in commercial virgin olive oil (cv. Arbequina) during storage, with special emphasis on the phenolic fraction. *Food Chem.*, 85, 357-364, 2004.

[50] Angerosa, F., Mostallino, R.; Basti, C.; Vito. R. Virgin olive oil odour notes: their relationships with volatile compounds from the lipoxygenase pathway and secoiridoid compounds. *Food Chem.*, 68, 283-287, 2000.

[51] Angerosa, F., Servili, M.; Selvaggini, R.; Taticchi, A.; Esposto, S.; Montedoro, G. Volatile compounds in virgin olive oil: occurrence and their relationship with the quality. *J. Chromatogr. A*, 1054,17-31, 2004.

[52] Salas, J. J.; Sanchez, C.; Garcia-Gonzalez, D. L.; Aparicio, R. Impact of the suppression of lipoxygenase and hydroperoxide lyase on the quality of the green odor in green leaves. *J. Agricu. Food Chem.*, 53:1648-1655, 2005.

[53] Morales, M.T.; Aparicio, R. Effect of extraction conditions on sensory quality of virgin olive oil. *J. Am. Oil Chem. Soc.*, 76: 295-300, 1999.

[54] Cimato, A., Dello Monacoa, D., Distante, C., Epifani, M., Siciliano, P., Taurino, A.M., Zuppa, M., Sani, A. Analysis of single-cultivar extra virgin olive oils by means of an electronic nose and HS-SPME/GC/MS methods. *Sensor Actuator B*, 114, 674-680, 2006.

[55] Sinesio, F.; Moneta, E.; Esti, M. The dynamic sensory evaluation of bitterness and pungency in virgin olive oil. Food Qual. Prefer., 16, pp. 557-564, 2005.

3

Combining Different Types of Prebiotic Plant Isolates Toward, Enhancing the Growth of Probiotic Organisms

Fathima Hajara Aslam[1,*], Jagath Wansapala[2], Manel Perera[3]

[1]Department of Food Science, Spectrum Institute of Science and Technology, Colombo, Sri-Lanka
[2]Department of Food Science and Technology, Faculty of Applied Sciences, University of Sri-Jayewardenepura, Gangodawila, Nugegoda, Sri-Lanka
[3]MicroChem Laboratories, Nugegoda, Sri-Lanka
*Corresponding author: fathimahajara@gmail.com

Abstract The growth stimulatory effect produced by combining different sources of prebiotics i.e; Fiber isolates from *Musa* sp pseudostem, polyphenol extracts from *Sesbania grandiflora* flower petal and non-digestible polysaccharide extracts from *Artocarpus heterophyllus* seed were assessed against probiotic organisms, *Lactobacillus acidophilus* and *Bifidobacterium animalis* subsp. *lactis* BB-12 *in vitro* in liquid cultures. Different combinations were formulated by integrating the three sources of prebiotics at two different levels i.e; fibre (0.2% and 2%), polyphenol extracts (0.2% and 0.6%) and non-digestible polysaccharide extracts (0.2% and 1.2%) to obtain eight treatments. The formulation which consisted 2% fibre, 0.2% polyphenol and 0.2% non-digestible polysaccharide was able to promote significant biomass increment in *Lactobacillus acidophilus*, while the treatment consisting 2% fibre, 0.6% polyphenol and 0.2% non-digestible polysaccharide demonstrated highest proliferation for *Bifidobacterium animalis* subsp. *lactis* BB-12 *in vitro* which were statistically different ($p<0.05$) than other formulations.

Keywords: *L. acidophilus, B. animalis subsp. lactis BB-12, S. grandiflora, Musa sp, A. heterophyllus*

1. Introduction

The increasing need and trend for functional food in the current society is causing a remarkable change in food pattern. The awareness which initiated interest between diet and health has increased the production of various neutraceutical and functional food ingredients that support health beyond supplying daily nutritional requirement [1]. To achieve a healthier gut system there are several type of probiotic as well as prebiotic supplements that are administrated by humans to increase the proliferation of beneficial gut microbes. This study focuses on integrating different source of prebiotics at suitable quantities to optimize the colonization of *Lactobacillus acidophilus* and *Bifidobacterium animalis* subsp *lactis* BB-12 *in vitro*.

L. acidophilus and *B. animalis* subsp *lactis*-BB-12 have been used in food industry for many years in the process of converting carbohydrates into lactic acid which is responsible to the sour taste in most of the fermented dairy food products such as yoghurt and kefir and due to the large array of beneficial physiological benefits they offer [2,3]. This lowers the pH to an extent where the growth of pathogenic and spoilage microorganisms are prevented there by assuring the prevention of various diseases transmissible through food such as gastrointestinal infections. In the meantime the probiotic resident microbiota in the host exerts a microbial barrier against microbial pathogen which is considered a valuable physiological function [4,5]. They are also capable of strengthening the immune system to overcome allergy conditions [6,7].

Under natural conditions a protective gut micro flora develops and there is no need for a probiotic or prebiotic supplement but consumption of large amount of processed food and sterile food prevent the access of natural colonization of certain strains [8]. Beyond that, antibiotic therapy, excessive alcoholic consumption, stress, exposure to toxic component, other diseases are capable of changing the natural equilibrium of intestinal flora which is not considered favorable to host. But the proper prebiotic intake is capable to re-establish the altered gut flora since they are potential heeling ingredients [9]. Processing conditions are also capable of altering the effect that is imposed by prebiotics, there by impairing the benefits imbibed by probiotic strains in order to bring out desirable physiological changes [10].

Fermentable carbohydrates (complex carbohydrates), polyphenols and oligosaccharides, which are not digested or poorly digested in the small intestine are passed to the lower gut where they become available for some colonic bacteria, but are not utilized by majority of the micro flora present in the colon [11]. The criteria as classifying a food as prebiotics are; It should resist digestion or absorption in the upper gastrointestinal tract, when it enters the colon it should be selectively metabolized by a limited number of beneficial bacteria and capable to change the colonic micro flora to a healthier bacterial flora rendering physiologic effect that is beneficial for host [9,12].

Fiber isolates of *Musa* sp (Banana) pseudostem, polyphenolic extracts of *Sesbania grandiflora* flower petal, and non-digestible polysaccharide extracts of *Artocarpus heterophyllus* (Jackfruit) seed are the major three prebiotic ingredients that were considered in this study. *S. grandiflora* is a widely available plant in family Fabaceae. The flowers of *S. grandiflora* have been reported to have anti-microbial activity against certain pathogenic microbial species [13]. Also recent studies show that the polyphenolic extract of *S. grandiflora* has significant growth promoting effect on *L. acidophilus [14]*. Dietary polyphenols are natural compounds which are chemically large heterogeneous group of compounds. Polyphenols can be classified into flavonoids and nonflavonoids based on their chemical structure and complexity. It has been found that only 5–10% of the total polyphenol intake is absorbed in the small intestine which can be considered low molecular weight polyphenols. The remaining polyphenols (90–95% of total polyphenol intake) may accumulate in the large intestine and be available to colon microbes since they consist oligomeric or polymeric form [15].

The other prebiotic source considered in this research is *Musa* sp pseudostem and jackfruit seed. Banana pseudo stems are rich sources of ligno-cellulosic fiber, which can promote beneficial health effects, but at the present considered as waste product of banana cultivation and not properly utilized [16].

Previous studies shows, jackfruit seed which is used in this study as a prebiotic ingredient, consist considerable amount of extractable indigestible polysaccharide and capable of selectively stimulating the growth of three probiotics, *Lactobacillus acidophilus*, *Lactobacillus plantarum* and *Bifidobacterium bifidum* [17]. All three phytonutrient isolates, selected in this study obey the prerequisites to be considered as prebiotics. Designing an integrated prebiotic combination which has far more potential to cause significant biomass increase of *L. acidophilus* and *B. animalis* subsp. *lactis* BB-12 was the core intention of this study. Since inducing proliferation of these organisms can harvest peak performance of probiotics. This sounds the importance of permanent implantation of probiotics in colon.

2. Materials and Methods

2.1. Extraction of Plant Prebiotics

2.1.1. Fiber from *Musa* sp

Fibers were extracted from the pseudostem of *Musa* sp which was purchased from local market in Sri Lanka. The suitable stalk of the plant was cut to a length of 150 cm and its outer sheath was removed, since fibers are located at the outer sheath. Then fibers was extracted from this section using roller drums [18]. Finally the fibers were completely cleaned in water to remove the waste materials and then dried at 60°C for three days to remove excess moisture, then powdered and stored at -18°C for further use.

2.1.2. Polyphenol Extract of *S. grandiflora* Flower Petals

The flower samples was collected from local market Sri Lanka. Then the petals were separated, sun dried and powdered and stored at -18°C for further use. Exactly eight grams of finely dried powder was weighted, and extracted with 80:20 of methanol:water at room temperature and filtered. The extraction was done until the filtrate was colorless. The extracts were pooled, concentrated in the rotary evaporator and freeze dried, then after stored in capped dark bottles at -18°C for further use [19].

2.1.3. Non-digestible Polysaccharide of *Artocarpus heterophyllus* (Jackfruit Seed)

Jackfruit seeds purchased from Sri Lankan local market, were washed, sliced, dried, milled and sieved for size of 2mm and stored at -18°C for further use. Sixty grams of jack fruit seed powder was extracted with 50 %(v/v) ethanol at 60°C shaken at 200 rpm in oil bath for 15 minutes where the liquid: solid ratio was maintained at 8. After extraction the solvent was evaporated using rotary evaporator and stored in dark cap bottle at -18°C for further use [20].

2.2. Bacterial Cultures

Lactobacillus acidophilus LA-5 (Chr. Hansen), *Bifidobacterium animalis* subsp. *lactis* BB-12(Chr. Hansen), purchased in lyophilized form, were used as probiotic test bacteria in this experiment. *Lactobacillus acidophilus* culture was activated in MRS broth (Oxoid) at 37°C under anaerobic conditions for 24 h [21].

Bifidobacterium animalis subsp. lactis BB-12 was activated in MRS broth (Oxoid) supplemented with 0.05% (w/v) L-cysteine HCl, under anaerobic conditions at 37°C for 48 h [22].

2.3. Experimental Design

Fiber (F), polyphenol extract (PE) and non-digestible polysaccharides (NDP) extracts were integrated as below in order to obtain eleven different treatment combinations.

Table 1. Prebiotic treatment combinations and composition

Treatment Combinations	Composition of F, PE, NDP (w/v)
A	0.2%, 0.2%, 0.2%
B	2%,0.2%,0.2%
C	0.2%,0.6%,0.2%
D	2%, 0.6%, 0.2%
E	0.2%,0.2%,1.2%
F	2%,0.2%,1.2%
G	0.2%,0.6%,1.2%
H	2%,0.6%,1.2%
I	**0.2%,0%,0%**
J	**0%,0.2%,0%**
K	**0%,0%,0.2%**

*The treatments in bold-face are designed using one prebiotic ingredient
°The treatments in non-bold-face are designed with three prebiotic ingredients combined at different concentrations.

2.3.1. Growth promoting property of integrated and non-integrated prebiotic treatments on *L. acidophilus*

The eleven treatments mentioned in Table 1, were introduced to fresh MRS broth (5 mL) separately and were inoculated with 0.1 mL *L. acidophilus* culture which contains 5.5×10^5 CFU/mL. In the positive control set *L. acidophilus* culture was grown in fresh MRS broth (5 mL) omitting prebiotic ingredients.

The cultures were incubated at 37°C for 24 h under anaerobic conditions. After incubation the samples were diluted in MRS broth and plated on MRS agar using pour plate technique and incubated at 37°C under anaerobic conditions. All steps were carried in triplicates. The effect of eleven treatments on *L. acidophilus* was determined comparing the number of CFU in the presence of different combinations of prebiotic ingredients, against those obtained from controls.

2.3.2. Growth Promoting Property of Integrated and Non-integrated Prebiotic Treatments on *B. animalis* subsp. *lactis* BB-12

The eleven treatments mentioned in Table 1 were introduced to fresh MRS broth enriched with 0.05% cysteine HCL (5 mL) separately and were inoculated with 0.1 mL *B. animalis* subsp. *lactis* BB-12 culture which contains 6.3×10^4 CFU/mL. In the positive control set *B. animalis* subsp. *lactis* BB-12 culture was grown in fresh MRS broth enriched with 0.05% cysteine HCL (5 mL) omitting prebiotic ingredients.

The cultures were incubated at 37°C for 24 h under anaerobic conditions. After incubation the samples were diluted in MRS broth and plated on MRS-cysteine agar using pour plate technique and incubated at 37°C under anaerobic conditions. All steps were carried in triplicates. The effect of eleven treatment on *B. animalis* subsp. *lactis* BB-12 was determined comparing the number of CFU in the presence of different combinations of prebiotic ingredients, against those obtained from controls.

2.3.3. Determination of Total Polyphenolic Content in *S. grandiflora* Flower Extract

S. grandiflora flower extract (2 mg) was initially dissolved in 40 µl of dimethyl sulfoxide (DMSO) and 2 and 1 mg/ml concentrations were made using distilled water. Series of gallic acid concentrations (1, 0.5, 0.25, 0.125, 0.0625, 0.03125 and 0.015625 mg/ml) were used to construct the standard curve (n=6). Total Polyphenolic Content was determined by method described by Singleton. The assay was carried out in a micro plate reader. Total polyphenolic content was expressed as mg gallic acid equivalents per g of *S. grandiflora* flower extracts [23].

2.4. Statistical Analysis

Statistical analysis was carried out in order to determine the significant difference among all treatments towards growth promoting ability, using Minitab- 16.1.1. One-way ANNOVA.

3. Results

3.1. Initial Count of *L. acidophilus* and *B. animalis* subsp. *lactis*

Table 2. Initial bacterial counts of inoculum

Organism	Log CFU/ mL
Lactobacillus acidophilus	5.73±0.0404
Bifidobacterium animalis subsp. *lactis* BB-12	4.79±0.0153

Values are expressed as log means of three trials (mean ±SD, n=3).

3.2. Total Polyphenolic Content in *S. grandiflora* Flower Petals

The polyphenolic extracts of S. grandiflora flower used in this study contain a total polyphenolic content of 28.31± 0.44 mg of GAE/g of dry weight (mean ± SD, n=6).

3.3. Growth Stimulation Caused by Prebiotic Treatments toward *L. acidophilus* and *B. animalis* subsp. *lactis*

The effect of eight integrated prebiotic treatments and three non-integrated prebiotic treatments were assessed in terms of increased CFU/mL on *L. acidophilus* and *B. animalis* subsp. *lactis* compared to control.

Table 3. Growth of probiotic species with integrated prebiotic treatments

	Viable counts (log CFU/mL)	
	L. acidophilus	*B. animalis* subsp. *lactis* BB-12
Treatments		
A	^{b}6.96 ± 0.0115	^{b}6.17 ± 0.0052
B	*7.49 ± 0.0049	^{c}6.08 ± 0.0361
C	^{b}7.00 ± 0.0058	^{d}5.94 ± 0.0173
D	^{b}6.86 ± 0.0096	*6.48 ± 0.0015
E	^{b}7.08 ± 0.0058	^{e}5.86 ± 0.0006
F	^{c}6.74 ± 0.046	^{f}6.05 ± 0.0101
G	^{d}7.18 ± 0.0125	^{e}5.89 ± 0.0058
H	^{d}7.26 ± 0.0053	^{g}6.35 ± 0.0189
Control	^{a}5.78 ± 0.2223	^{a}5.68 ± 0.0012

Values are expressed as log means of three trials (mean ±SD, $n=3$)
Value in the same column with the same superscript letter are not statistically significant (p< 0.05) than each other.
Superscript * indicates the treatment producing maximum growth proliferation.

Table 4. Growth of probiotic species with non-integrated prebiotic treatments

	Viable counts (log CFU/mL)	
	L. acidophilus	*B. animalis* subsp. *lactis* BB-12
Treatment		
I	*6.51 ± 0.0674	^{m}5.79 ± 0.0173
J	^{m}6.15 ± 0.0050	*5.84 ± 0.0029
K	^{m}6.12 ± 0.0125	^{n}5.76 ± 0.0051
Control	^{x}5.78 ± 0.2223	^{x}5.68 ± 0.0012

Values are expressed as log means of three trials (mean ± SD, $n=3$)
Value in the same column with the same superscript letter are not statistically significant (p< 0.05) than each other.
Superscript * indicates the treatment producing maximum growth proliferation.

4. Discussion

4.1. Prebiotic Potential of Plant Extracts

Functional food is one step ahead in delivering more health promoting effects beyond satisfying basic nutritional requirement. Prebiotics are phytonutrients that falls in to the category of functional food and will amend the gut micro biome in a potential way of delivering tremendous health benefits for the host by increasing the proliferation and growth of beneficial bacteria who are not pathogenic, especially *Lactobacilli*, and *Bifidobacteria* [1,9].

Table 1, indicates the eight prebiotic treatments, designed by integrating fiber, polyphenol extracts and non-digestible polysaccharide extracts at different levels and three prebiotic treatments designed by including fiber, polyphenol extracts and non-digestible polysaccharide extracts separately. The effect of these eleven treatments, toward increasing the growth of *L. acidophilus* and *B. animalis* subsp. *lactis* BB-12 were assessed in terms of increased CFU/mL in liquid MRS cultures when incubated with the prebiotic treatments under anaerobic conditions.

The results obtained in Table 3 and Table 4, clearly indicates all treatments are significantly ($p< 0.05$) capable to induce growth of *L. acidophilus* and *B. animalis* subsp. *lactis* BB-12 compared to the control. This demonstrate the prebiotic potential of ingredients used in this experiment.

Fibers of *Musa* sp, polyphenol extracts of *S. grandiflora* flower and non-digestible polysaccharide extract of *A. heterophyllus* seed, when tested integrated and separately showed statistically significant increase in biomass of *L. acidophilus* and *B. animalis* subsp. *lactis* BB-12 *in vitro* emphasise, they are qualified functional foods possessing prebiotic potential. Also the phytonutrient isolates are capable to, initiate favorable health benefits in host, because the action they undertake toward increasing the proliferation of probiotics.

4.1.1. Prebiotic Potential of *Musa* sp Fiber Isolates

There are tremendous health benefits that host obtain from consumption of fiber, out of all, the outstanding feature becomes maintaining good gastrointestinal health which this study focus more on. Fibre components are capable to undergo partial or complete fermentation by the gut micro flora in the large bowel (colon). The complex fiber which undergo fermentation produces gases such as hydrogen, methane and carbon dioxide and short chain fatty acids such as acetate in major proportion, propionate and butyrate. The produced short chain fatty acids are capable to decrease the pH in the colon and there by inhibit the growth of pathogenic organisms and also the formation of toxic breakdown products. Also the short chain fatty acids that are generated will be absorbed into the systemic circulation and account as an energy source while inducing beneficial effects on lipid metabolism. The availability of fermentable substrates in the colon results in an increase in the number of bacteria and thereby increase the stool weight which is attributed to the capability of some fiber components to absorb fluid and so increase stool weight [24]. Recent studies carried on *Musa* sp pseudostem has revealed that, the outer sheath consists higher ratios of insoluble fiber where soluble fibers are present in minor ratios [25]. Fibers can be classified as soluble and insoluble fibers, soluble fibers have gel forming property thereby increasing the viscosity in intestinal tract and undergo complete fermentation than insoluble fibers. However, not all soluble fibers are viscous and some insoluble fibers may also be well fermented [26]. Previous studies executed on fermentation of soluble and insoluble fibers by *Lactobacilli* spp and *Bifidobacterium* spp and production of short chain fatty acids, grandly complies with the results obtained in Table 3, and Table 4 [27,28]. The previous studies insist that *Musa* sp fibers can be fermented by *L. acidophilus* and *B. animalis* subsp. *lactis* BB-12 and thereby induce proliferation in liquid culture media.

4.1.2. Prebiotic Potential of Polyphenol Extracts of *S. grandiflora* Flower

The poor bioavailability of native polyphenols is the major reason to consider polyphenols as prebiotic candidates [29]. *S. grandiflora* flower petals consist of numerous antimicrobial properties towards most of the pathogenic microorganism mean while possessing growth stimulatory effect on *L. acidophilus* has been discussed in previous work carried out. HPLC analysis of the flavonoids extracted from the broth which was cultured with polyphenol extracts of *S. grandiflora* flower and *L. acidophilus*, before and after the incubation has showed that there is a reduction of rutin content and increment of biomass of the bacterium in the broth after incubation for 24h [14]. Rutin, absorbs more slowly since it requires the assistance of cecal microbial enzymes to be hydrolyzed. This implies the ability of rutin to resist digestion by digestive juices and be available in the large intestine for microbial digestion [30]. This is a clear evident that insists polyphenol extracts of *S. grandiflora* flower is a potential candidate to cause growth of *L. acidophilus.* Also the previous studies carried out shows that *Bifidobacterium* spp is capable to biotrasformate rutin and fermentation of polyphenols stimulates proliferation of *Bifidobacterium* spp [31]. These previous studies, dictate the capability of polyphenol extracts to cause growth stimulation against *L. acidophilus* and *B. animalis* subsp. *lactis* which are clearly shown in Table 3, and Table 4.

4.1.3. Prebiotic Potential of Non-digestible Polysaccharide Extracts of *A. heterophyllus*

Jackfruit (*A. heterophyllus*) seed are proven in previous studies to possess prebiotic properties and stimulate the growth of beneficial intestinal probiotics. The non-digestible polysaccharide of jackfruit seed which consist degree of polymerization 5, has confirmed their prebiotic capacity and selective fermentation by *Lactobacillus* spp and *Bifidobacterium* spp. Also the fermentation is associated with production of significant amount of butyric acid under anaerobic conditions [32]. This forms a link between the results obtained in current study, declaring *A. heterophyllus* seed extracts which contain non-digestible polysaccharides are capable to increase the growth of *B. animalis* subsp. *lactis* BB-12 and *L. acidophilus* under anaerobic conditions when integrated with other prebiotic ingredients and as separate entities, as shown in Table 3 and Table 4.

4.2. The Impact of Integrated Prebiotic Combinations on *L. acidophilus*

The results obtained in Table 3, remarkably dictates, treatment B which consist 2% fiber, 0.2%, polyphenol extracts and 0.2% non-digestible polysaccharide has the highest potential to cause statistically significant ($p<0.05$) growth of *L. acidophilus* than the other seven integrated prebiotic treatments and control in liquid culture media. Treatment H (2% F, 6% PE, 1.2% NDP) and G (0.2% F,

0.6% PE, 1.2% NDP) are capable to induce statistically significant growth of *L. acidophilus* than the control next to treatment B but there is no statistically significant difference among these two groups, which emphasise that the effect imposed by both the treatments are the same toward enhancing the growth of *L. acidophilus*. Treatment E (0.2% F, 0.2% PE, 1.2% NDP), C (0.2% F, 0.6% PE, 0.2% NDP), A (0.2% F, 0.2% PE, 0.2% NDP) and D (2% F, 0.6% PE, 0.2% NDP) stand next in place to cause significant growth than the control on *L. acidophilus* but there is no significant difference among these treatments. The contribution of Treatment F (2% F, 0.2% PE, 1.2% NDP) towards the growth of *L. acidophilus* is significantly effective than the control but, the effect is quite minute compared with other integrated treatment combinations.

4.3. The Impact of Non-integrated Prebiotic Treatments on *L. acidophilus*

The results obtained in Table 4, clearly shows that when fiber isolates of *Musa* sp (treatment I), polyphenol extracts of *S. grandiflora* (treatment J) and Non- digestible polysaccharide extracts of *A. heterophyllus* (treatment K) when tested separately with a concentration of 0.2% (w/v) are able to promote growth of *L. acidophilus* than the control in a statistically significant ($p < 0.05$) manner. Among all three non-integrated treatments treatment "I "which consisted 0.2% fiber extracts showed significant growth promotion than the other treatments [33,34]. Also there were no significant difference between treatment J and K in producing growth stimulation in the case of *L. acidophilus*. It is evident that the growth stimulation caused by the non- integrated prebiotic treatments are higher than the control but lower than the integrated prebiotic treatments, which is clear by Table 3 and 4.

4.4. The Impact of Integrated Prebiotic Treatments on *B. animalis* subsp. *lactis* BB-12

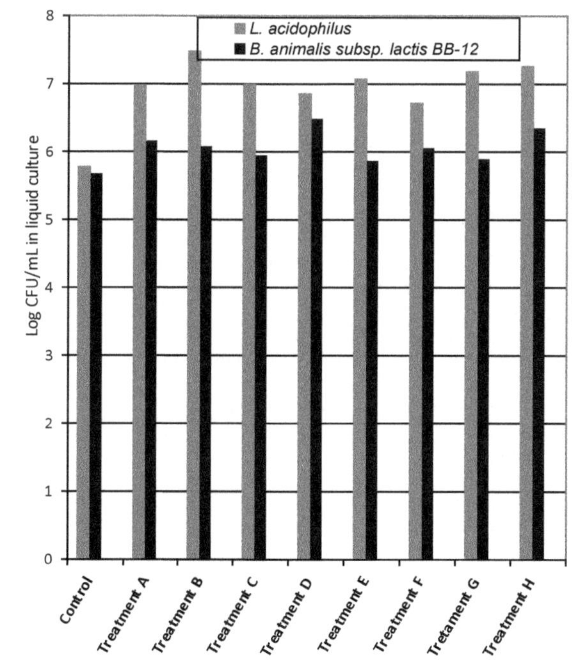

Figure 1. The effect imposed by integrated prebiotic treatments on *L. acidophilus* and *B. animalis* subsp. *lactis* BB-12 in liquid culture media

The results obtained in Table 3, remarkably dictates, treatment D which is constructed using 2% fiber, 0.6% polyphenolic extracts and 0.2% non-digestible extracts showed outstanding performance in promoting significant ($p<0.05$) biomass increment of *B. animalis* subsp. *lactis* BB-12, than rest of the seven integrated treatments and control in liquid culture media. The rest seven treatments are also capable to induce significant biomass increase of *B. animalis* subsp. *lactis* BB-12 than the control in the order of, treatment H (2% F, 0.6% PE, 1.2% NDP) > A (0.2% F, 0.2% PE, 0.2% NDP) > B (2% F, 0.2% PE, 0.2% NDP) > F (2% F, 0.2% PE, 1.2% NDP) > C (0.2% F, 0.6% PE, 0.2% NDP) > G (0.2% F, 0.6% PE, 1.2% NDP) > E(0.2% F, 0.2% PE, 1.2% NDP). But there is no significant difference between treatment G and E in relation to growth promotion.

4.5. The Effect of Non-integrated Prebiotic Treatments on *B. animalis* subsp. *lactis*

The results obtained in Table 4, clearly shows that when fiber isolates of *Musa* sp (treatment I), polyphenol extracts of *S. grandiflora* (treatment J) and Non- digestible polysaccharide extracts of *A. heterophyllus* (treatment K) when tested separately with a concentration of 0.2% (w/v) are able to promote growth of *B. animalis* subsp. *lactis* BB-12 than the control in a statistically significant ($p < 0.05$) manner. Among all three non-integrated treatments, treatment J which consisted 0.2% polyphenol extracts showed significant growth promotion than the other treatments[31]. Comparison made between Table 3 and 4, dictates growth stimulation caused by the non-integrated treatments are low than the growth induced by the integrated treatment combinations but higher than the control.

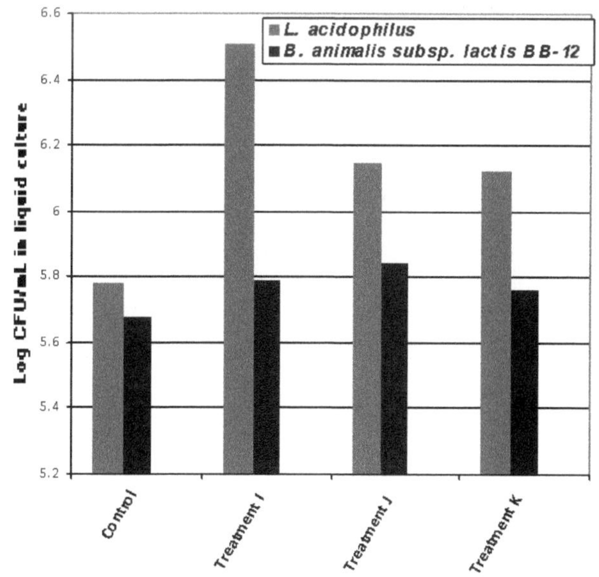

Figure 2. The effect imparted by non-integrated prebiotic treatments on *L. acidophilus* and *B. animalis* subsp. *lactis* BB-12 in liquid culture media

Further studies are required to understand the biotransformation mechanism and biotransformed products of all eight integrated prebiotic treatments when fermented by the test probiotic strains. Also detailed study is required to understand, the biotrasformated product mixture that is generated after fermentation e.g: short chain fatty acid production of fibers in the presence of polyphenolic extracts. Since, previous researchers have shown that it is not the initial component initiates proliferation of probiotics, but it is the biotransformed product environment that create the growth of probiotics [24]. The study demonstrates the ability of the prebiotic combination to enhance growth *in vitro*, but more work needed to be carried out to understand how these integrated prebiotic treatments work *in vivo* and the potential concentration to proliferate growth *in vivo*.

5. Conclusion

In conclusion we can say that the force, imposed by integrated prebiotic combinations toward causing significant increase of the biomass of *L. acidophilus* and *B. animalis* subsp. *lactis* BB-12 is greater than the force imparted by single prebiotic ingredients. But essentially not all integrated combinations are effective in causing maximum proliferation. Combination containing 2% fiber, 0.2% polyphenol extracts and 0.2% non-digestible polysaccharide extracts is significant in promoting growth of *L. acidophilus*, where 2% fiber, 0.6% polyphenol extracts and 0.2% non-digestible polysaccharide extracts combination is potential to cause remarkable growth of *B. animalis* subsp. *lactis* BB-12 *in vitro* in liquid culture. The effect caused by the integrated prebiotic combinations do not reflect reactions of the bacteria in the intestine, but the *in vitro* growth stimulatory effect on probiotic bacteria might indicate the *in vivo* interaction of the prebiotic combinations and probiotic organisms. This integrated prebiotic combinations can be employed in neutraceutical development, since the merged prebiotic combinations are capable to cause remarkable proliferation of probiotics which can render beneficial physiological health benefits for host.

Acknowledgement

The author acknowledge Mr. Ahamed Aslam, Mr. Abdussamad Ibrahim, Dr.Jagath Wansapala from Department of Food Science and Technology, University of Sri-Jayewardenepura and Manel Perera in their immense support and contribution in funding this project.

References

[1] Ogueke, C.C., et al., *probiotics and prebiotics* Parkistan journal of nutrition, 2010. 9: p. 833-843.

[2] Pyar, H. and K.K. Peh, *Characterization and Identlfication of Lactobacillus acidophilus using Biolog Rapid Identification System.* International Journal of Pharmacy and Pharmaceutical Sciences, 2014. Vol 6(Issue 1).

[3] Gorbach, S.L., *Lactic acid bacteria and human health.* Ann Med, 1990. 22(1): p. 37-41.

[4] Servin, A.L., *Antagonistic activities of lactobacilli and bifidobacteria against microbial pathogens.* FEMS Microbiology Reviews, 2004. 28 p. 405–440.

[5] Rossi, M. and A. Amaretti, *Probiotic properties of bifidobacteria.* 2010: Caister Academic Press: Norfolk, UK.

[6] Ishibashi, N., T. Yaeshima, and H. Hayasawa, *Bifidobacteria: their significance in human intestinal health.* Mal J Nutr, 1997. 3: p. 149-159.

[7] Mohd Shaufi, M.A., et al., *Deciphering chicken gut microbial dynamics based on high-throughput 16S rRNA metagenomics analyses.* Gut Pathog, 2015. 7(4): p. 015-0051.

[8] Fuller, R. and G. Peridigón, *Gut flora, nutrition, immunity and health.* 2008: John Wiley & Sons.

[9] Sousa, V.M.C., E.F. Santos, and V.C. Sgarbieri, *The Importance of Prebiotics in Functional Foods and Clinical Practice.* Food and Nutrition Sciences, 2011. Vol. 2 p. 4.

[10] Huebner, J., *Effect of processing conditions on the prebiotic activity of commercial prebiotics.* International Dairy Journal, 2008. 18: p. 287-293.

[11] Gibson, G.R., *Dietary modulation of the human gut microflora using prebiotics.* The British Journal of Nutrition, 1998. 4.

[12] Grajek, W., *Probiotics, prebiotics and antioxidants as functional foods.* on-line at: www.actabp.pl, 2005. 52: p. 665-671.

[13] Munde-Wagh, K.B., *Phytochemical, Antimicrobial Evaluation and Determination of Total Phenolic and Flavonoid Contents of Sesbania grandiflora flower extracts* International Journal of Pharmacy and Pharmaceutical Sciences 2012. 4(4).

[14] China, R., et al., *Antimicrobial activity of Sesbania grandiflora flower polyphenol extracts on some pathogenic bacteria and growth stimulatory effect on the probiotic organism Lactobacillus acidophilus.* Microbiological Research, 2012. 167: p. 500-506.

[15] Cardona, F., *Benefits of polyphenols on gut microbiota and implications in human health.* The Journal of Nutritional Biochemistry, 2013. 24(8): p. 1415-1422.

[16] Mukhopadhyay, S., et al., *Banana fibers–variability and fracture behaviour.* Cellulose, 2008. 31(3.61).

[17] Thammarutwasik, P., et al., *Prebiotics – A Review.* Songklanakarin J. Sci. Technol, 2009. 31: p. 401-408.

[18] Arthanarieswaran, V.P., A. Kumaravel, and M. Kathirselvam, *Evaluation of mechanical properties of banana and sisal fiber reinforced epoxy composites: Influence of glass fiber hybridization.* Materials & Design, 2014. 64: p. 194-202.

[19] China, R., et al., *In vitro antioxidant activity of different cultivars of banana flower (Musa paradicicus L.) extracts available in India.* J Food Sci 2011. 76(9): p. 1292-9.

[20] Veeraphong Bhornsmithikun, P.C., Ram Yamsaengsung, and a.K. Prasertsit, *Continuous extraction of prebiotics from jackfruit seeds.* Songklanakarin J. Sci. Technol, 2010. 6: p. 635-642.

[21] Bzducha, A. and M.W. Obiedziński, *Influence of two probiotic Lactobacillus strains on CLA content in model ripening cheese* polish journal of food and nutrition sciences, 2007. 57: p. 65-69.

[22] Shadan Abas, A.-W.A.A.A.-S., *Screening for Bacteriocins Production in Enteric Bifidobacterium Isolates and Study of Some Production Affecting Factors.* Medical Journal of Babylon, 2012. 9.

[23] Singleton, V.L., Orthofer, R. and Lamuela-Raventos, R.M, *Analysis of total phenols and other oxidation substrates and antioxidants by means of Folin-Ciocaltue reagent.* MethEnzymol, 1999. 299: p. 152-178.

[24] Buttriss, J.L. and C.S. Stokes, *Dietary fibre and health: an overview.* Nutrition Bulletin, 2008. 33(3): p. 186-200.

[25] Aziz, N.A.A., *Chemical and functional properties of the native banana (Musa acuminata × balbisiana Colla cv. Awak) pseudo-stem and pseudo-stem tender core flours.* Food Chemistry, 2011. 128(3): p. 748-753.

[26] Slavin, J., *Fiber and Prebiotics: Mechanisms and Health Benefits.* Nutrients, 2013. 5(4): p. 1417-1435.

[27] Farooq, U., et al., *Enhancement of short chain fatty acid production from millet fibres by pure cultures of probiotic fermentation.* Tropical Journal of Pharmaceutical Research, 2013. 12(2): p. 189-194.

[28] Wang, J., et al., *In vitro fermentation of xylooligosaccharides from wheat bran insoluble dietary fiber by Bifidobacteria.* Carbohydrate polymers, 2010. 82(2): p. 419-423.

[29] Kemperman, R.A., C. Bolca, and L.C.a.V. Roger, E.E, *Novel approaches for analysing gut microbes and dietary polyphenols: challenges and opportunities.* Microbiology, 2010. 156: p. 3224–3231.

[30] Manach, C., et al., *Bioavailability of rutin and quercetin in rats* FEBS Letters, 1997. 409: p. 12-16.

[31] Parkar, S.G., T.M. Trower, and D.E. Stevenson, *Fecal microbial metabolism of polyphenols and its effects on human gut microbiota.* Anaerobe 2013. 23: p. 12-19.

[32] Wichienchot, S., et al., *Extraction and analysis of prebiotics from selected plants from southern Thailand.* Songklanakarin J. Sci. Technol, 2011. 33(5): p. 517-523.

[33] Idrees, M., et al., *Enzymatic saccharification and lactic acid production from banana pseudo-stem through optimized pretreatment at lowest catalyst concentration.* EXCLI Journal, 2013. 12: p. 269-281.

[34] Montet, D., R.C. Ray, and N. Zakhia-Rozis, *Lactic Acid Fermentation of Vegetables and Fruits.* Microorganisms and Fermentation of Traditional Foods, 2014: p. 108.

4

Influence of Chemical Properties of Wheat-Lupine Flour Blends on Cake Quality

Abdelrahman R. Ahmed[1,2,*]

[1]Faculty of Specific Education, Home Economics Department, Ain Shams University, Cairo, Egypt
[2]Institute for Food Technology and Food Chemistry, Department of Food Rheology, Technical University of Berlin, Sekr, KL-H1, Königin-Luise-Str, Berlin / Germany
*Corresponding author: abdohat1@yahoo.com

Abstract Legumes have interesting nutritional properties and their inclusion in diet is encouraged; their incorporation into bakery products could be a good method for increasing consumption. In this study, the influence of the partial replacement of wheat flour by lupine flour on the quality characteristics of butter cake was analyzed. In the present study, three different concentrations (5, 10 and 15 % as a substitution of wheat flour) of lupine flour were used. Lupine flour showed higher levels of moisture, crude protein, ash, crude fat and dietary fiber than the wheat flour. The lupine flour showed higher levels of phenolic and flavonoids than the wheat flour. Conversely, wheat flour showed higher levels of total flavonols. Results clearly indicate that lupine flour exhibited higher antioxidant activity with DPPH and ABTS than the wheat flour. Essential amino acids (lysine, threonine, isoleucine, phenylalanine and tryptophane) in lupine flour were higher than those in wheat flour except methionine content which was higher in wheat flour (1.7 g/kg). A sensory acceptability of the cake is satisfactory up to 10 % concentration of lupine flour given. Even though deterioration in the structural formation in the batter system after the addition of lupine flour were detected that the blends have relatively good structure due to dominance of wheat to bake acceptable, protein enriched consumable cake. Finally, lupine flour up to 10% can be used successfully in bakery products. This could be utilized for the development of composite blends from locally produced lupine at small scale industry level as value-add products.

Keywords: *lupine, phenolic compounds, amino acids, cake quality*

1. Introduction

Cakes are important bakery products. Their worldwide market currently grows with about 1.5% a year. Challenges in the cake market include cost reduction, increased shelf life and quality control. Cake making consists of mixing the ingredients into a batter which, because of the high level of liquid phase in cake recipes, has a low viscosity, and baking such batter into cake [1].

Wheat-flours used in cake elaboration have lower protein content, and it is known that one of the most important characteristics of cake elaboration flours is particle size [2]. Gluten does not play an important role in this kind of product, which means that flours from other cereals [3,4,5] or even from pulses, such as chickpeas or lupine [6,7] can be used.

Pulses, also known as grain legumes, are essential to human diets in many parts of the world [8]. Pulses are unique for a human diet in terms of their nutritional profile. They are rich in protein, carbohydrates, dietary fiber, some minerals and vitamins and they are also low in fat [9,10,11].

Legume/pulse proteins, because of their composition, are considered a good supplement for cereal based foods, since both legume and cereal proteins are complementary with regard to their essential amino acids. Legume proteins are rich in lysine and deficient in sulphur-containing amino acids, whereas cereal proteins are deficient in lysine, but have adequate amounts of sulphur amino acids [12]. Apart from being nutritious, pulse proteins are highly functional and exhibit properties like solubility, gelation and water binding playing a crucial role in structure formation and mouth feel of the finished products. Among the legumes/ pulses tested as protein-enriching agents of bakery products, in the form of various protein preparations (e.g. flour, protein isolate, etc.), are soybean, chickpea, pea and lupine [6,13,14,15,16,17].

Research has indicated that consumption of pulses may have potential health benefits including reduced risk of cardiovascular disease, cancer, type-2 diabetes, osteoporosis, hypertension, gastrointestinal disorders, adrenal disease and reduction of LDL cholesterol [18-26].

The consumption could be higher if the food industry and professional organizations take up the challenge to incorporate grain legumes in novel, convenient and healthy food products [27]. Because of the nutritional and health promoting properties of the pulses, the development of value-added pulse based products for new market opportunities in the functional food and nutraceutical industry is being promoted [28].

Lupine flour is a novel food ingredient derived from the endosperm of lupine, a grain legume. It contains 40–45 % protein, 25–30 % fiber, and negligible sugar and starch [29]. It can be incorporated into high carbohydrate foods, resulting in significant increases in protein and fiber, reductions in refined carbohydrate, and little change in product acceptability [30]. Increasing protein at the expense of refined carbohydrate in the diet may benefit blood pressure. Therefore, the aim of this study was to determine the effect of the partial replacement of wheat flour by lupine flour on the quality of cakes.

2. Materials and Methods

2.1. Materials

Local Egyptian breeds of lupine (*Lupinus albus L.* variety Giza) were obtained from the Agricultural Research Centre, Giza, Egypt. Lupine flour was obtained after grinding lupine grains in a laboratory hammer mill (Retsch - Germany) until they could pass through a 250 µm screen. Commercial wheat flour type 405 was obtained from Lidl Market (Berlin-Germany). All other chemical reagents used in the experimental analysis were of analytical grade.

2.2. Chemical Analysis

2.2.1. Proximate Composition

Proximate composition was carried out according to ICC Standard Methods [31]. Moisture content was determined by drying the samples at 105°C to constant weight (ICC 109/01). Ash content was determined by calcinations at 900°C (ICC 104/1). Nitrogen content was determined by using Kieldahl method with factor of 5.7 to determine protein content (ICC 105/2). The total lipid content was determined by defeating in the Soxhelt apparatus with hexane (ICC 136). The determination of starch content was assessed using a polarimetric method according to Ewers, modified by [32]. All the measurements of analyzed samples were made in triplicate.

2.2.2. Determination of Total Phenolics

Total phenolic content was determined by the Folin–Ciocalteu micro-method [33]. A 20 µL aliquot of extract solution was mixed with 1.16 mL of distilled water and 100 µL of Folin–Ciocalteu's reagent followed by 300 µL of 200 g L^{-1} Na_2CO_3 solution. The mixture was incubated in a shaking incubator at 40°C for 30 min and its absorbance at 760 nm was measured. Gallic acid was used as standard for the calibration curve. Total phenolic content expressed as gallic acid equivalent (GAE) was calculated using the following linear equation based on the calibration curve:

$$Ab = 0.98C + 9.925 * 10^{-3}\left(R^2 = 0.9996\right) \quad (1)$$

where Ab is the absorbance and C is the concentration (mg GAE g^{-1} dry weight (DW)).

2.2.3. Determination of Total Flavonoids

Total flavonoid content was determined by the method of Ordoñez et al., [34]. A 0.5 mL aliquot of 20 g L^{-1} $AlCl_3$ ethanolic solution was added to 0.5 ml of extract solution. After 1 h at room temperature the absorbance at 420 nm was measured. A yellow color indicated the presence of flavonoids. Extract samples were evaluated at a final concentration of 0.1 mg ml^{-1}. Total flavonoid content expressed as quercetin equivalent (QE) was calculated using the following equation based on the calibration curve:

$$C = 0.0255 * Ab\left(R^2 = 0.9812\right) \quad (2)$$

where Ab is the absorbance and C is the concentration (mg QE g^{-1} DW).

2.2.4. Determination of Total Flavonols

Total flavonol content was determined by the method of Kumaran and Joel Karunakaran, [35]. To 2 mL of extract solution, 2 mL of 20 g L^{-1} $AlCl_3$ ethanolic solution and 3 mL of 50 g L^{-1} sodium acetate solution were added. The absorption at 440 nm was read after 2.5 h at 20°C. Extract samples were evaluated at a final concentration of 0.1mg mL^{-1}. Total flavonol content expressed as QE was calculated using the same equation of flavonoids.

2.3. Antioxidant Activity of Extracts

Because of the differences among the various test systems available, the results of a single method can provide only a limited assessment of the antioxidant properties of a substance [36]. For that reason, in this study the antioxidant capacity of each extract was determined through two complementary assay procedures.

2.3.1. DPPH Radical-Scavenging Activity

The DPPH assay according to Lee et al., [37] was utilised with some modifications. The stock reagent solution (1 x 10^{-3} mol L^{-1}) was prepared by dissolving 22 mg of DPPH in 50 mL of methanol and stored at −20°C until use. The working solution (6 x 10^{-5} mol L^{-1}) was prepared by mixing 6 mL of stock solution with 100 mL of methanol to obtain an absorbance value of 0.8±0.02 at 515 nm, as measured using a spectrophotometer. Extract solutions of different concentrations (0.1 mL of each) were vortexed for 30 s with 3.9 mL of DPPH solution and left to react for 30 min, after which the absorbance at 515 nm was recorded. A control with no added extract was also analysed. Scavenging activity was calculated as follows:

$$\text{DPPH radical} - \text{scavenging activity } (\%) = \left[\left(Ab_{\text{control}} - Ab_{\text{sample}}\right) / \ Ab_{\text{control}}\right] * 100 \quad (3)$$

where Ab is the absorbance at 515 nm.

2.3.2. ABTS Radical-Scavenging Activity

For the ABTS assay the method of Re et al., [38] was adopted. The stock solutions were 7 mmol L^{-1} ABTS solution and 2.4 mmol L^{-1} potassium persulfate solution. The working solution was prepared by mixing the two stock solutions in equal quantities and allowing them to react for 12–16 h at room temperature in the dark. Then 1 mL of the resulting $ABTS^+$ solution was diluted with 60 mL of methanol to obtain an absorbance of 0.706 ± 0.001 units at 734 nm, as measured using a spectrophotometer. $ABTS^+$ solution was freshly prepared for each assay. Extract solutions of different concentrations (1 mL of each) were allowed to react with 1 mL of $ABTS^+$ solution for 7 min, after which the absorbance at 734 nm was recorded. A control with no added extract was also analysed. Scavenging activity was calculated as follows:

$$\text{ABTS radical} - \text{scavenging activity } (\%) = \left[\left(Ab_{control} - Ab_{sample}\right) / Ab_{control}\right] * 100 \quad (4)$$

where $Ab_{control}$ is the absorbance of ABTS radical + methanol Ab_{sample} is the absorbance of ABTS radical + extract.

2.4. HPLC Analysis

A total of 20 mg grounded dried samples were extracted for 15 min using 750 µL 70 % methanol (v/v, pH 4, phosphoric acid) in an ultrasonic water bath (Sonorex digital 10 p, Bandelin) on ice. Samples were centrifuged for 5 min at 6000 rpm. The supernatants were collected and the pellets were re-extracted twice more with 500 µL 70 % methanol. The combined supernatants from each sample were reduced to near dryness in a centrifugation evaporator (Speed Vac, SC 110) at 25°C. Samples were then made up to 1 ml with 40 % acetonitrile. The samples were filtrated using 0.22 µm filter, and then analyzed with HPLC (Dionex Summit P680A HPLC-System), equipped with P680 pump, ASI-100 automated sample injector, a Narrow-Bore Acclaim PA C16-column (3 µm, 2.1 * 150 mm, Dionex) and PSA-100 photodiode array detector (Dionex) and software Chromeleon 6.8 (Dionex, USA). The column was operated at a temperature of 35°C.

The mobile phase consisted of 0.1 % (v/v) phosphoric acid in ultrapure water (eluent A) and of 40 % (v/v) acetonitrile in ultrapure water (eluent B). A multistep gradient was used for all separations with an initial injection volume of 40 µL and a flow rate of 0.4 mL/min. The multistep gradient was as follows: 1 min: 0.5 % (v/v) B; 1-10 min: 0.5-40 % B; 10-12 min: 40 % B; 12-18 min: 40-80 % B; 18-20 min: 80 % B; 20-24 min: 80-99 % B; 24-30 min: 99- 100 % B; 30-34 min: 100-0.5 % B; 34-39 min: 0.5 % B. Simultaneous monitoring was performed at 290, 330 and 254 nm at a flow rate of 0.4 mL/min. Phenolic acid quantity was calculated from HPLC peak areas at 290 nm against the internal standard and external standards. Identification and quantification of phenolic acids present was done by comparing retention time and area of the peaks in the extracts with that of the standard phenolic acids (chlorogenic acid, caffeic acid, cinnamic acid, coumaric acid, rosmarinic acid and sinapic acid).

2.5. Amino Acid Analysis

Amino acid content was determined as described by Moore et al., [39]. The analysis was performed in Central Service Unit, National Research Centre, Egypt using LC3000 amino-acid analyzer (Eppendorf-Biotronik, Germany). The technique was based on the separation of the amino acids using strong cation exchange chromatography followed by the ninhydrin colour reaction and photometric detection at 570 nm. Standard amino acids were used for comparison of resulting profiles, allowing quantitation of amino-acid residues. The defatted powdered seeds were hydrolyzed with 6 N HCl at 110°C in teflon capped vials for 24 h. After vacuum removal of HCl, the residues were dissolved in a lithium citrate buffer, pH 2.2. Twenty µL of the solution were loaded into the cation exchange column (pre-equilibrated with the same buffer), then four lithium citrate buffers with pH values of 2.2, 2.8, 3.3 and 3.7, respectively, were successively applied to the column at a flow rate of 20 mL/min. The ninhydrin flow rate was 10 mL/h under these conditions and a typical analysis required 160 min.

Methionine was determined as methionine sulfone, after oxidation with performic acid. An amino acid standards containing cysteine were treated parallel with the samples and used to quantify the methionine content. The amino acid content of the reference protein was taken from [40].

2.6. Cake Formulation and Baking

The recipes of butter cake used: 100 g (wheat flour or wheat flour substituted with 5, 10 or 15 % lupine flour), 80 g egg, 60 g sugar, 50 ml Milk (1.5 % fat), 50 g margarine and 4 g baking powder. A creaming mixing procedure was used. All ingredients, except for the flour and baking powder, were mixed for 2 min at speed 6 using a Kitchene Aid Professional mixer (KPM5). After the addition of the flour and baking powder, the mixing process continued for 3 min at speed 8. 200 g of cake batter were placed into 120 mm diameter and 45 mm height, metallic, lard coated pan, and were baked in an electric oven for 25 min at 200°C.

2.7. Loaf Volume Determination

Cake mass was weighted after 3 hours at room temperature. The volume (cm^3) was measured by rapeseed replacement method described in the [41]. The specific volume was obtained by dividing the volume by their weights.

2.8. Color Measurements

Crust and crumb color of fresh cake was measured with a Minolta Calorimeter (CR 200 Japan). Color readings were expressed by Hunter values for L*, a* and b*.

2.9. Sensory Evaluation

Evaluation of the cake quality characteristics was carried out following cooling to room temperature for 2 h. Sensory evaluation was performed by ten trained panelists who were graduate students and staff members of the Department of Rheology, Institute of Food Technology and Food Chemistry, Technical University, Berlin. Cakes were randomly assigned to each panelist. The panelists were asked to evaluate each sample for appearance, crumb texture, crumb grain, crust color, taste, odor and overall acceptability. A 10 point scale was used where 10 "excellent and 1" extremely unsatisfactory.

2.10. Statistical Analysis

Analysis of variance (ANOVA) was carried out using SAS program (Statistical Analysis System version. 9.1) SAS Institute Inc. [42]. The chemical composition and cake characteristics of wheat batter with or without lupine flour were analysed using ANOVA. When the treatment factor effect was found significant, indicated by a significant F-test ($p < 0.05$), differences between the respective means were determined using least significant difference (LSD) and considered significant when $p < 0.05$. Mean ± standard deviation of mean was used.

3. Results and Discussion

3.1. Chemical Composition of Wheat Flour, Lupine Flour and Their Blends

Lupine is a good source of nutrients, not only proteins but also lipids, dietary fiber, minerals, and vitamins [43]. The results for the chemical composition of wheat flour (WF), lupine flour (LF), and their blends are shown in Table 1. The lupine flour showed higher levels of moisture, crude protein, ash, crude fat and dietary fiber than the wheat flour. Conversely, wheat flour showed higher levels of starch. Highly significant differences ($P < 0.05$) were observed between the two types of flours. Mean protein and dietary fiber increased with increasing amount of lupine flour added. There was no significant difference between wheat flour and supplemented flour with different concentration of lupine for moisture content (Table 1).

The chemical properties of wheat flours have been studied previously by several researchers and they found that moisture content ranged between 12.5 to 14.6 % crude protein content 8.23 to 12.71 % and ash content 0.42 to 0.66 [44].

Table 1. Chemical composition of wheat flour (WF), lupine flour (LF) and their blends

Analysis	WF	LF	LF level (%)		
			5	10	15
Moisture	11.27 ± 0.09	12.37 ± 0.46	11.35 ± 0.19	11.42 ± 0.17	11.54 ± 1.16
Protein	12.1 ± 0.20	38.6 ± 0.87	13.73 ± 0.24	14.75 ± 0.27	16.28 ± 0.31
Ash	0.40 ± 0.02	3.41 ± 0.03	0.55 ± 0.01	0.70 ± 0.04	0.85 ± 0.03
Fat	1.62 ± 0.19	9.94 ± 0.16	2.04 ± 0.19	2.45 ± 0.17	2.87 ± 0.16
Starch	69.8 ± 1.96	0.98 ± 0.04	66.36 ± 3.94	62.92 ± 1.88	59.48 ± 2.84
S. D. F	1.1 ± 0.09	11.0 ± 1.36	1.60 ± 0.14	2.09 ± 0.27	2.59 ± 0.31
I. D. F	1.6 ± 0.23	30.8 ± 2.45	3.06 ± 0.31	4.52 ± 0.42	5.98 ± 0.53
T. D. F	2.7 ± 0.15	41.8 ± 3.08	4.66 ± 0.27	6.61 ± 0.43	8.57 ± 0.60

Mean ± standard deviation. S.D.F: Soluble Dietary Fiber. I.D.F: Insoluble Dietary Fiber. T.D.F: Total Dietary Fiber.

Table 2. Extract yield, total polyphenols content and antioxidant capacity of wheat flour (WF), lupine flour (LF) and their blends

Analysis	WF	LF	LF level (%)		
			5	10	15
Yield extract (%)	13.7 ± 3.99	36.2 ± 1.73	11.9 ± 0.35	13.6 ± 1.35	14.0 ± 1.21
Total phenolic (µg GAE/g DW)	126.63 ± 3.52	138.17 ± 8.35	132.17 ± 0.58	142.5 ± 7.10	156.53 ± 3.88
Total flavonoids (µg QE/g DW)	6.33 ± 0.15	8.93 ± 0.06	7.67 ± 1.27	7.93 ± 0.06	8.4 ± 0.52
Total flavonols (µg QE/g DW)	32.03 ± 6.13	31.60 ± 4.70	29.10 ± 2.48*	28.27 ± 2.96*	27.00 ± 1.08*
DPPH (%)	3.31 ± 0.35	20.62 ± 1.22	5.1 ± 0.10*	6.04 ± 0.77*	7.16 ± 0.26*
ABTS (%)	26.7 ± 0.21	43.42 ± 0.37	29.41 ± 0.37*	31.09 ± 0.00*	32.35 ± 0.37*

Mean ± standard deviation

Protein content of lupine (38.6 %) was higher than that of a lot of legumes. Favier et al., [45] reported that haricot bean, lentil and soy bean contain 28.8 %, 26.7 % and 40.5 % protein, respectively. Because of the high protein content, lupine flour could be used in the human diet. Also, temperature of denaturation of these proteins is higher than animal protein, so they are technologically easier to handle [46]. Lupine flour had a high amount of crude fiber (16.2 %). These fibers have many desirable properties, including white color, high water-holding capacity (7.1 g H_2O/g) and beneficial effects on human health [47]. Therefore, lupine flour can be incorporated into a wide range of foods to make dietary products.

3.2. Phenolic Compounds and Antioxidants Capacity of Wheat Flour, Lupine Flour and Their Blends

Phenolic compounds ubiquitous in plants are key phytochemical drivers of the health and functional foods and nutraceutical industry. Research with polyphenol compounds from various crops has created a growing market for polyphenol-rich ingredients, estimated to be worth around $ 99 million in Europe in 2003 [48].

Conventional solvent extraction has been reported in a laboratory scale using acetone, hexane, methanol and ethanol [49]. In this study, methanol was used for the extraction of antioxidant compounds from wheat, lupine flour and their blends (Table 2). The extraction yield 13.7 and 36.2 g/100 g dry weight for wheat and lupine flour respectively.

The lupine flour showed higher levels of total phenolic and total flavonoids than the wheat flour. Conversely, wheat flour showed higher levels of total flavonols. These results confirmed by statistical analysis, which highly significant differences ($P<0.05$) were observed between the two type of flours. Total phenolic and total flavonoids increased with increasing amount of lupine flour added to be 132.17 ± 0.58, 142.5 ± 7.10, 156.53 ± 3.88 (µg GAE/g DW) and 7.67 ± 1.27, 7.93 ± 0.06, 8.4 ± 0.52 (µg QE/g DW) for substituting wheat flour with lupine flour at 5, 10 and 15 %, respectively on dry weight basis. The contents of phenolic acids in lupine used in this study are comparable to levels reported previously [50], especially in cultivars of L. albus grown in Portugal. Phenolic content of lupins were higher than those of bean cultivars grown in Manitoba [51] probably as a result of relatively high flavonoid content. The methanolic extracts of lupine seed were analysed by high performance liquid chromatography figure 15.

As shown in figure 1, nine phenolic acids were separated and identified. This method is well reproducible and provides good separation in terms of migration time and resolution.

The antioxidant effects of extracts of various wheat flour (WF), lupine flour (LF) and their blends at different concentration (5, 10 and 15 %) were measured. Since the active substances of flour extracts tested are different, the antioxidant activities of these extracts cannot be evaluated by only a single method. Therefore, two different models were used in this study [52].

Free radicals which are involved in the process of lipid peroxidation are considered to play a major role in numerous chronic pathologies, such as cancer and cardiovascular diseases among others [53]. The DPPH radical has been widely used to evaluate the free radicals' scavenging ability of various natural products and has been accepted as a model compound for free radicals originating in lipids [54]. The effect of antioxidants on diphenyl-p-picryl hydrazyl (DPPH) radical scavenging was thought to be due to their hydrogen donating ability. DPPH is a stable free radical and accepts an electron or hydrogen radical to become a stable diamagnetic molecule. The assay is based on the reduction of DPPH. Because of its odd electron, DPPH gives strong absorption maxima at 515 nm (purple color) by visible spectroscopy. As the odd electron of the radical becomes paired off in the presence of a hydrogen donor, i.e., a free radical scavenging antioxidant, the absorption intensity is decreased, and the resulting decolorization is stochiometric with respect to the number of electrons captured [55].

Table 2 showed that the scavenging activity of methanolic extracts against DPPH for wheat flour (WF), lupine flour (LF) and their blends. Significant ($p < 0.05$) differences between wheat and lupine flour extracts were observed. Results clearly indicate that lupine flour exhibited higher antioxidant activity with DPPH and ABTS than the wheat flour. The antioxidant activity increased with increasing amount of lupine flour added to be 5.1 ± 0.10, 6.04 ± 0.77, 7.16 ± 0.26 in DPPH and 29.41 ± 0.37, 31.09 ± 0.00, 32.35 ± 0.37 in ABTS respectively, for substituting wheat flour with lupine flour at 5, 10 and 15 %, respectively on dry weight basis.

Wang et al., [56] found that some compounds which have $ABTS^+$ scavenging activity did not show DPPH scavenging activity. In this study, there was not the case. The $ABTS^+$ scavenging data suggests that the components within the extracts are capable of scavenging free radicals via a mechanism of electron/hydrogen donation and should be able to protect susceptible matrices from free radical-mediated oxidative degradation.

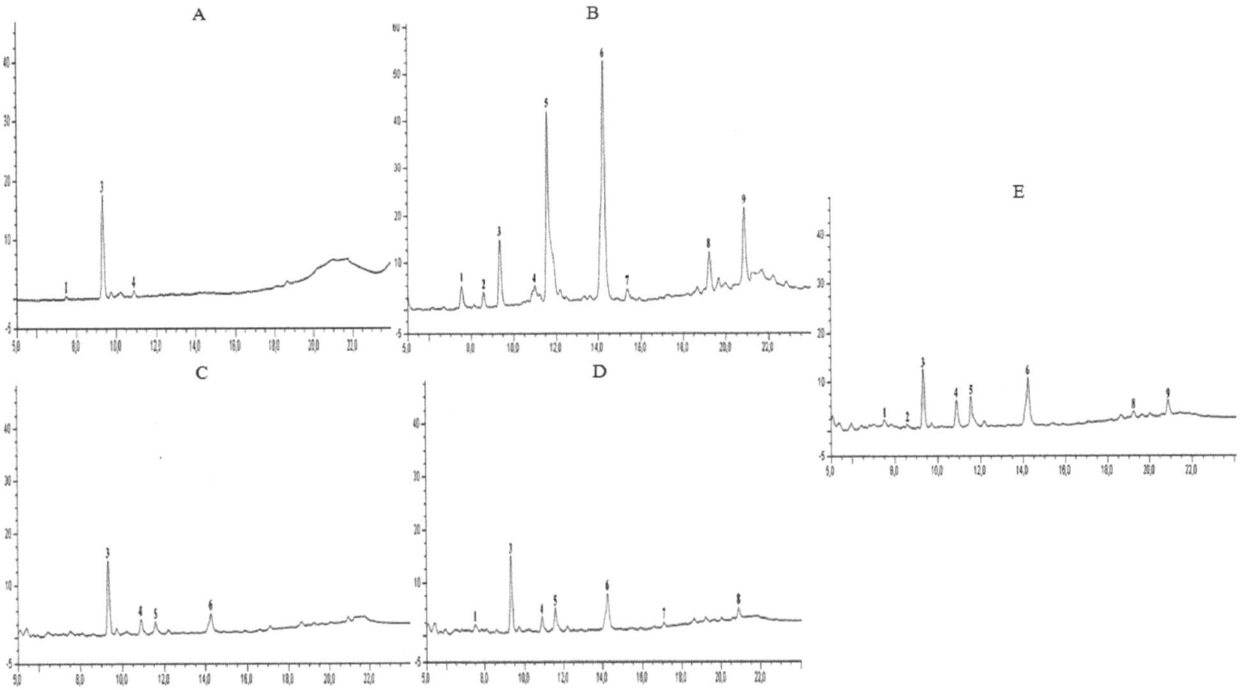

Figure 1. HPLC chromatogram of methanol extract of: wheat flour (A), lupine flour (B) and wheat flour supplemented with lupine flour at different concentration, 5 % (C). 10 % (D) and 15 % (E). 1. gallic, 2. procatechuic, 3. p-hydroxybenzoic, 4. vanillin, 5. P coumaric, 6. chlorogenic, 7. cinnamic 8. sinapine and 9. ferulic acid

3.3. Amino Acids Content

Lupine seeds represent a good balance of essential amino acids [57]. They are considered to be a good source of lysine, and are generally poor in the sulfur-containing amino acids (methionine and cysteín) [58] and threonine [59].

The results for the amino acid content of wheat flour (WF) and lupine flour (LF) are shown in Table 3. In contrast to plants, humans and animals are able to synthesize only 9 amino acids used in protein synthesis (non-essential amino acids). The biosynthesis of the remaining (essential) amino acids, thereby the protein synthesis, is not possible without their continuous supply through food consumption. In the case of low-protein diets, symptoms, such as delay in growth, negative nitrogen uptake or disturbances in protein synthesis, can take place. Therefore foods rich in exogenous amino acids are desirable.

Table 3. The total amino acids % dry matter, for wheat (WF) and lupine flour (LF)

Amino Acid g/kg	WF	LF	FAO/WHO**	Amino Acid g/kg	FAO/WHO	WF	LF
Lysin*	3,0 ± 0.13	16.35 ± 0.56	5.8	Leucin	6.6	6,7 ± 0.12	26.13 ± 1.08
Threonin*	2,7 ± 0.08	11.48 ± 0.47	3.4	Phenylalanin*	6.3	4,1 ± 0.16	14.20 ± 0.40
Valin	4,8 ± 0.28	16.65 ± 0.76	3.5	Tyrosin	6.1	2,6 ± 0.10	13.90 ± 0.53
Methionin*	1,7 ± 0.15	1.59 ± 0.16	2.5	Tryptophan*	1.1	1,4 ± 0.07	10.94 ± 0.70
Serine	2,3 ± 0.50	1.52 ± 0.60	-	Arginin	-	4,3 ± 0.18	36.13 ± 3.78
Isoleucin*	3,4 ± 0.20	14.85 ± 0.36	2.8	Histidin	1.9	2,5 ± 0.19	5.89 ± 0.33

*Essential Amino Acids **FAO/WHO, [40]

The results showed that the essential amino acids (lysine, threonine, isoleucine, phenylalanine and tryptophane) in lupine flour were higher than those in wheat flour except methionine content which was higher in wheat flour (1.7 g/kg). This result was confirmed by Lubowicki et al., [60]. Sujak et al., [61] reported that lupine seeds of different species representing diverse varieties of sweet lupine grown in Poland manifest a large deficiency of sulphur containing amino acids, for which the recommended level is 3.5 g/16 g N [62]. Methionine levels of 1.59 g/kg, found for the lupine flour was low but comparable to results reported previously for other lupins [63]. The recommended level of methionine is 2.5 g/kg [64]. Of great importance is the presence of sulphur containing amino acids, mainly methionine, which is necessary for the synthesis of cysteine, as well as phenylalanine needed for the synthesis of tyrosine [62].

The protein demand of different organisms depends on their physiological state stipulated mainly by age. For example, young and growing mammals (up to approximately two years in humans) need proteins rich in amino acids, such as arginine and histidine, as such amino acids are the source of the active centers of many enzymes. In contrast, adults show almost no physiological demand for these amino acids.

From the results we can noticed that lupine flour is rich with arginine and histidine (36.13 and 5.89 g/kg respectively). Protein quantity, as well as composition, is the limitation of protein quality [64]. For humans, adequate quantities of lysine, methionine and tryptophan are considered necessary in food of high nutritive value [62]. A number of approaches, based on the analysis of amino acids, have been considered for the estimation of protein quality in human and fodder foods. According to Alsmeyer et al., [65], the nutritional value of food should be expressed in terms of leucine and tyrosine contents, while other classifications are based on the chemical scores for 9–11 amino acids considered essential. Lupine flour showed high content of lysine (16.35 g/kg) more than wheat flour (3.0 g/kg).

3.4. Influence of Lupine Flour Incorporation on Cake Properties

According to Table 4, cake height diminished as the lupine flour percentage increased. During the baking process, baking powder generates gases, which should be retained in order to guarantee good cake volume, and in that respect flour quality has an important role to play. Another important factor is the gelatinization temperature of the flour, as Howard, [66], pointed out for layer cakes, whereas Mizukoshi et al., [67] reached the same conclusion for sponge cakes. The starch gelatinization at low temperatures would prevent the correct expansion of dough.

Table 4. Cake characteristics of wheat flour and lupine flour composite flours

Sample	Cake height cm	Cake weight g	Cake volume cm³	Specific volume cm³/g
WF	7.5 ± 0.53	265.1 ± 13	660 ± 56	2.49 ± 0.6
LF 5 %	7.0 ± 0.33	265.2 ± 38	670 ± 46	2.53 ± 0.45
LF 10 %	6.5 ± 0.23	264.5 ± 11	690 ± 68	2.61 ± 0.27
LF 15 %	6.0 ± 0.65	269.2 ± 8	700 ± 57	2.60 ± 0.30

Mean ± standard deviation of mean.

Lupine flour also presented higher protein content, and different amino acid composition than wheat flours which could affect cake characteristics, especially volume [68]. As far as lupine derivatives are concerned, flour gave cakes with higher volume. This fact can be explained by the higher fiber content of lupine flour and its effect on pasting behavior of starches [69].

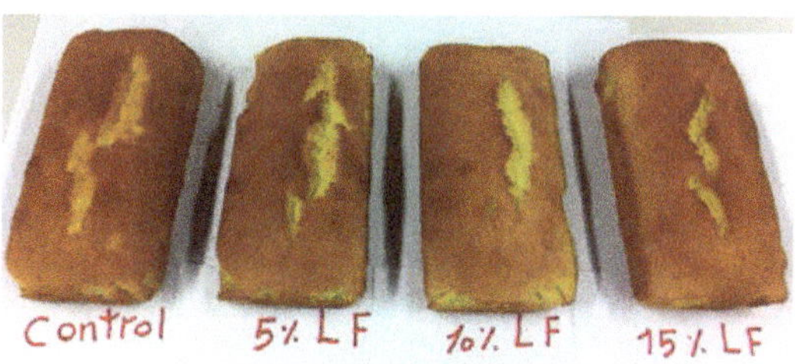

Figure 2. Comparison of the baking properties of wheat flour and mixes with lupine flour cake

The volume index is an indicator of cake volume and, as expected, followed a similar tendency as volume. Symmetry indicates the differences in height between the central zone and the lateral zone. Thus, a high symmetry suggests that cakes mainly rise in their central part, while a negative symmetry indicates that cake volume falls down at the end of the baking process. Hence, symmetry gives an idea about gas retention in the final baking phase. In both cases the incorporation of lupine flour at different levels had no effect on cake symmetry (Figure 2).

This result is not confirmed with Gômez et al., [6] who studied the influence of the total or partial replacement of wheat flour by chickpea flour on the quality characteristics of two kinds of cake. They reported that in both cases the incorporation of chickpea flour reduced their symmetry, and in the case of layer cakes with 100 % of chickpea flour it became negative. As for cake weight, no significant differences were found. Therefore, the water retention capacity was not affected by the substitution of wheat flour by lupine flour.

Crust color in cakes varied with the quantity and the kind of lupine flour addition. This influence was more important in cakes. The crust color data of cakes are shown in Table 5 Cakes became darker (lower L*) as the lupine flour quantity increased. With regard to the wheat flour, it produced the brightest cakes. No considerable differences in crust yellowness were found among the different lupine flour samples, while all lupine flour gave higher b values than the control (wheat flour). The crust color of cakes was generated in the baking process due to the Maillard reactions between sugars and amino acids, and the caramelization process of sugars. Therefore, the differences observed when the quantity of lupine flour increased could be attributed to the high protein content and the different amino acid composition of the composite flours compared to the wheat flour.

Table 5. Color measurements of cake from wheat flour and lupine flour composite flours

Color parameters	Crust			Crumb		
	L*	a*	b*	L*	a*	b*
WF	90.53 ± 3.79	1.24 ± 0.69	4.92 ± 2.74	93.62 ± 3.74	0.90 ± 0.60	2.31 ± 1.58
LF 5 %	76.48 ± 8.24	4.19 ± 1.85	11.88 ± 6.20	92.39 ± 1.61	1.32 ± 0.41	2.55 ± 1.59
LF 10 %	74.38 ± 10.43	4.71 ± 2.09	14.22 ± 3.13	92.15 ± 2.31	3.00 ± 0.50	7.78 ± 1.82
LF 15 %	73.44 ± 4.50	4.94 ± 4.05	15.95 ± 1.76	91.48 ± 1.55	4.89 ± 30	11.99 ± 1.73

Mean ± standard deviation of 10 different points on Crust and Crumb.

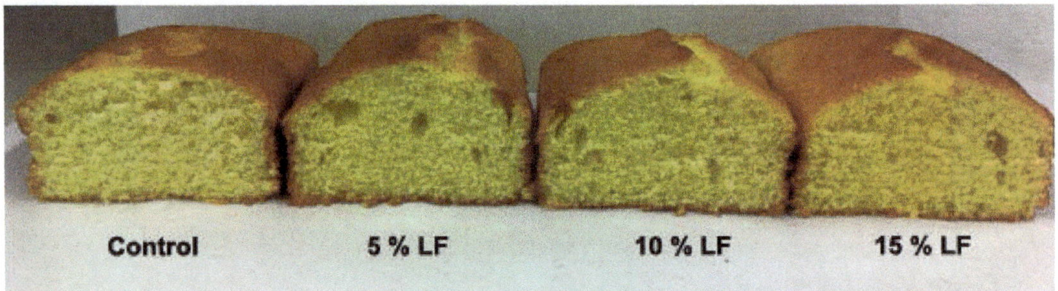

Figure 3. Crumb color of wheat flour and mixes with lupine flour cake

Table 6. Sensory evaluation of cake from wheat flour and lupine flour composite flours

Sample	Appearance	Crumb texture	Crumb grain	Crust color	Taste	Odor	Overall acceptability
WF	7.9 ± 1.5	7.9 ± 1.5	7.1 ± 1.6	7.6 ± 1.8	7.7 ± 2.1	8.1 ± 1.9	8.2 ± 1.3
LF 5 %	7.7 ± 2.0	8.1 ± 1.1	7.9 ± 0.9	8.3 ± 1.6	7.8 ± 1.8	8.2 ± 1.6	8.1 ± 1.5
LF 10 %	8.0 ± 1.4	8.0 ± 1.5	8.3 ± 0.9	8.1 ± 1.5	7.4 ± 2.2	7.9 ± 1.8	8.1 ± 1.2
LF 15 %	8.6 ± 1.3	7.7 ± 2.1	8.9 ± 1.5	8.7 ± 1.3	8.2 ± 2.0	8.2 ± 1.1	8.7 ± 1.3

Mean ± standard deviation of ten panellists

Cake crumb does not reach temperatures above 100°C, so the Maillard or caramelization reactions by sugars fail to take place. Therefore, crumb color must be the result of the raw materials colors and their interactions. The crumb color data of cakes are shown in Table 5 and Figure 3 Cake crumb of lupine flour samples were lightly darker than control, and the lupine flour addition reduced its luminosity, but this effect was not significant. These results agree with those obtained by Dodok et al., [70], who observed that chickpea flour triggered the change in the crumb color when they studied the addition of chickpea flour to bread dough.

Sensory evaluation results of fortified cake with lupine flour are presented in Table 6. It was found that cake which was fortified with different levels of lupine flour had non significant difference in taste, odour and overall acceptability as compared to unfortified cake (control).

The present results came in agreement with Sabanis et al., [71] who reported that, organoleptic properties (colour, flavour and overall acceptability) improved with a low proportion of chickpea flour, especially for 5 % w/w substitution. Alabi and Anuonye, [72] indicated that up to 50 % of some legume products could be added without significant loss in palatability.

4. Conclusion

This study demonstrates that it is possible to incorporate lupine flour into cake formulations up to 10%. The substitution percentage lupine flour used led to the pursued objectives (nutritional improvement, free-gluten products, special organoleptic characteristics, etc.). This could be utilized for the development of composite blends from locally produced lupine at small scale industry level as value-add products.

References

[1] Cauvain SP, Young L: Baked products. in Science, technology and practice. Oxford, UK: Blackwell Publishing, 2006.

[2] Gómez M, Ruiz-París E, Oliete B: Original article: Influence of flour mill streams on cake quality. International Journal of Food Science & Technology 2010; 45: 1794-1800.

[3] Oliete B, Pérez GT, Gómez M, Ribotta PD, Moiraghi M, León AE: Use of wheat, triticale and rye flours in layer cake production. International Journal of Food Science & Technology 2010; 45: 697-706.

[4] Ronda F, Gomez M, Caballero PA, Oliete B, Blanco CA: Improvement of quality of gluten-free layer cakes. Food Science and Technology International 2009; 15: 193-202.

[5] Turabi E, Sumnu G, Sahin S: Rheological properties and quality of rice cakes formulated with different gums and an emulsifier blend. Food Hydrocolloids 2008; 22: 305-312.

[6] Gômez M, Oliete B, Rosell CM, Pando V, Fernandez E: Studies on cake quality made of wheat_chickpea flour blends. LWT - Food Science and Technology 2008; 41: 1701-1709.

[7] Paraskevopoulou A, Chrysanthou A, Koutidou M: Characterisation of volatile compounds of lupin protein isolate - enriched wheat flour bread. Food Research International 2012; 48: 568-577.
[8] Roy F, Boye JI, Simpson BK: Bioactive proteins and peptides in pulse crops: Pea, chickpea and lentil. Food Research International 2010; 43:4 32-442.
[9] Iqbal A, Khalil IA, Ateeq N, Khan MS: Nutritional quality of important food legumes. Food Chemistry 2006; 97: 331-335.
[10] Rochfort S, Panozzo J: Phytochemicals for Health, the Role of Pulses. Journal of Agricultural and Food Chemistry 2007; 55: 7981-7994.
[11] Tosh SM, Yada S: Dietary fibres in pulse seeds and fractions: Characterization, functional attributes, and applications. Food Research International 2010; 43: 450-460.
[12] Eggum BO, Beame RM: The nutritive value of seed proteins. in Muller WGPH (ed): Seed protein biochemistry, genetics and nutritive value. Junk: The Hague, 1983, 499-531.
[13] Doxastakis G, Zafiriadis I, Irakli M, Tananaki C: Lupin, soya and triticale addition to wheat flour doughs and their effect on rheological properties. Food Chemistry 2002; 77: 219-227.
[14] Kiosseoglou V, Paraskevopoulou A: Functional and physicochemical properties of pulse proteins. in B. Tiwari AG, & B. McKenna (ed): Pulse foods: Processing, quality and nutraceutical application. Burlington: Elsevier Inc. Academic Press, 2011, 57-90.
[15] Paraskevopoulou A, Provatidou E, Tsotsiou D, Kiosseoglou V: Dough rheology and baking performance of wheat flour-lupin protein isolate blends. Food Research International 2010; 43: 1009-1016.
[16] Ribotta PD, Arnulphi SA, Leôn AE, Anôn MC: Effect of soybean addition on the rheological properties and breadmaking quality of wheat flour. Journal of the Science of Food and Agriculture 2005; 85: 1889-1896.
[17] Sadowska J, Blaszczak W, Fornal J, Vidal-Valverde C, Frias J: Changes of wheat dough and bread quality and structure as a result of germinated pea flour addition. Eur Food Res Technol 2003; 216: 50.
[18] Anderson JW, Major AW: Pulses and lipaemia, short- and long-term effect: potential in the prevention of cardiovascular disease. Br J Nutr 2002; 88 S 263-271.
[19] Bazzano LA, He J, Ogden LG, Loria C, Vupputuri S, Myers L, Whelton PK: Legume consumption and risk of coronary heart disease in US men and women: NHANES I Epidemiologic Follow-up Study. Arch Intern Med 2001; 161: 2573-2578.
[20] Bazzano LA, Thompson AM, Tees MT, Nguyen CH, Winham DM: Non-soy legume consumption lowers cholesterol levels: a meta-analysis of randomized controlled trials. Nutr Metab Cardiovasc Dis 2011; 21: 94-103.
[21] Campos-Vega R, Loarca-Piña G, Oomah BD: Minor components of pulses and their potential impact on human health. Food Research International 2010; 43: 461-482.
[22] Cao JJ, Gregoire BR, Sheng X, Liuzzi JP: Pinto bean hull extract supplementation favorably affects markers of bone metabolism and bone structure in mice. Food Research International 2010; 43: 560-566.
[23] Jacobs DR, Jr., Gallaher DD: Whole grain intake and cardiovascular disease: a review. Curr Atheroscler Rep 2004; 6: 415-423.
[24] Mathers JC: Pulses and carcinogenesis: potential for the prevention of colon, breast and other cancers. Br J Nutr 2002; 88 S 273-279.
[25] Tharanathan RN, Mahadevamma S: Grain legumes - a boon to human nutrition. Trends Food Sci. Technol. 2003; 14: 507.
[26] Venn BJ, Mann JI: Cereal grains, legumes and diabetes. Eur J Clin Nutr 2004; 58: 1443-1461.
[27] Schneider AV: Overview of the market and consumption of pulses. Br. J. Nutr. 2002; 88: S 243.
[28] Anton AA, Ross KA, Lukow OM, Fulcher RG, Arntfield SD: Influence of added bean flour (Phaseolus vulgaris L.) on some physical and nutritional properties of wheat flour tortillas. Food Chemistry 2008; 109: 33-41.
[29] Evans AJ, Cheung PCK, Cheetham NWH: The carbohydrate composition of cotyledons and hulls of cultivars of Lupinus angustifolius from Western Australia. Journal of the Science of Food and Agriculture 1993; 61: 189-194.
[30] Lee YP, Mori TA, Sipsas S, Barden A, Puddey IB, Burke V, Hall RS, Hodgson JM: Lupin-enriched bread increases satiety and reduces energy intake acutely. Am J Clin Nutr 2006; 84: 975-980.
[31] ICC: ICC Standard Methods. In, International Association for Cereal Chemistry http://www.icc.or.at/, Access date 28/5/2011 2001.
[32] Davidek J, Hrdlicka J, Karvanek M, Pokorny J, Velisek J: Laboratory handbook for foodstuffs. Prague, CR, SNTL-Alfa, 1981.
[33] Arabshahi-Delouee S, Urooj A: Antioxidant properties of various solvent extracts of mulberry (Morus indica L.) leaves. Food Chemistry 2007; 102: 1233-1240.
[34] Ordoñez AAL, Gomez JD, Vattuone MA, Isla MI: Antioxidant activities of Sechium edule (Jacq.) Swartz extracts. Food Chemistry 2006; 97: 452-458.
[35] Kumaran A, Joel Karunakaran R: In vitro antioxidant activities of methanol extracts of five Phyllanthus species from India. LWT - Food Science and Technology 2007; 40: 344-352.
[36] Sacchetti G, Maietti S, Muzzoli M, Scaglianti M, Manfredini S, Radice M, Bruni R: Comparative evaluation of 11 essential oils of different origin as functional antioxidants, antiradicals and antimicrobials in foods. Food Chemistry 2005; 91: 621-632.
[37] Lee SC, Kim JH, Jeong SM, Kim DR, Ha JU, Nam KC, Ahn DU: Effect of far-infrared radiation on the antioxidant activity of rice hulls. J Agric Food Chem 2003; 51: 4400-4403.
[38] Re R, Pellegrini N, Proteggente A, Pannala A, Yang M, Rice-Evans C: Antioxidant activity applying an improved ABTS radical cation decolorization assay. Free Radic Biol Med 1999; 26: 1231-1237.
[39] Moore S, Spackman DH, Stein WH: Chromatography of amino acids on sulphonated polystyrene resins: An improved system.. Anal. Chem. 1958; 30: 1185-1190.
[40] FAO/WHO: Energy and protein requirements. InGeneva, Nutrition Report Series No. 935, 2007.
[41] AACC: Cereal Laboratory Methods, Method 46-09. In(Am. Assoc. Cereal Chem., St. Paul, MN), 7th Ed., American Association of Cereal Chemists, 1983.
[42] SAS: SAS Online Doc_ 9.1. InCary, NC, SAS Institute Inc., 2004.
[43] Martínez-Villaluenga C, Zieliński H, Frias J, Piskuła MK, Kozłowska H, Vidal-Valverde C: Antioxidant capacity and polyphenolic content of high-protein lupin products. Food Chemistry 2009; 112: 84-88.
[44] Ahmad I, Anjum FM, Butt MS: Quality characteristics of wheat varieties grown in Pakistan from 1933-1996. Pak. J. Food Sci. 2001; 11: 1-4.
[45] Favier JC, Ripert JI, Toque C, Feinberg M: Reportoire general des aliments (composition tables). ed 2nd, Paris, Inra Editions, 1995.
[46] Chapleau N, de Lamballerie-Anton M: Improvement of emulsifying properties of lupin proteins by high pressure induced aggregation. Food Hydrocolloids 2003; 17: 273-280.
[47] Huyghe C: White lupin (Lupinus albus L.). Field Crops Research 1997; 53: 147-160.
[48] Nutraingredients: Nutraceutical offers high polyphenol extracts. In, http://www.nutraingredients.com/news/ng.asp?id=62581, 2005.
[49] Kosar M, Dorman HJ, Can Baser KH, Hiltunen R: Screening of free radical scavenging compounds in water extracts of Mentha samples using a postcolumn derivatization method. J Agric Food Chem 2004; 5 2: 5004-5010.
[50] Ricardo-da-Silva J, Laureano O, Beir˜ao da Costa M: Total phenol and proanthocyanidin evaluation of Lupinus species. In Proceedings of the VIIth International Lupin Conference,. Evora, Portugal, 1993:250-254.
[51] Oomah BD, Cardador-Martínez A, Loarca-Piña G: Phenolics and antioxidative activities in common beans (Phaseolus vulgaris L). Journal of the Science of Food and Agriculture 2005; 85: 935-942.
[52] Huang D, Ou B, Prior RL: The chemistry behind antioxidant capacity assays. J Agric Food Chem 2005; 53: 1841-1856.
[53] Dorman HJD, Koşar M, Kahlos K, Holm Y, Hiltunen R: Antioxidant Properties and Composition of Aqueous Extracts from Mentha Species, Hybrids, Varieties, and Cultivars. Journal of Agricultural and Food Chemistry 2003; 51: 4563-4569.
[54] Da Porto C, Calligaris S, Celotti E, Nicoli MC: Antiradical properties of commercial cognacs assessed by the DPPH (.) test. J Agric Food Chem 2000; 48: 4241-4245.
[55] Yamaguchi F, Ariga T, Yoshimura Y, Nakazawa H: Antioxidant and antiglycation of carcinol from Garcinia indica fruit rind. J. Agric. Food Chem. 2000; 48: 180-185.

[56] Wang M, Li J, Rangarajan M, Shao Y, LaVoie EJ, Huang T-C, Ho C-T: Antioxidative Phenolic Compounds from Sage (Salvia officinalis). Journal of Agricultural and Food Chemistry 1998; 46: 4869-4873.

[57] Drakos A, Doxastakis G, Kiosseoglou V: Functional effects of lupin proteins in comminuted meat and emulsion gels. Food Chemistry 2007; 100: 650-655.

[58] Gulewicz P, Martínez-Villaluenga C, Frias J, Ciesiołka D, Gulewicz K, Vidal-Valverde C: Effect of germination on the protein fraction composition of different lupin seeds. Food Chemistry 2008; 107: 830-844.

[59] Pisariková B, Zraly Z, Bunka F, Trckova M: Nutritional value of white lupine cultivar Butan in diets for fattening pigs. Veterinárni Medicina 2008; 53: 124-134.

[60] Lubowicki R, Petkov K, Kotlarz A, Jaskowska I: Chemical composition and nutritive value of protein of lupins seeds. Folia Universitaet Agricultura Zootechnica, Szczecin 2000; 39: 9-104.

[61] Sujak A, Kotlarz A, Strobel W: Compositional and nutritional evaluation of several lupin seeds. Food Chemistry 2006; 98: 711-719.

[62] Molvig L, Tabe LM, Eggum BO, Moore AE, Craig S, Spencer D, Higgins TJV: Enhanced methionine levels and increased nutritive value of seeds of transgenic lupins (Lupinus angustifolius L.) expressing a sunflower seed albumin gene. Proceedings of the National Academy of Sciences 1997; 94: 8393-8398.

[63] El-Adawy TA, Rahma EH, El-Bedawey AA, Gafar AF: Nutritional potential and functional properties of sweet and bitter lupin seed protein isolates. Food Chemistry 2001; 74: 455-462.

[64] Tabe L, Higgins TJV: Engineering plant protein composition for improved nutrition. Focus in Protein Science 1998; 3:2 82-285.

[65] Alsmeyer RH, Cunningham AE, Happich ML: Equations predict PER from amino acid analysis. Food Technology 1974; 28: 34-40.

[66] Howard NB: The role of some essential ingredients in the formation of layer cake structures. Bakers Digest 1972; 46: 28-37.

[67] Mizukoshi M, Maeda H, Amano H: Model studies of cake baking, II. Expansion and heat set of cake batter during baking. Cereal Chem 1980; 55: 352-355.

[68] Mohamed S, Hamid NA: Effects of ingredients on the characteristics of rice cakes. Journal of the Science of Food and Agriculture 1998; 76: 464-468.

[69] Sasaki T, Yasui T, Matsuki J: Influence of non-starch polysaccharides isolated from wheat flour on the gelatinization and gelation of wheat starches. Food Hydrocolloids 2000; 14:2 95-303.

[70] Dodok L, Modhir AA, Hozova B, Halasova G, Polacek I: Importance and utilization of chickpea in cereal technology. Acta alimentaria 1993; 22: 119-129.

[71] Sabanis D, Makri E, Doxastakis G: Effect of durum flour enrichment with chickpea flour on the characteristics of dough and lasagne. Journal of the Science of Food and Agriculture 2006; 86: 1938-1944.

[72] Alabi MO, Anuonye JC: Nutritional and sensory attributes of soy-supplemented cereal meals. Nigerian Food J. 2007; 25: 100-110.

5

Vinification By-products and Their Phenolic Compounds

Milene Teixeira Barcia[1,*], Paula Becker Pertuzatti[2], Vivian Caetano Bochi[1], Isidro Hermosín-Gutiérrez[3], Helena Teixeira Godoy[1]

[1]Departamento de Ciência dos Alimentos, Universidade Estadual de Campinas, Campinas, Brazil
[2]Instituto de Ciências Exatas e da Terra, Engenharia de Alimentos, Universidade Federal de Mato Grosso, Barra do Garças, Brazil
[3]Instituto Regional de Investigación Científica Aplicada, Universidad de Castilla-La Mancha, Ciudad Real, Spain
*Corresponding author: milenebarcia@ig.com.br

Abstract The waste production by grape industrialization has been growing because about 60% of the harvested fruits are used in vinification. Grape marc (skins and seeds) and lees (sediment solids) are the most abundant by-products of the winemaking process, because the former represents approximately 5% of the total grapes processed and the second, 4% of the total volume of wine produced. Some studies have shown that winery residues contain phenolic compounds, potent antioxidants that were not fully transferred to the wine during the winemaking process, thus have economic and functional interest.

Keywords: *by-products, grape marc, lees, phenolic compounds, grape, winemaking*

1. Introduction

The generation of waste by the wine industry has grown because part of the planted grapes are destined for wine production, which generates about 5% waste in the form of grape marc, in relation to produced grapes, and 4% as lees, in relation to the produced wine. One of the wine residues generated in greater amounts is the grape marc. It consists mainly of skin and seed that are generated after concomitant maceration, alcoholic fermentation, and from grape pressing (in red wine or after pressing of the previously crushed grapes in the production of white wine). After fermentation of the must, the lees are obtained by decantation consisting of fine particles of grape residue and dead yeast. To date, viticulture industry data shows that these residues are being used as animal feed, as fertilizer, in as a source for alcohol distillation, as raw material for pharmaceutical and food industries. This explains the growing interest in exploring nobler forms of winemaking usage, as in the extraction of phenolic compounds [1,2,3].

Thus, this review aims to demonstrate the main phenolic compounds present in winery residues, which include phenolic acids, flavan-3-ols, flavonols and anthocyanins, Due to their importance to health and by the potential to be used as active compounds by pharmaceutical and food industry, this work provide relevant background information to future researches in the feasibility of winery waste as raw material.

2. Viticulture

Thousands of grape varieties exist in the world. Most of them belong to the species *Vitis vinifera*, of European origin, used for the production of fine wines. Moreover, others species classified as *Vitis labrusca* and *Vitis bourquina* are American grapes that are available *in natura* for fresh consumption or to be used as raw material for production of table wines. Thus, viticulture is considered as having great diversity with cultivars of *Vitis vinifera,* American grapes (*Vitis labrusca* and *Vitis bourquina*), and interspecific hybrid grapes. Some of them are consecrated for the wide adaptability and for the characteristics of the wines produced; others, have a more restricted adaptation, remaining in their regions of origin, catering for the preparation of typical and unique products [4].

3. Winery Waste

The grape wine industries produce large amounts of agro-industrial waste. With this problem, the search for alternatives in the use of the generated organic matter is growing within several research centers [5].

Figure 1. Grape marc BRS Lorena (A) and grape marc Cabernet franc (B)

The grape marc (Figure 1) is the most abundant of residues, produced after the pressing of the grapes in the production of white wine, or after the maceration/alcoholic fermentation step during the production of red wine. The lees consist of fine particles of grape residue and inactive yeasts that are obtained by decantation after the process of must fermentation [2,3].

Lees (Figure 2) are regarded as the residues that accumulate in wine vessels after fermentation. This phenomenon occurs during storage or after authorized treatment and is separated by filtering and/or centrifuging this product. This residue is mainly composed of micro-organisms (yeast), poorly soluble salts of tartaric acid (such as potassium and calcium bitartarate), inorganic matter and phenolic compounds [6].

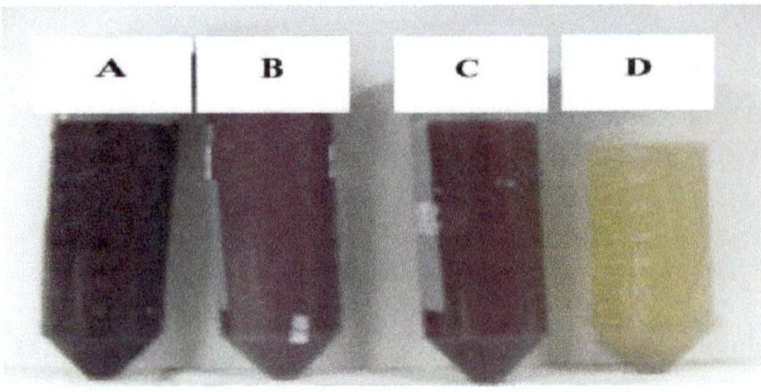

Figure 2. Lees BRS Violeta (A), Cabernet sauvignon (B), Cabernet franc (C) and BRS Lorena (D)

The quantities of lees obtained depend on several factors; those inherent of the grape constitution, ripeness, and hygiene state of the berries. Climatic conditions and the winemaking techniques adopted arenot able to establish a precise relation with the amount of lees generated [7].

According to industry data, the waste generated in wine production is usually used as animal feed, even with the inconvenience of the presence of alcohol, and as vineyard fertilizer, with the grape marc also being used in distilling in the alcohol winery. However, most of this waste is still disposed without any previous treatment resulting in damage to the environment as, for example, the contamination of surface and ground water [8,9].

However, the use of these wine industry residues appropriately could provide a balance in the environment. The large amounts of waste produced by agro-industries cause serious damage, as already mentioned [5,8,10].

The disposal of these residues, as is done today, causes an economic deficit in the supply chain, since many of these residues are rich in bioactive compounds, i.e. compounds with antioxidant powers among other benefits like antimicrobial activity and cardiovascular health benefits. Therefore they have a high commercial value, awakening scientific and economic interest. Among the bioactive compounds, the waste generated by the wineries is a rich source of phenolic compounds since a significant amount results from the processing. The grape marc (skins and seeds), rachis and lees all together represent an average of about 30% of the volume of grapes used for wine production, which makes them a promising source of natural bioactive substances with high added value [11]. This explains the increasing interest in exploring the phenolic compounds such as antioxidants present in wine waste by virtue of their functionality.

Previous studies with the grape marc have demonstrated that it contains compounds which were not completely removed during the winemaking process, because it has a high amount of secondary metabolites of grape, including phenolic acids, flavan-3-ols and anthocyanins [12].

A vast majority of studies in the literature are conceptualized in the study of residual phenolic compounds. A part of these compounds is transferred to wine, however a greater part remains in waste, because these compounds are present in the solid parts of the grape and its extraction depends mainly on the technological conditions used during winemaking such as: maceration time, temperature, intensity and pressure duration, use of enzymes, type of yeast and SO_2 concentration [12,13].

Amico et al. [14], Arvanitoyannins et al. [10] and Amendola, Faveri and Spigno [15] presented a study reporting the functional properties of the wine residue, specifically the analyzed grape marc, and the main compounds found were phenolic compounds, anthocyanins, catechins, tartaric acid, malic acid, sugars, fatty acids and minerals. Kammerer, Claus Carle and Schieber [16] characterized the phenolic compounds of grape marc produced in Germania, and identified 13 anthocyanins, 11 phenolic acids, catechins and stilbene. Karvela, Makris, Kalogeropoulos, Karathanos and Kefalas [17] found gallic acid, tannic acid, catechin and dimers and flavanols in wine seed residues in Greece.

Studies on wine grape marc demonstrate large variations in the content of phenolic compounds found, ranging from 0.1 to 7.0 g of gallic acid equivalent/100 g of dry grape marc. This oscillation is mainly due to the different varieties of grapes used, the diverse winemaking techniques, and different methods of analysis for this type of waste [5,9,15,18].

4. Phenolic Compounds

The vine (*Vitis* spp.), among many other vegetables, stands out due to the high levels of phenolic compounds present in the fruit tissues, leaves and seeds, as well as the variability of the chemical structures found. A growing number of studies have studied the qualitative/quantitative aspects of these compounds in biomass of various species and varieties of vines and their products, such as wines and juices and other products [19,20,21].

Chemically, the phenolic compounds belong to a class of molecules with large structural diversity, characterized by presenting an aromatic ring, having one or more hydroxyl groups directly linked with the ring structure, and may range from simple molecules to highly polymerized phenolic compounds [22].

Phenolic compounds are secondary metabolites that do not participate in metabolic pathways responsible for growth and reproduction, so, their nature and concentration vary greatly.

They can be classified into two groups: flavonoids and non-flavonoids.

4.1. Flavonoids

Flavonoids are those which have the chemical structure described as C_6-C_3-C_6 consisting of two phenolic rings (A and B), linked by a central pyran (containing oxygen) ring

(C). Variations in replacing the C ring result in important classes of flavonoids such as flavonols, flavones, flavanones, flavanols (or catechins), isoflavones and anthocyanidins. The replacements of the A and B rings originate from different flavonoid compounds within each class [23].

Flavonoids are widely synthesized in the endoplasmic reticulum cell before being translocated and stored in the central vacuole of the producer cell. Their alleged role in grapes (and other plants) is like a line of defense against microbial pathogens, pests and herbivores and against changes in water content, light, UV radiation and mineral deficiency [24,25].

Flavonoids are formed from the combination of synthesized derivatives of phenylalanine (shikimic acid metabolic pathway) and acetic acid (acetyl-CoA). The group of flavonoids is also known as polyphenolic and usually occurs in plants in the form of glycosides, one of the classes of substances responsible for allocating the sensory profile. The distribution of flavonoids in the plant depends on many factors and can vary according to the order and the plant family, as well as the variation of species. As an example, one can mention the red grapes variety that are characterized by having higher contents of anthocyanins in the skin tissue, than the white and pink varieties [26].

Among the flavonoids, occurrence is mainly seen in the skins of red grapes, and to a lesser extent in the pulp of the dye varieties, being responsible for blue, violet, purple and red colors. Besides the possibility of being used as substitutes for synthetic dyes in food, the relationship between antioxidant capacity and possible beneficial health effects is also studied [27].

Anthocyanins are included in naturally occurring pigments group (Figure 3). Structurally, anthocyanins are glycosylated derivatives of 2-phenyl benzopirilium cation, also called flavilic cation. The substitution at position 3 of the C ring and in position 5 and 7 of the A ring, by one or more sugar units in anthocyanidin (aglycone) results in anthocyanin. Structural variations of anthocyanins denote different linked sugars, of polymerization and of modes or positions of hydroxylation and methylation. In some cases, the sugars are shown to be acylated by *p*-coumaric and caffeic acids [28].

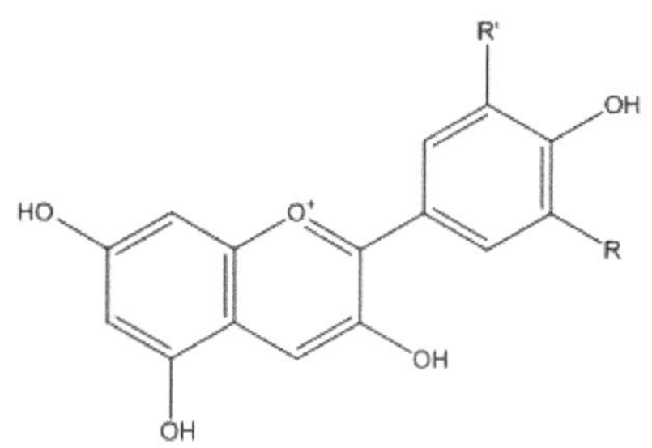

Figure 3. Basic structure of anthocyanidins [29]

In literature, 23 anthocyanidins have been mentioned in grapes, these differ from each other by the number and position of hydroxyl/methoxyl groups, although there is a predominance of 5 anthocyanins: cyanidin, delphinidin, peonidin, petunidin and malvidin, present both in wine and in non-wine grapes. Small amounts of perlagonidin were also found in wine grapes. Their relative amounts vary with the variety, however malvidin is usually the most abundant in quantity. It is characteristic of *Vitis vinifera* to find only one glucose molecule attached at position 3 (3-glycoside), since other species of the genus *Vitis* also contain diglycosides at positions 3 and 5 (3,5-diglicosides), as seen in *Vitis labrusca* (for example, Concord), *Vitis rupestris*, *Vitis riparia*, and the Asian species *Vitis amurensis* which contain both 3-glycosides and 3,5-diglycosides, while the *Vitis rotundifolia* (Muscadine) grapes contain only 3,5-diglycosides [27,30]. The *V. vinifera* hybrids with non-wine grapes also contain a mixture of the anthocyanidins 3-glycoside and 3,5-diglycosides,

Unlike grapes, phenolic composition of wine, shows that about 25% anthocyanins may have been polymerized with flavonoids and other phenolic compounds during fermentation. This level can rise to over 40% within one year and may reach the level of 100% after several years [25].

In red wine made from *Vitis vinifera* grape cultivars, the pyranoanthocyanins are formed in the early stages of fermentation, mainly derived from the reaction of pyruvic acid with anthocyanins (anthocyanidin-3-glycosides) and acetaldehyde, two intermediary yeast metabolites [31]. So it is not surprising to find a kind of pyranoanthocyanin in the winemaking by-products, as reported for Sicilian cultivar grape marc Nerello Mascalese [14].

Barcia et al. [32] found 19 anthocyanins in the skin (from grape marc) of Cabernet sauvignon and Cabernet franc cultivars, along with 6 and 5 pyranoanthocyanins respectively, while in the lees 9 pyranoanthocyanins were identified along with 16 and 14 anthocyanins respectively for Cabernet sauvignon and Cabernet franc. These had free aglycones and acetylated and/or glycosylated aglycones. Barcia et al. [33] also found 27 anthocyanins in the skin (from grape marc) and 26 anthocyanins and 3 pyranoanthocyanins in the lees of BRS Violeta.

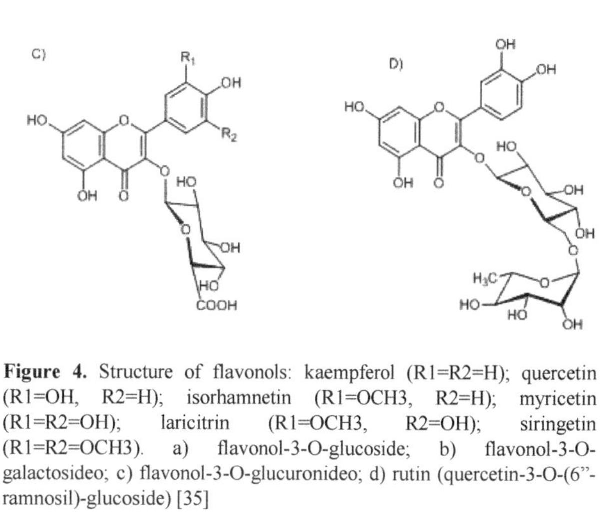

Figure 4. Structure of flavonols: kaempferol (R1=R2=H); quercetin (R1=OH, R2=H); isorhamnetin (R1=OCH3, R2=H); myricetin (R1=R2=OH); laricitrin (R1=OCH3, R2=OH); siringetin (R1=R2=OCH3). a) flavonol-3-O-glucoside; b) flavonol-3-O-galactosideo; c) flavonol-3-O-glucuronideo; d) rutin (quercetin-3-O-(6”-ramnosil)-glucoside) [35]

Castillo-Munoz, Gómez-Alonso, García-Romero and Hermosín-Gutiérrez [34] reported that the occurrence of flavonols in red grape cultivars is represented by six structures: kaempferol, quercetin, isorhamnetin, myricetin, laricitrin and siringetin (Figure 4).

A few years ago it was stated that only white grapes were methoxylated derivatives of flavonols disubstituted from ring B, such as kaempferol and quercetin. However, it has been demonstrated that the white grapes are all possible derivatives of flavonols with one or two substituents on ring B (kaempferol, quercetin, and isorhamnetin) with all tri-substituents (myricetin, laricitrin and siringetin) being absent in ring B [36,37]. Moreover, the predominant flavonols in *Vitis vinifera* are quercetin and myricetin on red grapes and quercetin only for white grapes, while in *Vitis labrusca* grapes it is quercetin [35].

The conjugates in grape flavanols are mainly 3-O-glycosides and the presence of sugars in other flavonol skeletal positions has never been reported. In *V. vinifera* grapes, other two complete minor series of glycosides are found as well, derivatives 3-galactoside and 3-glucuronides. 3-glucuronide of quercetin is an exceptional case since it is usually one of the most important flavonols, both in red and white grapes. Furthermore, in the case of quercetin, small proportions of the derivative 3-rhamnosylglucoside (3-rutinoside, a disaccharide) are found, which for several years was confused with the 3-glucuronide quercetin, even affirming that it was the most abundant of flavonols [27]. In the case of non-wine varieties of red grapes, like isorhamnetin in grapes, only glucose derivatives have been identified, though myricetin and quercetin may also occur as glucuronide and galactoside, and myricetin derivatives are often in these cases the majority [38].

In Brazilian winery wastes, about 18 flavonols were found in the skins of Cabernet sauvignon and Cabernet franc cultivars and on average 15-16 flavonols, in the lees, respectively [32]. In the wastes of hybrid cultivars (BRS Violet and BRS Lorena) an average of 12 flavonols were identified [33].

The flavan-3-ols are compounds found in higher quantities in grape seeds, and also importantly in skins. The combination of several monomeric units of catechins and epicatechins are called proanthocyanidins or condensed tannins, these are not easily hydrolysable. In the case of varieties of *Vitis vinifera*, the main elements are monomeric forms of (+) - catechin and (-) - epicatechin and mostly, their oligomers and polymers [30]. Proanthocyanidins in skins and seeds show different structural features. The proanthocyanidins from the skin have a higher degree of polymerization and contain prodelfinidins (structural units with 3 substituents on ring B, such as gallocatechin and epigallocatechin), while the seeds have a lower degree of polymerization and do not contain prodelfinidins and have relevant amounts of gallic esters (such as 3-gallate of epicatechin) [39]. Among the phenolic compounds, high molecular weight tannins have the ability to combine with proteins or other polymers such as polysaccharides, causing the sensation of astringency. The low molecular weight tannins on the other hand tend to provide a bitter taste [40].

4.2. Non-Flavonoids

The non-flavonoids consist of phenolic acids, hydrolysable tannins, and stilbenes (C_6-C_2-C_6) [5,23].

The small amounts of phenolic material found mainly in white wines are not flavonoids (hydroxycinnamic), such as caftaric, *p*-coumaric and ferulic acid [25], due to limited maceration with the solid parts of the grape (skins in this case) occurring when obtaining the must, that leads to the end of the fermentation.

The hydroxybenzoic acids include gallic acid, *p*-hydroxybenzoic, protocatechuic, vanillic and syringic, which have the common C_1-C_6 structure; while the hydroxycinnamic acids have a side chain with three carbon atoms (C_6-C_3), they are vinyl-analogs of hydroxybenzoic acid, such as caffeic, ferulic, sinapic and *p*-coumaric acids [41].

Hydroxycinnamic acids (HCA) are representatives of the classes of phenolic acids found in grapes and wine. The main HCA found in grapes and wine are combined with tartaric acid (hydroxycinnamoyl-tartaric acids), which are called caftaric acid, *p*-cutaric acid and fertaric acid, derived from caffeic, *p*-coumaric and ferulic acids, respectively (Figure 5).

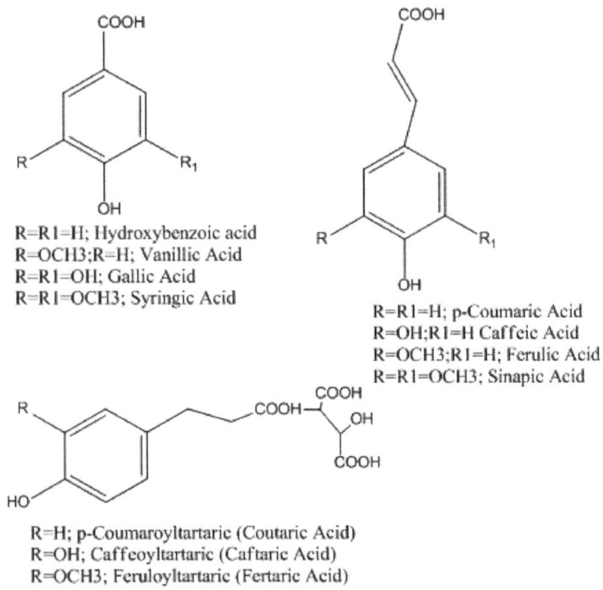

Figure 5. Structure of phenolics acids [42]

Typically caftaric, cutaric and fertaric acids are found in the pulp of grapes in the *trans* form, and are released during grape pressing, while p-cutaric acid is presented in an appreciable manner in *cis*-form, with the *cis* isomers of caftaric and tartaric acids being practically non-existent. Thus, they remain in the wine residues and are associated with the wine darkening process and are precursors of volatile phenolic compounds. During the wine fermentation process, the partial hydrolysis of these types of esters yields the free hydroxycinnamic acids. These are converted into ethyl ester, or ethyl caffeate and ethyl coumarate [42], and can also be decarboxylated by yeasts and bacteria (4-forming vinilfenols), and even later transformed into volatile ethylphenols of unpleasant aroma by Brettanomyces type yeast.

In Cabernet sauvignon and Cabernet franc cultivars, 4 to 5 derivatives of hydroxycinnamic acids (HCA) were identified from the skin, and 5 to 6 HCA in the lees, respectively [32]. And in hybrid cultivar wastes (BRS

Violet and BRS Lorena) 7 to 8 HCA derivatives were found [33].

Besides these compounds present in the grape, resveratrol, a polyphenol belonging to the class of stilbenes is also found. Resveratrol (3,5,4'-trihydroxystilbene) is a phenolic compound that comprises two aromatic rings linked by an ethylene bridge. It is synthesized in the plant using an enzyme, stilbene synthase, which combines a molecule of hidroxicinnamoil-Coenzyme A (CoA) and three molecules of malonyl-CoA in two isomeric forms: *trans*-resveratrol and *cis*-resveratrol (Figure 6).

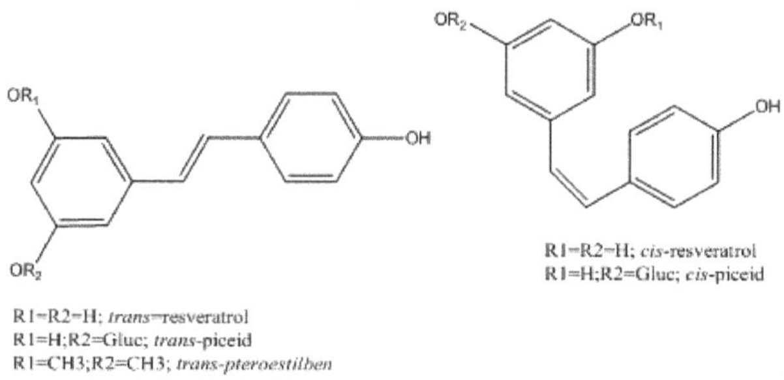

Figure 6. Structure of resveratrol [42]

The *trans* form is the true form produced in a biosynthetic pathway and is glycosylated at position 3 (3-glucoside *trans*-resveratrol, also known as *trans*-piceid) [43,44]. The *cis* isomer may already be formed in grapes by induction of UV radiation. The piceid is also hydrolyzed in the grape and later in wine releasing resveratrol. Resveratrol has been identified in wine, the vine leaves and the skin of grapes, and its concentration decreases significantly during grape ripening [42].

Modifications in resveratrol, such as the already described glycosylation, methylation and polymerization, produce piceid, and Pterostilbene and viniferins. The *trans*-resveratrol has attracted special attention because of their spatial conformation that can activate calcium absorption by the blood vessels, triggering vasodilation, which contributes to lowering blood pressure, a fact demonstrated by epidemiological studies, which claim inverse correlation between moderate wine consumption and the incidence of cardiovascular disease [40].

In Brazilian wineries, waste such as the lees of Cabernet sauvignon and Cabernet franc, only 3 stilbenes were identified, among them: *trans*-piceid, *cis*-piceid and *trans*-resveratrol [32]. The same stilbenes were identified in wine residue shells of BRS Lorena [33].

The hydrolysable tannin result from binding of a sugar, usually glucose, to a phenolic compound, particularly gallic acid or ellagic acid, thereby forming gallotannins and ellagitannins, respectively. These compounds do not contain flavonoid molecules and are not found naturally in grapes. However, they are present in the wood used in storage or aging wines and therefore can be transferred to the wine and waste during these processes.

5. Trends and Prospects

With the data presented in this study, it was observed that there is concern about the destination of the waste generated by the wine industry. This waste has been studied by virtue of the presence of bioactive compounds of great value.

Previous studies have shown that the grape marc, consisting of skins and seeds, and lees present considerable amounts of phenolic compounds with potential health promoting properties.

Because of this, the recovery of phenolic compounds from industrial waste is gaining attention, especially attributed to the functional properties that these compounds exert, such as anti-inflammatory, antimutagenic and anticancer.

With the work currently being performed, and with the results obtained that show that the wine by-products have the potential to be applied in food, pharmaceutical and chemical industries, the recovery of by-products generated is expected to grow in coming years and that may solve the problem of the generation of industrial waste and provide consumer health benefits.

References

[1] Toscano, G., Riva, G., Duca, D., Pedretti, E.F., Corinaldesi, F. and Rossini, G. "Analysis of the characteristics of the residues of the wine production chain finalized to their industrial and energy recovery". *Biomass and Bioenergy,* 55, 260-267, 2013.

[2] Maragkoudakis, P.A., Nardi, T., Bovo, B., D'Andrea, M., Howell, K.S., Giacomini, A. and Corich, V. "Biodiversity, dynamics and ecology of bacterial community during grape marc storage for the production of grappa". *International Journal of Food Microbiology,* 162, 143-151, 2013.

[3] Cortés, S., Rodríguez, R., Salgado, J.M. and Domínguez, J.M. "Comparative study between Italian and Spanish grape marc spirits in terms of major volatile compounds". *Food Control,* 22, 673-680, 2011.

[4] Camargo, U.A., Tonietto, J. and Hoffkmann, A. "Progressos na viticultura brasileira". *Revista Brasileira de Fruticultura,* volume especial, 144-149, 2011.

[5] Cataneo, C.B., Caliari, V., Gonzaga, L.V., Kuskoski, E.M. and Fett, R. "Atividade antioxidante e conteúdo fenólico do resíduo agroindustrial da produção de vinho". *Semina: Ciências Agrárias, Londrina,* 29, 1, 93-102, 2008.

[6] Pérez-Serradilla, J.A. and Luque de Castro, M.D."Role of lees in wine production: A review". *Food Chemistry,* 111, 447-456, 2008.

[7] Silva, L.M.L.R. "Caracterização dos subprodutos da vinificação". *Revista Millenium,* 28, 123-133, 2003.

[8] Lafka, T., Sinanoglou, V. and Lazos, E.S. "On the extraction and antioxidant activity of phenolic compounds from winery wastes". *Food Chemistry,* 104, 1206-1214, 2007.

[9] Llobera, A. and Cañellas, J. "Dietary fibre content and antioxidant activity of Manto Negro red grape (Vitis vinifera): pomace and stem". *Food Chemistry,* 101, 659-666, 2007.

[10] Arvanitoyannis, I.S., Ladas, D. and Mavromatis, A. "Potencial uses and aplications of treated wine waste: a review". *International Journal of Food and Technology,* 41, 475-487, 2006.

[11] Melo, P.S., Bergamaschi, K.B., Tiveron, A.P., Massarioli, A.P., Oldoni, T.L.C., Zanus, M.C., Pereira, G.E. and Alencar, S.M. "Composição fenólica e atividade antioxidante de resíduos agroindustriais". *Ciência Rural,* 41, 6, 1088-1093, 2011.

[12] Makris, D.P., Boskou, G. and Andrikopoulos, N.K. "Polyphenolic content and in vitro antioxidant characteristics of wine industry and other agri-food solid waste extracts". *Journal of Food Composition and Analysis,* 20, 125-132, 2007.

[13] Gallego, M.A.G., García-Carpintero, E.G., Sánchez-Palomo, E., Viñas, M.A.G. and Hermosín-Gutiérrez, I. "Oenological potential, phenolic composition, chromatic characteristics and antioxidant activity of red single-cultivar wines from Castilla-La Mancha". *Food Research International,* 48, 7-15, 2012.

[14] Amico, V., Napoli, E.M., Renda, A., Ruberto, G., Spatafora, C. and Tringali, C. "Constituents of grape pomace from the Sicilian cultivar 'Nerello Mascalese'". *Food Chemistry,* 88, 599-607, 2004.

[15] Amendola, D., Faveri, D.M. and Spigno, G. "Grape marc phenolics: Extraction kinetics, quality and stability of extracts". *Journal of Food Engineering,* 97, 384-392, 2010.

[16] Kammerer, D., Claus, A., Carle, R. and Schieber, A. "Polyphenol Screening of Pomace from Red and White Grape Varieties (*Vitis vinifera* L.) by HPLC-DAD-MS/MS". *Journal of Agriculture and Food Chemistry, 52,* 4360-4367, 2004.

[17] Karvela, E., Makris, D.P., Kalogeropoulos, N., Karathanos, V.T. and Kefalas, P. "Factorial design optimisation of grape (Vitis vinifera) seed polyphenol extraction". *European Food Research Technology,* 229, 731-742, 2009.

[18] Shrikhande, A.J. "Wine by-products with health benefits". *Food Research International,* 33, 469-474, 2000.

[19] Cadot, Y., Chevalier, M. and Barbeau, G. "Evolution of the localization and composition of phenolics in grape skin between veraison and maturity in relation to water availability and some climatic conditions". *Journal of the Science of Food and Agriculture,* 91, 1963-1976, 2011.

[20] Lago-Vanzela, E.S., Rebello, L.P.G., Ramos, A.M., Stringheta, P.C., Da-Silva, R., García-Romero, E., Gómez-Alonso, S. and Hermosín-Gutiérrez, I. "Chromatic characteristics and color-related phenolic composition of Brazilian young red wines made from the hybrid grape cultivar BRS Violeta ("BRS Rúbea" × "IAC 1398-21")". *Food Research International,* 54, 33-43, 2013.

[21] Bindon, K., Varela, C., Kennedy, J., Holt, H. and Herderich, M. "Relationships between harvest time and wine composition in Vitis vinifera L. cv. Cabernet Sauvignon 1. Grape and wine chemistry". *Food Chemistry,* 138, 1696-1705, 2013.

[22] Zhu, L., Zhang, Y. and Lu, J. "Phenolic contents and compositions in skins of red wine grape cultivars among various genetic backgrounds and originations". *International Journal of Molecular Science,* 13, 3492-351, 2012.

[23] Del Rio, D., Rodriguez-Mateos, A., Spencer, J.P.E., Tognolini, M., Borges,G. and Crozier, A. "Dietary (Poly) phenolics in Human Health: Structures, Bioavailability, and Evidence of Protective Effects Against Chronic Diseases". *Antioxidants & Redox Signaling,* 18, 1818-1892, 2013.

[24] He, F., Liang, N.N., Mu, L., Pan, Q.H., Wang, J., Reeves, M.J. and Duan, C.Q. "Anthocyanins and Their Variation in Red Wines I. Monomeric Anthocyanins and Their Color Expression". *Molecules,* 17, 1571, 2012.

[25] Jackson, R. *Wines Science. Principles and applications.* (3th ed.). USA: Elsevier, 2008.

[26] Cordenunsi, B.R., Nascimento, J.R.O., Genovese, M.I. and Lajolo, F.M. "Influence of cultivar on quality parameters and chemical composition of strawberry fruits grown in Brazil". *Journal Agriculture Food Chemistry,* 50, 9, 2581-2586, 2002.

[27] Castillo-Muñoz, N., Gomez-Alonso, S., Garcia-Romero, E., Gomes, M.A., Velders, A.H. and Hermosín-Gutiérrez, I. "Flavonol 3-0-Glycosides Series of Vitis vinifera Cv. Petit Verdot Red Wine Grapes". *Journal Agriculture Food Chemistry,* 57, 209-219, 2009.

[28] Okumura, F., Soares, M.H.F.B. and Cavalheiro, E.T.G. "Identificação de pigmentos naturais de espécies vegetais utilizando-se cromatografia em papel". *Química Nova,* 25, 4, 680-683, 2002.

[29] Ribéreau-Gayon, J., Glories, Y., Maujean, A. and Dubourdieu, D. *Handbook of Enology: The chemistry of wine stabilization and treatments.* (2th ed.). John Wiley & Sons, Ltda, 2006.

[30] Blouin, J. and Guimberteau, G. "Principales componentes de la uva y evolución durante la maduración". In: J. Blouin and G. Guimberteau (Ed.). *Maduración y madurez de la uva.* (pp.57-100). Madri: Mundi-Prensa, 2002.

[31] Blanco-Vega, D., López-Bellido, F.J., Alía-Robledo, J.M. and Hermosín-Gutiérrez, I. "HPLC-DAD-ESI-MS/MS characterization of pyranoanthocyanins pigments formed in model wine". *Journal of Agricultural and Food Chemistry,* 59, 9523-9531, 2011.

[32] Barcia, M.T., Pertuzatti, P.B., Rodrigues, D., Gómez-Alonso, S., Hermosín-Gutiérrez, I. and Godoy, H.T. "Occurrence of low molecular weight phenolics in Vitis vinifera red grape cultivars and their winemaking by-products from São Paulo (Brazil)". *Food Research International.* 62, 500-513, 2014.

[33] Barcia, M.T., Pertuzatti, P.B., Gómez-Alonso, S., Godoy, H.T. and Hermosín-Gutiérrez, I. "Phenolic composition of grape and winemaking by-products of Brazilian hybrid cultivars BRS Violeta and BRS Lorena". *Food Chemistry,* 159, 95-105, 2014.

[34] Castillo-Muñoz, N., Gómez-Alonso, S., García-Romero, E. and Hermosín-Gutiérrez, I. "Flavonol profiles of vitis vinifera red grapes and their single-cultivar wines". *Journal Agriculture Food Chemistry,* 55, 992-1002, 2007.

[35] Hermosín-Gutiérrez, I., Castillo-Muñoz, N., Gómez-Alonso, S. and García-Romero, E. Flavonol profiles for grape and wine authentication. In S.E. Ebeler, G.R. Takeoka and P. Winterhalter, *Progress in authentication of food and wine* (pp. 113-129). Washington, DC: American Chemical Society, ACS Symposium Series, 2011.

[36] Castillo-Muñoz, N., Gómez-Alonso, S., García-Romero, E. and Hermosín-Gutiérrez, I. "Flavonol profiles of Vitis vinifera white grape cultivars". *Journal of Food Composition and Analysis,* 23, 699-705, 2010.

[37] Hermosín-Gutiérrez, I., Gómez-Alonso, S., Castillo-Muñoz, N., Fernández-González, M., Martínez-Gascueña, J. and García-Romero, E. Chapter 5. *Screening the phenolic composition of the wide diversity of autochthonous grape cultivars from La Mancha (Spain),* 2012. In *Grapes: Production, Phenolic Composition and Potential Biomedical Effects.* J. C. Câmara, Ed. Nova Science Publishers, Inc.: New York, 135-164, 2014.

[38] Rebello, L.P.G., Lago-Vanzela, E.S., Barcia, M.T., Ramos, A.M., Stringheta, P.C., Da-Silva, R., Castillo-Muñoz, N., Gómez-Alonso, S. and Hermosín-Gutiérrez, I. "Phenolic composition of the berry parts of hybrid grape cultivar BRS Violeta (BRS Rubea × IAC 1398-21) using HPLC-DAD-ESI-MS/MS". *Food Research International,* 54, 354-366, 2013.

[39] Cheynier, V., Dueñas-Paton, M., Salas, E., Maury, C., Souquet, J.M., Sarni-Manchado, P. and Fulcrand, H. "Structure and Properties of Wine Pigments and Tannins". *American Journal of Enology and Viticulture,* 57, 298-305, 2006.

[40] Abe, L.T., Mota, R.V., Lajolo, F.M. and Genovese, M.I. "Compostos fenólicos e capacidade antioxidante de cultivares de uvas *Vitis labrusca L.* e *Vitis vinifera* L". *Ciência e Tecnologia de Alimentos,* 27, 2, 394-400, 2007.

[41] Balasundram, N., Sundram, K. and Samman, S. "Phenolic compounds in plants and agri-industrial by-products: Antioxidant activity, occurrence, and potential uses". *Food Chemistry,* 99, 191-203, 2006.

[42] Garrido, J. and Borges, F. "Wine and grape polyphenols. A chemical perspective". *Food Research International,* 44, 3134-3148, 2011.

[43] Poutaraud, A., Latouche, G., Martins, S., Meyer, S., Merdinoglu, D. and Cerovic, Z.G. "Fast and local assessment of stilbene content in grapevine leaf by in vivo fluorometry". *Journal of Agricultural and Food Chemistry,* 55, 13, 4913-4920, 2007.

[44] Fernández-Mar, M.I., Mateos, R., García-Parrilla, M.C., Portas, B., Cantos-Villar, E. "Compostos bioativos em vinho: Resveratrol, hidroxitirosol e melatonina: Uma revisão". *Food Chemistry,*130, 4, 797-813, 2012.

Preparation and Characterization of Microwave-made Oat and Fruit Cupcakes

Paula Becker Pertuzatti[1], Denise Soares Dorneles de Oliveira[2], Kelly Cristina Campagnolo Port[2], Larissa Alves Vieira[2], Mariana Bento da Cruz[1], Gabrielle Lázara Ribeiro Rodrigues de Barros Vinhal[2,*]

[1]Universidade Federal de Mato Grosso, Instituto de Ciências Exatas e da Terra, Curso de Engenharia de Alimentos, Barra do Garças, MT, Brazil

[2]Instituto Federal de Educação, Ciência e Tecnologia de Mato Grosso, Curso Técnico em Alimentos, Barra do Garças, MT, Brazil

* Corresponding author: gvinhal@hotmail.com

Abstract The development of new products is increasingly being studied to meet the demands of today's consumer market which seeks products with sensory and nutritional quality associated with health benefits. In this sense, fruits, vegetables and cereals have stood out, since in addition to preventing some diseases when consumed properly combined with healthy habits, they can be used as ingredients in various products such as those of breadmaking. The objective of this study was to develop and evaluate chemical, physicochemical and sensory characteristics of microwave-made cupcakes of oats and fruits. Two cupcake formulations were made, one with oats, banana and apple and the other with oats, pineapple and mint. The analyses were moisture, protein, lipids, ash, carbohydrates, pH, titratable acidity, soluble solids, reducing, non reducing and total sugars, as well as sensory analysis performed by the emotional test of acceptance and purchase intent using the hedonic scale. The results indicated that the products are presented as a good source of energy and protein, low-fat and good acceptance and purchase intent, especially the oat cupcake with banana and apple suggesting an alternative to the baking industry.

Keywords: *cupcake, oats, chemical and physicochemical properties, sensory evaluation*

1. Introduction

The demand for foods that help in promoting health and thereby potentiate the prevention of some diseases is currently on the increase. Accordingly, foods considered functional have gained more and more space in the meals of consumers. Although the term functional food does not have a universal definition, according to the American Dietetic Association (ADA) all foods are functional in some physiological level [1], because they provide nutrients or other substances that furnish energy, sustain growth, or maintain/repair vital processes. However, functional foods move beyond necessity to provide additional health benefits that may reduce disease risk and/or promote optimal health. Functional foods include conventional foods, modified foods (ie, fortified, enriched, or enhanced), medical foods, and foods for special dietary use [2].

Some of the foods that have these functionality are fruits, vegetables and cereals. Oat constitutes a healthy basis for food products [3]. Oat has gained relevance in human nutrition because it is one of the few cereals with a high content of soluble fibre -namely β-glucan-, and is a good source of proteins, vitamins, and minerals [4]. In addition oats contain large amounts of antioxidants such as vitamin E, various phenolic compounds and avenanthramides [5]. Current public health recommendations aimed at reducing the risk of coronary heart disease in the United Kingdom (UK) suggest including oats and oat-based products as part of a healthy diet [6].

Furthermore, there is a demand for products that not only provide functionality but also convenience, and microwave foods are increasingly available on the market, since they require less cooking time compared to conventional ovens.

The present study aimed to process and characterize microwave-prepared cupcakes of oat and fruits, aiming to produce a product that adds functionality, practicality and is well accepted by consumers.

2. Materials and Methods

2.1. Obtaining Raw Materials

The ingredients used in the formulation of cupcakes were oats in thin flakes, wheat flour, eggs, vegetable oil, brown sugar, baking soda, baking powder, chia (*Salvia hispanica* L.), banana (*Musa paradisiaca* L.), apple

(*Malus domestica Borkh*), pineapple (*Ananas comosus* (L.) *merril*) and mint (*Mentha* sp.), all purchased at local shops in the city of Barra do Garças-MT and taken to the Food Technology Laboratory of the Federal Institute of Mato Grosso, Barra do Garças Campus, where they were stored until processing time.

2.2. Processing Cupcakes

Two types of oat and fruit cupcakes were prepared, one with banana and apple and the other with pineapple and mint. The rest of the ingredients were the same for the two assays. The formulation is shown in Table 1.

Table 1. Ingredients used in the cupcakes preparation

Ingredients	Quantity (g)
Wheat flour	65
Oats in thin flakes	65
Brown sugar	120
Oil	60
Eggs	150
Chia	5
Baking powder	5
Sodium bicarbonate	3
Banana	450
Apple	450
Pineapple	450
Mint	20

The ingredients were mixed manually in stainless steel container with the aid of spatula during 5 minutes. The homogeneous dough was placed on silicone shapes with dimensions of 7×6×3 cm. They were then taken to the high power microwave oven (Electrolux-MTD30) where they remained for five minutes. The products were cooled at room temperature, removed from the shapes and properly stored until analysis.

2.3. Chemical and Physicochemical Analysis

Analysis were performed in the Food Analysis Laboratory, belonging to the Institute of Exact and Earth Sciences, Federal University of Mato Grosso. Samples were crushed and homogenized for further analysis, performed in triplicate. All reagents used were of appropriate degree of purity for analysis (P.A).

The contents of moisture, lipids, ash, pH, titratable acidity and soluble solids were carried out in accordance with Instituto Adolfo Lutz [7]. Analysis of protein, reducing sugars, non-reducing and total sugars were carried out according to A.O.A.C. [8].

2.4. Sensory Analysis

Sensory analysis was performed in the Sensory Analysis Laboratory of Food of the Institute of Extact and Earth Sciences, of the Federal University of Mato Grosso and followed the rules established by Instituto Adolfo Lutz [7]. Affective tests of acceptance by hedonic scale and purchase intent tests were carried out with 100 untrained judges aged 18 to 50 with availability and interest in participating in the tests.

For the affective test of acceptance, a structured hedonic scale of 9 points, ranging from "extremely liked" (9) to "extremely disliked" (1), was used to evaluate the following parameters: color, taste, aroma, texture and overall acceptance. Consumer purchase intent was assessed using a scale of 5 (five) points ranging from "would definitely buy" (5) to "would definitely not buy" (1).

2.5. Statistical Analysis

Data were subjected to analysis of variance (ANOVA) and the study of differences between means detected by Tukey test ($p < 0.05$).

3. Results and Discussion

3.1. Chemical and Physicochemical Analysis

Table 2 presents the values of the chemical analysis of cupcakes. Results were expressed as mean and standard deviation.

Table 2. Chemical analysis carried out on the products (%)

	Product	
	BA	PM
Moisture	42.82±1.0[b]	51.51±1.6[a]
Proteins	2.1±0.1[a]	2.0±0.3[a]
Lipids	2.80±0.4[a]	2.53±0.8[a]
Ash	1.55±0.2[a]	1.23±0.1[a]
**Carbohydrates	50.73	42.73

* In the same row, means with the same letters do not differ significantly at $p < 0.05$. BA = oat cupcake with banana and apple; PM = oat cupcake with pineapple and mint.

** Obtained by difference.

The moisture content found in the cupcake with banana and apple was 42.82% while that of pineapple with mint was 51.51%, with a significant difference between them at $p < 0.05$. In a study of preparation of chocolate cake with yacon and inulin flour, [9] found values close to the studied (42.57%) for standard formulation. The determination of moisture is one of the most important measurements and is used in food analysis because it is related to its stability, quality and composition, and can affect the storage, packaging and processing [10].

The amount of protein found in the samples were 2.1% for BA and 2.0% for PM, with no significant difference at $p < 0.05$. These values were similar to those found by [11], who found 2.3% protein when preparing carob gluten-free cupcakes.

The values obtained in the analysis of lipids were 2.80% for the BA and 2.53% for the PM, with no significant difference at $p < 0.05$. However, in a study of the chemical composition of oat cake with addition of amaranth, [12] found values above the present study, 9.89% lipids for formulation with 22% amaranth, according to the authors this is due mainly to the Amaranth lipid content being 8% [13]. Thus it can be concluded that the products of this study have relatively low lipid levels when compared with those of other authors.

The ash content found in BA was 1.55% and 1.23% in PM, with no significant difference between them ($p < 0.05$). Studying the chemical characteristics of cupcakes added with different concentrations of banana peel flour, [14] found values above those in the present study (2.49 and 2.41%) in ash content, according to the authors, different results may be found due to the different varieties of bananas used.

Table 3 shows the values of physicochemical analysis of cupcakes.

Table 3. Physicochemical parameters of cupcakes

Product	pH	Acidity (%)	Soluble Solids (°Brix)
	Parameters		
BA	6.85±0.03[a]	0.55±0.05[a]	9.0±0.01[b]
PM	6.56±0.06[a]	0.55±0.06[a]	10.0±0.01[a]

*In the same column, means with the same letter do not differ significantly at $p < 0.05$. BA = oat cupcake with banana and apple; PM = oat cupcake with pineapple and mint.

The values found in the analysis of pH and acidity were 6.85 and 0.55 for BA and 6.56 and 0.55 for PM, respectively, with no significant difference between them ($p < 0.05$). These values were similar to those found by [15], who on studying cookies made with *Moringa oleifera* leaf found 6.75 and 6.67 for analysis of pH and values above those of this study for acidity, obtaining 1.56 and 1.65 for standard formulation and a formulation with 9.09% of *Moringa oleifera* leaf powder. For the soluble solids analysis, the values were 9 and 10 °Brix, respectively, for the BA and PM, with a significant difference at $p < 0.05$.

Table 4 shows the values of the analysis of reducing, non-reducing and total sugars of cupcakes.

Table 4. Reducing, non-reducing and total sugars

Product	Sugars (g glucose.g sample-1)		
	Reducing	Non-reducing	Total
BA	2.63±0.09[a]	2.67±0.21[a]	5.44±0.14[a]
PM	2.54±0.01[a]	2.57±0.15[a]	5.25±0.17[a]

* In the same column, means with the same letter do not differ significantly at $p < 0.05$. BA = oat cupcake with banana and apple; PM = oat cupcake with pineapple and mint.

Studies on reducing, non-reducing and total sugars of cupcakes are scarce in literature. The amount of reducing, non reducing and total sugars found in BA were 2.63; 2.67 and 5.44 g glucose.g $sample^{-1}$ respectively. In PM, the values were 2.54; 2.57 and 5.25 g glucose.g $sample^{-1}$ respectively. There was no significant difference between the samples in all the analysis performed at $p < 0.05$.

3.2. Sensory Analysis

Table 5 shows the means of the attributes evaluated in cupcakes such as color, aroma, texture, flavor, global acceptance, as well as purchase intent. Averages above 7.0 were obtained for all attributes, which corresponds to "Like Moderately" in a structured scale of 9 points, highlighting the attribute flavor which reached a value of 8.01 for BA which corresponds to "Like Very Much" on the hedonic scale. To be considered a product with good acceptance, the averages should exceed 6 on the hedonic scale of 9 points, which is the quality and marketing threshold [16].

Table 5. Sensory evaluation of oat cupcakes with fruits

	Product	
	BA	PM
Color	7.59±1.5[a]	7.50±1.4[a]
Aroma	7.77±1.3[a]	7.41±1.5[a]
Texture	7.82±1.4[a]	7.34±1.6[b]
Flavor	8.01±1.4[a]	7.33±1.8[b]
Global Acceptance	7.92±1.2[a]	7.36±1.6[b]
Purchase Intent	4.24±0.9[a]	3.86±1.1[b]

*In the same line, means with the same letter do not differ significantly at $p < 0.05$. BA = Oat cupcake with banana and apple; PM = oat cupcake with pineapple and mint.

Attributes such as aroma and taste are probably the most important characteristics that influence the sensory properties of food products containing different added ingredients [17].

In evaluating purchase intent, there was no significant difference between BA and PM samples at $p < 0.05$. The BA had the highest value (4.24) corresponding to "probably buy" in a structured 5-point scale, while PM obtained 3.86 corresponding to "might buy/probably not buy" on the 5-point scale.

Figure 1 shows the percentages for acceptance analysis and purchase intent of cupcakes. For the attributes color, aroma, texture, flavor and overall acceptance between both samples, the values exceeded 80% acceptance. According to [18], for a product to be accepted in terms of their sensory properties, it is necessary to obtain an acceptability rate of at least 70%.

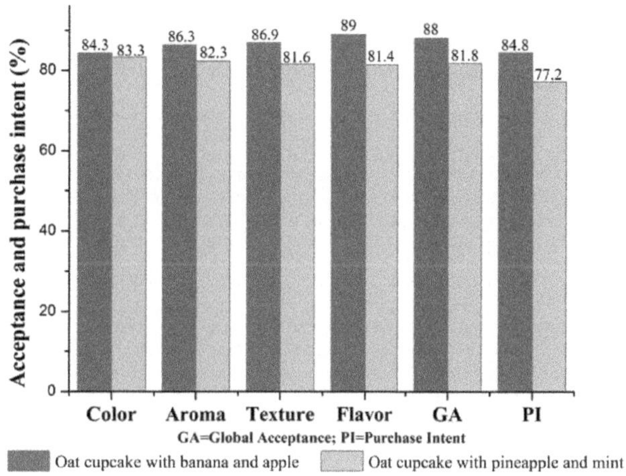

Figure 1. Acceptance and purchase intent of cupcakes.

Therefore, when you look at Figure 1, it can be seen that BA can be considered more accepted by the judges, because its values were higher when compared to PM, both for acceptance, as well as purchase intent.

To develop a new product, one of the key points is to assess its acceptability in order to predict its behavior on the consumer market [19].

4. Conclusions

The development of these products made it possible to observe that both have the potential to be used as functional foods, due to the ingredients present such as oats, fruit and chia seed, furthermore, they are practical and easy to prepare.

The cupcakes presented considerable protein content and low lipid content.

Through sensory analysis, it was observed that cupcakes were well accepted by the judges, and presented a good purchase intent as well, with the oat cupcake with banana and apple standing out, suggesting the marketing of this new product, which becomes an alternative for bakery industries.

Acknowledgement

The authors thank the Food Engineering degree course at the Federal University of Mato Grosso, for the availability to perform the analysis in Food Analysis and Sensory Food Analysis laboratories, and the Federal

Institute of Education, Science and Technology of Mato Grosso for the research incentive.

References

[1] Position of The American Dietetic Association: Functional foods. Journal of the American Dietetic Association. 104, 814-826, 2004.

[2] Hasler, C. M.; Brown, A. C. Position of the American dietetic association: Functional foods. *Journal of the American Dietetic Association*, 109 (4), 735-746, 2009.

[3] Londono D. M., Gilissen L.J.W.J., Visser R.G.F., Smulders M. J.M., Hamer R. J. Understanding the role of oat b-glucan in oat-based dough systems. *Journal of Cereal Science* 62, 1-7, 2015.

[4] Butt, M.S., Tahir-Nadeem, M., Khan, M.K.I., Shabir, R., Butt, M.S. Oat: unique among the cereals. *European Journal of Nutrition*. 47, 68-79, 2008.

[5] Peterson, D.M., Hahn, M.J., Emmons, C.L. Oat avenanthramides exhibit antioxidant activities in vitro. *Food Chemistry* 79, 473-478, 2002.

[6] British Heart Foundation, 2015. Accessed February 05, 2015 from http://www.bhf.org.uk

[7] Instituto Adolfo Lutz. Normas analíticas do Instituto Adolfo Lutz: métodos químicos e físicos para análise de alimentos. São Paulo: Instituto Adolfo Lutz, edição IV, 1ªedição digital, 2008.

[8] A.O.A.C. Association of Official Analytical Chemists: Official Methods of Analysis of AOAC Internacional. v.2, 17.ed. Gaithersburg-EUA: AOAC, 2000.

[9] Moscatto, J.A; Prudêncio-Ferreira, S.H; Hauly, M.C.O. Farinha de yacon e inulina como ingredientes na formulação de bolo de chocolate. *Ciência e Tecnologia de Alimentos*, Campinas, 2004.

[10] Cecchi, H. M. Fundamentos teóricos e práticos em análise de alimentos. 2ª ed.rev. Campinas-SP. Editora Unicamp, 2003.

[11] Silva, B. C.; Carpenedo, E.; Silva, L. O.; Lavinas, F. C.; Ribeiro-Alves, M. A. Elaboração de cupcake de alfarroba isento de gluten. *Almanaque multidisciplinar de pesquisa*. Universidade Unigranrio. Ano I, Volume 1, Número 1, p. 79-93, 2014.

[12] Santos, J. L.; Dourado, C. P.; Bonfim, A. C. P.; Santos, E. F.; Manhani, M. R.; Novello, D. Bolo de aveia com adição de amaranto: composição físico-química e avaliação sensorial entre crianças. *Revista UNIABEU*, v.7 Número 16, p. 142-156, 2014.

[13] Capriles, V.D.; Coelho, K.D.; Matias, A.C.G.; Arêas, J.A.G. Efeito da adição de amaranto na composição e na aceitabilidade do biscoito tipo cookie e do pão de forma. *Alimentos e Nutrição, Araraquara*, v.17, n.3, p. 269-274, 2006.

[14] Carvalho, K. H.; Bozatski, L. C.; Scorsin, M.; Novello, D.; Perez, E.; Dalla Santa, H. S.; Scorsin, G.; Batista, M. G. Desenvolvimento de cupcake adicionado de farinha da casca de banana: características sensoriais e químicas. *Alimentos e Nutrição, Araraquara*, v. 23, n. 3, p. 475-481, jul/set. 2012.

[15] Baptista, A. T. A.; Silva, M. O.; Bergamasco, R.; Vieira, A. M. S. Avaliação físico-química e sensorial de biscoitos tipo *cookies* elaborados com folha de *Moringa oleífera*. *B.CEPPA*, Curitiba, v. 30, n. 1, p. 65-74, jan./jun. 2012.

[16] Muñoz, A.M.; Civille, V.G.; Carr, B.T. Sensory evaluation in quality control. New York: Van Nostrand Reinhold, 1992.

[17] Alamanou, S.; Bloukas, J.G.; Paneras, E.D.; Doxastakis, G. Influence of protein isolate from lupin seed (Lupinus albus. ssp. Graecus) on processing and quality characteristics of frankfurters. *Meat Science, Champaign*, v. 1, n. 42, p. 79-93, 1996.

[18] Teixeira, E.; meinert, E. M.; & Barbetta, P. A. Análise Sensorial de Alimentos. Série Didática. Florianópolis: Editora UFSC, p. 18-102, 1987.

[19] Robertson, J.A. Aplication of plant-based by products as fiber supplements in processed foods. *Recent Research Development in Agricultural & Food Chemistry*, v. 2, p. 705-717, 1998.

7

Effects of Pre-treatments and Drying Methods on Chemical Composition, Microbial and Sensory Quality of Orange-Fleshed Sweet Potato Flour and Porridge

Fana Haile[1], Shimelis Admassu[2,*], Abrehet Fisseha[3]

[1]Arbamich University
[2]Food, Beverage and Pharmaceutical Industry Development Institute, Ministry of Industry, Ethiopia
[3]Hawassa University, School of Nutrition, Food Science and Technology, Hawassa, Ethiopia
*Corresponding author: shimelisemire@yahoo.com

Abstract This study was aimed to evaluate the effect of pre-treatments and drying methods on the chemical composition (proximate composition, β-carotene content and phytochemicals), functional properties, microbial and sensory of orange-fleshed sweet potato flour and porridge. Physical (blanching) and chemical (1% salt, 0.5% citric acid solution soaked for 20 min) of the sliced OFSP followed by drying (sun, solar and fluidized bed dried) with complete randomized design for chemical composition, functional properties, and microbial load analysis and randomized complete block design was used for sensory analysis. The moisture (4 - 8%), protein (4 - 8%), fat (0.9 - 2.5), ash (4 - 8%), fiber (3.5 - 7%), total carbohydrates (80 - 84%) measured as a function of treatment and drying. The energy contribution of OFSP was determined by difference of the proximate excluding the fiber (352.9-365.6kcal/100g). The salt treated and FB dried had significantly lowest moisture content and highest ash content, citric acid treated and FB dried had significantly highest protein content and gross energy, control and FB dried had significantly highest fiber content, control and sun dried had significantly highest fat content and blanched and FB dried had significantly highest total carbohydrates. The β-carotene (82 -127μg/g), tannin (74 -108mg/100g) and phytate content (51 - 98mg/100g) was measured. Blanched and FB dried was observed better in retention of β-carotene and reduction of phytochemicals (tannin and phytate). The salt treated and FB dried was measured increment with WAC, viscosity and lowest LGC, blanched and FB had highest OAC and high bulk density in the control. The microbial analysis showed OFSP porridge within microbiological acceptable limit. The sensory acceptability showed OFSP was accepted in all sensory attributes. In conclusion, the blanching and FB drying techniques were the best approach in retaining the nutrients.

Keywords: *β-carotene, chemical composition, drying method, orange-fleshed sweet potato, pre-treatment, microbial quality, sensory acceptability*

1. Introduction

Sweet potato (*Ipomoea batatas*L.) ranks seventh among the most important food crops in the world [43]. Sweet potatoes are a good source of minerals [22], carbohydrates, fiber, antioxidants, starch and vitamins [4,28]. Orange-fleshed sweet potato (OFSP) is rich in beta-carotene, which is a precursor to vitamin A Contributes in alleviating vitamin A deficiency. Vitamin A deficiency is a major public health problem in developing countries of the most vulnerable groups [21]. The short shelf life of sweet potato due to its pershiblity of sweet potato [41] is a challenge in developing countries [46]. In Ethiopia sweet potato cultivation is currently being encouraged because of its numerous potential for food security; even though most of the Ethiopian farmers have the challenge of sweet potato tuberous roots storage after harvest like other parts of developing countries [41].

Drying methods are commercially accepted and economic of the raw and intermediate potato products to preserve by reducing the bulk weight [10,42,52]. Blanching, soaking in solution inactivate the enzyme activities (quality reduction and deterioration reduced), and improve product acceptability [9,12]. The futures of Orange-fleshed sweet potato in the developing country for target consumers with micronutrient supplemented porridge, bread, traditional doughnut, chapatti and cake is promising [31]. OFSP is a source of energy and nutrients, natural sweetener, color, flavor and dietary fiber [48].

2. Materials and Methods

The deep orange colored Tula variety (Voucher number CIP420027) orange-fleshed sweet potato was used to evaluate the nutritional and functional properties, microbial and sensor quality of OFSP flour and porridge. The tuber was Blanched in a water bath at 60°C for 5 min according to the method described by Emmanuel [13], soaked in salt solution of 1% concentration and citric acid solution (0.5%) for 20 min based on the method described by Bechoff [10] with slight modification and dried(sun, solar and fluidized bed drier at 65°C), Figure 1. The dried OFSP was milled in to flour and analyzed nutritional profile, functional properties, microbial and sensory quality using a standard method. Sensory analysis was conducted for the porridge (10gm Orange-fleshed sweet potato flour, 200ml water, about 1g salt and 3g sugar cooked for 8 min according to USAID [45].

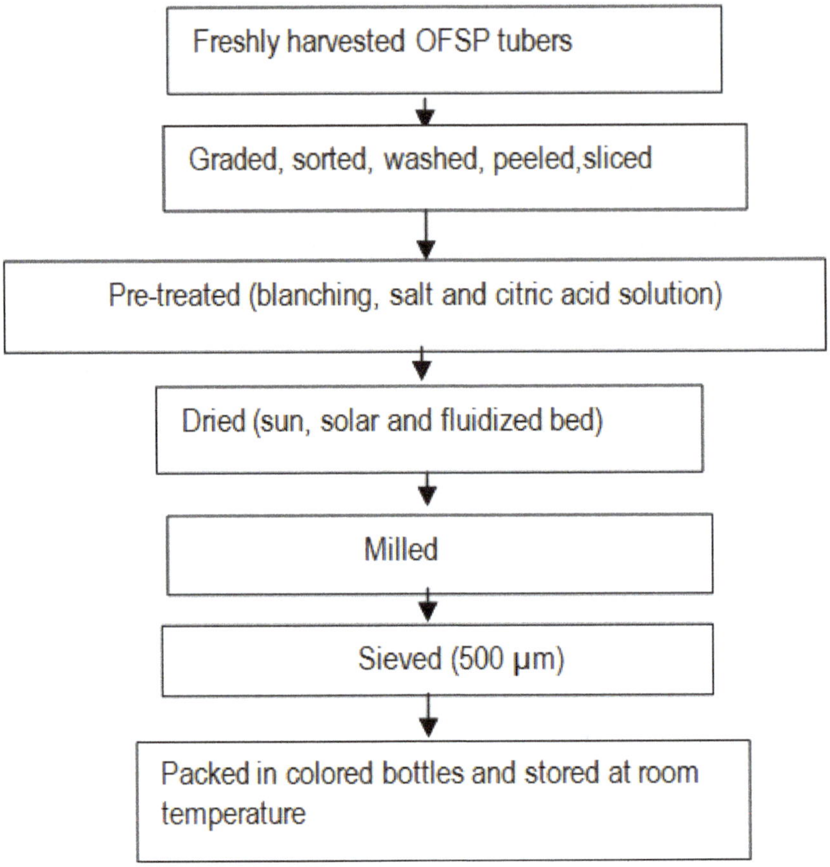

Figure 1. Flow chart for OFSP flour preparation; Source: [13]

The nutritional profile; Proximate composition [15,16], β-carotene [36], pH-value [15,16], water and oil absorption capacity [8], bulk density [8], least gelation capacity [2], phytochemicals such as tannins and phytate content were determined according to Maxson and Rooney [24] and Latta and Eskin [20]; respectively. The total plate count, yeasts and mold count analysis was conducted for the porridge [29]. The sensory quality of porridge was studied using the nine hedonic points.

The experiment was consisted of two factors three pre-treatments and three drying methods. Untreated OFSP sample was taken as a control. The experimental design was completely randomized design (CRD) for chemical composition, functional properties and microbial load and Randomized Complete Block Design (RCBD) for the sensory analysis. The data was subjected to two factor-analysis of variance (ANOVA) used SAS (Statistical Analysis System, version 9.1; SAS Institute, 2001). All the samples were analyzed in duplicates and results were expressed as mean ± standard deviation. The Fisher's least significant difference (LSD) test was used to test differences between means and the significant was set at ($P<0.05$).

3. Results and Discussions

3.1. Proximate Composition

The proximate composition (crude fat, crude protein, crude fiber, carbohydrate, ash, moisture) was analyzed and a significant difference was observed on the OFSP flour as affected by pre-treatments and drying methods (Table 1). The moisture content of OFSP flour was measured between 4 to 8%. The lower moisture contents the better shelf life and denser nutrient composition [19,26,51]. The salt treated in fluidized bed dried product revealed lowest moisture content. This is due to the fact that salt brining causes in the dehydration of the OFSP tuber and the high temperature of fluidized bed applied at a moderately short time compared to solar and sun dryer. Protein content was measured between 4 to 6%. The citric acid treatment with all drying method, salt and control with fluidized bed dried was observed increment in the crude protein content. The blanched sundried was observed reduced crude protein content due to the protein denaturation, leaching out and prolonged drying in uncontrolled environment [14,23,26,27,30,38]. Ünlüsayin et al. [44] describes salt induces the change in protein structure and denaturation up to 20-25%. The controlled and blanched fluidized bed dried was observed increment in its fiber content. The fat content (0.9 to 2.55%); Control sun dried measured increased fat content while the blanched with all drying methods was decreased fat content decrement. Different studies showed sun dried root and tuber crops had high fat than other drying methods [3,16,23,27]. The reduced fat content of the blanched OFSP was due to oxidation at increased temperature. Ash is important food constituent and inorganic residue after combustion at high temperature - prolonged time [18]. The ash content was measured between 4-7.5%. Salt treated fluidized bed dried and solar dried detected high due to salt absorption during soaking and lower moisture content [26] and blanched sun dried flour were measured with lowest ash content due the sensitivity of specific minerals to temperature at prolonged time during blanching and drying [3,16,23].

Table 1. The Proximate Composition of OFSP flour

Sample	Moisture (%)	Protein (%)	Fiber (%)	Fat (%)	Ash (%)	Carbohydrate (%)
Control - Sun dried	7.36 ± 0.08^{b}	4.39 ± 0.05^{f}	4.12 ± 0.26^{def}	2.55 ± 0.06^{a}	5.14 ± 0.06^{ef}	80.62 ± 0.05^{f}
Control - Solar dried	6.35 ± 0.02^{e}	4.63 ± 0.00^{e}	5.59 ± 0.02^{b}	2.04 ± 0.08^{b}	5.43 ± 0.06^{de}	81.56 ± 0.03^{e}
Control - Fluidized bed dried	5.11 ± 0.06^{g}	5.06 ± 0.15^{d}	7.30 ± 0.43^{a}	1.48 ± 0.01^{e}	5.68 ± 0.10^{cd}	82.68 ± 0.10^{c}
Blanched - Sun dried	7.79 ± 0.06^{a}	4.10 ± 0.07^{g}	3.84 ± 0.09^{fg}	1.17 ± 0.06^{h}	4.15 ± 0.07^{h}	82.79 ± 0.15^{c}
Blanched - Solar dried	6.46 ± 0.04^{e}	4.21 ± 0.05^{g}	5.30 ± 0.06^{bc}	0.99 ± 0.01^{i}	4.67 ± 0.06^{g}	83.69 ± 0.07^{b}
Blanched - fluidized bed dried	5.46 ± 0.03^{f}	4.34 ± 0.03^{f}	7.07 ± 0.07^{a}	0.90 ± 0.01^{i}	5.10 ± 0.14^{f}	84.20 ± 0.16^{a}
Salt -Sun dried	6.78 ± 0.12^{d}	4.45 ± 0.05^{f}	4.63 ± 0.07^{de}	1.42 ± 0.01^{e}	6.78 ± 0.40^{b}	80.01 ± 0.58^{f}
Salt -solar dried	5.09 ± 0.05^{g}	4.73 ± 0.03^{e}	4.17 ± 0.29^{efg}	1.32 ± 0.00^{fg}	7.31 ± 0.08^{a}	81.56 ± 0.06^{e}
Salt - fluidized bed dried	4.04 ± 0.05^{i}	5.18 ± 0.08^{d}	4.45 ± 0.19^{cd}	1.26 ± 0.00^{g}	7.51 ± 0.11^{a}	82.01 ± 0.15^{d}
Citric acid-sun dried	7.13 ± 0.04^{c}	5.44 ± 0.03^{c}	3.73 ± 0.04^{g}	1.89 ± 0.06^{c}	5.25 ± 0.01^{ef}	80.30 ± 0.05^{f}
Citric acid – solar dried	5.40 ± 0.09^{f}	5.61 ± 0.01^{b}	4.32 ± 0.26^{def}	1.62 ± 0.05^{d}	5.59 ± 0.04^{d}	81.78 ± 0.08^{de}
Citric acid-fluidized bed dried	4.38 ± 0.02^{h}	5.76 ± 0.01^{a}	4.78 ± 0.49^{de}	1.40 ± 0.04^{ef}	5.96 ± 0.02^{c}	82.50 ± 0.01^{c}

$^{a\text{-}g}$All values are means±SD

All values within the same column with different superscript letters are significantly different from each other at ($p<0.05$).

3.2. Beta Carotene Content

The β-carotene content was measure between 82 to 127μg/g as shown below (Figure 2). Blanching with fluidized drying measured higher β-carotene content as the enzyme (peroxidases, and lipoxygenases) activities which have a capacity in degrading β-carotene [35] was inactivated and β-carotene degradation was delayed. Salt and citric acid treated OFSP flour shows no significant difference but increment was observed relative to the control. Salting reduces osmotic tension of cells increasing the stress on bacteria and enzymes [47].

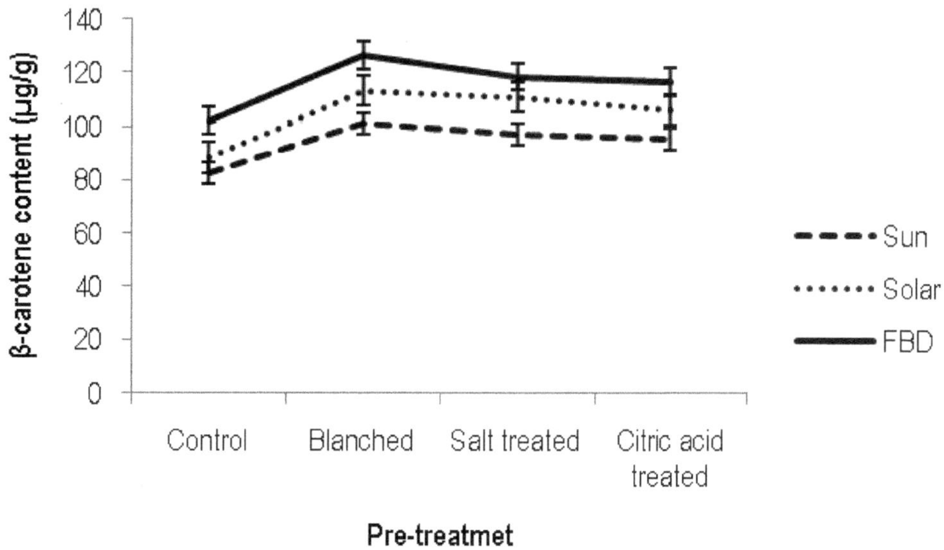

Figure 2. The beta-carotene content (μg/g) of OFSP flow as affected by pre-treatment and drying

A combination of blanching and fluidized bed drying techniques had higher retention on the beta-carotene content of OFSP flour samples compared to the other. Pre-treatments and drying methods had significant effect on the beta-carotene content of OFSP flour.

3.3. Phytochemical Composition and pH Value

Tannin and phytate of OFSP flour was detected between 74 to 108 and 51 to 98/100g respectively. Control sun dried OFSP flour was measured high in tannin content while the blanched and fluidized bed dried OFSP flour shows lowest in tannin and phytate content. Pre-treatments reduce levels of both tannin and phytate content in different dried products [25,34,49]. The decrease in tannin and phytate content in fluidized bed dried OFSP might have due to more controlled conditions of temperature and air velocity and also with the fluidized nature of drying particles. Pareek and Kaushik [32] reported highest tannin content was recorded in solar dried aonla flour followed by oven dried, microwave dried and lowest in fluidized bed. The pH value was measured between 4.5 to 7.0 for all pre-treatments and drying methods. The citric acid treated and sun dried OFSP flour was observed lower pH level. The higher acidity value was due to the added citric acid in the solution or the acidifying property of citric acid.

Table 2. effect of pre-treatments and drying methods on the phytochemicals and pH value of OFSP flour (mg/100g)

Sample	Tannins	Phytate	pH
Control - Sun dried	108.46±1.33[a]	98.39±1.57[a]	5.53±0.03[f]
Control - Solar dried	97.16±0.12[c]	94.27±0.40[b]	5.69±0.01[e]
Control - Fluidized Bed dried	92.49±0.62[d]	80.61±0.54[c]	5.73±0.07[de]
Blanched -Sun dried	90.75±0.13[e]	61.78±0.22[g]	5.59±0.02[f]
Blanched - Solar Dried	82.89±1.29[g]	59.21±0.21[h]	5.71±0.07d[e]
Blanched - Fluidized Bed dried	74.15±0.83[h]	50.78±0.22[j]	5.79±0.06[d]
Salt - Sun dried	98.11±0.04[c]	66.28±0.94[de]	6.48±0.01[c]
Salt - Solar dried	92.88±0.13[d]	64.19±0.35[f]	6.63±0.02[b]
Salt - Fluidized Bed dried	86.33±0.67[f]	56.73±0.64[i]	6.71±0.02[a]
Citric acid - Sun dried	100.05±0.62[b]	66.78±1.65[d]	4.63±0.02[h]
Citric acid - Solar dried	96.72±0.04[c]	64.74±0.85[ef]	4.74±0.03[g]
Citric acid- Fluidized Bed dried	91.78±0.12[de]	57.23±0.07[i]	4.78±0.01[g]

[a-h]All values are means±SD

All values within the same column with different superscript letters are significantly different from each other at (p<0.05)

3.4. Functional Properties

Pre-treatments and drying have a significant effect on the bulk density, water absorption capacity, oil absorption capacity, viscosity of OFSP flour. Bulk density is important for the selection of packaging material [18] was measured between 0.70 to 0.75g/ml. The control fluidized bed dried OFSP flour was measured higher bulk density. However, the pre-treatments have no significant effect [2,11]. Olapade and Ogunade [30] observed higher bulk density in the blanched cabinet dried yellow and cream fleshed sweet potato flour. Therefore, different bulk density was observed different for different root and tuber crops due to the difference of variety, pre-treatment, blanching time and temperature and drying method used. The water absorption capacity of OFSP flour was measured 2.81 to 3.9ml/g. The salt treated fluidized bed dried OFSP flour were measured higher due to high polar amino acid residue of protein having affinity for water [50] while blanched sun dried flour had lowest water absorption capacity. The oil absorption capacity of OFSP flour was measured from 0.55 to 1.03ml/g. The Sun dried flour had lowest oil absorption capacity compared to solar and fluidized bed dried in all pre-treatments and the blanched - fluidized bed dried OFSP flour was observed with high oil absorption capacity. Viscosity is the ability of starch to form paste after cooling and the unstable starch accompanied break down the bonding and damaged starches [1,39]. The viscosity of OFSP flour was measured between 72 to 371.2cps. The salt treated fluidized bed dried OFSP flour and porridge had measured high highest viscosity value. The higher viscosity value was due to the lowest moisture content and higher water absorption capacity of the fluidized bed dried flour and form a thick porridge with desirable attribute for industrial uses as a thickening power is required [7]. The least gelation capacity of OFSP flour samples ranged from 7.0 to 20.5% (Table 3). Citric acid treatment and Sun drying was observed with increased gelatin capacity in treatments. Salt and fluidized bed drier and their combination showed that lowest gelation property as shown in below (Table 3).

Table 3. The Functional properties of OFSP as affected by pre-treatments and drying process

Sample	BD(g/ml)	WAC(ml/g)	OAC(ml/g)	Viscosity(cps)	LGC (%)
Control - Sun dried	0.71±0.00[d]	2.90±0.00[ef]	0.55±0.01[h]	79.4±0.28[k]	18.80±0.28[b]
Control - Solar dried	0.72±0.00[b]	3.14±0.01[d]	0.63±0.02[fg]	144.5±0.71[g]	17.45±0.07[c]
Control - Fluidized Bed dried	0.74±0.00[a]	3.43±0.04[c]	0.75±0.01[cd]	290.7±0.85[c]	11.10±0.14[g]
Blanched -Sun dried	0.71±0.00[d]	2.81±0.15[f]	0.71±0.00[de]	95.55±0.49[j]	17.65±0.35[c]
Blanched - Solar Dried	0.71±0.00[cd]	3.06±0.04[de]	0.89±0.00[b]	155.0±1.41[f]	16.3±0.14[d]
Blanched - Fluidized Bed dried	0.73±0.00[ab]	3.38±0.01[c]	1.03±0.06[a]	311.70±1.13[b]	9.4±0.28[h]
Salt - Sun dried	0.71±0.00[cd]	3.42±0.06[c]	0.66±0.04[efg]	119.50±0.71[i]	16.05±0.07[d]
Salt - Solar dried	0.72±0.00[c]	3.68±0.08[b]	0.74±0.00[cd]	242.85±0.92[e]	14.0±0.71[e]
Salt - Fluidized Bed dried	0.73±0.00[ab]	3.9±0.04[a]	0.94±0.06[b]	371.15±0.92[a]	7.0±0.28[i]
Citric acid - Sun dried	0.71±0.00[d]	3.18±0.11[d]	0.6±0.04[gh]	72.65±0.49[l]	20.5±0.71[a]
Citric acid - Solar dried	0.71±0.00[cd]	3.41±0.15[c]	0.71±0.02[def]	134.0±1.41[h]	18.65±0.07[b]
Citric acid- Fluidized Bed dried	0.73±0.00[ab]	3.78±0.01[ab]	0.81±0.01[c]	267.0±0.99[d]	12.35±0.21[f]

[a-l]All values are Means±SD

All values within the same column with different superscript letters are significantly different from each other at (p<0.05)

BD= Bulk density, WAC= Water absorption capacity, OAC= Oil absorption capacity, LGC=Least gelation capacity.

3.5. Microbial Quality

The porridge of OFSP was analyzed for the microbial quality to guarantee the quality at the time of consumption after two months storage. Total plate count (TPC), yeast and mold was detected (1.53-2.88*10^2 and 0.43-1.35*10^2cfu/ml; respectively) for the pr-treated and dried OFSP porridge. The salt treated OFSP porridge had lowest TPC and the control had highest as the salt solution inhibits the microbial growth at the time of raw material preparation [33]. The citric acid treatment dried with fluidized bed drier shows lowest yeast and mold and TPC count. The microbial dose was observed in the acceptable limits as International microbiological standards recommended units for foods total bacterial plate counts should be less than 10^3cfu/g and the International Commission for Microbiological Specification for Foods [17] states that ready-to-eat foods with plate counts between 0-10^3 are acceptable, between 10^4 - ≤ 10^5 is tolerable and 10^6.

Table 4. Microbial load (10^2cfu/ml) of OFSP flour as affected by pre-treatment and drying process

Sample	Total plate count	Yeast and Mold count
Control - Sun dried	2.88±0.04^{a}	1.35±0.71^{a}
Control - Solar dried	2.78±0.04^{ab}	1.00±0.73^{c}
Control - Fluidized Bed dried	2.58±0.04^{cde}	0.85±0.85^{de}
Blanched -Sun dried	2.73±0.04^{abc}	1.15±0.95^{b}
Blanched - Solar Dried	2.58±0.04^{cde}	0.88±1.06^{cde}
Blanched - Fluidized Bed dried	2.43±0.18^{e}	0.65±1.41^{fg}
Salt - Sun dried	2.53±0.04^{de}	0.98±0.35^{cd}
Salt - Solar dried	2.18±0.04^{f}	0.75±0.65^{ef}
Salt - Fluidized Bed dried	1.53±0.04^{h}	0.58±0.35^{g}
Citric acid - Sun dried	2.63±0.04^{bcd}	0.75±0.71^{ef}
Citric acid - Solar dried	2.43±0.18^{e}	0.58±0.35^{g}
Citric acid- Fluidized Bed dried	1.98±0.04^{g}	0.43±0.35^{h}

$^{a-g}$All values are means±SD

All values within the same column with different superscript letters are significantly different from each other at ($p<0.05$).

3.6. Drying Effect on Product Quality

The OFSP raw tuber was dried; (Sun, solar and fluidized bed) after cleaning, sorting and sliced in to uniform pieces to maintain uniform drying process. After drying the dried OFSP was grounded into flour and then the moisture content was measured. The lower the moisture content the better shelf life and nutritional profile. The moisture content of OFSP flour was measured below 8% for all drying techniques. FBD was observed lower moisture content as a result better nutrient density and lower phytochemical (phytate and taninins) composition as shown in Table 1 and Table 2; respectively. The FBD was better in β-carotene retention and functional properties comparedto solar and sun drying (Figure 2 and Table 3). This is due to the fact that the β-carotene degrading enzymes and the volatile components of OFSP were delayed in a controlled manner [35]. Drying has a significant effect on the total microbial load of the OFSP porridge (Table 4). Porridge prepared from the fluidized bed dried orange fleshed sweet potato powder was better in its total microbial quality due to the saving in the cross contamination and control in growth of microorganisms. Drying shows insignificant effect on the sensorial quality attributes of porridge. However, product made from sun dried was liked by the consumers. This might be due to adaptation of the traditional products prepared by sun drying process.

3.7. Sensory Quality

Consumer based sensory quality of porridge was conducted using the nine-hedonic point of the porridge's color, taste, aroma, mouth feel and overall acceptability as it affects the purchasing behavior of consumers [40] Table 5. All the sensory attributes were liked slightly - moderately. The color of blanched with sun and solar dried were liked more due to the inactivation of endogenous enzymes reduce pigment (yellowish and brown color) degradation. Citric acid treated with fluidized dried porridge was liked more for its taste and no significant difference was observed for aroma. The citric acid treated sun dried; salt treated fluidized bed dried and blanched fluidized bed dried were score high for the mouth feel of OFSP porridge. The mouth feel depends on moisture content, viscosity, bulk density and taste difference. Blanched sun dried and citric acid treated sun dried was scored higher value for the overall acceptability. The study revealed sensory qualities were affected by pre-treatments and drying methods.

Table 5. Sensory acceptability score of OFSP porridge as affected by pre-treatments and drying methods

Sample	Color	Taste	Aroma	Mouth feel	Overall acceptability
Control - Sun dried	7.09±1.21^{abcd}	6.31±1.48^{f}	6.11±1.47^{b}	5.83±1.44^{d}	6.70±1.04^{cd}
Control - Solar dried	7.15±1.34^{abcd}	7.13±1.26^{ab}	6.18±1.62^{b}	5.75±1.31^{d}	6.48±1.15^{d}
Control - Fluidized Bed dried	6.95±1.77^{cd}	7.08±1.63^{abc}	6.18±1.68^{b}	6.23±1.61^{c}	6.58±1.42^{cd}
Blanched -Sun dried	7.49±1.58^{a}	7.03±1.36^{abc}	6.74±1.68^{a}	6.05±1.45^{c}	7.20±1.12^{a}
Blanched - Solar Dried	7.48±1.97^{a}	6.60±1.77^{def}	6.53±1.54^{ab}	6.30±1.63^{bc}	6.78±1.32^{bcd}
Blanched - Fluidized Bed dried	7.35±1.20^{abc}	6.68±1.29^{cdef}	6.78±0.81^{a}	6.76±1.18^{a}	6.90±0.95^{abc}
Salt - Sun dried	7.03±0.94^{bcd}	6.98±1.14^{abcd}	6.85±0.97^{a}	6.58±0.74^{abc}	7.05±0.78^{ab}
Salt - Solar dried	6.93±1.22^{d}	6.80±0.93^{bcde}	6.83±1.10^{a}	6.58±0.78^{abc}	6.88±0.72^{abc}
Salt - Fluidized Bed dried	6.85±0.92^{d}	7.08±1.38^{abc}	6.70±1.13^{a}	6.83±1.19^{a}	6.90±0.87^{abc}
Citric acid - Sun dried	7.25±0.95^{abcd}	7.25±1.12^{a}	6.83±1.14^{a}	6.89±1.11^{a}	7.05±0.78^{ab}
Citric acid - Solar dried	7.38±1.49^{ab}	7.03±1.20^{abc}	6.63±1.42^{a}	6.58±1.29^{abc}	6.80±1.26^{bcd}
Citric acid- Fluidized Bed dried	6.35±1.24^{e}	6.43±1.12^{ef}	6.65±1.16^{a}	6.33±1.09^{bc}	6.53±0.98^{d}

$^{a-f}$All values are means±SD

All values within the same column with different superscript letters are significantly different from each other at ($p<0.05$).

4. Conclusions

The orange-fleshed sweet potato flour was prepared by pre-treating of the OFSP tuber with 0.5% citric acid, 1% salt solution and blanching. The pre-treated and control OFSP were dried with sun, solar and fluidized bed drying methods. In the conversion of OFSP to flour the pre-treatments and drying methods and their combination had a significant effect on the chemical composition, functional, microbial and sensory quality of OFSP flour and porridge. Blanching and fluidized bed drying method had a significant effect on the nutrient retention and a best

Combination processing method of OFSP flour and porridge is the best approach for preparing OFSP flour and porridge, enriched with pro-vitamin A. OFSP is rich is β-carotene (pro-vitamin A) recommended to supplement stable food lucked with vitamin A and vitamin A deficiency reduction will be granted. The porridge prepared from OFSP flour was slightly too moderately acceptable to all sensory attributes and within the acceptable microbial limit according to the International Commission for Microbiological Specification for Foods. The post-harvest loss (physiological and nutritional losses) of OFSP is manageable with the cost-effective pre-treatment and controlled drying (fluidized bed drying) methods use.

Acknowledgements

The authors thanks all who have put their efforts from the start to end of the research experiments, expressly Arbaminch University, Hawassa University, Irish Aid – CIP, Addis Ababa Institute of Technology, Food, Beverage and Pharmaceutical Industry Development Institute and Ethiopian Public Health Institute (EPHI).

References

[1] Adebayo-Oyetoro A., Olatidoye O., Ogundipe O., Balogun I. and Apara T. 2012. Effect of Local cassava fermentation methods on functional pasting and sensory properties of Lafun. *Continental Journal of Agricultural Sciences*; 6(2): 1-8.

[2] Akubor, I. 2013. Effect of ascorbic acid and citric acid treatments on the functional and sensory properties of yam flour. *International Journal of Agricultural Policy and Research;* 1 (4), pp. 103-108.

[3] Akubor, I. and John, I. 2012. Quality evaluation and cake making potential of sun and oven dried carrot fruit. *International Journal of Biosciences;* 2, No. 10(2), p. 19-27.

[4] Anderson, L. and Gugerty, K. 2013. Sweet potato Value Chain: Ethiopia. Evans School Policy Analysis and Research. University of Washington. EPAR Brife No. 219. 11pp.

[5] AOAC. 2000. Association of Official Analytical Chemists. Official Methods of Analysis, 15th ed. Arlington, VA., USA.

[6] AOAC. 2000. Association of Official Analytical Chemists. Official Methods of Analysis, 16thEdn. Washington, DC.

[7] Aprianita A, Purwandari U, Watson B, Vasiljevic T. 2009. Physico-chemical properties of flours and starches from selected commercial tubers available in Australia. *International Food Research Journal*; 16: 507-520.

[8] Aremu, O. Olonisakin, A. Atolaye, O. and Ogbu, F. 2007. Some Nutritional Composition and Functional Properties of *Prosopis Africana.Journal of Scientific and Industrial Research;* 42(3): 269-280.

[9] Babajide, J., Obadina, O., Oyewole, B. and Ugbaka, N. 2006. Microbial quality of dry yam ''gbodo'' parboiled with / without adjuncts. *African Journal of Biotechnology;* 5: 278-281.

[10] Bechoff, A. 2010. Investigating carotenoid loss after drying and storage of orange-fleshed sweetpotato. PhD thesis. Natural Resources Institute. University of Greenwich. 260pp.

[11] Bentil, B. 2011. Assessment of three different drying technologies (sun, solar and bin) used for the production of cassava (Manihotesculenta Crants) Chips in Ghana. MSc. Thesis, School of research and graduate studies, kwamenkrumah University.

[12] Doymaz, I. 2007. Air-drying characteristics of tomatoes. *Journal of Food Engineering,* 78(4): 1291-1297.

[13] Emmanuel, N. 2012. Application of Windhexe Dehydration Technology for producing ß-carotene Rich Flours From Sweet potatoes. MSc. Thesis, North Carolina State University .Food Science. Raleigh, North Carolina; 102 pp.

[14] Eric, D. 2013. Effect of drying methods on nutrient quality of Basil (*Ocimumviride*) leaves cultivated in Ghana.*International Food Research Journal* 20(4): 1569-1573.

[15] Eric, K. 2012. The effect of drying methods (catalytic flameless infrared - CFIR, oven and solar drying) on the quality of sweet potato flour.MSc.Thesis, College of science, department of biochemistry and biotechnology, Kwame Nkrumah University of Science and Technology, Kumasi.

[16] Eze, J. and Akubor, I. 2012. Effect of Drying Methods and Storage on the Physicochemical Properties of Okra.*Journal Food Process and Technology;* 3:177.

[17] International Commission on Microbiological Specifications for Foods (ICMSF). 1996. Microorganisms in Foods: Microbiological Specifications of Pathogens.

[18] Idowu, O., Olaoye, O., Sogotemi, C., Ajayi, B. 2013. Quality assessment of flours and amala produced from three varieties of sweet potato ipomeabatatas. International journal of food and nutritional sciences, 2(4): 2320-7876.

[19] Kikafunda, J., Abenakyo, L., Lukwago, F. 2006. Nutritional and sensory properties of high energy/nutrient dense composite flour porridges from germinated maize and roasted beans for child-weaning in developing countries: a case for Uganda. Ecol. Food Nutr., 45, 279-294.

[20] Latta and Eskin. 1980. Determination of Phytate. *Journal of Cereal chemistry.*

[21] Low, J., Arimond, M., Osman, N., Cunguara, B., Zano, P. and Tschirley, T. 2007. A Food based approach introducing Orange-fleshed sweet potatoes increased vitamin A intake and serum retinol concentrations in young children in rural Mozambique. *Journal of Nutrition*; 137: 1320-1327.

[22] Luis, G., Rubio, C., Gutiérrez, A., González-Weller, D., Revert, C., Hardisson, A. 2013. Evaluation of metals in several varieties of sweet potatoes (Ipomoea batatas L.): comparative study. 186(1):433-440.

[23] Malomo, O., Jimoh, O., Adekoyeni, O., Soyebi, E, Alamu, A . 2013. Effect of blanching and unblanching on rheological properties of sweet-potato bread. College of Food Sciences, Bells University of Technology, Ota Ogun State, Nigeria. *Academic Research International*; Vol. 4 No. 3.

[24] Maxson and Rooney (1972). Condensed Tannin of Burns. *Journal of Cereal chemistry.*

[25] Mosha, T, Gaga, H, Pace, R, Laswai, H, Mtebe, K. 1995. Effect of blanching on the content of antinutritional factors in selected vegetables. Plant Foods Hum.Nutr. (Formerly QualitasPlanturum), 47(4): 361-367.

[26] Muyanja, C., Kyambadde, D. and Namugumya, B. 2012. Effect of Pretreatments and Drying Methods on Chemical Composition and Sensory Evaluation of Oyster Mushroom (*PluerotusOestreatus*) Powder and Soup. *Journal of Food Processing and Preservation*; 38, Issue 1, pages 457-465.

[27] Ngozi, A., Aminat, A., Adediran, A. and Ebunoluwa, O. 2013. The Effect Of Pretreatment Of Plantain (Musa Parasidiaca) Flour On The Pasting And Sensory Characteristics Of Biscuit. *International Journal of Food and Nutrition Science*; vol. 2 No.1.

[28] Odongo, B., Mwanga, R., Owori, C., Niringiye, C., Opio, F., Ewell, P., Berga, L, Agwaro, G., Sunjogi, L., Abidin, E., Kikafunda., J. and Mayanja, R. 2002. Development and promotion of orange fleshed sweet potato to reduce vitamin A deficiency in Uganda. Kampala, Uganda, 7pp.

[29] Olaoye, O., Onilude, A. and Idowu, O. 2006. Quality characteristics of bread produced from composite flours of wheat, plantain and soybeans. *African Journal of Biotechnology; 5 (11), pp. 1102-1106.*

[30] Olapade, A. and Ogunade, O. 2014 Production and evaluation of flours and crunchy snacks from sweet potato (*Ipomeabatatas*) and maize flours. *International Food Research Journal;* 21(1): 203-208.

[31] Owori, C., Lemaga, R., Mwanga, A., Namutebi and Kapinga, R. 2007. Sweet potato recipe book: sweet potato processed products from Eastern and Central Africa. *African Crop Science Society;* Kampala, Uganda. 93pp.

[32] Pareek, S. and Kaushik, A. 2012. Effect of drying methods on quality of Indian gooseberry (*EmblicaofficinalisG*aertn.) poweder during storage. *Journal of Scientific and Industrial Research;* Vol.71. pp. 727-732.

[33] Pepe, O., Blaiotta, G., Moschetti, G., Greco, T., and Villani, F. 2003.Rope-producing strains of *Bacillus* spp. from wheat bread and strategy for their control by lactic acid bacteria. *Applied Environmental Microbiology;* 69:2321–2329.

[34] Prajapati, K., Prabhat, K., Nema, S., Rathore. 2011. Effect of pretreatment and drying methods on quality of value-added dried

aonla (*Emblicaofficinalis*Gaertn) shreds. *Journal of Food Science and Technology;* 48(1):45–52.

[35] Prakash, S., Jha, S. and Datta, N. 2004. Performance evaluation of blanched carrots dried by three different driers. *Journal of Food Engineering;* 62, 305-313.

[36] Rodriguez- Amaya, D., Kimura, M., Godoy, H. and Arima, H. 1988. Assessment of provitamin A determination by open column chromatography/visible absorption spectrophotometry. *Journal of Chromatography Science;* 26:624-629.

[37] SAS Institute. 2001. Statistical Analysis System (SAS) user's guide, Release 8.1 1999- 2000. North Carolina, U.S.A.

[38] Seidu, J., Bobobee, E., Kwenin, W., Tevor, W., Mahama, A. and Agbeven, J. 2012. Drying of sweet potato (*Ipomoea batatas*) (chipped and grated) for quality flour using locally constructed solar dryers. *Journal of Agricultural and Biological Science*; vol. 7, no. 6.

[39] Shimelis, Admassu, E., Meaza, M., and Rakshit, S.K. 2006. Physico-chemical properties, pasting behavior and functional characteristics of flour and starches from improved Beans in East Africa. *GIGR Journal*; 8(1): 1-18.

[40] Surkan S, Albani O, Ramallo L. 2009. Influence of storage conditions on sensory shelf-life of yerba mate. *Journal of Food Quality*. 32: 58-72.

[41] Teye, E. 2010. Developing appropriate storage technology for sweet potatoes in the coastal savannah zone of Ghana. Thesis, University of Cape Coast, Cape Coast, Ghana.

[42] Truong, V. and Avula, R. 2010. Sweet potato purees and powders for functional food ingredients. In R. C. Ray, Post-Harvest Aspects In Food. USA: Nova Science Publishers. 117-161pp.

[43] Tumwegamire, S., Kapinga, R., Mwanga, R., Niringiye, C., Lemaga, B. and Nsumba, J. 2007. Acceptability studies of orange –fleshed sweet potato verities in Uganda. In: proceeding of the 13th ISTRC Symposium, pp.807-813, Kampala, Uganda.

[44] Ünlüsayin, M., Erdilal, B., Gümüş, E. and Gülyavuz, H. 2010. The effects of different salting methods on extract loss from rainbow trout. *Pakistan Vet.Journal ;* 30(3), 131-134.

[45] USAID. 2011. Maternal, Infant, and Young Child Nutrition in Malawi,United States Agency for International Development (USAID),*Community Nutrition Workers.*

[46] Van Hal, M. 2000. Quality of sweet potato flour during processing and storage. *Food Review International;* 16: 1-37.

[47] Wijinker, J., Koop, G. and Lipman, L. 2006. Antimicrobial properties of salt (NaCl) used for the preservation of natural casings. *Journal ofFood Microbiology;* 23: 657-662.

[48] Woolfe, J. 1992. Sweet potato: an untapped food resource, Cambridge University. Press and the International Potato Center (CIP). Cambridge, UK.

[49] Yakubu, N., Amuzat, A. and Hamza, R. 2012. Effect of processing methods on the nutritional contents of bitter leaf (*Vernoniaamygdalina*).*American journal of food and nutrition*; 2(1): 26-30.

[50] Yusuf, A., Ayedun, H., Sanni, L. 2008. Chemical composition and functional properties of raw and roasted Nigerian benniseed (*Sesamumindicum*) and Bambara groundnut ((*Vigna subterranean*).*Food Chemistry*; 111: 277-282.

[51] Zakpaa, H, Al-Hassan, A, Adubofour, J. 2010. An investigation into the feasibility of production and characterization of starch from 'apantu' plantain (giant horn) grown in Ghana. *African Journal of Food Science;* 4(9): 571-577.

[52] Zhang, M., Tang, I., Mujumdar, A. and Wang, S. 2006. Trends in microwave-related drying of fruits and vegetables. *Trends in Food Science & Technology;* (17):524 - 534.

8

Constraints and Compliances of Traceability in Low Grown Orthodox Black Tea Manufacturing Process

ChandanaVindika Kumara Lokunarangodage*, Indira Wickramasinghe, Kamburawala Kankanamge Don Somathilaka Ranaweera

Department of Food Science and Technology, University of Sri Jayewardenepura, Nugegoda, Sri Lanka
*Corresponding author: vindikal@gmail.com

Abstract Traceability practices and their compliances in low grown orthodox black tea manufacturing process were examined, while proposing possible solutions for identified major drawbacks. The physical traceability in supply chain was considered one step forward and one step backward from the point of manufacturing, starting from auction/buyer back to supplier. Randomized stratified sampling was used. The traceability was evaluated using a checklist, end product sampling, open ended interviews, observations and internal document studies. The orthodox process was more complicated unlike other production processes due to the different separation techniques employed for grading and variety of grades produced because the sifting/grading was the key to number of different tea varieties. Major traceability issues were observed in leaf collection and grading operations due to complexity of separation through Myddleton, Chota, Michie and Winnower, which reduced the specific amounts produced, where bulking and blending process further extended complexity, while increasing the mixing of different made tea together with increased number of suppliers. Considering 1st, 2nd, 3rd dhool and big bulk with given separation techniques during grading; a single tea leaf could pass many paths before it end up in a specific product due to weight, size and shape of the leaf of a shoot based on the way it was rolled in orthodox rollers, where traceability up to tea bush, grading, blending and traceability of sample back to supplier was not fully complying. Nevertheless, supplier records, traceability after packing, traceability at dispatch and after dispatch were in full compliance, and other factors had varying degree of traceability compliances which make the compliances unachievable. Alternatively, if made tea is considered as bulk material, use of emerging technologies like Radio Frequency Identification (RFID) tags or/and DNA barcoding may be potential tools in rectifying such drawbacks and further research is needed to assess their efficacy in the field.

Keywords: *orthodox black tea, manufacturing, traceability, compliance, Myddleton shifter, supplier, grading, dhool*

1. Introduction

Tea was known as a beverage at least for over 5000 years, it was initially considered as a medicine and later grew into a beverage which finally became the most popular beverage in the world or the second most important drink after water [1]. According to the Chinese legends, tea was first discovered in the time of Second Emperor and herbalist, Shennong in 2737 B.C.E, accidently when a dried tea leaf was fallen into the boiling water which was intended for quenching emperor's thirst; who drank boil water as a habit, while he was travelling to another region.

The commercial planting of tea in Sri Lanka was introduced by a Scotsman, James Taylor in 1867 [2]. Since then the Ceylon Tea has been the world's number one brand for over a century and the local tea industry has made a significant growth over the time while securing its position in the global market as a leading producer and exporter of high quality black tea. In terms of international trade, tea is one of the major export revenue earners for the country, where thousands of lives are depending on it directly or indirectly. In 2007, Sri Lanka was the fourth-largest tea-producing country according to global production statistics and the country has produced 318,470 tons of tea which contributed nearly 9.1% of the world's total tea production [3].

Nevertheless, country has dedicated over 221,000 hectares or approximately 4% of the total land area for tea cultivation [4]. Accordingly, Sri Lankan tea plantations can be categorized as large plantations as well as smallholdings and there are approximately 118,274 hectares of smallholdings with 397,223smalhodings [5] out of the total land cultivated; around 43% is managed by the corporate sector, with a production of about 35%, while the balance of 57% is in the smallholder sector, with a production of 65% of the total. The average productivity in the smallholder sector is substantially higher, at about 1,853 kg/ha compared to the corporate sector productivity of 1,459 kg/ha [6]. The production grows at an annual rate of around 10% [7] while current average production amounted around 328 million kilograms in 2012 with majority of black tea. The annual production contributed from 14 tea growing districts of the country with figures of low grown 62%, mid-grown 16% and high grown 22% respectively, while 99% of the produce was black tea out of which 92% of the produce were accounted as orthodox black tea and 7% was accounted as CTC (cut, tear and curl) [5].

Considering current trends in export market, tea is moving worldwide as a healthy beverage which has advantages of availability but disadvantage of complex supply chains. Today global food competition is more intense where stakeholders started to adapt their mind sets toward a more holistic approach on supply chain management while focusing on food safety and traceability in a farm-to-fork perspective [8] because food industry has drastically changed during recent decades [9]. Nevertheless, many governments are improving food safety measures to safeguard their citizens by increasing control at all stages of food production, processing and distribution with hazard analysis critical control point (HACCP) and traceability based food safety management systems (FSMS) such as ISO 22000: 2005. According to Food and Agriculture Organization of the United Nations, traceability requirements are gaining significance in the global tea economy, with the potential for significant impacts on the world's tea producers, which is expected that the demand for traceability will result in widespread industry restructuring, which will require institutional responses within each of the major tea-producing regions [10]. The traceability is used as a tool to track product movement through manufacturing and distribution chain up to end user and backward to raw material supplier. Today consumer safety is the firm concern, which is a matter of cooperation between all actors involved in food supply chain rather than a confrontation; with the sharing of information, use of common standards and languages to pinpoint the food safety issues [11].

Concern over food safety is becoming central to supply chain restructuring in tea. Due to the costs, sampling and methodological constraints associated with the monitoring of finished food products, there is a trend towards preventing contamination at source through monitoring of estate and factory processes. This is associated with traceability based food safety management systems such as Good Agricultural Practices (GAP), Good Manufacturing Practices (GMP), ISO 22000 and HACCP certification. The second driver of traceability is consumer demands for independent verification that teas have been produced in the absence of abusive labour practices, by respecting worker rights and providing a living wage to smallholders, and by not polluting the environment or threatening biodiversity [10].

According to the ISO 9001:2000 standard, traceability is defined as "The ability to trace the history, application or location of product or service that is under consideration"[12], which is a process that makes it possible to find the traces of the various steps (manufacturing steps, sources of its raw materials with relevant suppliers, controls and tests carried out) and locations (storage locations, equipments used to manufacture or handle it, direct customers and end users) a product has passed through from its creation through to its final disposal [11]. Traceability requirements in food supply chain due to consumer concerns were started to rise in the later part of nineteenth century with the discoveries of microbes as well as vitamins which were further increased due to the development of supply chains followed by the improved production methods [13]. The product traceability is the ability to follow the food movement through specified stages of production, processing and distribution in one step forward and backward at any given place of the food supply chain, while facilitating efficient product recall in the event of hazard left the production facility. As a result of global, integrated partner oriented approach throughout food supply chain, every operator must identify their product in a unique way while recording their destinations with the links between incoming and outgoing products as well as end users on databases [11].

Further, product traceability helps to determine the origin of a food safety problem and to comply with legal requirements while meeting consumers' expectations for the safety and quality of purchased products [14]. The aim of all these measures is to safeguard consumers from biological, chemical and physical hazards that may be present in food while implementing full traceability in food supply chains. According to the Kelepouris et al. [15], Morrison [16], Van Dorp [17] Viaene&Verbeke [18] full traceability or traceability throughout entire supply chain is essential for ensuring food safety and quality which was mainstream requirement to regain or maintain consumer confidence, where traceability cannot improve food safety or quality itself although it can provide necessary information and keep track of product movements [19].

Considering the complex nature of tea supply chain as well as the complex manufacturing operations involved in the tea industry, it is very difficult to identify the movement of a given supplier's raw material throughout the production process at a single glance. It was rarely discussed subject in many researches, where black tea's health benefits, chemical structures, disease prevention etc., has been extensively studied by many researchers. The existing manufacturing processes are also very old when compared to most of the modern food industries where traceability was not a major concern over a century. Hence the propose of this study was to find out the major drawbacks in the area of traceability by analyzing factors affecting the traceability process in orthodox black tea manufacturing and to find out possible solutions to the issues identified in the study.

2. Materials & Methods

The project designed to evaluate the problem areas of the tea manufacturing process related to the food safety applications and then to find out reliable solutions with minimizing or eliminating the existing complications through Pareto principle. The sampling plan was randomized stratified sampling, with the use of Factory Information.xls provided by the Sri Lanka Tea Board which was used to select tea factories based on the availability or non-availability of food safety or quality assurance certifications from the low grown orthodox black tea manufacturing industries. The most prominent area was the Southern province while Galle, Matara and Rathnapura districts were set as major target areas for the project execution due to the fact that most of the orthodox black tea factories were located in these areas and the easy access as well as shorter distances between factories, where concentration of factories were high. The project has two phases for execution and sample size for each phase was 30 factories where phase II was focused mostly on the factories which have ISO 22000/ISO 9001/HACCP systems or Good Manufacturing Practices with Japanese 5S implementations. However, there are number of tea factories which have abandoned the implemented food safety management systems due to various reasons, where additional priority was given to understand the problems they faced, that led to abandoning as well as to find out the remaining practices and their efficacy.

The first stage was completed with gap analysis to evaluate the basic major issues which has more impact on food safety and the product quality with the evaluation of 25 tea factories that have not implemented ISO 22000/HACCP food safety management system (FSMS). The second stage was designed based on the output of the first stage audits, where another set of 20 factories were analyzed and which have ISO 22000 FSMS/HACCP or previously implemented but currently abandoned. The audits had two dimensions where first major audit was targeted to identify the issues prevailing in the food safety management systems and their supporting documents while second objective was to find out traceability of the manufacturing process as a part of generic model development. The internal audits were designed based on ISO 22000 food safety management system requirements and root cause analysis carried out at the gap analysis stage which pointed out the issues in industry as well as strategic approaches to the problems identified while traceability was a major concern from the gap analysis.

In addition to the gap analysis and internal audits conducted in stage I and stage II, separate traceability study was carried out in parallel to the stage I and II to understand the traceability applications and practices in tea industry using over 40 tea factories with or without ISO 22000 or HACCP FSMS or ISO 9001 QMS. The selected factories were evaluated based on the visits made in parallel to gap analysis carried out at phase I and internal audits conducted in phase II. Required data was collected through standard check list followed by open ended discussions. A traceability exercise was conducted for a selected made tea sample, practically trying to trace back green leaf supplier for selected sample with the help of available documents/records and employees working on relevant processing areas.

The two testing criteria were common for all the factories evaluated and special attention was given to the factories which were currently practicing ISO 22000/ISO 9001/HACCP systems or previously followed even though the certifications were expired. The check list was prepared considering the various standard requirements of traceability.

The results of the study were analyzed using MINITAB 17 Version and 1-Propotion test was applied for the results at 0.05 significant levels. Hypotheses were built based on the proportion of factories which were complied with traceability factors and results were separately analyzed for every factor.

3. Results and Discussion

According to the evaluations carried out, over 45 factories were observed and complete data was observed on 40 factories with or without ISO 22000, HACCP or ISO 9001 certifications. The traceability was evaluated using a check list, end product sampling, open ended interviews, observations, and internal document studies in all the tea factories. As to the results, traceability was found in place up to a certain extent in all the tea manufacturing processes from one step forward. It was mostly limited to the made tea up to auctioneer or the wholesale buyer after completion of the manufacturing process, where made tea could be traced from manufacturer through product coding, manufacturing date, brand, and name of the manufacturer up to auction and through auction records purchaser can be located.

According to the CBI, the consumer markets were dominated by the popular blended brands which can contain up to 36 different types of tea varieties that were blended in the consuming country to ensure the unique taste of their brand at a competitive price where tea buyers source different teas from around the world mainly China, India, Kenya and Sri Lanka [1]. Accordingly, buyers blend tea after purchasing according to their brand's requirements depending on the market and cost. Therefore, the process was complex and the traceability will lead to a bunch of manufacturers rather than a single manufacturer.

Thus rest of the traceability was not studied because most of the purchases were export oriented and it was beyond the research objectives. However, that can be traced up to the consumer since no processing was involved other than blending, repacking or value addition according to the consumer preferences. Thus traceability of a product complains can be elaborate as consumer claim to retailer, who claim to wholesaler and who will then inform the buyer through auctioneer and the manufacturer. If there is a product recall, the same channel can be used vice versa.

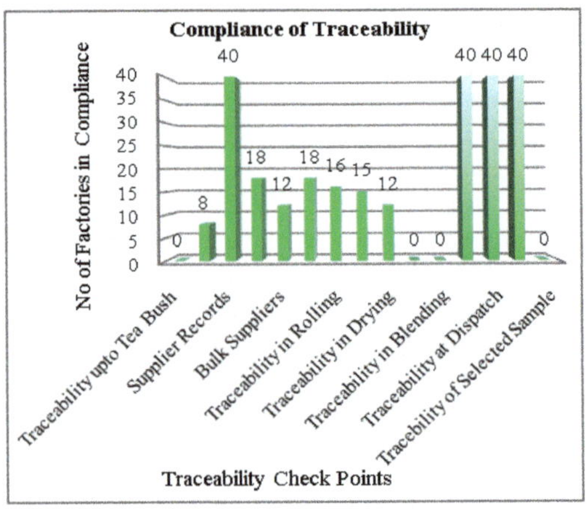

Figure 1. Constraints and Compliances of Traceability in Low Grown Orthodox Black Tea Supply Chain

The results of the study was given in the bar chart above (Figure 1), which illustrate the traceability practices as well as the achievements in the factory flow level. Further to the results obtained traceability up to tea bush, traceability in grading, traceability in blending and traceability of sample back to the supplier wasn't fully complying according the study sample ($p < 0.05$). On the other hand, supplier records, traceability after packing, traceability at dispatch and traceability after dispatch was fully traceable or all the tea factories were in compliance with these four factors ($p = 1$) . As a rule of thumb, other factors had varying degree of traceability compliances, which indicated that those areas were more or less neglected but it was not impossible. The Figure 2 below shows the probability of implementation and it further prove that what were already implemented and what was not achieved due to explained reasons or neglected according to the representative sample.

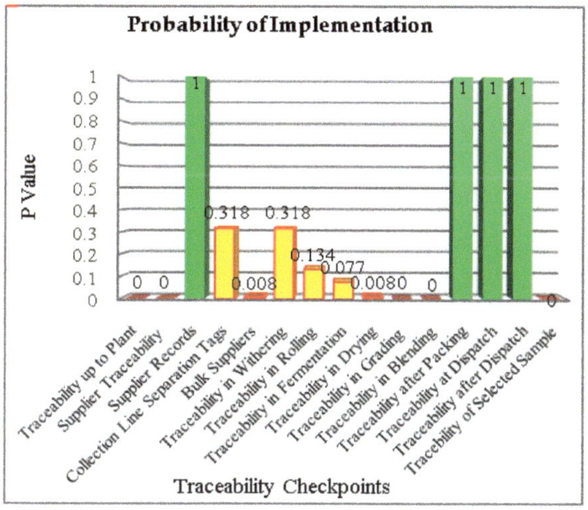

Figure 2. Implementation Probability of Traceability in Low Grown Orthodox Black Tea Supply Chain

When considering traceability in one step backward, none of the brought leaf based tea factories were able to exactly locate the farmer or the field where green leaves were harvested. Large estates were able to locate the field of the harvesting carried out, but they also unable to locate the records of exact tea bush or the labourer who harvested a given green leaves which mostly process in their own production facility. On the other hand, smallholder suppliers were more critical because their leaves were mixed each other while collecting and transport, where traceability was only possible up to a bunch of suppliers.

Bearing in mind the traceability up to tea bush which was the primary objective, a quantity of 4.50 – 4.65 kg of green leaves were required to manufacture 1kg on made tea (final product) [20]. 1kg of green leaves were to be harvested from 20 - 30 tea bushes with average yield where it is extremely difficult or almost impossible to practically locate the exact tea bush in the case of a product complain or hazard occurrence. The most possible paper based system could reach the leaf collectors or bunch of suppliers approximately, but it was impossible to locate the exact supplier or tea bush who contributes to the issue. Thus one-step backward traceability in orthodox black tea manufacturing was rarely achieved due to following reasons in smallholder suppliers. Most of the suppliers were smallholders who had less than one hectare of lands, which produce less than a single roll charge in a one supply of green leaves. It was usually varying around 3 – 15 suppliers at a time. Tea cultivation requires a relatively small investment, and the risk of complete crop failure is low. Problematic issues for smallholders producing bulk tea products include low farm gate prices, poor extension services and limited marketing channels [1]. As a result, their tea leaves were mostly collected through leaf collectors. According to keleporis et al, successful implementation of supply chain traceability calls for co-operation among all the stakeholders of the supply chain [15]. According to the interviews conducted, leaf collectors were not educated enough by manufacturers to keep traceability records. On the other hand, cost of additional labour and transport was high, where leaf collectors were operating at little margins. Nevertheless, leaf collectors as well as factory employees also believe that tea leaves are a uniform and homogenous raw material; origin in similar fields which cannot be distinguished each other very easily. Thus it was very difficult to motivate them to practice traceability in supply chain, where extreme competition for leaves in the market has the bargaining power for them to change the manufacturer at any time without notice.

After withering during release to the rolling process, small quantities were bulked or large quantities segregate to make a single batch of size about 250kg depending on the dryer capacity, which was call the charge, [21] and this activity increased the mixing of direct smallholder suppliers as well as leaf collectors. According to the study sample (as shown in the Figure 1 – traceability compliance), 18 factories maintain collector bulks with identification tags ($p > 0.05$), thus the leaf could be traced with time records backward to leaf collectors as well as bunch of smallholder suppliers at a given date but it was impossible to trace up to the exact supplier or the tea bush. The same leaves could be traced forward through rolling, fermentation, drying up to fiber removal, easily with time records when the production was arranged in such a way.

Typically, orthodox black tea manufacturing process was much more complicated unlike other production processes (e.g. CTC) due to the different separation techniques employed as well as the number of different varieties produced. Hence, most critical part of the tea traceability systems was to trace the product within the manufacturing process. The sifting/grading was the key to number of different tea varieties and ungraded dry tea leaves initially pass through Myddleton after fiber removal. According to Samaraweera, the Myddleton bubble tray stalk extractor has an oscillation motion of 200 to 220 oscillations per minute of two aluminum bubble trays with perforations where bubbles impart a string action at the bottom of the layer which makes long particles such as fibers or stalk to float. It helps small and round particles to settle at the bottom of the layer while top particles to fall through perforations. To achieve this action, it must be feed with a continuous thick layer without exposing the bubbles, which will sort the stalky long leaf portion by passing through a tray that is used to separate long leaf particles and stalks at the beginning of the sorting process [21].

Thereafter, Chota and Michie shifters were used to separate uniform sizes and then to different varieties. Michie oscillatory shifters were equipped with one or two trays stacked one on top the other with a slight slopes towards forward direction that gives oscillatory motion on the vertical plane which help particles to jump forward at 250 to 260 oscillations per minute. Michichi shifter was used to separate long and wiry particles which were basically used to produce long leaf and wiry grades [21]. After separations, semi graded tea was then passed through a winnower to separate according to the weight. The winnower is used to separate light tea particles such as flakes, fibers, tea fluff and extraneous matters such as sand or dirt [21].

However, this process was very complicated and very difficult to understand easily for an average person. Considering 1st dhool, 2nd& 3rd dhool as well as the big bulk, given generic process flowcharts (Figure 3, Figure 4 and Figure 5 below) were drawn where it was obvious that single tea leaf could pass many paths before it end up in a specific product due to the weight, size and the shape of the leaf of a shoot based on the way it was rolled in the orthodox rollers.

On the other hand, consider the same leaf if it broken to small sizes and large sizes they normally end up in two different grades. The same shoot's tender leaves will mostly end up in first grades while matured low weight leaves will mostly end up in second grades based on the weights. Considering the complexities involved, the Figure 3, Figure 4, and Figure 5 shows the number of different grades produced in a given dhool. It was almost 32 for a first dhool (Figure 3) considering the generic processes without 2nd or 3rd grade products where their sizes were different from one another as well as different parts of the tea shoot.

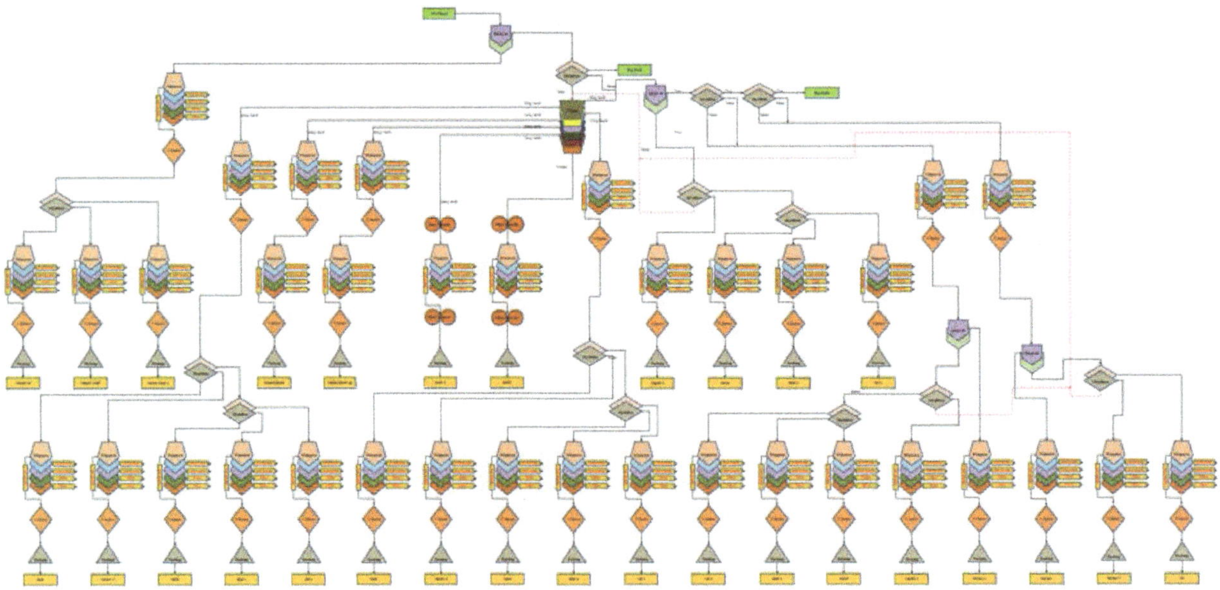

Figure 3. Graphical representation ofgeneric process flow chart for the 1stDhoolsifting program of orthodox black tea

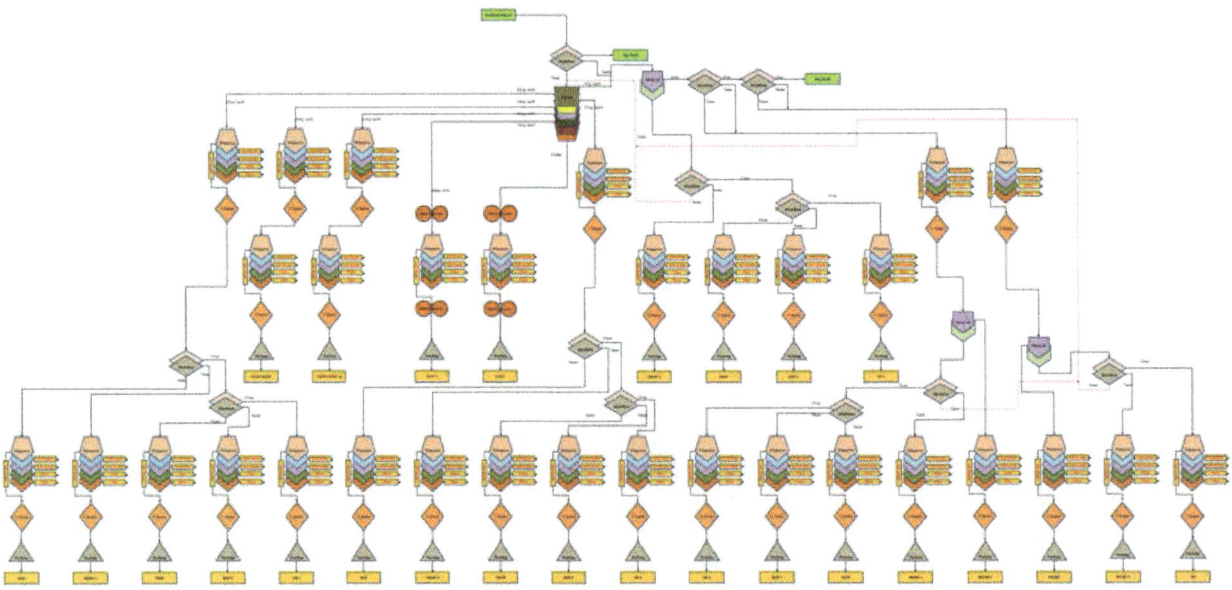

Figure 4. Graphical representation ofgeneric process flow chart for the 2nd& 3rdDhool sifting program of orthodox black tea

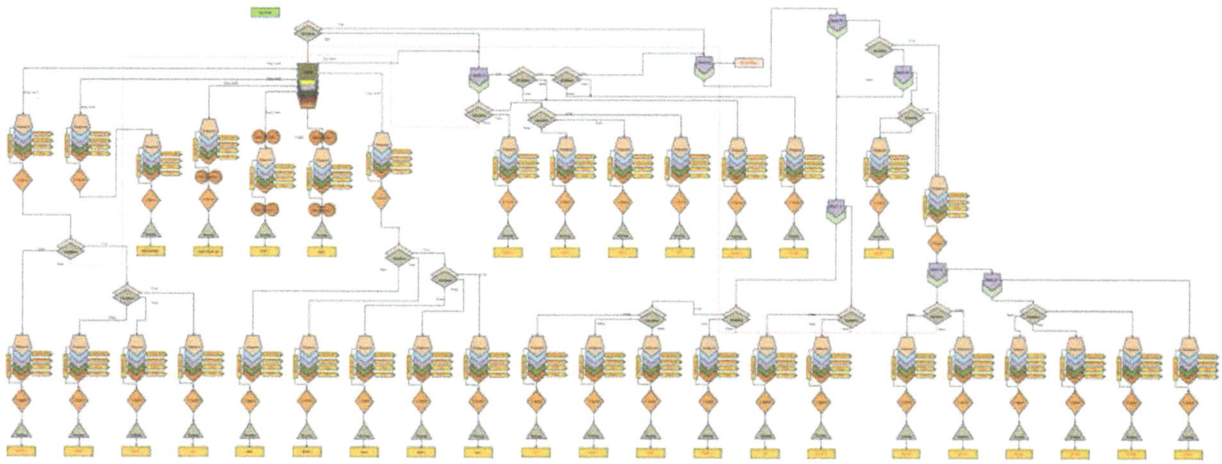

Figure 5. Graphical representation of generic process flow chart for the Big Bulk sifting program of orthodox black tea

This phenomenon was almost same for all the dhools where it wasn't a very easy process to keep records of the product movement. Nevertheless, the second grades were much less discussed in the orthodox black tea manufacturing process but those amounts were also massive and their movements were never recorded in the industry on accurate manner, where the existing traceability practices were not complete. The existing systems had introduced various documents to keep traceability tracks on the product movements, but none of them were able to comply with the theoretical traceability requirements according to the definition of traceability[12].

According to the given process flow diagrams below, one tea leaf might pass many times through Myddleton, Michie and Winnower as well as the Colour Sorter before it reached the final grade. Some leaves may recycle the same path again and again with mixing new coming ungraded teas such as throughput of 5mm Myddlton which go back again to Chota (red line – Figure 3, Figure 4 and Figure 5) and passing through same sieves before end up in a final grade. On the other hand, each time the product went through winnower, based on the weight; it separates the product in to 1st grade, 2nd grade, 3rd grade and fiber while removing stones and other extraneous particles and reducing the bulk into small quantities. 1st grade was proceed with next steps of manufacturing process, while 2nd and 3rd grades were followed separate pathways which generates second quality tea products which were usually called second selling mark products. Considering 1st dhool, 2nd dhool, 3rd dhool and big bulk, when each of these process flowcharts were equally complicated and largely distributed, it was very difficult to follow a paper based or even a digital method to track down the traceability of a specific tea leaf.

Due to the continuous feeding and requirement of uniform thickness in Middleton shifter; batches couldn't be separated accurately rather than guessing. Thus there must be number of documents and complex storing systems to identify different batches as well as batch wise separation techniques which will add more cost to the production. While dried tea was passing through shifters within grading room operations, the same bulk was separated into small different grades based on leaf size, type, dhool, weight and buyer requirements where orthodox black tea has around 30 different grades basically which further creates issues in reaching the initial leaf supplier.

There are various other processes involved, but these processes were the major critical points for the traceability issues in orthodox black tea manufacturing unlike the cut, tear and curl (CTC) process where it is easily traceable due to non-complex process which give less varieties as well as large quantities. In orthodox process, the output of a single grade was small where several batches were required to prepare a standard quantity for packing or as to buyer requirements, thus bulking is needed.

Figure 6. A Typical Blending Operation

The blending (Figure 6) was carried out to give specific product characteristics to a given tea product where different grades are mixed together. Hence this process further increase the traceability issues while increasing the number of raw material suppliers as well as individual tea plants contributing to a given tea product. Nevertheless, these areas still can be traceable with a proper organization of production flow and better recording systems which was not available in any of the systems evaluated. The cost is one of the major factors which decide the implementation of such measures, because even though the factories are fully dedicated and implemented such systems, the buyers do not pay attention to such efforts while they were involved in purchasing. Thus it is inappropriate to request complete treatability from manufacturers without providing solutions to the areas mentioned in the study.

On the other hand, if orthodox black tea manufacturing was considered as a bulk material, the traceability can be achieved with less hazardous work load where bunch of suppliers or fields were the primary objective. Then paper based traceability systems can be developed to trace single day's production which can be more easily segregated while storing and recording where it can be traced through grading and blending up to packing. Nevertheless, it can be trace one step backward practically up to bunch of suppliers through existing records while adding few new records within grading operation from drier out to finished product packing. Rest of the areas of manufacturing namely receiving, withering, rolling, fermenting and drying can be managed through addition of time records or codes to the existing food safety management systems.

Finally, if the cost is bearable to the consumer, DNA barcoding is a widely used molecular-based system, which can identify biological specimens, and is used for the identification of both raw materials and processed food, which can be possible to employ to identify exact issue concerning samples [22]. On the other hand, Radio Frequency Identification (RFID) tags also can be integrated to track down the raw material movements as well as finish product by using the passive RFID tags with the bulk storing bins from field to the factory and then after drier moth to the packing since raw material and finish product both moved within plastic crate which can be continually used with a reader. The passive RFID tags are designed to depend on the tag reader as its power source where reader unit can communicate with the tag within 6 meters which are low in product cost and it is manufactured for disposal with disposable consumer products [23]. Considering current developments in RIFD tags technology, it possibly can be used in the complicated operations in orthodox back tea manufacturing process to track down intermediate product movement within the factory environment where the production should plan according to the requirements while synergizing with existing paper based systems to make it workable. However given model will be able to identify the traceability on the bulk movement where individual supplier and the end user can be located. But these technologies are not currently applied in tea trade where it will not guarantee the accurate identification of exact tea bush in an event of hazard occurrence which needs further research to explore the possibilities of exact application of the given technologies.

Accordingly, the research recommends considering tea as a bulk material like edible oils in the event of processing where it is really practical to consider traceability after made tea left the factory and if there is any food safety issue manufacturer has to address all the suppliers in the supply chain and check their plantations for identification of abnormalities, if necessary. If the made tea is considered as a bulk material, it will be helpful to develop user friendly, realistic and practical paper based traceability models which will not increase the product cost or unnecessary documentation works for the quality assurance system operators.

4. Conclusion

As to the results, traceability was found in place up to a certain extent in all the tea manufacturing processes from one step forward. The research was limited to the made tea up to auctioneer or the wholesale buyer after completion of the manufacturing process, where made tea can be traced from manufacturer through product coding, manufacturing date, brand, and name of the manufacturer up to auction and through auction records, purchaser can be located.

When considering traceability in one step backward, none of the brought leaf tea factories were able to exactly locate the farmer or the field where green leaves were harvested. Large estates were able to locate the field of the harvesting carried out, but they also unable to locate the records of exact plant or the labourer harvested a given green leaves which mostly process in their own production facility. On the other hand, smallholder suppliers were more critical because their leaves are mix each other while collecting and transport where only possible up to a bunch of suppliers. Bearing in mind the traceability up to tea bush which was the primary objective, a quantity of 4.50 – 4.65kg of tea leaves were required to manufacture 1kg on made tea. 1kg of green leaves were to be harvested from 20 - 30 tea bushes with average yield where it was extremely difficult or almost impossible to locate the exact tea bush in the case of a product complain or hazard occurrence. The most possible paper based system could reach the leaf collectors or bunch of suppliers approximately, but it was impossible to locate the exact supplier who contributes to the issue or tea bush.

Typically, orthodox black tea manufacturing process was much more complicated unlike other production processes due to the different separation techniques employed as well as the number of different grades produced. Thus most critical part of the tea traceability systems was to trace the product within the manufacturing process. The sifting/grading was the key to number of different tea varieties. Considering 1st dhool, 2nd dhool, 3rd dhool and big bulk, when each of these process flowcharts are equally complex and largely distributed, it is very difficult to follow a paper based or even a digital method to track down the traceability of a specific tea leaf. According to the given results and the observations it was obvious that orthodox black tea production process was very complicated, where it was suitable to consider as a bulk material in the event of traceability which help to develop user friendly traceability systems in orthodox black tea manufacturing process. On the other hand, use of emerging technologies like Radio Frequency Identification (RFID) tags or/and DNA barcoding may be potential tools in rectifying such drawbacks and further research is needed to assess their efficacy in the field.

Considering all these factors, it was obvious that tea traceability practices were very complicated to the manufacturer and product was not priced considering these critical factors. Thus requesting traceability from the consumer's tea cup back to the tea bush that supplied green leaves to produce the given made tea sample was not justifiable as well as very difficult in the manufacturing process. However, educating all the stake holders may help to improve the traceability practices up to certain extent, but if the manufacturer can increase the hygienic practices to improve the food safety of the product, while educating all the stakeholders in supply chain that will be more meaningful than considering the traceability as a critical buying factor due to the complex processing nature of the product.

References

[1] CBI Market Channels and Segments for Tea 'Your trade route through the European market' http://www.cbi.eu/system/files/marketintel_platforms/2013_market_channels_and_segments_tea_-_coffee_tea_and_cocoa.pdf.

[2] Elliott, E. C. and Whitehead F.J: *Tea planting in Ceylon*. 1928.

[3] Top 10 Tea Producing Countries. Available: http://www.toptenofcity.com/commerce/top-10-largest-tea-producing-countries.html. [Accessed 4th April 2014].

[4] Department of Census and Statistics, 2005 Census of Tea Small Holdings. Sri Lanka

[5] Karunaratne S., "*Statistical Information on Plantation Crops - 2012*" Ministry of Plantation Industries, Colombo, November 2013, p 1-68.

[6] Munasinghe M., Deraniyagala Y and Dasanayake N.: *SCI Research Study Report. : Economic, Social and Environmental Impacts and Overall Sustainability of the Tea Manufacturing Industry in Sri Lanka*. May 2013. Available: http://www.sci.manchester.ac.uk/sites/default/files/Economic,%20Social%20and%20Environmental%20Impacts%20and%20Overall%20Sustainability%20of%20the%20Tea%20Manufacturing%20Industry%20in%20Sri%20Lanka_0.pdf.

[7] Department of Census and Statistics, 2012.

[8] Helena L., Christina S. and Annika O.: Traceability in Food Supply Chain: Towards Synchronized Supply Chain. : *Lund University Publications*, 2008 September. Available at http://lup.lub.lu.se/record/1165966.

[9] Stadig, M., Breg, B., Bergström, B., Janson, C.-G., Karlsson, R., Wiik, L., &Johnsson, M. Spårbarhet i Livsmedelskedjan, SIK, SIK-Dokument 161. 2002.

[10] FAO Traceability, Supply Chains and Smallholders: Case-Studies from India and Indonesia. *Committee on Commodity Problems. Intergovernmental Group on Tea* Seventeenth Session. 2006. p 1-21.

[11] Hand Book 7.1:*Traceability and Labelling*, Available at: http://pip.coleacp.org/files/documents/edes/publications/EDES%20fascicule%207-1_EN_web.pdf. [Accessed 04th January 2014].

[12] ISO 9001:2000 standard.

[13] Popper, D.E., "Traceability: Tracking and privacy in the food system", *Journal of Geographical Review*, Vol. 97, No.3, 2007. p. 365-389.

[14] UNEP. "Green Economy and Trade": Green Economy Report's Agriculture chapter 2. Available: http://www.unep.org/greeneconomy/Portals/88/GETReport/pdf/Chapitre%202%20Agriculture.pdf, 2013.p 45-88.

[15] Kelepouris, T., Pramatari, K., &Doukidis, G. "RFID-enabled traceability in the food supply chain", *Industrial Management + Data Systems*, vol. 107, no. 2, 2007. p. 183.

[16] Morrison, C. "Traceability in food processing: an introduction," in Food authenticity and traceability, M. Lees, ed., Woodheadpublishing limited, Cambridge, 2003. p. 459-471.

[17] Van Dorp, K.-J. "Beef labeling: The emergence of transparency", *Supply Chain Management*, vol. 8, no. 1, 2003. p. 32-40.

[18] Viaene, J. &Verbeke, W. "Traceability as a key instrument towards supply chain and quality management in the Belgian poultry meat chain", *Supply Chain Management*, vol. 3, no. 3, 1998 p. 139.

[19] Wang, X. & Li, D. "Value Added on Food Traceability: a Supply Chain Management Approach", *2006 IEEE International Conference on Service Operations and Logistics, and Informatics*2006. p. 493-498.

[20] Institute of Social Developments: *SustainabiliTea*, Report on Sri Lankan Tea Industry. Available: somo.nl/publications-en/Publication_3095/at_download/fullfile, January 2008.

[21] Samaraweera D. S. A., Chapter 20.: Technology of Tea Processing. : "Hand Book on Tea", 2008. p 265-322,.

[22] Galimberti A., Mattia F D., Losa A., Bruni I., Federici S., Casiraghi M., Martellos S, Labra M.: "DNA barcoding as a new tool for food traceability".: *Food Research International* 50 (2013) 55-63.

[23] Kevin Bonsor and Wesley Fenlon, How RFID Works: Available: http://electronics.howstuffworks.com/gadgets/high-tech-gadgets/rfid3.htm. [Accessed 25th June 2015].

Nutritional and Functional Characterization of Undecorticated Groundnut (*Arachis hypogaea* L.) Seeds from Bosso Market, Minna, Nigeria

Mustapha S.[1,*], Mohammed U.M.[2], Adeosun N.O.[3], Mathew T.J.[1], Muhammed S.S.[4], Ibn-Aliyu A.[1]

[1]Department of Chemistry, Federal University of Technology, Minna, Nigeria
[2]Department of Chemistry, Federal College of Education, Kontagora, Niger State, Nigeria
[3]Department of Chemistry, Federal College of Education (Technical), Asaba, Nigeria
[4]Department of Biochemistry, Federal University of Technology, Minna, Nigeria
*Corresponding author: saheedmustapha09@gmail.com

Abstract The undecorticated groundnut (*Arachis hypogaea* L.) seeds were analyzed for proximate compositions, some valuable minerals and functional properties using standard method of analysis. The results showed the seeds samples are within the range of 40.10 to 42.13 % fat, 24.48 to 26.37 % crude protein, 4.24 to 6.30 % carbohydrate, 3.02 to 5.08 % moisture, 17.32 to 22.70 % crude fibre and 1.51 to 2.33 % ash. Minerals content include; sodium, calcium, phosphorus, potassium, manganese, zinc, iron and copper were found to be present in the seeds flours. The dominant macro mineral element (potassium) could be recommended as source of dietary supplement. Functional properties ranged from 198.45 to 201.70 % water absorption capacity, 170.30 to 190.50 % foaming absorption capacity, 25.20 to 31.50 % emulsion capacity, 9.35 to 13.24 % emulsion stability, 3.50 to 4.50 % foaming capacity and 0.62 to 0.80 g/cm^3 bulk density. However, the raw groundnut seed flour had higher values of functional properties than the toasted and sundried groundnut seed flours. The high emulsion stability and activity of the seed flours signify their potential applications in food systems.

Keywords: *groundnut, nutritional properties, functional properties, proximate compositions, minerals*

1. Introduction

Groundnut, or peanut (*Arachis hypogaea* L.), is a species in the legume family Fabaceae, native to Mexico, South and Central America [1]. In Nigeria, among the leading producing states of groundnut is Niger state [2]. In this Northern part of Nigeria, groundnuts are processed into various products which include groundnut paste popularly known as peanut butter, groundnut cake, salted groundnut and groundnut soup. Groundnut has contributed extremely to the development of the Nigerian economy through the sales of seeds, cakes, oil and haulms [3]. A substantial part is eaten in the cooked form, while fewer people indulge in eating the seeds raw [4]. Therefore, good nutrition is very essential to humans. Healthy population promotes development of the people through the relation between food, nutrition and health which needs to be reinforced in order to satisfy the needs of the increasing population [5].

However, the seeds serve as an important diet in many countries. Owing to its richness in high quality dietary protein and oil it helped to reduce malnutrition in the developing Countries [6] and help to supplement the nutrients of the basic carbohydrate food of the needy who cannot afford enough protein food of animal origin Achu *et al.* [5]. Apart from this, cake made from groundnut could be used in infant food formulations. Groundnut provides considerable amounts of mineral elements like phosphorous, calcium, magnesium and potassium and vitamin E, K, and B group to supplement the dietary requirements of humans and farm animals [7]. Groundnut contained 44 to 56% oil, 22 to 30% protein, and 9.5 to 19.0% total carbohydrates as both soluble and insoluble carbohydrate [8,9,10,11]. According to Aremu *et al.* [12] seed proteins possess the essential functional properties for their successful utilization of some food products. Functional properties are essential physicochemical characteristics which affect the behaviour of proteins in food systems during manufacturing, storage and preparation. Some of these properties can be as a result of the thermodynamically favorable protein water interactions such as wettability, swelling, water retention, solubility, foaming and emulsification [13]. In previous study as reported by Yu *et al.* [14] the functional properties of defatted groundnut flour such as emulsification, bulk density, viscosity, and water and oil absorption were essential in food processing and formulation of food product.

A survey of the literature showed that research had been done on the proximate and mineral analysis of groundnut seeds, but there is little or no information as regards the chemical composition and functional properties of groundnut from a market in Minna, Niger state. It is on this basis that this research was conducted so as to establish the base line data in the chemical compositions, minerals and functional properties of groundnut seeds in this region.

2. Materials and Methods

2.1. Sample Collection and Treatment

Groundnuts (*Arachis hypogaea* L.) seeds were purchased from Mobile markets, Minna, Nigeria and were transported with polythene bag to the laboratory. The seeds were undecorticated, pre-treated by washing with distilled water and then divided into three portions: the first part was air dried, second part sun dried and the third part was toasted in an oven for 3 hours at 105°C. The toasted seeds and the other two portions of sample were grounded into a paste to pass through mesh sieve.

2.2. Proximate Analysis

The standard AOAC [15] methods were used to determine proximate composition of the groundnut samples.

2.3. Analysis of Mineral Elements

The wet-ashing method was employed for the digestion of the seed sample; 1 g of sample was digested with 20 cm^3 of concentrated HNO_3 and perchloric acid (1:1) and thereafter transferred to a 100 cm^3 volumetri flask. It was diluted to volume with distilled water and stored in sample bottle. The mineral element contents were determined using an atomic absorption spectrophotometer (Perkin–Elmer model 703) and a flame photometer as described by Onyeike and Acheru [16].

2.4. Functional Properties

2.4.1. Bulk Density

The weighed sample was transferred into 10 cm^3 graduated. The bottom of the cylinder was gently tapped on a laboratory bench for several times until no further diminution of the sample level was observed after it was filled up to the calibrated mark. Bulk density is expressed as:

$$\text{Bulk density}(g/cm^3) = \frac{\text{Weight of the sample(g)}}{\text{Volume of the sample}(cm^3)}$$

2.4.2. Emulsifying Properties

Emulsifying capacity (EC) was determined according to the method of described by Pedroche *et al.* [17] with little modification. 1.0 g of each sample was homogenized at a speed of 160 rpm for 1 min at room temperature in 25 cm^3 distilled water. The protein solution was mixed with 25 cm^3 of vegetable oil followed by homogenization at 160 rpm for 1 min and finally, the emulsion was centrifuged for 5 min. Emulsifying activity was determined by:

$$\text{Emulsifying capacity}(\%) = \frac{\text{Height of emulsion layer}}{\text{Height of the content in the tube}} \times 100$$

Emulsion stability (ES) was measured by re-centrifugation followed by heating at 80°C for 30 min. Emulsion stability is expressed as follow:

$$\text{Emusifying stability}(\%) = \frac{\text{Height of remaining emulsion layer}}{\text{Height of original emulsified layer}} \times 100$$

2.4.3. Foaming Capacity (FC) and Foam Stability (FS)

Foaming capacity (FC) and stability (FS) were determined by the method described by Sze-Tao and Sathe [18] with certain modification. 1 g of each sample was isolated in 50 cm^3 of distilled water. The solution was stirred at a speed of 160 rpm for 10 min. The blend was immediately transferred into a 100 cm^3 graduated cylinder. The volumes were recorded before and after stirring.

$$\text{Foam capacity}(\%) = \frac{\begin{pmatrix}\text{Volume after whipping} \\ -\text{Volume before whipping}\end{pmatrix}\text{ml}}{(\text{Volume before whipping})\text{ml}} \times 100$$

For the determination of foam stability, foam volume changes in the graduated cylinder were recorded at 25 min of storage.

$$\text{Foam stability}(\%) = \frac{\begin{pmatrix}\text{Volume after standing} \\ -\text{Volume before whipping}\end{pmatrix}\text{ml}}{(\text{Volume before whipping})\text{ml}} \times 100$$

2.4.4. Fat Absorption Capacity (FAC)

Fat absorption capacity was determined using the method described by Lin and Zayas [19] with minor modifications. 1 g of each sample was weighed into 20 cm^3 pre-weighed centrifuge tube and thoroughly mixed with 5 cm^3 vegetable oil. The emulsion was incubated at room temperature for 25 min and then centrifuged for 15 min and the supernatant was carefully removed, and the tube was reweighed. FAC was determined by:

$$\text{Foam absorption capacity} = \frac{F_2 - F_1}{F_0}$$

Where F_o is the weight of the sample (in gram), F_1 is the weight of the tube plus the sample (in gram), and F_2 is the weight of the tube plus the sediment (in gram).

2.4.5. Water Absorption Capacity (WAC)

The modified procedure as described by Rodriguez-Ambriz *et al.* [20] was used to determine the water absorption capacity. 1 g of each sample was weighed into 20 cm^3 pre-weighed centrifuge tube. 10 cm^3 of distilled water was added in to the tube under continuous stirring with a glass rod at room temperature for 25 min. The tube was centrifuged for 20 min and the amount of added distilled water resulting in the supernatant liquid in the test tube was recorded. WAC expressed as grams of water per gram of sample, was calculated by:

$$\text{Water absorption capacity} = \frac{W_2 - W_1}{W_0}$$

Where W_o is the weight of the sample (in gram), W_1 is the weight of the tube plus the sample (in gram), and W_2 is the weight of the tube plus the sediment (in gram).

2.5. Statistical Analysis

Results are expressed as the means and standard deviations of three separate contents. The data were statistically analyzed by 2-way analysis of variance (ANOVA). Standard deviations were compared by Duncan's multiple range tests at 5% level of significance ($P \leq 0.05$).

3. Results and Discussion

3.1. Proximate Composition of Undecorticated Groundnut Seed

Results of the proximate compositions of the groundnut seeds are shown in Table 1.

Table 1. Proximate Composition (%) of Undecorticated Groundnut Seed

Parameter	Raw	CV	Toasted	CV	Sundried	CV
Ash	1.51±0.01[a]	0.66	2.33±0.01[c]	0.43	2.03±0.01[b]	0.43
Moisture	5.08±0.02[b]	0.39	3.02±0.03[a]	0.99	3.05±0.01[a]	0.33
Crude Protein	26.37±0.03[c]	0.11	24.48±0.02[a]	0.08	25.32±0.02[b]	0.08
Crude Fiber	22.70±0.02[c]	0.09	17.32±0.01[a]	0.06	22.06±0.01[b]	0.05
Fat	40.10±0.01[a]	0.03	42.13±0.02[c]	0.05	41.21±0.02[b]	0.05
Carbohydrate	4.24±0.03[a]	0.71	9.16±0.02[c]	0.22	6.30±0.03[b]	0.48
Fatty acid	32.08		33.70		32.95	
Energy(KJ/100g)	1936.27		1984.76		1960.82	

Values are means ± standard deviation of duplicate determination. Calculated fatty acid = 0.8 × crude fat. Calculated soluble energy (KJ/100)= protein × 17 + fat × 37 + carbohydrate.Coefficient of variation (CV)% = $\frac{\text{Standard deviation}}{\text{Mean}} \times 100$.

The moisture content of the raw groundnut seeds was significantly high compared to the toasted and the sundried groundnut since it was not exposed to any heat. The moisture contents of the groundnut seeds were lower when compared to the reported work of Ibeabuchi [21] on the proximate and functional properties of raw and fermented bottle gourd seed (*Lagenaria sicerari*). As a result of this, it would enhance storage stability by preventing mould growth and also suitability and adaptability for further use in food formulation. Ash content is an indicator for mineral elements in seed. The ash contents (1.51-2.33 %) of the groundnut seed flours are higher than 1.38-1.48 % of *Arachis hypogaea* seed flour reported by Ayoola and Adeyeye [22]. The increase in the ash content of toasted groundnut may be as a result of pretreatment since volatile compounds are lost. According to Pomeranz and Clifto [23] ash contents of seeds should be in the range 1.5-3.5 % for suitable consumption. In this study, the ash contents fall within this range thus it can be recommended for human consumption and animal feeds.

The protein contents in both the raw and sun dried groundnuts were low while in the toasted groundnut, the heat applied showed a synergistic effect. The carbohydrate content was higher in the toasted and the sundried than in the raw groundnut. The crude protein contents (24.48-26.37 %) are greater than that of benniseed (16.4–18.1 %) as reported by Yusuf *et al.* [24]. Therefore, the seeds can be used as alternative sources of protein in diets. High fiber content in the samples is an advantage, and hence this may have potential uses as food supplement. The coefficient of variation (CV) % of the groundnut samples range from 0.08 to 0.99 % indicating the results were not critically varied.

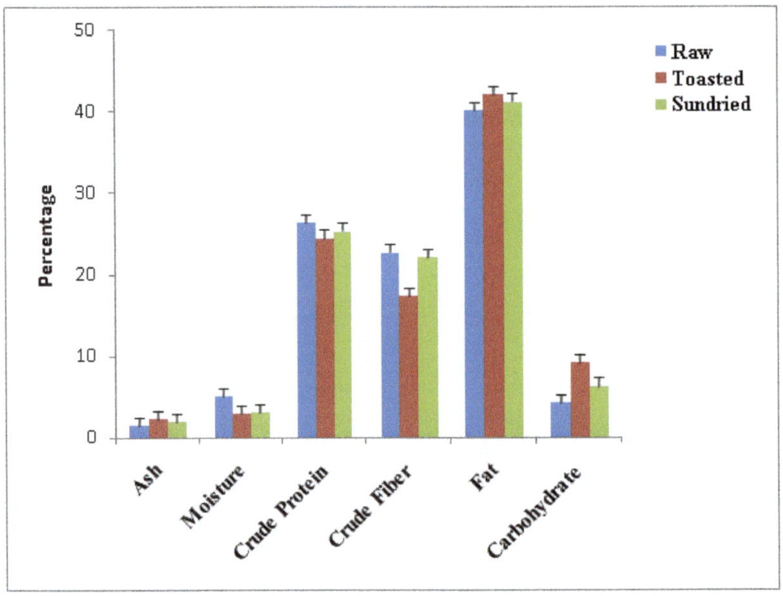

Figure 1. Graphical Representation of Proximate Composition of Groundnut Samples

3.2. Mineral Composition of Undecorticated Groundnut Seed

From "Table 2" and "Figure 2", the toasted groundnut does not lead to reduction in levels of mineral contents but rather increases in the levels since volatile and organic matter are lost by heat which corresponds with the observation of Ayoola and Adeyeye [22]. Therefore, groundnut is a good source of oil protein and minerals which can be used in diets to prevent against some mineral deficiencies

Table 2. Mineral Composition (mg/100g) of Undecorticated Groundnut Seed

Mineral	Raw	Toasted	Sun-dried
Sodium	64.23±0.25[b]	66.13±0.15[c]	62.80±1.06[a]
Potassium	501.00±1.00[c]	535.67±0.11[a]	521.33±0.58[b]
Calcium	90.00±1.00[a]	122.33±1.53[c]	114.80±0.27[b]
Iron	4.53±0.58[a]	5.10±0.10[c]	4.72±0.20[b]
Zinc	4.24±0.06[a]	5.35±0.05[c]	4.81±0.01[a]
Copper	4.61±0.01[a]	5.59±0.80[b]	4.80±0.27[a]
Lead	ND	ND	ND
Manganese	5.50±0.10[a]	6.03±0.15[b]	5.50±0.10[a]
Phosphorus	5.83±0.21[a]	6.80±0.10[b]	6.14±0.15[a]

ND = Not detected. Values are means ± standard deviation of triplicate determination.

Na, K, Cu, Fe and Mn contents were found to be higher than conophor nut, cashew and bean seed [25,26]. Potassium was the highest mineral element. The highest value of potassium agreed with the study that potassium was the most major mineral in Nigerian agricultural products [27]. The Na/K ratios obtained in raw, toasted and sundried samples are 0.12, 0.13 and 0.12 respectively. The values were less than 1 which signifies that the groundnut seeds flour would probably be a good source for electrolyte balance and controls high pressure in the body [28,29]. The raw, sun dried and the toasted seeds contained appreciable amounts of calcium. Calcium is of greater amount when compared to *Abelmoschus esculentus* reported by Ndangui *et al.* [30]. This is very important to health of humans and is required for formation of bones and teeth, formation of blood clot, and body mechanisms [31]. The intake of the seeds flour which contains zinc as a micro element will play a role in the body in the healing of wound. The presence of phosphorous in the samples indicates that the soil nutrient where the samples were planted is rich in the mineral. So therefore, intake of the mineral would help in bone growth, kidney function, cell growth and in maintaining the body's acid-alkaline balance [32]. The absence of lead in the samples indicated that the investigated samples were free from the toxic metal.

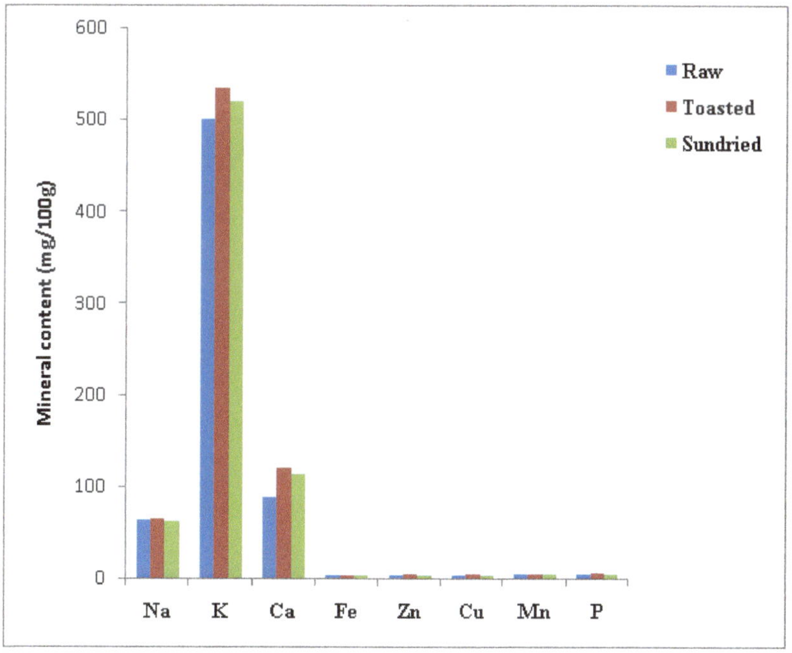

Figure 2. Graphical Representation of Percentage Mineral Content of Groundnut Samples

3.3. Functional Properties of Undecorticated Groundnut Seed

The toasted groundnut showed decrease in its functional properties compared to the raw and sundried seeds (Table 3). The fat absorption capacity (FAC) of raw groundnut was higher than that of toasted and sundried groundnut. This suggests that raw groundnut may have more hydrophobic proteins [33].

The water absorption capacity (WAC) (201.7 %) of raw groundnut seeds was the highest. The high water absorption capacity may be due to the high polar amino acid residues of raw groundnut proteins have strong affinity for water molecules [34]. The foam stability of toasted seeds decreased and other functional properties showed similar trends. Foam stability is dependent on protein content and environmental temperature. The higher the protein content the higher the foam stability [35]. Heat processing diminished the nitrogen solubility of proteins by denaturation and reduced their foaming capacity [36]. High temperature and prolong toasting time would not be desirable for processing the seeds where water or oil absorption are importance. Heat has been reported to inactivate the anti-nutritional factors [37] and also increase the antioxidant activity [38] seed flour. The application of heat during the pretrearment of the toasted and sundried groundnut seeds may contribute to the decreased in their foaming capacity and foaming stability leading to protein denaturation. Similar trends have been reported by Rahma and Mustafa [39] for peanut flour.

Table 3. Functional Properties of Undecorticated Groundnut Seed

Functional property	Raw	Toasted	Sun-dried
Water absorption capacity (%)	201.70	185.20	198.45
Fat absorption capacity (%)	190.50	170.30	183.45
Emulsifying capacity (%)	31.50	25.20	28.72
Emulsifying stability (%)	13.24	9.35	11.30
Foam capacity (%)	4.50	3.50	3.80
Bulk density (g/cm^3)	0.80	0.62	0.72

The emulsion activity shows the ability and capacity of a protein to help in the formation of an emulsion and is related to the protein's ability to absorb to the interfacial area of oil and water in an emulsion [40]. The emulsion stability normally reflects the ability of the proteins to impart strength to an emulsion for resistance to stress and changes [41]. Raw seed flour has the highest emulsion capacity (31.50 %) because the seeds contain low amount of carbohydrate and this may be attributed to the high fat content in the groundnuts. This may be of great important in food emulsion such as yogurt, mayonnaise and ice-creams because of it high emulsion capacity.

The bulk densities of groundnut seed flour of raw, toasted and sundried were found to be in the range of 0.72-0.80 g/cm^3. The bulk density values in this study for the groundnut seed flour are not comparable with 0.54 – 0.57 g/cm^3 for different chickpea varieties as reported by Kaur *et al.* [42]. Higher bulk density is advantageous since it helps to reduce the paste thickness which is an essential factor in child feeding [43,44]. However, the groundnut seed flours could be use as food supplement for children.

4. Conclusion

The undecorticated groundnut seeds contained high level of oil and protein this makes it a potential source of edible oil. The sources of protein and essential minerals can serve as various food and economic purposes. Functional properties of the groundnut seeds flour could be a good replacement in food formulation. This work has helped us know the nutritive values of groundnut from a part of Nigeria which can be considered as a good source of protein and promote the incorporating groundnut to improve protein level in reducing malnutrition in the developing countries. However, more studies should be carried out to investigate the possibility of using the seeds as an ingredient on other food products in order to increase the application of such value added food ingredient.

References

[1] Gibbon, D. and Pain, A, *Crops of the drier region of the tropics*. longman group Ltd UK, 1985, pp 146.

[2] NAERL, Agricultural survey for 2011 wet seasons National Agricultural Extension Research and Liaisons Service 2011.

[3] Olorunju, P. E. Alabi, O. and Tanimu, B, Priorities and strategies for groundnut research in Nigeria. Paper presented at the national workshop on groundnut rehabilitation in Nigeria, Kano, 11-12 May, pp15, 1999.

[4] Oderinde, R. A. and Ajayi, I. A, Metal and oil characteristics of Terminalia catappa, *Riv.Ital Sostarvze Grasse,* 75, 361–362, 1998.

[5] Achu, M. B., Fokou, E., Tchengang, C., Fotso, M. and Tchouanguep, F. M, Nutritive value of some cucurbitaceae oil seed from different region in Cameroon, *African Journal of Biotechnology,* 4, 1329-1334, 2005.

[6] Asibuo, J. Y., Akromah, R. O., Safo- Kantanka, O. S. Osei Adu-Dapaah, H. K. and Agyeman, A, Chemical composition of groundnut, *Arachis hypogaea* (L) land races. *African Journal of Biotechnology,* 7, 2203-2208, 2008.

[7] Savage, G. P. and Keenan, J. I, *The composition and nutritive value of groundnut kernels,* In: smart J (ed). The Groundnut crop: scientific basis for improvement. London: Chapman and Hall, 1994, pp: 173-213.

[8] Crocker, W. and Barton, L. V, *Physiology of seed,* chronic Botanica, Waltham, Massacusetts, 1957, p267.

[9] Oke, O. L, Chemical studies on some Nigerian pulses. *West African Journal of Biological and Applied Chemistry,* 9, 52-55, 1967.

[10] Abdel Rahman, A. H, Compositional Study on some Egytain Peanut varieties, *Rivista Halinna Aelle Sostanze Grasse,* 59, 287-288, 1982.

[11] Woodroof, J. G, Peanuts production, processing products 3rd edition Avi plublishing Company Inc. Westport, Connecticut. 1983.

[12] Aremu, M. O., Olaofe, O. and Akintayo, E. T, Functional properties of some Nigerian varieties of legume seed flours and flour concentration effect on foaming and gelation properties, *Journal of Food Technology,* 5 (2), 109–115, 2007.

[13] Moure, A., Sineiro, J., Domínguez and Parajó, J. C, Functionality of oilseed protein products: A review. Food *Research International* 39, 945-963, 2006.

[14] Yu, J., Ahmedna, M. and Goktepe, I, Peanut protein concentrates: production and functional properties as affected by processing, *Food Chemistry,* 103, 121-129, 2007.

[15] AOAC, *Official methods of Analysis,* 15th edu. Association of Official Analytical chemist. Washington, DC, 2000.

[16] Onyeike, E. N. and Acheru, G. N, Chemical composition of selected Nigerian oil seeds and physiochemical properties of oil extracts. *Food Chemistry,* 77, 431-437, 2002.

[17] Pedroche, J., Yust, M., Lqari, H., Giron-Calle, J., Alaiz, M., Vioque, J. and Millan, F, Brassica carinata protein isolates: chemical composition, protein characterization and improvement of functional properties by protein hydrolysis, *Food Chemistry,* 88, 337-346, 2004.

[18] Sze-Tao, K. and Sathe, S, Functional properties and *in vitro* digestibility of almond (*Prunus dulcis* L.) protein isolate, *Food Chemistry,* 69, 153-160, 2000.

[19] Lin, C. and Zayas, J, Functionality of defatted corn germ proteins in a model system: fat binding capacity and water retention, *Journal of Food Science,* 52, 1308-1311, 1987.

[20] Rodriguez-Ambriz, S., Martinez-Ayala, A., Millan, F. and Davila-Ortiz, G, Composition and functional properties of *Lupinus campestris* protein isolates, *Plant Foods and Human Nutrition*, 60, 99-107, 2005.

[21] Ibeacbuchi, J. C, Proximate and functional properties of raw and fermented bottle groud seed (*Lagenania siceraria*), *International Journal of Biotechnology and Food Science,* 2 (4), 82-87, 2014.

[22] Ayoola, P. B. and Adeyeye, A, Effect of Heating on the Chemical Composition and Physico-Chemical properties of Arachis hypogae (Groundnut) Seed Flour and Oil, *Pakistan Journal of Nutrition,* 9(8), 751-754, 2010.

[23] Pomeranz, Y. and Clifto, D, *Food Analysis Theory and Practices.* In: Melon EE (ed) Chemical Food Analysis Practical Manual – Chemistry. Avi Publishing Company, Westport, C.T. 1981.

[24] Yusuf, A. A., Ayedun, H. and Sanni, L. O, Chemical composition and functional properties of raw and roasted Nigeria benniseed (Sesamum indicum) and bambara groundnut (vigina subterranean), Food Chemistry, 111, 277-282. 2008.

[25] Akinhanmi, T. F., Atasie, V. N. and Akintokun, P. O, Chemical Composition and Physicochemical Properties Of Cashew nut (Anacardium occidentale) Oil and Cashew nut Shell Liquid, *Journal of Agricultral Food and Environmental Science,* 2 (1), 2008.

[26] Aremu, M. O., Olonisakin, O., Bako, D. A. and Madu, P. C, Compositional studies and physicochemical characteristics of

cashew nut (Anarcadium occidentale) flour, *Pakistan Journal of Nutrition,* 5 (4), 328-333, 2006.

[27] Olaofe, O. and Sanni, C. O, Mineral contents of grain and baby foods, *Journal of Science and Food Agriculture,* 45, 191-194, 1988.

[28] Yusuf, A. A., Mofio, B. M. and Ahmed, A. B, Nutrient Contents of Pride of Barbados (*Caesalpinia pulcherrima* Linn.) Seeds, *Pakistan Journal of Nutrition,* 6 (2), 117-121, 2007.

[29] National Research Council NRC, Recommended dietary Allowances 10th ed., Nat. Academy Press, Washington DC, USA. (1989).

[30] Ndangui, C. B., Kimbonjuila, A., Nzikou, J. M., Matos, L., Pambou-Tobi, N. O. G., Abena, A. A., Siluo, Th., Scher, J. and Desboy, S, Nutrtitive composition and properties physicochemical of gurbo (Abelomoschos esculentus L.) seed and oil, *Research Journal of Environmental and Earth Sciences,* 2 (1), 49-54, 2010.

[31] Olusanya, J.O, *Proteins, In: Essentials of food and nutrition,* Apex Books Limited, Lagos, 2008, pp.13-21.

[32] Fallon, S. and. Enig, M.G, *Nourishing Traditions:* The Cookbook that Challenges Policitally Correct Nutrition and the Diet Dictocrats, 2001, 40-45.

[33] Lawal, O. S. and Adebowale, K. O, Effect of acetylation and succinylation on solubility profile, water absorption capacity, oil absorption capacity and emulsifying properties of muncuna bean (Mucuna pruriens) protein concentrate, *Nahrung/Food,* 48 (2), 129-136, 2004.

[34] Narayana, K. and Narasinga Rao, M. S, Functional properties of raw and heat processed winged bean (Psophocarpus tetragonolobus) flour, *Journal of Food Science,* 47, 1534-1538, 1985.

[35] Ocloo, F. C. K., Darfour, B., Wilson, D. D. and Ofosu, D. O, Physical, proximate, functional and pasting properties of flour produced from gamma irradiated cowpea (Vigna unguiculata,L. Walp), *Radiation Physics and Chemistry,* 81, 450-457, 2012.

[36] Yasumatsu, K., Sawada, K., Moritaka, S., Misaki, M., Toda, J., Wada, T. and Ishii, I, Whipping and emulsifying properties of soybean products, *Journal of Agriculture and Biological Chemistry,* 36, 719-727, 1972.

[37] Thangadurai, D. Chemical composition and nutritional potential of Vigna Unguiculata ssp Cylindrica (Fabaceae), *Journal of Food Biochemistry,* 29, 88-98, 2005.

[38] Jeong, S. M., Kim, S. Y., Nam, K. C., Ahn, D. U. & Lee, S. C, Effect of seed roasting conditions on the antioxidant activity of defatted sesame meal extracts, Journal of Food Science, 5(69), 377-381, 2004.

[39] Rahma, F. H. and Mustafa, M. M, Functional properties of peanut flour as affected by different heat treatments, *Journal of Food Science and Technology,* 25, 11-15, 1988.

[40] Du, S. Jiang, H. Yu, X. and Jane, J, Physicochemical and functional properties of whole legume flour, *Food Science and Technology,* 55, 308-313, 2014.

[41] Singh, N., Kaur, N., Rana, J. C. and Sharma, S. K, Diversity in seed and flour properties in field pea (Pisum sativum) germplasm, *Food Chemistry,* 122, 518-525, 2010.

[42] Kaur, M. and Singh, N, Studies on functional, thermal and pasting properties of flours from different Chickpea (*Cicer arietinum* L.) cultivars, *Food Chemistry,* 91, 403-411, 2005.

[43] Padmashree, T.S., Vijayalakshmi, L. and Puttaraj, S, Effect of traditional processing on the functional properties of cowpea (*Vigna Catjang*) flour, *Journal of Food Science and Technology,* 24, 221-224, 1987.

[44] Milán-Carrillo, J., Reyes-Moreno, C., Armienta-Rodelo, E., CaráH bez-Trejo, A. and Mora-Escobedo, R, Physicochemical and nutritional characteristics of extruded flours from fresh and hardened chickpeas (*Cicer arietinum* L), *Lebensmittel-Wissenschaft und-Technologie,* 33, 117-123, 2000.

10

Biotransformation in Temperate Climate Fruit: A Focus on Berries

Paula Becker Pertuzatti[1,*], Milene Teixeira Barcia[2]

[1]Instituto de Ciências Exatas e da Terra, Engenharia de Alimentos, Universidade Federal de Mato Grosso, Barra do Garças, Brazil
[2]Escola de Química e Alimentos, Universidade Federal do Rio Grande, Rio Grande, Brazil
*Corresponding author: paulapertuzatti@yahoo.com.br

Abstract Berry fruits, cultivated in temperate zones, are popularly used in the human diet. With the increase in production and consumption, the development of food industry is oncoming in the sense of developing new products that can meet market needs, in addition to the development new formulations, enhancing the industrial waste and bio-transforming it to generate high value-added products is fundamental. The use of fermentation processes to increase the level of functional compounds and the production of enzymes, antibiotics and flavorings is a good example.

Keywords: *bioactive compounds, flavour, antibiotics, enzymes, solid-state fermentation*

1. Introduction

Berry fruits, cultivated in temperate zones, such as blueberry, blackberry, raspberry, strawberry and cranberry are popularly used in human diet [1].

The cultivation of these fruits is becoming more popular in many countries. Brazil has a privileged climatic condition for such cultivation since it has temperate regions in four states. Thus, the country has potencial production potential for all year. In addition, another economically important aspect is that the Brazilian production of fruits occurs from December to February, the off-season in the United States and the European Union that are the main consumer centers.

Currently most of the Brazilian production is consumed *in natura* or exported. However, in other countries where the consumer has already entered these fruits in his daily diet, several products are processed and the waste generated by industries has been used as fermentation substrate obtaining a lot of value-added by-products.

This review aimed to present the main characteristics of strawberry, blackberry, raspberry, cranberry and blueberry, regarding the growing potential in Brazil, the biologically active compounds present and the use of industrial waste of these fruits as a substrate for biotransformation and production of phenolic compounds, enzymes, antibiotics, organic acids and flavorings.

2. Berries

The *small fruit* designation is used in the international literature to refer to several crops like strawberry, blackberry, raspberry, blackcurrant, blueberry, etc., and is limited to the name – *e.g.* berries - and to the size of the crop [2].

According to reference [3], the cultivation of berries has attracted the attention not only of consumers, but also the interest of producers and traders. In general, the cultivation of these species is characterized by low cost of deployment, affordable cost of production for small producers, economic return, adaptation to socio-economic conditions and the local environment, unskilled manpower demand, possibility of cultivation in an organic system and greater demand than supply [4]. The profitability of this type of fruit per hectare is high, exports have grown in recent years, as well as the planted area and in many regions constitute an important alternative for both small farms as well as for large companies.

The great marketing appeal of these fruits is due to the high content of antioxidants and anti-cancer substances that attract the consumer, as in the case of the blueberry, which is regarded as one of the richest in natural antioxidants among small fruits [5,6].

Berries are a natural source of substances which have antioxidant activity. Extracts of various fruits such as blackberry, cherry, blackcurrant, blueberry and raspberry have been found to be effective in combating free radicals [7]. Recently, due to association of the small fruits to bioactive properties such as high content of antioxidants and anti-cancer substances, fruit consumption demand by the population has increased, in search of food supplementation from diet diversification based on fruits.

The changes in the eating habits of population, observed in recent years, coupled with the increased purchasing power of the low income population, has provided ample market opportunity for the production of fresh and industrialized fruits in Brazil, with the southern states, São Paulo and Minas Gerais standing out, especially considering that climate conditions in these regions allow the supply of temperate fruit species at different times of the year [2,8].

2.1. Strawberry

Strawberry is a perennial, creeping, herbaceous belonging to the Rosacea family and genus *Fragaria* [9]. The edible part is the strawberry, which is a non-climacteric pseudofruit [10] of bright red coloring, engaging odor, texture and slightly acidified taste [11]. The color of strawberry is due to anthocyanins, and its unique flavor is due to the citric and malic acids, and sugars [12].

The strawberry is the most popular and consumed type of berry fruit for its sweet flavour. There are more than 600 strawberry cultivars; they differ in flavour, size and texture. In addition to cultivated strawberries, some cultivars grow wild or come from natural breeding. These can be usually smaller in size, but feature a more intense flavor [13].

2.2. Cranberry

Cranberry is a red fruit belonging to the Ericaceae family native of the United States [14]. Due to the great bitter taste, the fruit is not consumed *in natura*. According to data from the National Agricultural Statistic Service (NASS) [15] the fruit is marketed in three basic categories: fresh (5%); sauce, concentrated and various value-added applications (35%); and juice (60%) [14]. There are no reports yet, about cultivation of this species in Brazil.

Cranberries, like other small fruits, are rich in phenolic phytochemicals, such as phenolic acids (especially ellagic acid), tannins and flavonoids. The consumption of this fruit has historically been associated with health benefits, especially decreases in the incidence of urinary tract infections and intestinal disorders caused by *E. coli* O157:H7, its consumption has also been associated with the ability to inhibit peptic ulcer associated with *Helicobacter pylori* bacteria. The processing of this fruit to elaborate juices generates pomace as a byproduct. The pomace is mainly composed of peel, pulp and fruit seed. This is high in fiber and has relatively little amount of carbohydrates and protein [1,14,16].

2.3. Blackberry

The blackberry belongs to *Rosaceae* family and the *Rubus genus* that contains about 740 species divided according to some authors, into 12-15 subgenres [17,18]. This fruit has an erect or creeping posture, which produces aggregated fruit formed by minidrupes of about four to seven grams of black color and a sweet to sweet-acid flavor [19]. There are numerous blackberry cultivars known and marketed, but currently the two most important are the Brazos and Tupi [20].

In Brazil there are five native species of blackberries - *R. urticaefolius*, *R. erythroclados*, *R. brasiliensis*, *R. sellowii* and *R. imperialis* - white, pink, red or black colors. None of the Brazilian species was domesticated. The cultivars of blackberries used in the country are the result of introductions, crosses and selections of American cultivars [8]. Rio Grande do Sul state is Brazil's main producer, but the fruit has a high potential for cultivation in temperate regions such as Santa Catarina, Paraná, São Paulo and Minas Gerais [18].

2.4. Raspberry

The raspberry belongs to the family Rosaceae, genus Rubus and subgenus Idaeobatus. This subgenus comprises about 200 species, distinguished by the ease with which their mature fruits are separated from the receptacle. Many of these species have been improved, but only the red and black raspberry are produced on a large scale [8,21].

The raspberry is native to central and northern Europe and parts of Asia [8]. However the *Idaeobatuss* subgenre, occurs on five continents, but has its distribution centered mainly in the Northern Hemisphere, with a focus on Asia, Europe and North America [22]. In Brazil, the raspberry crop was introduced in the region of Campos do Jordão - SP, and the major producing states are Rio Grande do Sul, São Paulo and Minas Gerais [3].

2.5. Blueberry

Blueberries are a small fruit native of the United States, belonging to the genus *Vaccinium*, and the *Ericaceae* family. According to statistics from the Food and Agriculture Organization (FAO), from 1970 to 2011, the world production of blueberries increased approximately 7 times [23]. According to the National Agricultural Statistics Service (NASS) of the United States Department of Agriculture (USDA), the United States is the leading producer of blueberries, which makes the blueberry the second most produced and marketed small fruit in the country, only behind strawberry [24].

In South America, blueberry cultivation (*Vaccinium* spp.) is booming in countries like Chile, Argentina and Uruguay [25]. In Brazil, the cultivation of blueberries has not been well known. However, Brazil has the potential to produce blueberries throughout allyear. So, an important economic aspect is to produce Brazilian blueberries at least from December through February, which is between harvest seasons in the main consumer centers in the USA and European Union [26].

In the world there are three main groups of blueberries commercially produced: the lowbush, the highbush and the rabbiteye [8].

3. Bioactive Compounds

Fruit extracts showed antioxidant capacity, which helps slow down or prevent the development of several degenerative diseases, due to their abilities to react with free radicals. This protects the tissues in the human body against oxidative stress and pathologies due to cancers, cardiovascular diseases and inflammatory processes [27,28]. These beneficial effects are most probably related to the presence of bioactive compounds, such as carotenoids, tocopherols, vitamin C and phenolic compounds [29,30].

A large number of *in vitro* and *in vivo* studies have shown that berries stand out as fruits with high anticarcinogenic properties, which in turn are attributed to the high content of bioactive compounds such as flavonoids (anthocyanins, flavonols and flavanols), tannins (proanthocyanidins, ellagitannins and gallotannins) and phenolic acids (hydroxybenzoic acids, hydroxycinnamic and derivatives) (Figure 1) [31]. However the reference [26] found small amounts of carotenoids and ascorbic acid, along with elevated levels of tocopherols in blueberries, which was also reported by others [18,32] in fruits such as blackberry, blueberry and goldenberry (*Physalis peruviana*). So, in other study, [6] verified that compounds other than phenolics can affect the antioxidant activity of blueberries.

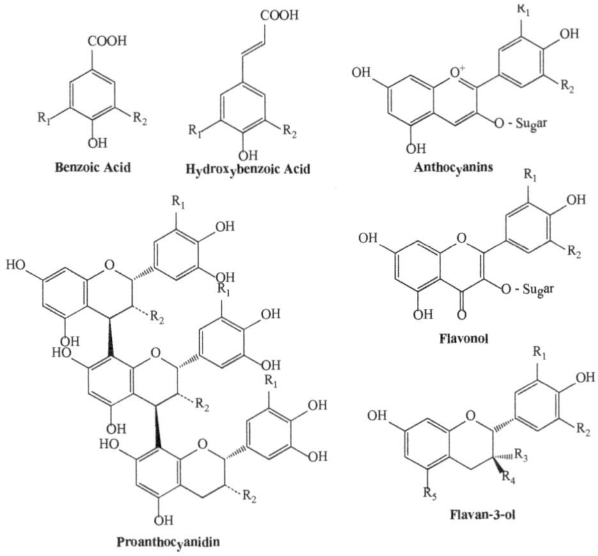

Figure 1. Structure of flavonoids found in Berries

Among the bioactive compounds present in berries, phenolic compounds stand out, which comprise more than 8,000 substances that have already been identified [33]. These compounds are classified according to the structure of the aromatic ring, the number of carbon atoms substituting the ring and connecting it together. Thus, the metabolic changes that phenolic compounds undergo during the metabolic processes in plants make them show a structural and functional diversity and can be hydroxylated and methoxylated in various positions [34].

Table 1 presents the content of phenolic phytochemicals in strawberry, raspberry, blackberry, cranberry and blueberry.

With regard to other types of pigments that can be found, in a study by [39] determining the content of lutein, zeaxanthin, lycopene, β-carotene, α-carotene and β-cryptoxanthin in strawberry, raspberry, blackberry and blueberry, it was found that blackberry had the highest carotenoid content (440 $\mu g.100g^{-1}$), followed by raspberry (370 $\mu g.100g^{-1}$) and blueberry (290 $\mu g.100g^{-1}$), the strawberry had the lowest levels (26 $\mu g.100g^{-1}$). However, when the carotenoid content of berries is compared to other fruits, the values are considered low, this is because in fruits whose main compounds belong to the anthocyanin class, berries for example, the carotenoid content is reduced during ripening [40].

As for the tocopherol content, in a study of 33 fruits by [41] it was found that berries (strawberry, raspberry, blueberry, blackberry and cranberry) had amounts greater than most fruits. However, it is important to note that while the tocopherols and tocotrienols are found in photosynthetic organisms the presence of these compounds in plant foods occurs mainly in dark green vegetables, in oil seeds, vegetables oils and wheat germen [42].

Table 1. Phenolic compounds values in strawberry, cranberry, raspberry, blackberry and blueberry ($mg.100g^{-1}$ of fresh fruit)

Compound	Strawberry	Cranberry	Raspberry	Blackberry	Blueberry
Ellagic acid	34.4 - 63.0 [35,36]	12.0 PS [36]	150 [36] PS	ND [18]	2.7 [26]
Galic acid	0.5 - 4.4 [36]	ND [36]	1.9 - 3.8 [36]	350.5 [18]	58.7 [26]
Ferulic acid	0.0 [36]	ND [36]	0.3 - 1.7 [36]	35.7 [18]	9.8 [26]
Caffeic acid	<0.05 -1.4 [36]	ND [36]	0.6 - 1.0 [36]	15.4 [18]	0.7 - 7.7 [26]
Chlorogenic acid	*	0.5 [37]	*	*	186 - 208 [38]
p-coumaric acid	0.7-17.0[35,36]	0.0 - 2.0 [36,37]	0.7 - 2.0 [36]	ND [18]	0 - 3.9 [18,26,37]
p-hydroxybenzoic	1.0 - 3.6 [36]	ND [36]	1.5 - 2.7 [36]	0.01 [18]	0.03 [26]
Quercetin	0.3 - 3.9 [35]	7.3 - 25.0 [36]	2.9 [36]	20.2 [18]	2.2 - 15.6 [26,35]
Myricetin	ND [36]	1.1 - 7.7 [36]	0.0 [36]	ND [18]	0.8 - 36.0 [26,35]
Kaempferol	0.2 - 9.9 [35]	0.0 - 0.3 [36]	<0.01 [36]	ND [18]	6.2 [26]
Catechin	*	41.700 DM [36]	*	ND [18]	ND [26]
Epicatechin	*	44.700 DM [36]	*	120.2 [18]	ND [26]

DM: Dry Matter; * Not found in literature; - ND: Not Detected

In a study by [26] with pulp, peel and the whole fruit of six blueberry cultivars, it was found that the peeling of the fruit has 10-58 times more tocopherols than the fruit flesh itself. Similar results were found for carotenoids, vitamin C, anthocyanins, flavonols, flavanols and phenolic acids (with the exception of gallic acid), all found in larger quantities in the peel of the fruit, resulting in a higher antioxidant activity of peel.

4. Metabolism of Berries Waste via Semi-Solid Fermentation

The food industry produces large volumes of waste, both solids and liquids resulting from the production of food preparation and consumption. The disposal of such waste results in both economic and environmental problems, because in addition to being highly pollutant, they represent a loss of biomass and valuable nutrients [43,44]. Because of this, the biotransformation of food processing waste has received increased attention as it represents a possible resource of waste conversion into useful and value-added products [44]. These residues have long been used as substrates for solid-state fermentation, production of fertilizers, animal feeds, ethanol production, organic acids and the production of various types of enzymes [14].

4.1. Increased Content of Phenolic Compounds and Antioxidant Activity

Among the strategies used for the enrichment of phenolic antioxidants in fruits, the bioprocessing of waste fruits and vegetables using solid state fermentation with microorganisms that have GRAS (substances generally recognized as safe for use in food) to generate phytochemical profiles that make the increase of health benefits possible [14,43].

The solid state fermentation in small fruits can be done with juice, pulp or by using the resulting byproduct of the extraction juice or production of berry wine [14,45]. There are reports in the literature using byproduct from processing of cranberry juice as substrate for semisolid fermentation using fungi such as *Rhizopus oligosporus* and *Lentinus edodes* in order to increase the production of phenolic compounds and thereby increase the antioxidant action of the fruit [14,16]. These studies have shown that during fermentation there is an increase in the production of phenolic compounds reaching levels found in fruit and juices. [16] found that on 10th day of fermentation of cranberry byproduct, an increase in the production of phenolic compounds occurred resulting in an increase of 11 to 25% in antioxidant capacity compared with the initial values.

The increase of phenolic acids, such as ellagic acid in cranberry-product due to solid state fermentation has also been reported [16,43,46]. This may be due to ellagitannins hydrolysis by enzymes that hydrolyze tannins produced by the fungi used in fermentation. These fungal hydrolases such as glycosidase and fructofuranosidase break the glycosidic bonds existing between the phenolics and sugar, resulting in an increase in antimicrobial activity with *V. parahaemolyticus*, *E. coli* O157:H7 and *H. pylori* being the microorganisms considered to be more sensitive to presence ellagic acid. This sensitivity of micro-organisms could be due to disruption of membrane integrity, blockade of ion channels in the plasma membrane or inhibition of ATP synthesis by inhibiting the flow of electrons through the electron transport chain in the bacterial membrane [16].

The reference [47] found that after 50 days of fermentation using the fungus *Lentinus edodes,* the yield of the free phenolic achieved a maximum 0.5 $mg.g^{-1}$ of byproduct, the authors also identified the major free phenolic through High Performance Liquid Chromatography (HPLC).

4.2. Enzyme Production

Different strains of fungi can be used in solid state fermentation (SSF) process, since the majority can be easily grown on the surface of solid substrates. The objective of the process may be the production of biomass, waste products such as organic acids, production of flavor compounds and flavoring or enzyme production. Since agricultural wastes are rich sources of natural polymers (cellulose, lignin, pectin and starch), large amounts of enzymes are produced [48].

Among the fungi used *Lentinus edodes*, also known as *Shiitake* or Chinese mushroom grows very well in apple pomace and other lignocellulosic waste as a byproduct of processing cranberries or strawberries [43]. This fungus has been used to produce enzymes such as laccase and manganese peroxidase [48] and to produce polygalacturonase [43].

Polygalacturonase is widely used in the food industry, assisting in processing, maceration, liquefaction, extraction, clarification and filtration of juices and fruit wines. For the most common industrial uses, fungal polygalacturonase is more advantageous because it has a higher enzymatic activity. In addition, it presents its optimum activity at low pH, which facilitates its application in processing of fruits and vegetables. Commercial preparations of polygalacturonase used in the food industry are typically derived from fungal sources especially *Aspergillus niger* and the yeast *Kluyveromyces marxianus* [43]. In recent studies the same authors found that high levels of polygalacturonase activity were produced by *L. edodes* during SSF in strawberry pulp. The strawberry pulp was presented as the best substrate for the production of polygalacturonase, while the byproduct of cranberry was not a good substrate.

4.3. Production of Aromas

In 2012, the world market of aromas, flavors and fragrances saw transactions of 23.4 billion dollars, and there are chances to accelerate this segment by 2017 [49]. Currently, there is a great interest from consumers for these products thus increasing the economic value of bioflavors [44].

Natural flavors are chemicals that have properties that are produced from raw materials of vegetable or animal origin by physical, enzymatic or microbiological treatments [50].

The microbial synthesis of these natural flavorings is generally carried out by submerged fermentation [44]. Its results in high costs in natural flavors, for example, synthetic raspberry ketone costs US $ 58.00 per kilogram while the natural costs US $ 3,000.00 per kilogram (Sigma, 2000). Due to high cost of this technology now used on an industrial scale, there is a need to develop inexpensive processes, which make bioflavors economically viable. This could be achieved by exploiting the metabolic pathways and alternative technologies such as solid state fermentation.

4.4. Antibiotic Production

According to [51], production of antibiotics by fungi have some similar characteristics to the production of other secondary metabolites, such as production of antibiotics in a specific lineage; there is instability in the biosynthetic process, which tends to decrease with successive samplings used for the maintenance of the lines; the production of antibiotics follows fungal growth kinetics associated with a specific culture medium; increased production of a secondary metabolite occurs frequently in the sporulation stage of the microorganism; the structural variety of biosynthesized metabolites can be increased by small variations of culture medium.

New techniques such as solid state fermentation have emerged in order to promote a faster fungal growth and a differentiated production of secondary metabolites [52]. The solid-state fermentation employs a natural substrate as carbon source in the presence of a minimal amount of water. The solid state fermentation for production of antibiotics has been considered very advantageous, with an increased yield and shorter fermentation period on parallel procedures performed in submerged cultures [51].

There are studies that use of raspberry seed powder for production of the antibiotic Neomycin by *Streptomyces marinensis* using solid fermentation [53,54]. This antibiotic is an important aminoglycoside widely used in pharmaceutical preparations because it is effective against gram positive bacteria, gram negative bacteria and mycobacteria [54]. It was first produced by *Streptomyces fradial*, and a mutant strain (*Streptomyces marinensis*) was later developed and identified as a producer of Neomycin, [53].

The reference [53] observed that the concentration of raspberry powder used as substrate has a direct relation with production of Neomycin and may act as a limiting nutrient, since small variations in concentration alter the growth rate of fungi and/or product formation. It was also found that dextrin and raspberry seed powder are key nutrients that control the biosynthesis of the antibiotic.

5. Trends and Prospects

From the studies presented in this work, it was observed that there is a global trend of cultivation of small fruits.

Care with improved food quality is currently on the rise. Many research groups in several fields as food science, biochemistry, medicine and nutrition among others have developed products aiming at the ingestion of larger amounts of bioactive compounds that prevent the development of diseases. The matrices studied, have excellent functional characteristics and can become a good alternative.

The main trends observed are the increasing number of consumers of these fruits and, consequently, the expansion of agricultural area devoted to these crops.

With increasing production and consumption, the development of food industry is urgent in developing new products that can meet market needs. In this regard, many countries already have a variety of juices, jellies, creams, fillings and the like.

In addition to the development of new formulations attractive to consumer, it is necessary to advance in the way of performing the treatment of industrial effluents. In Brazil, the most common is the use of food industry waste in the manufacture of animal feed, ethanol production or fertilizers. However, a challenging thought was presented in this review. This consists of enhancement of the industrial waste, bio-transforming it to generate high value-added products. As an example, the use of fermentation processes to increase the level of functional compounds, making the residue rich in these matrix components much as the fruit itself. Thus, the enhanced residue can be used for other purposes such as the development of concentrated extracts. The use of bio-transformation for the production of enzymes, antibiotics and fragrances also presents a potential market. In short, in Brazil, a major challenge is to develop a new way of seeing the industrial waste, not as a means to increased costs for treatment, but as an alternative of income and development.

This paper presents an important perspective to Brazil, and let taking advantage of its climate potential can become a producer, consumer and exporter of wide range of small temperate fruits. This strategy promotes the development of small agriculture, strengthens the economy and improves the dietary quality of the population.

References

[1] Nile, S.H, and Park, S.W, Edible berries: Bioactive compounds and their effect on human health, *Nutrition*, 30, 134-144, 2014.

[2] Antunes, L.E.C, Amora-preta: Nova opção de cultivo no Brasil, *Ciência Rural*, 32 (1), 151-158, 2002.

[3] Pagot, E, and Hoffmann, A, Produção de pequenas frutas no Brasil, In: Seminário Brasileiro sobre pequenas frutas, 1, 64p, 2003.

[4] Poltronieri, E, Alternativa para o mercado interno de pequenas frutas, In: Seminário Brasileiro sobre pequenas frutas, 1, 37-40, 2003.

[5] Prior, R.L, Cao, G, Martin, A, Sofic, E, Mcewen, J, O'brien, C, et al. Antioxidant capacity as influenced by total phenolic and anthocyanin content, maturity, and variety of Vaccinium species, *Journal of Agricultural and Food Chemistry*, 46, 2686-2693, 1998.

[6] Pertuzatti, P.B, Barcia, M.T, Rodrigues, D, Cruz, P.N, Hermosín-Gutiérrez, I, Smith, R, and Godoy, H.T, Antioxidant activity of hydrophilic and lipophilic extracts of Brazilian blueberries, *Food Chemistry*, 164, 81-88, 2014.

[7] Pantelidis, G.E, Vasilakakis, M, Manganaris, G.A, and Diamantidis, G, Antioxidant capacity, phenol, anthocyanin and ascorbic acid contents in raspberries, blackberries, red currants, gooseberries and Cornelian cherries, *Food Chemistry*, 102, 777-783, 2007.

[8] Raseira, M.C.B, and Antunes, L.E.C, A cultura do mirtilo, Embrapa Clima Temperado, 29-36, 2004.

[9] Gomes, P, *Fruticultura brasileira*, 13ed., Nobel, São Paulo, 2007, 342-348.

[10] Chitarra, M.I.F, and Chitarra, A.B, *Pós-colheita de frutos e hortaliças: Fisiologia e manuseio*, ESAL/FAEPE, Lavras, 2005, 735p.

[11] Henrique, C.M, and Cereda, M.P, Utilização de biofilmes na conservação pós-colheita de morango (Fragaria Ananassa Duch) cv IAC, *Ciência e Tecnologia de Alimentos*, 19 (2), 270-276, 1999.

[12] Silva, P.A, *Qualidade de morangos cultivados na região de Lavra-Mg, armazenados em temperatura ambiente*, Universidade Federal de Lavras, Lavras, 2006, 71p.

[13] Padula, M.C, Lepore, L, Milella, L, Ovesna, J, Malafronte, N, Martelli, G, and Tommasi, N, Cultivar based selection and genetic analysis of strawberry fruits with high levels of health promoting compounds, *Food Chemistry*, 140, 639-646, 2013.

[14] Vattem, D.A, Ghaedian, R, and Shetty, K, Enhancing health benefics of berries through phenolic antioxidant enrichment: focus on cranberry, *Asia Pac. Journal Clinical Nutrition*, 14 (2), 120-130, 2005.

[15] NASS, *National Agricultural Statistics Service Agricultural Statistics board. Cranberries*, Annual Reports, Washington D.C, USDA, 2001.

[16] Vattem, D.A, Lin, Y.T, Labbe, R.G, Shetty, K, Antimicrobial activity against select food-borne pathogens by phenolic antioxidants enriched in cranberry pomace by solid-state bioprocessing using the food grade fungus Rhizopus oligosporus, *Process Biochemistry*, 39, 1939-1946, 2004.

[17] Daubeny, H.A, *Brambles*, Fruit Breeding, New York, John Wiley & Sons, 1996, 109-190.

[18] Jacques, A.C, Pertuzatti, P.B, Barcia, M.T, Zambiazi, R.C, and Chim, J.F, Estabilidade de compostos bioativos em polpa congelada de amora-preta (Rubus fruticosus) cv. Tupy, *Química Nova*, 33 (8), 1720-1725, 2010.

[19] Mota, R.V, Caracterização física e química de geleia de amora-preta, *Ciência e Tecnologia de Alimentos*, 26 (3), 539-543, 2006.

[20] Sousa, M.B, Curado, T, Negrão, F, and Trigo, M.J, Amora-qualidade pós colheita, *Folhas de divulgação da Agro*, 556 (7), 1-31, 2007.

[21] Sousa, M.B, Curado, T, Negrão, F, and Trigo, M.J, Framboesa-qualidade pós colheita, *Folhas de divulgação da Agro*, 556 (6), 1-35, 2007.

[22] Oliveira, P.B, Valdiviesso, T, Mota, M, and Fonseca, L.L, A planta da framboesa-morfologia e fisiologia, *de divulgação da Agro*, 108 (655), 2005.

[23] FAOSTAT, *Agriculture Data*, Food and Agriculture Organization of the United Nations, Statistics Division, 2012. Available: http://faostat.fao.org/

[24] USDA-NASS, *Blueberries Production*, 2013. Available: http://www.nass.usda.gov/Statistics_by_Subject/index.php?sector=CROPS.

[25] Bañados, M.P, Blueberry production in South America, *Acta Horticulturae*, 715, 165-172, 2006.

[26] Pertuzatti, P.B, Barcia, M.T, Jacques, A.C, Vizzotto, M, Godoy, H.T, and Zambiazi, R.C, Quantification of several bioactive compounds and antioxidant activities of six cultivars of Brazilian blueberry, *The natural Products Journal*, 2, 188-195, 2012.

[27] Wang, H, Cao, G, and Prior, R, Oxigen radical absorbing capacity of anthocyanins, *Food Chemistry*, 45, 304-309, 1997.

[28] Tapiero, H, Tew, K.D, Ba, G.N, and Mathé, G, Polyphenols: do they play a role in the prevention of human pathologies?, *Biomed Pharmacother*, 56 (4), 200-207, 2002.

[29] Faria, A.F, Marques, M.C, and Mercadante, A.Z, Identification of bioactive compounds from jambolão (Syzygium cumini) and antioxidant capacity evaluation in different pH conditions, *Food Chemistry*, 126 (4), 1571-1578, 2011.

[30] Barcia, M.T, Pertuzatti, P.B, Jacques, A.C, Godoy, H.T, and Zambiazi, R. C, Bioactive compounds, antioxidant activity, and percent composition of jambolão fruits (Syzygium cumini), *The natural products journal*, 2, 2012.

[31] Seeram, N.P, Adams, L.S, Zhang, Y, Lee, R, Sand, D, Scheuller, H.S, and Heber, D, Blackberry, black raspberry, blueberry, cranberry, red raspberry, and strawberryextracts inhibit growth and stimulate apoptosis of human cancer cells in vitro, *Journal of Agricultural and Food Chemistry*, 54, 9329-9339, 2006.

[32] Barcia, M.T, Jacques, A.C, Pertuzatti, P.B, and Zambiazi, R.C, Determination by HPLC of ascorbic acid and tocopherols in fruits, *Semina: Ciências Agrárias*, 31 (2), 381-390, 2010.

[33] Vinson, J.A, Su, X.H, Zubik, L, and Bose, Phenol antioxidant quantity and quality in foods: Fruits, *Journal of Agricultural and Food Chemistry*, 49, 5315-5321, 2001.

[34] Beer, D, Joubert, E, Gelderblom, W.C.A, and Manley, M, Phenolic Compounds: A Review of Their Possible Role as In Vivo Antioxidants of Wine, *S.Afr. J. Enol.Vitic.*, 23 (2), 48-61, 2002.

[35] Häkkinen, S, and Törrönen, A.R, Content of flavanols and selected phenolic acids in strawberries and *Vaccinium* species: Influence of cultivar, cultivation site and technique., *Food Res. Intern*, 33, 517-524, 2000.

[36] Häkkinen, S, Heinonen, M, Kärenlampi, S, Mykkänen, H, Ruuskanen, J, and Törrönen, R, Screening of selected flavonoids and phenolic acids in 19 berries. *Food Research International*, 32, 345-353, 1999.

[37] Ehala, S, Vaher, M, and Kaljurand, M, Characterization of phenolic profiles of northern European berries by capillary electrophoresis and determination of their antioxidant activity, *Journal Agricultural and Food Chemistry*, 53, 6484-6490, 2005.

[38] Skrede, G, Wrolstad, R.E, and Durst, R.W, Changes in Anthocyaninas and Polyphenolics During Juice Processing of Highbush Blueberries (*Vaccivium corymbosum L.*). *Journal of Food Science*, 65 (2), 357-364, 2000.

[39] Marinova, D, and Ribarova, F, HPLC determination of carotenoids in Bulgarian berries, *Journal of Food Composition and Analyses*, 20,. 370-374, 2007.

[40] Jacques, A.C, Pertuzatti, P.B, Barcia, M.T, and Zambiazi, R.C, Nota científica: compostos bioativos em pequenas frutas cultivadas na região sul do estado do Rio Grande do Sul. *Braz. J. Food. Technology*, 12 (2), 123-127, 2009.

[41] Chun, J., et. al. Tocopherol and tocotrienol contents of raw and processed fruits and vegetables in the United States diet. *J. Food Comp. Analysis*, 19, 196-204, 2006.

[42] Batista, E.S, Costa, A.G.V, and Pinheiro-Sant'ana, H.M, Adição da vitamina E aos alimentos: implicações para os alimentos e para a saúde humana. *Rev. Nutr.* 20 (5), 525-535, 2007.

[43] Zheng, Z, and Shetty, K, Solid state production of polygalacturonase by *Lentinus edodes* using fruit processing wastes. *Process Biochem.* 35, 825-830, 2000.

[44] Laufenberg, G, Kunz, B, and Nystroem, M, Transformation of vegetable waste into value added products: (A) the upgrading concept; (B) practical implementations. *Bioresource Technology*, 87, 167-198, 2003.

[45] Su, M.S, and Silva, J.L, Antioxidant activity, anthocyanins, and phenolics of rabbiteye blueberry (Vaccinium ashei) by-products as affected by fermentation. *Food Chemistry*, 97, 447-451, 2006.

[46] Vattem, D.A, and Shetty, K, Solid-state production of phenolic antioxidants from cranberry pomace by *Rhizopus oligosporus*. *Food Biotecnology*, 16 (3), 189-210, 2002.

[47] Zheng, Z, andShetty, K, Solid-State Bioconversion of Phenolics from Cranberry Pomace and Role of *Lentinus edodes β*-Glucosidase. *Journal of Agricultural and Food Chemistry*, 48,895-900, 2000.

[48] Hatvani, N, and Mécs, I, Production of laccase and manganese peroxidase by *Lentinus edodes* on malt-containing by-product of the brewing process, *Process Biochem.* 37, 491-496, 2001.

[49] Bain &Company, *Potencial de diversificação da indústria química brasileira*, 2014, Available: http://www.bndes.gov.br/SiteBNDES/export/sites/default/bndes_pt/Galerias/Arquivos/produtos/download/aep_fep/chamada_publica_FEPprospec0311_Quimicos_Relat4_aromas.pdf

[50] Kosseva, M, and Webb, C, *Food Industry Wastes*, Elsevier, Academix Press, 2013, 312 p.

[51] Takahashi, J.A, and Lucas, E.M.F, Occurrence and structural diversity of fungal metabolites with antibiotic activity. *Quimica Nova*, 31 (7), 1807-1813, 2008.

[52] Asagbra, A.E., Sani, A.I, and Oyewole, O.B, Solid state fermentation production of tetracycline by Streptomyces strains using some agricultural wastes as substrate. *World Journal of Microbiology and Biotechnology*, 21, 107-114, 2005.

[53] Adinarayana, K, Prabhakar, T, Srinivasulu, V, Anitha, V, and Jhansi L.P, Optimization of process parameters for cephalosporin C production under solid state fermentation from Acremonium chrysogenum, *Process Biochem*, 39, 171-177, 2003.

[54] Ellaiah, P. et al. Production of lipase by immobilized cells *Aspergilus niger*, *Process Biochemistry*, 39 (5), 525-528, 2004.

11

Effect of Honey and Sugar Solution on the Shelf Life and Quality of Dried Banana (*Musa paradisiaca)* Slices

Md. Sultan Mahomud[1], Md. Keramot Ali[2], Md. Mostafijur Rahman[3], Md. Hafijur Rahman[3], Tajnuba Sharmin[1], Md. Jiaur Rahman[4,5,*]

[1]Department of Food Engineering and Technology, Hajee Mohammad Danesh Science and Technology University, Dinajpur, Bangladesh
[2]Department of Food Technology and Rural Industries, Bangladesh Agricultural University, Mymensingh, Bangladesh
[3]Faculty of Agriculture, Hajee Mohammad Danesh Science and Technology University, Dinajpur, Bangladesh
[4]Department of Food Processing and Preservation, Hajee Mohammad Danesh Science and Technology University, Dinajpur, Bangladesh
[5]Department of Biochemistry (MSc Student), Memorial University of Newfound, St. John's, NL, Canada
*Corresponding author: jia.hstu@yahoo.com

Abstract The main purpose of the study was to investigate the effect of solution on the shelf life and quality of banana slices and development of high quality dehydrated banana products. Ripe banana collected from local market cut into 3, 5 and 7 mm slices. Solution effect was assessed using honey, sugar and mixed (honey plus sugar, 1:1) solution. The osmosis samples were dried in a mechanical drier at 65°C for 24 hour up to moisture content 14.6%. Fresh and dehydrated banana were analyzed for their chemical composition. The effect of pre-treatment (4 min steam blanching plus 20 min sulphyting) and nutrient content also evaluate. Percent solid gain was assumed as indicator of solution of solution effect. Percent solid gain slightly decreased (per unit weight) with increasing thickness of banana slices at constant immersion time (3 hour) and concentration (72% TSS). This gain was higher in honey solution followed by sugar solution. The response of taste panel revealed that banana slices prepared by 4 min steaming plus 20 min sulphyting (0.5% KMS) and subsequently dipping in honey solution gave better colour and flavor. It has been also revealed that pre-treatment by 4 min steaming plus 20 min sulphyting (0.3% KMS) and subsequently dripping in honey was gave better colour and flavour of banana slices. So these pre-treated dehydrated products showed highest degree of acceptability. Studies on the effect of various packaging materials showed that the single layer polythene plus keep in tin can gave the best result for storing the dried banana slices. Among the different storage condition of dried banana slices (75% RH, 80%RH and 90% RH), the 75% RH was found most effective for storing the processed slices. However, in all packaging systems and storage conditions, the slices absorbed moisture over the storage period and lost its quality. To maintain better quality of slices for longer shelf-life, the low relative humidity and moisture proof packaging system (polythene plus kept in tin can or laminated aluminum foil) may be required.

Keywords: *banana, pretreatments, processing, sugar & honey solution, shelf life, quality*

1. Introduction

Ripe bananas have been considered as a part of human diets for many years. Many people are involved with production and consumption of ripe banana around the world. The moisture content in ripe banana is about 80% and therefore very susceptible to post-harvest losses and considerable weight loss during transportation and storage [1]. Post-harvest losses are a major challenge for tropical products such as mango, pineapple, banana etc. Generally, Deterioration occurs in a fully ripened banana within 4-7days [1].

Banana (*Musa paradisiaca)* is one of the cheapest, most plentiful and nourishing of all fruits [2]. This is one of the superior fruits of Bangladesh with respect to popularity, availability, production and consumption [3]. It is also delicious and seedless. It is the most delicious fruit in the world. As a matter of fact, in Bangladesh, banana is a popular and economic fruit which constitutes 42% of fruit consumption [3]. The importance of this fruit is due to its calorific and nutritive value and of its versatile use to the consumer. It contains appreciable amount of vitamin B and certain amount of vitamin A and C.

Banana contains a considerable amount of minerals such as K, P, Ca and Fe [4]. Various products like banana chips, banana figs, flour, powder, jam, confectionary, dehydrated slice etc. could be prepared from banana [2]. Banana is not seasonal in nature like many other fruit crop and is available in large quantity throughout the year. But banana is highly perishable food item and could not be preserved for long time after harvesting. Most of the world's bananas are eaten either raw, in the ripe state, or as a cooked and proportion are processed in order to obtain a storable product. At present few varieties of banana (e. g. Sagar) has been cultivating commercially could be encouraged for more production and spoilage could be prevented by proper preservation.

Drying is one of the oldest methods for the preservation of food products. Heated air drying method has been developed as a newer technique of drying food and also considered as more hygienic and economic [5,6]. Dried products could be used directly or treated as secondary raw material [7]. The hot air drying of food material has advantages s such as control product quality achievement of hygienic conditions and on reduction of production loss [8]. Shrinkage of the cells browning loss of redrying ability, wettability and case hardening are some common problems associated with drying of tropical fruits, which reduce their market value and general acceptability [9,10].

Among processed products, dried ripe banana slices contribute significantly, since it can be stored and used easily. To decrease the drying cost and increase the product quality several pre-treatment before drying are conducted. Among these immersion in high concentrated aqueous or sweet (honey) solution has potential to give a higher quality products. Before drying immersion in sugar solution prior to drying reduced drying time upto 1.6 times [11]. One of the main constrains of the dried banana products is flavour deterioration and discoloration which resulted in less marking demand. To overcome these constrains, besides steam blanching, immersion of the slices in different solution before drying might give improved colour and flavour. Keeping the above views in consideration, the present investigation were undertaken to fulfill the following objectives: To assess the solution effect on qualitative and nutritive changes in ripe banana slices. To develop high quality dehydrated banana slices. To assess the overall acceptability of the processed slices. Packaging study of the processed slices. And storage study of the processed slices.

2. Materials and Method

2.1. Sample Collection and Banana slice Preparation

The banana samples were collected from local markets and cut into slices of 3 mm, 5 mm and 7 mm thickness using stainless knives and the peels were removed manually.

2.2. Preparation of Sugar and Sugar Honey Mixed Solution

The natural honey was collected from local market of Dinajpur, Bangladesh. The total soluble solid (TSS) of honey examined by refractometer was found 72%. The sugar solution was prepared by mixing by water with heating and TSS was adjusted until it reached to the same TSS of honey (72%). The mixed solution was made by mixing (ratio is 1:1) sugar solution and honey and TSS was adjusted to that of honey (72%).

2.3. Pre-treatments and Processing of Banana Slices

Table 1. List of the pretreatments

Pre-treatments	Time
Hot Water Blanching	4 min
	2 min
Steam Blanching	3 min
	4 min
Steam Blanching + Sulphyting (0.3% KMS Solution)	4+20 min
Steam Blanching + Sulphyting (0.5% KMS Solution)	4+20 min

The banana slices were blanched in steam (100°C and atmospheric pressure) for 4 minutes to prevent the enzymatic activity which may causes undesirable changes during processing. The blanched banana slices were dipped in 0.3% and 0.5% KMS solution for 20 minutes to avoid non-enzymatic browning which act as ant browning chemicals. Then the sliced banana was immerged in sugar solution, honey solution and mixed solution (sugar + honey; 1:1) for 3 hours. The solution concentration was 72% (W/W) and the solution to product ratio was 7:1.

Ripe banana
↓
Washing/Peeling
↓
Slicing
↓
Washing/Peeling
↓
Blanching (Steam blanching at 100^{0}C for 4 minutes)
↓
Sulphiting (0.3% KMS and 0.5% KMS for 20 minutes)
↓
Osmotic dehydration
(i. Dipping in honey for 3 hrs, ii. Dipping in sugar solution for 3 hrs, iii. Dipping in mixed solution (honey + sugar) for 3 hrs)
↓
Removing from solution and rinsing
↓
Dried in cabinet drier (at 60-65^{0}C) for 24 hrs)
↓
Cooling and packaging and sealing
↓
Storage

Figure 1. Flow Chart of Banana Slices Processing

Samples were removed from the solution at the end of 3 hours, quickly rinsed to remove surface sugar and moisture for the gravimetric determination of the water loss and solids uptake. The pre-treated slices were dried in mechanical drier by spreading the slices on the trays immediately. Tray loading is kept at $^1/_2$ – 1 kg/m^2 of drying area. The drying temperature was maintained at 60-

65°C. After drying the dried products was cooled, packed in polythene bas (100 g/bag) and sealed the bag by an electrical sealing machine and then stored.

2.4. Chemical Analysis

The fresh banana and the dried slice were analyzed for moisture, ash, protein, fat, ascorbic acid and acidity. The total soluble solids (TSS) were read directly from an Abbe's refractometer. All the determination was done in triplicates and the results were expressed as the average value. Moisture, Ash, Fat and protein content were determined by [12] methods.

2.5. Total Carbohydrate

Total carbohydrate content of the samples were determined as total carbohydrate by difference, that is by subtracting the measured protein, fat, ash, moisture from 100.

2.6. Quality and Sensory Evaluation of Banana Slices

The sensory and quality attributes test of banana slices was carried out according to [13,14] with some modifications. A taste- testing panel consisting of 10 experts evaluated the consumer's acceptability of developed products. The panelists (10) were requested to assign appropriate score of characteristics colour, flavour, texture and overall acceptability of various banana slices. To note the score a hedonic scale having a highest score of 9 for 'like extremely' and lowest of 1 for 'dislike extremely' was used. The hedonic scale was arranged such that: 9 = like extremely, 8 = like very much, 7 = like moderately, 6 = like slightly, 5 = neither like or dislike, 4 = dislike slightly, 3 = dislike moderately, 2 = dislike very much and 1 = dislike extremely.

2.7. Shelf Life of Pretreated Banana Slices

To assess the shelf-life of dried banana slices packed in various packaging system, moisture uptake was determined gravimetrically. The packaging systems used were: (a) single layer high density polythene, (b) double layer high density polythene, (c) single layer high density polythene plus kept in tin can.

Each packing unit contains 100g of samples. Moisture uptake by samples in the above packaging systems was determined at every three days interval gravimetrically from known initial moisture content of samples.

It was also essential to observe the storage behaviour of banana slices at various storage conditions. For observing shelf life of banana slices at various storage conditions each 100 g sample was packed in single layer high density polythene and stored at 75%, 80% and 90% relatively humidity at room temperature. The moisture uptake by slices at every three days interval was determined gravimetrically from known initial moisture content of samples. The above mentioned content relative humidity were maintained in dessicator by using saturated NaCl (reagent grade) solution for 75% RH, saturated $(NH_4)_2SO_4$ solution for 80% RH and saturated KNO solution for 90% RH. Samples in pre weighed Petri dishes were kept in wire-mesh support structure above the saturated salt solution at the bottom of the dessicator were airtight using grease.

2.8. Data Analysis

All means of the values and standard deviations were calculated from obtained data and analyzed statistically by the help of MSTAST-C. Meanwhile, DMRT was applied to determine differences among different samples.

3. Results and Discussion

3.1. Composition of Fresh Banana

The fresh banana pulp was analyzed for moisture, sugar, protein, fat, ash and total carbohydrate. The results are presented in Table 2. The fresh banana pulp contained moisture 71%, ash 0.99%, sugar 5.20%, fat 0.2%, protein 1.3%. The composition of fresh ripe banana under study was more or less similar to these reported by [15]. Anon [15] reported that the composition (per 100 g) of fresh ripe banana is as follows: moisture 70.0%, carbohydrate 22.0%, crude fat 0.5%, protein 1.2%, far 0.3%, ash 0.9%, phosphorus 290 ppm, calcium 80.0 ppm, iron 6.0 ppm, β-carotene 0.5 ppm, riboflavin 0.5 ppm, niacin 7.0 ppm, ascorbic acid 120 ppm. The variation in composition of banana pulp might be due to varietal difference, variation in stage of maturity, time elapsed between harvesting and analysis and growing condition of banana.

Table 2. Composition of fresh ripe banana

Components	Quantity (%)
Moisture	71.00
Total carbohydrate	26.54
Ash	0.96
Sugar	5.20
Fat	0.20
Protein	1.30

3.2. Effect of Pretreatments on the Colour of Banana Slices

Immediately after slicing, banana slice became dark (core colour) due to enzymatic reaction. To eliminate this problem, banana slices were subjected various pretreatment just after slicing. The treatment was hot water blanching, steam blanching plus 20 minutes sulphyting (0.3% KMS) and 4 minutes steam blanching plus 20 minutes sulphyting (0.5% KMS). The samples evaluated visually and the result shown in Table 3.

Table 3. Effect on pretreatment in colour

Pretreatments	Time	Colour (considered only core colour of banana)
Hot Water Blanching	4 min	Dark
	2 min	Slightly dark
Steam Blanching	3 min	Almost natural
	4 min	Almost natural
(a) Steam Blanching + Sulphyting (0.3 % KMS Solution)	4 + 20 min	Attractive
(b) Steam Blanching + Sulphyting (0.5 % KMS solution)	4 + 20 min	More attractive

It is observed from Table 3 that hot water blanching and short time steam blanching yielded banana slices of unattractive. Treatment of 4 minutes steam blanching plus 20 minutes sulphyting (0.5% KMS) resulted in attractive colour of banana slices and they used for further studies.

3.3. Effects of Sugar and Honey Solution on the Pre-Treated Banana Slices

The samples of three different thicknesses were dipped in natural honey solution, sugar solution and mixed (honey plus sugar, 1:1) solution. The concentration of sugar and mixed solution was maintained equal to that of natural honey solution (72%). After three hours of immersion the weight loss of the banana slices was measured. The weight loss and the solid grain were considered as indicator of solution effect. The results showed that the weight loss increased with increasing thickness. After recording weight loss, the samples immediately dried in a mechanical drier. The solid grain and sugar content of the samples is shown in Table 4 and Table 5. Since rapid water loss took place during first 3-4 hours. Hope and Vitalthe [16] immersion time was selected for 3 hours.

Table 4. Effect of solution on the pretreated banana slices

Thickness of banana slices (mm)	Solution Effect as % Solid Gain		
	Honey	Sugar	Mixed
3	12.11	11.21	11.42
5	11.25	10.75	10.95
7	10.08	9.98	10.20

Table 5. Effect of solution as percentage (%) of total sugar on the pre-treated banana slices

Thickness of slice (mm)	Solution effect as % total sugar (sugar grain per unit weight		
	honey	sugar	mixed
3	52.49	54.22	53.78
5	50.29	52	50.97
7	49.19	51.29	50.25

It is observed from Table 4 that the percent solid grain per unit weight slightly decreased with increasing thickness slices. Since the solution concentration was remained the same, the solid grain and the sugar content showed slight variation. The higher solid and sugar content is due to infusion of sucrose during immersion in different solution. The solid grain is higher in honey solution than other solution. It can also be seen from Table 5 that sugar content of banana slices also increase with increasing thickness. During experiment it is seen that 3 mm thickness slices could not retain its shape in the solution rather they broke into small pieces while removing from it and also they become hard after drying. On the other hand, 7 mm thick slices took more time during drying and their quality also not acceptable. In this regard 5 mm thick banana slices was found better during pre testing, dipping in solution and drying. So 5 mm thick slices gave good products.

3.4. Effects of Sugar and Honey Solution on the Nutrients Status Banana Slices

The development products were analyzed for their nutrients content and the result are compared to those of fresh banana as shown in Table 6. It is seen that there is noticeable difference in ascorbic acid content between fresh bananas and processed as development banana slices. But the development products contained more total sugar and protein than fresh banana when processed products are comparing to fresh sample. It is shown that all the nutrients accept vitamin C apparently shown higher value in processed products. It should be noted that the nutrients content presented are based on wet weight basis and that (in solution) sample contained additional sucrose due to infusion during osmosis.

Due to this additional sucrose, the dried products showed higher total sugar a lower fat content. Here should be describing that the 3 type of solution treatment when applied to observed effect on the nutritional quality of processed product. The solution one is honey two is sugar solution and other is mixed solution. The nutrient content of processed production is shown in Table 6.

Table 6. Nutrients content of fresh and processed banana slices

Components	Fresh	Processed (dried)		
		Honey	Sugar	Mixed
Sugar (%)	5.2	50.29	52	50.97
Fat (%)	0.2	0.005	0.002	0.0035
Protein (%)	1.3	3.08	2.10	2.96
Ascorbic acid (%)	6.93	4.27	2.31	3.95

3.5. Sensory Quality Evaluation of Dried Banana Slices

The samples of processed banana slices produced from banana slices with pretreatment were subjected to the sensory evaluation. The colour, flavour, texture and overall acceptability of 6 samples of processed slices were evaluated by a panel of 10 judges. Samples no. 1 was prepared by 4 min. steam blanching and subsequently dipped in 0.5% KMS for 4 min. and then dipped in honey solution for 3 hrs. Sample 2 was prepared by 4 min. steam blanching and dipped 0.5% KMS for 20 min. and subsequently dipped in sugar solution. Sample 3 was prepared by steam blanching then dipped in 0.5% KMS solution for 20 min. and subsequently dipped in mixed solution (sugar and honey) for 3 hours. Sample 4, 5 and 6 prepared by same above procedure but different in percent of KMS which percent was 0.3%. Then mean scores for colour, flavour, texture and overall acceptability of different sample are present in Table 7.

A two way analysis for variance (ANOVA) was carried out for colour preference and result revealed that there were ($p<0.01$) difference in colour acceptability among the chips. The calculated value F (16.0164) was greater than the tabulated value F (4.8692). Then indicates that colour of different samples of banana slices were not equally acceptable.

As shown in Table 7, the sample 5 is secured the lowest score for the colour preference than other five samples. However sample 1 and 2 were equally acceptable at 1% level of statistical significant. Similar sample 1, 2, 4, 5 and 6 were equally acceptable at 1% level of statistical significant in colour preference and significant different from sample 3 which secured the highest score (6.90 out of 9).

Table 7. Mean score for colour, flavour, texture and overall acceptability of various pretreated banana slices

Sample type	Sensory attributes			
	Colour	Flavour	Texture	Overall acceptability
A	5.40± 0.100[b]	5.20±0.992[b]	5.40±0.132[b]	6.20±0.164[b]
B	5.30± 0.082[b]	5.10±0.962[b]	5.30±0.097[b]	5.90±0.078[b]
C	6.90± 0.131[b]	7.00±0.154[b]	7.00±0.085[b]	7.10±0.142[b]
D	4.80±0.062[b]	4.80±0.062[b]	5.30±0.090[b]	5.20±0.040[b]
E	4.00±0.054[b]	3.90±0.035[b]	5.10±0.082[b]	5.10±0.043[b]
F	4.50±0.058[b]	4.20±0.046[b]	4.90±0.075[b]	4.40±0.053[b]
LSD P<1.01	0.3486	0.7312	0.8358	0.7290

All values are expressed as mean ±SD

Mean followed by different superscript letter in each column are significantly different (p<1.01).

Sample A: Pretreatment by 4 minutes steam blanching plus 20 minutes sulphyting (0.5% KMS solution) then dipping in honey for 3 hours plus 24 hours dried in cabinet drier at 60-65°C, Sample B: Pretreatment by 4 minutes steam blanching plus 20 minutes sulphyting (0.5% KMS solution) then dipping in sugar for 3 hours plus 24 hours dried in cabinet drier at 60-65°C, Sample C: Pretreatment by 4 minutes steam blanching plus 20 minutes sulphyting (0.5% KMS solution) then dipping in mixed solution (sugar + honey, 1:1) for 3 hours plus 24 hours dried in cabinet drier at 60-65°C, Sample D: Pretreatment by 4 minutes steam blanching plus 20 minutes sulphyting (0.3% KMS solution) then dipping in honey for 3 hours plus 24 hours dried in cabinet drier at 60-65°C, Sample E: Pretreatment by 4 minutes steam blanching plus 20 minutes sulphyting (0.3% KMS solution) then dipping in sugar for 3 hours plus 24 hours dried in cabinet drier at 60-65°C and Sample F: Pretreatment by 4 minutes steam blanching plus 20 minutes sulphyting (0.3% KMS solution) then dipping in mixed solution (sugar + honey, 1:1) for 3 hours plus 24 hours dried in cabinet drier at 60-65°C.

In case of flavour preference among the slice samples a two analysis of variance (ANOVA) showed that the pretreatment has significant influence on flavour acceptability (P<0.01). Since the calculated F (6.6454) was greater than the tabulated value F (4.0878). The result showed that (Table 7) sample 1, 2 and 4 were equally acceptable securing 5.2, 5.1 and 4.8 respectively. Sample 4 is significantly different from sample 5 but equally acceptable to sample 6. It is noted that sample 3 secured the highest score (7.00 out of 9) while sample 5 is the lowest score (3.90).

In case of texture of produced banana samples, the panelist's detected significant differences among the samples tasted at 1% level of statistical significance as calculated F (18.000) was higher than the tabulated value of F (4.5393). As shown in Table 7 (DMRT) the samples were not equally acceptable. The result showed that the sample 3 was the highest acceptable among the samples 1, 2, 4 and 5 securing 7 out of 9 while the sample 6 secured the lowest score (4.90).

A two way analysis of variance (ANOVA) was carried out for and result rebuild that there were 1% level of statistical significant (p<0.01) overall acceptability among the slices. The calculated value F (8.82539) was greater than the tabulated value F (1.1719). This indicates that the overall acceptability of different sample of slices were not equally acceptable. As shown in table (DMRT) the sample 6 secured the lowest score among the samples for overall acceptability preference. However sample 1, 2 and 4 were equally acceptable at 1% level of statistical significant. Sample 5 secured the second lowest score and was significantly different from other sample. The result showed that sample 3 was the highest acceptable among the samples tested in overall acceptability preference than the sample 1, 2, 4, 5 and 6 and securing the highest score (7.10).

3.6. The Shelf Life of Dried Banana Slices

3.6.1. Shelf Life of Dried Banana Slices at Various Packaging System

Studies were conducted to determine storage stability in terms of moisture grain by using different packaging materials and equilibrium relative humidity. Dried banana slices with 14.6% moisture content were packed in single layer polythene package, double layer polythene package and packed by single layer polythene plus kept in tin can and kept in the laboratory at room temperature and humidity. The moisture content by banana slices were determined gravimetrically at (from initial known moisture content) every 3 days, interval. Results are shown in Table 8.

Table 8. Moisture content of dried banana slices at various packaging system

Days of storage	Moisture Content %		
	Packed in single layer polythene	Packed in double layer polythene	Packed in single layer polythene plus kept in tin can
3	18.0±0.430[ab]	16.2±0.244[a]	15.0±0.283[b]
6	20.5±0.243[b]	18.0±0.303[d]	16.0±0.294[c]
9	22.0±0.362[d]	18.8±0.273[ab]	17.0±0.374[a]
12	23.0±0.342[abc]	20.5±0.334[ad]	18.0±0.523[bc]
15	23.8±0.095[bd]	20.8±0.560[bc]	19.0±0.952[b]
18	24.7±0.370[a]	21.3±0.351[b]	20.0±0.981[a]
21	25.5±0.082[d]	21.9±0.311[a]	21.0±0.241[cd]
24	26.0±0.221[d]	22.5±0.233[d]	21.5±0.490[bc]
27	26.4±0.153[abc]	22.7±0.241[ab]	22.0±0.219[abc]
30	26.8±0.263[a]	22.9±0.129[c]	22.2±0.649[c]

All values are expressed as mean ±SD

Mean followed by different superscript letter in each column are significantly different (p<0.05).

The results showed that percentage of moisture content by samples, packaged in polythene package and subsequently kept in tin can was the lowest among packaging systems tested, while slices packed in the single layer polythene package percentage of water was the highest among the systems. Slices packed in double layer polythene showed second highest moisture content. The above trend continued all throughout the investigation period up to 30 days and a linear relationship of moisture was found with storage period of 10 days for all the packaging systems, investigated. Moisture content of dried banana slices for various packaging system are illustrate in Figure 2.

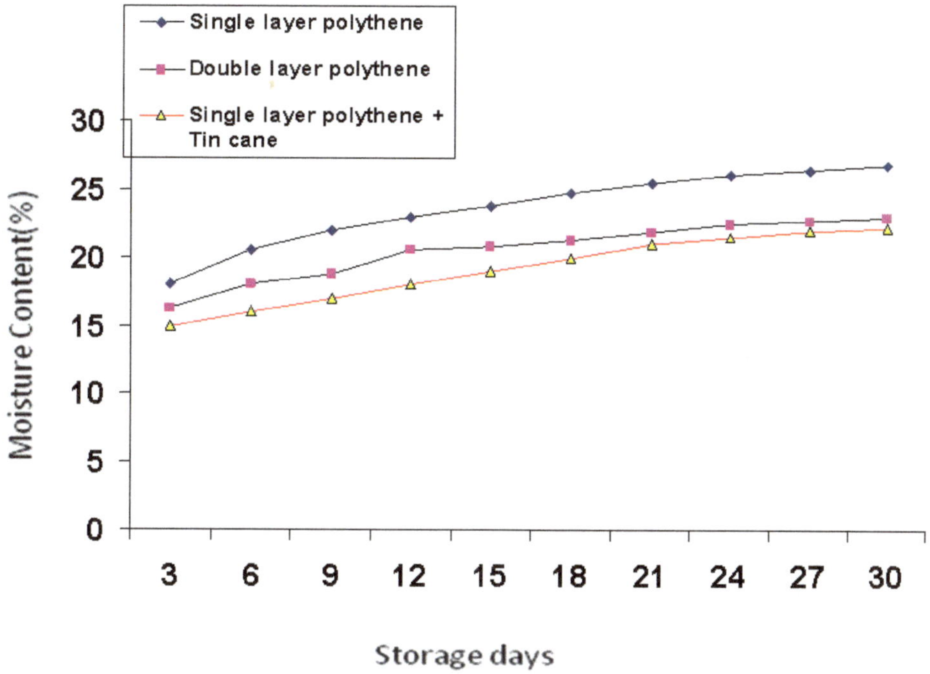

Figure 2. Moisture content of dried banana slices for various packaging systems such as single layer polythene, double layer polythene and single layer polythene plus kept in tin can

3.6.2. Shelf Life of Dried Banana Slices at Various Storage System

Studies were conducted to determine storage stability in terms of moisture grain by storing in different relative humidity. Dried banana slices with 14.6% moisture content were packed in single layer polythene packages and kept in tin can 75%, 80% and 90% relative humidity using different saturated salt solution as mentioned by Islam (1999). The moisture content of slices was measured gravimetrically at every three days interval. Results are shown in Table 9.

Table 9. Moisture uptake by dried banana slices at various storage conditions (at room temperature)

Days of storage	Moisture Content %		
	Stored in 75% RH	Stored in 80% RH	Stored in 90% RH
3	18.3±0.254^{a}	19.3±0.321^{bc}	20.3±0.549^{abc}
6	20.3±0.423^{ab}	21.2±0.450^{ab}	22.5±0.704^{bc}
9	22.3±0.287^{b}	22.8±0.196^{a}	23.8±0.327^{cd}
12	24.0±0.472^{c}	23.6±0.309^{b}	24.5±0.307^{abc}
15	24.5±0.371^{bc}	24.5±0.134^{d}	25.2±0.209^{a}
18	25.0±0.412^{abc}	25.0±0.809^{bc}	25.8±0.476^{c}
21	25.5±0.324^{bc}	25.8±0.421^{bc}	26.3±0.654^{b}
24	25.8±0.413^{a}	26.5±0.388^{d}	27.1±0.543^{ab}
27	26.3±0.542^{b}	27.2±0.623^{c}	28.3±0.342^{bc}
30	26.6±0.278^{cd}	27.9±0.554^{a}	28.9±0.432^{a}

All values are expressed as mean ±SD
Mean followed by different superscript letter in each column are significantly different ($p<0.05$).

The results showed that during first three days the increase in moisture content was high (from 14.6% to 18.3%-20.3%) for all cases and after that the rising moisture content was quite slow. The dried banana slices stored at 90% RH, the product moisture content continued to increase up to 30th days of storage. The dried banana slices stored at 90% RH reached moisture content 28.9% on 30th days of storage. Similarly the product stored at 80% RH reached moisture content 27.9% on 30th days of storage, while the product stored at 75% RH attained 26.6% moisture content on 30th days of storage period. Up to 30th days of storage, the higher the RH, the higher was the moisture content of banana slices. The above results indicated that the initial rate of moisture uptake was very high for all cases. The equilibrium moisture content of banana slices corresponding to 75% RH, 80% RH and 90% RH was not attained at 30th days of storage like banana slices. The behavior may attribute low percentage of fat contained in dried banana slices.

The above study clearly indicated that moisture content of dried banana slices during storage was dependent on relative humidity of the storage structure. Storage at low relative humidity and for short period would be helpful to preserve the dried banana slices. Banana slices packed in single layer high density polythene can be kept for longer period. While packed in single layer polythene plus kept in tin can could be kept for 15 days if 15% moisture content could be allowed without much loss is crispness. In case of dried slices packed in single layer polythene plus kept in tin can could be kept for 15 days if 20% moisture content could be allowed. Therefore, packaging of both slices in double layer high density polythene and single layer polythene plus kept in tin can of in laminated aluminum foil is recommended.

4. Conclusion

The study on storage of processed or dried banana slices showed that the moisture uptake by single layer polythene plus kept in tin can was the lowest. Moisture uptake was highest in polythene and was lowest in polythene and kept in tin cane. Moisture uptake was highest in the slices stored in only single layer polythene. Therefore, single layer polythene plus kept in tin cans are considered is better packaging aids. In case of dried banana slices single layer polythene plus kept in tin can or laminated aluminum foil is recommended. The finding of this study may help in developing commercial processing techniques of banana to reduce post harvest losses of banana in Bangladesh.

References

[1] Abano E. E. and Sam-Amoah L. K. (2011): Effect of different pretreatments on drying characteristics of Banana slices. ARPN Journal of Engineering and Applied Sciences, VOL. 6, NO. 3, PP. 121-129.

[2] Khader, J.B. M.A. Chellapan, K. Alagica and P.K. Challopadhaya. 1985. Banana In: Fruit of India Trop. A sub-trop (ed. Bose Tk.) Longmans Ltd. Calcutta.

[3] Haque, M.A. 1988. Kolar Bagan, Banana Research Project, BAU, Mymensingh, 24 p. Bangladesh.

[4] Ketiku, A.O. 1976. Chemical evaluation of the nutritive value of varieties of banana fruits grown in Nigeria. J. Assoc. Adv. Agril. Sci. AI; 3(2): 53-56.

[5] Das I., S.K. Das and S. Bal. 2004. Specific energy and quality aspects of infrared (IR) dried parboiled rice. J. Food Eng. 62: 9-14.

[6] Motevali A., S. Minaeiy, M.H. Khoshtaghazaz, M.Kazemi and A.M. Nikbakhtyy. 2010. Drying of pomegranate arils: Comparison of predictions from mathematical models and neural networks. Int. J. Food Eng. 6(3): Article 15.

[7] Menges H.O. and C. Ertekin. 2006. Mathematical modeling of thin layer drying of Golden apples. J. Food Eng. 77: 119-125.

[8] Corzo O., N. Bracho A. Pereira and A. Vasquez. 2008. Weibull distribution for modeling air drying of coroba slices. LWT-Food Sci. Technol. 41(6): 1108-1115.

[9] Dalgleish J.N. and N.M. Andy 1988. Dehydration and Dried Products. Food Industries Manual. M.D. Ranken (Ed). Blackie Glasgow.

[10] Singh U., S.K. Jain, A. Doshi, K.H. Jain and K.V. Chahar. 2008. Effects of Pretreatments on Drying Characteristics of Button Mushroom. International Journal of Food Engineering. The Berkeley Electronic Press. 4(4).

[11] Islam, M.N. 1999. Influence of process parameters on the effectiveness of osmotic dehydration of some fruits and vegetables. Bangladesh J. Agril. Engg. 4 (1 and 2):65-73.

[12] AOAC. 1975. Official methods of Analysis of the Association of Official Analytical Chemistry.15th Ed. Washington, D.C.

[13] Rahman M.J, Talukder M.A.I., Rani I., Saha K.C. and Nahid M.S.I (2013). The effect of processing techniques on the shelf life, nutritional and sensory quality of ginger (zingiber officinale) powder and paste. Journal of innovation and development strategy. 7(3), 66-66.

[14] Tummala J., Galla N.R. and Dubashi (2008). Physiochemical changes during processing and storage of greenchili (Capsicum annum L.). Journal of processing and preservation. 32, 868-880.

[15] Anon. 1992. Trends in world cassava production. Journal of Cassava Newsletter. [Cited from Food Science and Technology Abstracts 25(1), 1993].

[16] Hope, G.W. and D.G. Vitale. 1972. Osmotic dehydration. Intl. Res. Centre. Monographs IDRC-004, Ottawa, Ontario, Canada.

12

Amino Acid and Fatty Acid Profile of Twenty Wild Plants Used as Spices in Cameroon

Armand Abdou Bouba[1,*], Roger Ponka[2], Goudoum Augustin[2], Nicolas Njintang Yanou[3], Mehanni Abul-Hamd El-Sayed[4], Didier Montet[5], Joël Scher[1], Carl Moses Mbofung[6]

[1]Laboratory of Engineering and Biomolecules (LiBio), Nancy-University, Nancy, France
[2]Department of Agriculture, Livestock and Derivate Products, University of Maroua, Maroua, Cameroon
[3]Department of Biological Sciences, Faculty of Science, University of Ngaoundere, Ngaoundere, Cameroon
[4]Department of Food Science and Nutrition, Faculty of Agriculture, Sohag University, Sohag, Egypt
[5]CIRAD, Montpellier, France
[6]Department of Food Science and Nutrition, University of Bamenda, Bamenda, Cameroon
*Corresponding author: abdouarmand@yahoo.fr

Abstract The amino acids and fatty acids of twenty wild plants used as spices which were collected in Cameroon were investigated. The most important of total non-essential amino acids (TNEAA) contents have been found in *A. daniellii* (52.04%), *C. frutescens* (50.53%), *D. psilurus* (56.22%), *H. gabonii* (fruit) (58.81%), *P. brazzeana* (59.11%), *S. melongena* (59.78 %) and *X. aethiopica* (56.78%), respectively. Concerning the levels of the essential amino acids (EAA), all had EAA contents of at least 30% except *E. giganteus* (17.85 %), *S. melongena* (27.05 %), and *T. tetraptera* (15.85 %). For WHO protein standard, Lysine was considered as the first limiting amino acid (LAA) for *A. daniellii* (1.54%), *P. brazzeana* (16.35%) and S. *striatinux* (40.19 %). In addition, Meth+Cys were the second LAA for *D. psilurus, E. giganteus, S. melongena, S. zenkeri* (fruit), *T. tetraptera, X. aethiopica* (non detected); *H. gabonii* (fruit) (29.2%), *M. myristica* (6%), *M. whitei* (8.8 %) and *P. guineense* (39.2%), respectively. Spices which contained a large amounts of the essential fatty acids (ω – 3) were found in *C. frutescens* (31.98%), *F. leprieuri* (40.07 %), *H. gabonii* (fruit) (34.84 %), *M. whitei* (54.77%), *P. umbellatum* (39.25%), *S. melongena* (34.23%), *S. zenkeri* (fruit) (35.68%), and *T. tetraptera* (33.64%), respectively. Furthermore, (ω – 6) were found in *D. psilurus* (49.29%), *H gabonii* (bark) (47.91%) and *F. xanthoxyloides* (43.79%). In addition, (C22: 6n3) were *S. zenkeri* (bark) (20.45%) and *X. aethiopica* (30.10 %). These results show that tropical edible plants could contribute significantly to the diets of indigenous population in sub-Saharan Africa and should provide their public health.

Keywords: *spices, amino acid, fatty acid, wild plant, WHO*

1. Introduction

Balance diet intake can provide our needs for energy and body building; all power is based on the principle of food metabolism for our well-being. Furthermore, the habits can to take diet renders the population of developing countries to consume a balanced family dishes made with a basic food (cereal, tuber) rich in energy and sauce. Different parts of plants are widely used as spices and / or ingredients in the preparation of many dishes in many countries like Cameroon. Besides its primary role which is the flavoring, spices can be a significant source in nutrients like proteins, amino acids, minerals and vitamins [1,2,3]. Associated with water, proteins constitute the major part of our body weight. Essential amino acid (EAA) deficiency induces the slowdown of growth and development in children, gives diseases and causes the destruction of cells in adults. Among these amino acids, some are bioactive (histidine, lysine, isoleucine, leucine, valine, threonine, and phenylalanine + tyrosine) and other are antioxidants (histidine, methionine + cysteine) [4]. Fats are essential for a balanced diet because of the physiological role play in the nervous system and in the transportation of fat-soluble vitamins (vitamins A, D, E and K). Essential to the growth of children, vegetable oils are also taken into consideration, fats are also necessary for the maintenance of good health among others. These oils are composed largely of unsaturated fatty acids USFA (ω-3 and ω-6) that help to reduce the risk which incur the cardiovascular disease [5,6,7,8]. In a context of fighting against deficiencies in food, it is important to estimate the nature and amounts of micronutrients ingested by the people of traditional foods [9]. Hence the purpose of this part of the study is to determine amino acid and fatty acid profile of some wild plants used as spices in Cameroon.

2. Material and Methods

2.1. Plant Material

Spices were collected in West Cameroon or either purchased in local markets of Bafoussam and Douala cities in Cameroon, then transported to Ngaoundere. Twenty different spices were harvested and purchased (see Table 1). Once taken to laboratory, the spices were cleaned and dried using a solar dryer shell for seven days. Dried spices were then powdered using an electric grinder (Culatti, Polymix, France) through a 500 µm sieve.

Table 1. Name and parts of the spices used in the study

Botanical name	Common name	French name	Family	Used parts
Aframomum daniellii	Bastered melegueta	Maniguette sucrée	Zingiberaceae	Fruit
Capsicum frutescens	Bird pepper	Petit pigment rouge	Solanaceae	Fruit
Dichrostachys glomerata	Sickle bush	oreilles de souris	Mimosaceae	Fruit
Dorstenia psilirus		Remèdes des serpents	Moraceae	Root
Echinops giganteus	Giant japanese butterbur	Racines tubéreuse	Asteraceae	Root
Fagara leprieuri	Prickly ash	Grappe odoriférante	Rutaceae	Fruit
Fagara xanthoxyloïdes		Bouche béante	Rutaceae	Fruit
Hua gabonii	Garlic tree	Fruit de l'arbre de l'ail	Huacaceae	Fruit
Hua gabonii	Garlic tree	Ecorce de l'arbre de l'ail	Huacaceae	Bark
Mondia whitei	White ginger	Racine sucrée	Periplocaceae	Root
Monodora myristica	Calabash nutmeg	Fausse noix de muscade	Annonaceae	Almond
Pentadiplandra brazzeana	Joy perfume tree	Liane blanche	Pentadipandraceae	Root
Piper guineense	Black pepper (Ashandi)	Poivre sauvage	Piperaceae	Fruit
Piper umbellatum	Cordoncillo		Piperaceae	Flower
Scleria striatinux		Racine de chaume	Cyperaceae	Fruit
Scorodophleus zenkeri	Divida (African)	Fruit de l'arbre à l'ail	Caesalpinaceae	Fruit
Scorodophleus zenkeri	Divida (African)	Ecorce de l'arbre à l'ail	Caesalpinaceae	Bark
Solanum melongena	Aubergine	Aubergine	Solonaceae	Fruit
Tetrapleura tetraptera	Aidan tree	Fruit à 4 ailes	Mimosaceae	Fruit
Xylopia aethiopica	Ethiopian pepper (African)	Poivre d'Éthiopie	Annonaceae	Fruit

The obtained powders were finally sealed in polyethylene bags for better conservation they were stored at 4°C in refrigerator until analysis is carried out.

2.2. Amino Acid Determination

The samples were hydrolyzed under vacuum at 110 °C for 24 h in a Pico-Tag apparatus (Waters, Milford, MA, USA) in the presence of 6N HCl containing 0.1% Phenol, then recovered by water ultra-pure and deposited on an auto tap-parser-420 trademark Applied Biosystems (Applera Corp., Foster City, CA, USA). The derivation to phenyl isothiocyanate (PITC) was made in pre column. The separation of amino acids was made on a channel high performance liquid chromatography (HPLC) (Applied Biosystems model 172 A: Corp, Foster City, CA, USA) with a column of phase reverse PTC RP-18 (2, 1 * 220 mm) series 238638, lot no. 02 M 9-40531 Brownlee (Applied Biosystems model 172 A: Corp, Foster City, CA, USA). Solvents used for this analysis are acetate Na 45 mM pH 5.90 (A) and 30% acetate Na 105 mM pH 4.60 and 70% acetonitrile (B), and the detection was made at 254 nm. The acquisition and exploitation of the results were made by the Model 600 Data Analysis System software (Applied Biosystems, Applera Corp., Foster City, CA, USA). Tryptophan was not determined.

2.3. Chemical Score

This index is one of the parameters for assessing the nutritional value. The chemical index was calculated according to the method used by Chavan *et al.* [10] using the following equation:

$$\text{Chemical score (CS)} = \frac{(\text{mg of essential amino acid per g of test protein})}{\begin{pmatrix}\text{mg of the same essential amino acid} \\ \text{per g of protein in requirement pattern}\end{pmatrix}} \times 100.$$

The requirement pattern suggests by the FAO/WHO/UNU [11] for children of 1-2 years old was used for this purpose: Histidine = 1.8 g/100g protein; Isoleucine = 3.1 g/100g protein; Leucine = 6.3 g/100g protein; Lysine = 5.2 g/100g protein; Methionine + cysteine = 2.5 g/100g protein; Phenylalanine + tyrosine = 4.6 g/100g protein; Threonine = 2.7 g/100g protein ; and Valine = 4.1 g/100g protein.

2.4. Fatty Acids Determination

Fatty acid methyl esters were prepared by the method of Ackman [12] using the reagent to the concentration of 8% boron trifluoride in methanol (BF_3/MetOH). About 100 mg of oil sample were weighed in a screw test tube 10 mL. Then 1.5 mL of hexane and 1.5 mL of BF_3/MetOH were added. Then the tube tightly closed under nitrogen and has been stirred vigorously, and heated to 100 °C for 1 h. After cooling to ambient temperature, 2 mL of distilled water and 1 mL of hexane were added and stirred under nitrogen. After resting, the upper phase was recovered in a tube under nitrogen. The lower phase extraction was repeated twice on using 1 mL of hexane.

Phase hexane (methyl esters) have been collected and washed with 2 mL of distilled water, dried on anhydrous sodium sulfate. The solvent was evaporated under nitrogen. Hexane has been added to bring up concentration of methyl esters in a suitable concentration (5 mg/mL) analysis by gas chromatography. This analysis was performed on a chromatogram Peri 2000 (Perichrom, Saulx-les-Chartreux, France) equipped with a flame ionization detector. The separation was made on a capillary column (25 mm in length, diameter 0.25 mm, film thickness is 0.5 µm) lined with polyethylene glycol (Perichrom) Terephthalic acid-doped. The detector and injector temperatures were set at 260 °C. The program used for analysis as follows: maintained at the start at 70 °C for 2 min, the column has been to 180 °C (39.9 °C/min) and maintained during 8 min at this temperature, then it undergoes a second phase of warming up to 220 °C (3 °C/min) for 45 min. The cooling is carried out at the rate of 39.9 °C/min. The identification of the peaks was carried out using a standard of fatty acid supplied by Supelco (Bellefonte, USA): PUFA 1 (vegetable source). Winlab 3 (Perichrom, Saulx-les-Chartreux, France) software has allowed the integration of the chromatograms.

2.5. Statistical Analysis

All experiments were conducted in triplicate. The variances of the data analysis and the comparison of means using Duncan multiple comparison test conducted in using statistics software Statgraphics 5.0 (Manugistics, Rockville, Maryland, USA, 1997) [13]. Statistical significance was accepted at a level of $p<0.05$.

3. Results and Discussion

3.1. Total Proteins and Non-essential Amino Acids Composition

Table 2 shows the total proteins and their different compositions in non -essential amino acids. These results indicated that all the analyzed parameters varied significantly ($p < 0.05$) between the spices. Furthermore, it may be remarked that the total protein levels ranged from 4.82% (*D. psilurus*) and 13.61% (*M. myristica*) [14]. These values are lower than those found in the wild sea bass (*Dicentrarchus labrax*) which varies from 18.74% to 21.79% [15]. This table presents also the results of the non-essential amino acid (Aspartate + Aspargine, Glutamate + Glutamine, Serine, Alanine, Proline, Arginine and Glycine). It appears also that, serine in spices *E. giganteus*, *M. myristica*, and *S. zenkeri* "bark", - alanine in spices *E. giganteus*, *S. zenkeri* "fruit" and *T. tetraptera*, - proline in *F. leprieuri* and *P. umbellatum*, - arginine in *F. leprieuri*, *P. umbellatum*, *P. brazzeana* and *T. tetrapleura* were unable to reach the limit of detection. On the other hand, the aspargine and aspartate values ranged from 0.88% (*S. zenkeri* "fruit") to 17.65% (*S. melongena*). Aspartic acid, which occurs as a physiological compound in the mammalian pituitary and testis, has a role in the regulation of the release and synthesis of LH and testosterone [16].

Table 2. Total protein content (g/100g DM) and Non-essential amino acid composition (g/100g protein) of Spices

Spices	Total Protein * (NX6.25)*	Aspartate	Glutamine	Serine	Alanine	Proline	Arginine	Glycine	TNEAA
A. daniellii	8.53±0.99^{cd}	11.62±0.04^{e}	19.51±0.03^{p}	6.92±0.03^{f}	6.00±0.06^{d}	4.53±0.03^{i}	3.46±0.04^{i}	10.6±0.07^{k}	52.04
C. frutescens	9.44±0.75^{def}	14.98±0.04^{k}	11.47±0.03^{d}	6.39±0.08^{d}	8.33±0.03^{g}	7.85±0.03^{p}	1.51±0.06^{d}	9.20±0.04^{h}	50.53
D. glomerata	8.61±0.44^{cde}	0.89±0.03^{a}	1.28±0.06^{b}	14.59±0.06^{n}	13.17±0.04^{n}	4.11±0.03^{h}	6.96±0.03^{m}	18.67±0.04^{o}	41
D. psilurus	4.82±0.15^{a}	15.70±0.06^{l}	21.37±0.06^{r}	7.13±0.04^{h}	7.47±0.03^{e}	3.19±0.03^{g}	1.36±0.07^{c}	11.78±0.03^{l}	56.22
E. giganteus	5.88±0.31^{b}	15.02±0.01^{k}	27.73±0.04^{s}	ND	ND	0.44±0.03^{b}	0.23±0.04^{a}	38.71±0.08^{r}	43.42
F. leprieuri	6.41±0.25^{b}	13.86±0.03^{i}	20.82±0.08^{q}	8.57±0.04^{l}	ND	ND	ND	21.32±0.04^{p}	43.25
F. xanthoxyloïdes	7.98±0.10^{c}	9.17±0.01^{b}	13.29±0.04^{g}	5.55±0.04^{c}	9.93±0.01^{l}	8.25±0.03^{q}	2.70±0.06^{g}	8.50±0.06^{f}	48.89
H. gabonii (fruit)	6.15±0.65^{b}	17.06±0.03^{m}	16.44±0.04^{l}	7.30±0.03^{h}	8.23±0.04^{f}	7.35±0.03^{n}	2.43±0.04^{e}	10.31±0.06^{j}	58.81
H. gabonii (bark)	10.16±0.40^{f}	12.85±0.01^{h}	12.99±0.06^{e}	7.25±0.03^{h}	9.28±0.04^{j}	2.53±0.03^{f}	4.60±0.04^{l}	9.28±0.06^{h}	49.5
M. myristica	13.61±0.16^{h}	15.69±0.04^{l}	18.49±0.03^{o}	ND	6.93±0.03^{c}	5.90±0.06^{k}	4.05±0.03^{k}	7.30±0.06^{c}	51.06
M. whitei	4.73±0.05^{a}	11.13±0.03^{d}	7.16±0.03^{c}	4.18±0.07^{b}	8.69±0.03^{i}	7.58±0.06^{o}	3.09±0.03^{h}	7.72±0.04^{d}	41.83
P. brazzeana	11.27±0.14^{g}	9.32±0.04^{c}	14.75±0.04^{i}	9.86±0.03^{m}	0.77±0.03^{a}	1.10±0.03^{d}	ND	7.17±0.04^{b}	35.8
P. gunineense	8.52±0.23^{cd}	15.05±0.04^{k}	16.86±0.03^{n}	7.06±0.03^{g}	9.68±0.08^{k}	6.82±0.03^{l}	3.64±0.04^{j}	9.66±0.03^{i}	59.11
P. umbellatum	11.29±0.06^{g}	11.68±0.03^{e}	17.31±0.06^{n}	6.56±0.03^{e}	1.16±0.04^{b}	ND	ND	27.87±0.03^{q}	36.71
Scleria striatinux	9.50±0.21^{ef}	12.01±0.06^{f}	14.19±0.06^{h}	6.62±0.04^{e}	8.55±0.06^{h}	7.03±0.03^{m}	1.11±0.04^{b}	8.99±0.03^{g}	49.51
S. melongena	4.93±0.21^{a}	17.65±0.06^{o}	16.33±0.03^{k}	8.48±0.06^{k}	8.37±0.04^{g}	8.78±0.03^{r}	0.17±0.03^{a}	13.18±0.06^{n}	59.78
S. zenkeri (fruit)	12.17±0.74^{g}	12.25±0.03^{g}	15.82±0.01^{j}	3.32±0.03^{a}	ND	0.73±0.01^{c}	14.17±0.03^{n}	8.09±0.04^{e}	46.29
S. zenkeri (bark)	13.97±0.12^{h}	0.88±0.01^{a}	0.31±0.03^{a}	ND	12.93±0.04^{m}	1.36±0.01^{e}	1.06±0.03^{b}	3.64±0.04^{a}	16.54
T. tetraptera	4.96±0.28^{a}	14.47±0.04^{j}	19.46±0.03^{p}	7.64±0.04^{i}	ND	0.24±0.04^{a}	ND	42.34±0.04^{s}	41.81
X. aethiopica	7.92±0.30^{c}	17.52±0.04^{n}	16.78±0.03^{m}	8.11±0. 03^{j}	6.98±0.01^{c}	4.79±0.06^{j}	2.60±0.07^{f}	12.25±0.06^{m}	56.78

*Source: Abdou Bouba et al. (2012). Mean values that have the same superscripts letters in the same column are not significantly different at the 5% level; Mean ± standard deviation; n = 3. ND = Non Detected

Asparagine provides for a particular type of amine transamination in the liver and participates in the control of the metabolic functions of brain and central nervous system cells as well as the treatment of its [17]. Glutamate and glutamine were most abundant in all studied spices. The same observation was made by Bei-Zhong *et al.* [18] on nine samples of soybean in China. In our samples, concentrations ranged from 0.31% (*S. zenkeri* "bark") to 27.73% (*E. giganteus*). Glutamic acid is an exciting neurotransmitter for the central nervous system, the brain and the spinal cord. It serves as fuel for the brain and helps to correct the physiological imbalances in the body. Glutamine is the amino acid which is more abundant in muscles; its presence allows the building and maintaining muscle tissue. Glutamine struggles against the muscle hypertrophy which may appear after a period of prolonged rest or diseases such as cancer and AIDS [19]. The content in serine, in other spices varies from 3.32% (*S. zenkeri* "fruit") to 14.59% (*D. glomerata*). It is an amino acid necessary for the development of the muscles and the maintenance of the immune system. It is important in the formation of RNA and DNA from the cells. Alanine values ranged from 0.77% (*P. brazzeana*) and 13.17% (*D. glomerata*). It plays an important role in the transfer of nitrogen in the body; as well as glucose that the body uses as energy and strengthens the immune system producing antibodies. It also regulates toxic substances discharged in their inherent energy need muscles. The percentage of proline fluctuates between 0.24% (*T. tetraptera*) to 8.78% (*S. melongena*). This amino acid is to improve the texture of skin by facilitating the production of collagen and slowing the aging process and is used in curative treatment to avoid the problems of cartilage, tendons, muscle of the heart [20]. Arginine varies with contents that are set between 0.17% (*S. melongena*) and 14.17% (*S. zenkeri* "fruit"). Arginine is considered to be the "natural Viagra" that increases blood flow in the penis [20]. This acid also delays the growth of cancerous tumors by increasing the immune system and helps the detoxification of the liver by neutralizing ammonia and reduces the toxicity of alcohol. Also, it is frequently used in the treatment of infertility in men. The content of glycine in spices varied from 3.64% (*S. zenkeri* "fruit") to 42.34% (*T. tetraptera*). Furthermore, Glycine retards muscle degeneration, it improves the glycogen storage and thereby releases glucose to meet the energy needs. It is also developing an excellent scaring function. Contrary to the results obtained by Chavan *et al.* [10] on pea Beach to the Canada (54.1% to 51, 7%) we note that our values are low in general.

3.2. Essential Amino Acids Composition

The essential amino acid content (EAA) of each 20 spices summarized in Table 3. The total amounts of EAA ranged from 15.85% (*T. tetraptera*) and 79.81% (*S. zenkeri "fruit"*). Furthermore, these results were higher than those obtained by Jiyu *et al.* [21] on certain cultivars of *Myrica rubra* (27.83 to 29.91%) in China and on *Riconodendron heudelotii* and *Tetracarpidium conophorum* (up to 33 %) in Cameroon by Tchiegang *et al.* [22]. So, it concluded from these values that the spice *S. zenkeri* (fruit) could be a good source of protein supplement in essential amino acids.

Table 3 also shown that the distinction of the spices on the basis of their total namely; essential amino acid content as follows: - spices with a very high content of total amino acids have a value of 79.81% (*S. zenkeri "fruit"*), - the total essential amino acids spice relatively high with *P. brazzeana* (57.94%), *M. whitei* (50.46%) and *S. zenkeri* (bark), - spices with intermediate levels represented by *F. xanthoxyloides* (42.63%), *S. melongena* (41.51%), *M. myristica* (41.67%), *H. gabonii "bark"* (41.31%), *C. frutescens* (40.27%), *D. glomerata* (40.33%), *A. daniellii* (37.35%), *P. umbellatum* (35.42%) and *F. leprieuri* (35,33%), - Spices with average grades of *D. psilurus* (32%), *P. guineense* (31.93%), *H. gabonii* (fruit) (30.84%), *X. aethiopica* (30.96%) and *S. striatinux* (27.05%), - and spices to low levels of essential amino acids by *E. giganteus* (17.85%) and *T. tetraptera* (15.85%). In addition, when comparing the levels than the sources of plant proteins such as safflower seeds (15.13%) [23], pea (42.18% and 41.3%) [21] and the nuclei of *Myrica rubra* (27.83 to 29.21%), respectively [10], its realized that certain levels of proteins of our spices have essential amino acids with lower percentages and others high.

Table 3. Essential amino acid composition (g/100g protein) of spice

Spices	E A A	Histidine	Valine	Met + Cys	Threonine	Isoleucine	Leucine	Phe + Tyr	Lysine	TEAA/TAA
A. daniellii	37.35 ± 50.18^{g}	ND	9.81 ± 0.04^{m}	20.69 ± 0.04^{l}	1.94 ± 0.04^{a}	4.83 ± 0.04^{d}	ND	ND	0.08 ± 0.01^{a}	0.72
C. frutescens	40.27 ± 0.14^{h}	2.06 ± 0.01^{e}	2.23 ± 0.04^{h}	4.38 ± 0.07^{e}	6.39 ± 0.04^{f}	6.82 ± 0.06^{i}	7.68 ± 0.07^{i}	5.63 ± 0.04^{h}	5.08 ± 0.06^{f}	0.8
D. glomerata	40.33 ± 0.28^{h}	ND	ND	ND	ND	9.73 ± 0.03^{m}	3.89 ± 0.06^{d}	11.45 ± 0.04^{l}	15.32 ± 0.06^{q}	0.98
D. psilurus	32 ± 0.05^{e}	0.86 ± 0.03^{c}	7.92 ± 0.04^{l}	ND	5.95 ± 0.01^{e}	8.8 ± 0.07^{l}	ND	3.92 ± 0.06^{c}	4.55 ± 0.03^{e}	0.57
E. giganteus	17.85 ± 0.02^{b}	ND	ND	6.72 ± 0.06^{f}	2.35 ± 0.03^{b}	ND	ND	ND	8.78 ± 0.03^{n}	0.41
F. leprieuri	35.33 ± 0.14^{f}	0.72 ± 0.01^{b}	ND	7.18 ± 0.03^{g}	ND	12.9 ± 0.01^{o}	4.83 ± 0.03^{e}	ND	9.8 ± 0.03^{o}	0.82
F. xanthoxyloïdes	42.63 ± 0.07^{k}	1.61 ± 0.04^{d}	0.83 ± 0.04^{f}	7.72 ± 0.07^{i}	3.14 ± 0.04^{c}	6.69 ± 0.04^{g}	11.83 ± 0.03^{o}	5.33 ± 0.04^{g}	5.47 ± 0.04^{g}	0.87
H. gabonii (fruit)	30.84 ± 0.13^{d}	ND	2.57 ± 0.03^{i}	0.73 ± 0.04^{b}	ND	6.07 ± 0.03^{e}	7.84 ± 0.06^{j}	5.77 ± 0.08^{i}	7.91 ± 0.04^{l}	0.53
H. gabonii (bark)	41.31 ± 0.14^{i}	4.54 ± 0.04^{i}	0.65 ± 0.06^{e}	3.31 ± 0.01^{d}	9.56 ± 0.07^{h}	7.7 ± 0.04^{k}	3.27 ± 0.03^{c}	5.1 ± 0.03^{f}	7.08 ± 0.07^{k}	0.83
M. myristica	41.67 ± 0.31^{j}	9.55 ± 0.06^{k}	2.15 ± 0.03^{g}	0.15 ± 0.01^{a}	ND	6.51 ± 0.03^{h}	7.89 ± 0.01^{j}	8.9 ± 0.03^{k}	6.49 ± 0.06^{i}	0.82
M. whitei	50.46 ± 0.07^{m}	1.58 ± 0.05^{d}	0.22 ± 0.03^{c}	0.22 ± 0.01^{a}	6.44 ± 0.07^{f}	4.06 ± 0.01^{c}	6.4 ± 0.07^{f}	14.44 ± 0.04^{m}	17.1 ± 0.04^{r}	1.21
P. brazzeana	57.94 ± 0.06^{n}	7.55 ± 0.06^{j}	0.41 ± 0.04^{d}	36.22 ± 0.04^{n}	ND	10.45 ± 0.03^{n}	ND	2.46 ± 0.07^{b}	0.85 ± 0.03^{b}	1.62
P. gunineense	31.93 ± 0.08^{e}	0.08 ± 0.01^{a}	6.09 ± 0.06^{j}	0.98 ± 0.04^{c}	4.16 ± 0.03^{d}	6.45 ± 0.04^{h}	8.08 ± 0.04^{k}	ND	5.4 ± 0.06^{g}	0.53
P. umbellatum	35.42 ± 0.28^{f}	ND	ND	7.28 ± 0.03^{h}	6.79 ± 0.06^{g}	0.14 ± 0.01^{a}	0.085 ± 0.01^{a}	17.31 ± 0.03^{n}	3.81 ± 0.07^{d}	0.96
Scleria striatinux	41.51 ± 0.14^{ij}	2.43 ± 0.03^{f}	0.13 ± 0.04^{b}	20.33 ± 0.03^{k}	2.31 ± 0.03^{b}	3.92 ± 0.04^{b}	2.85 ± 0.03^{b}	7.45 ± 0.04^{j}	2.09 ± 0.04^{c}	0.84
S. melongena	27.05 ± 0.11^{c}	3.73 ± 0.04^{h}	0.09 ± 0.01^{ab}	ND	ND	7.24 ± 0.04^{j}	9.68 ± 0.04^{m}	0.53 ± 0.03^{a}	5.78 ± 0.06^{h}	0.45
S. zenkeri (fruit)	45.62 ± 0.28^{l}	10.7 ± 0.06^{l}	ND	ND	ND	8.76 ± 0.03^{l}	9.79 ± 0.03^{n}	4.04 ± 0.03^{d}	12.33 ± 0.06^{p}	0.98
S. zenkeri (bark)	79.81 ± 0.14^{o}	3.18 ± 0.04^{g}	ND	11.11 ± 0.03^{j}	39.41 ± 0.04^{i}	13.59 ± 0.04^{p}	6.64 ± 0.04^{g}	ND	5.88 ± 0.06^{h}	4.83
T. tetraptera	15.85 ± 0.07^{a}	ND	ND	ND	ND	ND	7.23 ± 0.03^{h}	ND	8.62 ± 0.04^{m}	0.38
X. aethiopica	30.96 ± 0.02^{d}	3.17 ± 0.03^{g}	7.27 ± 0.08^{k}	ND	ND	ND	9.1 ± 0.04^{l}	4.45 ± 0.06^{e}	6.97 ± 0.07^{j}	0.55

Mean values that have the same superscripts letters in the same column are not significantly different at the 5% level; Met = methionine; Cys = cysteine, Phe = phenylalanine; Tyr = tyrosine; Mean ± standard deviation; n = 3. ND = Non Detected.

Table 3 also was shown that the levels of essential amino acids of the different studied spices. Except lysine, others amino acids could not be detected in certain spices because of the lower sensitivity of the device used for this purpose. Also, the hisditine could not be detected in the spices *A. daniellii, D. glomerata, T. Tetraptera, H. gabonii «fruit», E. giganteus and P. umbellatum,-* valine in *D. glomerata, E. giganteus, F. leprieuri, P. umbellatum,*

S. zenkeri 'fruit and bark' and T. tetraptera,- also, methionine and cysteine not detected in *D. glomerata, D. psilurus, P. umbellatum, S. melongena, S. zenkeri "fruit", T. tetraptera and X. aethiopica*, - threonine in *D. glomerata, F. leprieuri, H. gabonii «fruit», M. myristica, P. brazzeana, S. melongena, S. zenkeri "fruit", T. tetraptera and X. aethiopica*, - isoleucine in *E. giganteus, T. tetraptera* and *X. aethiopica*, - leucine in *A. daniellii, D. psilurus*, and *P. brazzeana*, - and the phenylalnine and tyrosine in *A. daniellii, F. leprieuri* and *E. giganteus, P. guineense, S. zenkeri* "bark" and *T. tetraptera*.

However, the values of the hisditine in the other various spices range from 0.72% (*F. leprieurii*) to 10.7% (S. zenkeri "fruit"). Furthermore, Histidine is used in the treatment of cardiovascular disease with physiological antioxidant role it plays on the free radicals (hydroxyl radical and Singlet oxygen) [24]. According to WHO, the daily need in this amino acid is 12 mg/Kg or 840 mg to 70 Kg of body weight. Also, Methionine plus cysteine, have values that lie between 0.15% (*M. myristica*) and 36.22% (*P. brazzeana*). It is important to mention also that methionine acid is an antioxidant with high sulphur content [25], that helps prevent deficiencies of hair cells, skin, and nails. Furthermore, this amino acid protect against greasy clusters around the liver and the arteries that cause obstructions. It promotes detoxification of harmful agents such as lead and other heavy metals. Also, Cysteine, works as a powerful antioxidant to eliminate harmful toxins [25]. Therefore, these two sulfur amino acid deficiency would cause physiological disturbances, even the risks of contracting degenerative diseases [4]. Valine values oscillate between 0.09% (S. *melongena*), and 9.81% (*A. daniellii*). Medically, this amino acid is used in the treatment of liver diseases, gall bladder and promotes the intellectual liveliness [25]. The results also indicated that the value of threonine varied from 1.94% (*A. daniellii*) to 39.41% (*S. zenkeri* "bark"). Also, this amino acid permits to maintain the balanced intake of proteins in the body and is also an important part in the formation of dental enamel, the collagen and elastin. The amounts of isoleucine in the studied spices range from 0.14% (*P. umbellatum*) to 13.59% (*S. zenkeri* "bark"). In the deficiency, the isoleucine causes physical and mental disorders. For leucine, values were included between 0.08% (*P. umbellatum*) and 11.83% (*D. glomerata*). Leucine plays with isoleucine and valine to promote muscle function, skin and bones [20]. Phenylalanine plus tyrosine have values that fluctuate between 0.53% (*S. melongena*), and 17.31% (*P. umbellatum*). In addition, it is recognized that the phenylalanine is used by the brain to produce norepinephrine, a chemical that transmits signals between nerve cells. Also, it promotes alertness and vitality, regulates human mood and reduces the pain. This acid also used in the treatment of arthritis, depression, painful menstruation, migraine, obesity, the Parkinson's disease and schizophrenia [20,26]. The levels of lysine, which is an amino acid in different spices ranged from 0.08% (*A. daniellii*) and 15.32% (*D. glomerata*). These results revealed that individual amino acid varies in general from 10.7% (histidine) to 39.41% (threonine). The proteins are considered good nutritive value if their essential amino acid profile is higher than the reference level required by at least pre-school children (2-5 years that need most of these amino acids) due to inability of the human body to synthesize FAO/WHO/UNU [11] has recommended that pre-school children need a minimum of 34, 35, 25, 28, 66, 63 and 58 mg/g protein for Thr, Val, (Met+Cys), Ile, Leu, (Phe+Tyr) and Lys, respectively, in their protein diet [27]. In this respect, taking alone, all spices were deficient of some essential amino acids. For instance, it is recommended to use these spices in the mixture to equilibrate this balance.

3.3. Chemical Score

The chemical score proves to be a powerful tool to assess the nutritional value of a protein because it reflects the ability of the analyzed protein or food protein to contain of the essential amino acids, i.e. not synthesizable by the human body. The results cited in Table 4, shown that the chemical score index varied significantly ($p < 0.05$) between various essential amino acids for each of the spices. Furthermore, the chemical index of the histidine of the studied spices varied from 0.57 (*A. daniellii*) to 545.55 (*S. zenkeri* "fruit"). The valine oscillated between 0.00 (*D. glomerata*) and 203.75 (*A. daniellii*), methionine + cysteine, it fluctuated between 0.20 (*S. zenkeri* "fruit") to 564.44 (*P. brazzeana*), then for threonine, with values ranged between 0.22 (*S. zenkeri* "fruit") and 537.92 (*S. zenkeri* "bark"). The chemical score index of isoleucine varied from 0.25 (*X. aethiopica*) to 253.77 (*S. melongena*), while that of the leucine fluctuated between 0.10 (*P. brazzeana*) and 271.57 (*S. zenkeri* "bark"), and as the phenylalanine, more tyrosine which also ranged from 0.06 (*S. zenkeri* "bark") to 268.97 (*P. umbellatum*) and finally the chemical index of lysine whose values lie between 1.56 (*A. daniellii*) and 397.78 (*T. tetraptera*).

Furthermore, it is important in the same table to noted that: - histidine was the first limiting amino acid for some spices (*A. daniellii, D. glomerata, Echinops gigateus, P. umbellatum* and *T. tetraptera*), and also the second limiting amino acid for others (*D. psilurus, F. leprieuri, F. xanthoxyloides* and *P. guineense*); - While Valine was the first limiting amino acid for some spices (*C. frutescens, D. glomerata, Echinops gigateus, F. leprieuri, F. xanthoxyloïdes, H. gabonii (bark), M. whitei, P. guineense, S. Striatinux* and *S. zenkeri* (fruit and bark) and it was considered as the second limiting amino acid for: *P. brazzeana* and *S. melongena;* - In addition the other spices were *D. glomerata, D. psilurus, S. melongena* and *S. zenkeri* (fruit), Meth+Cys were the first amino acids and the second for others spices (*H. gabonii* (fruit), *M. myristica* and *M. whitei*); - Concerning threonine, it is the first limiting amino acid for *D. glomerata, F. leprieuri* and *H. gabonii* (fruit) and represented the second for *A. daniellii* and *Echinops gigateus;* - In respect of the isoleucine, it was the major limiting amino acid for the spices: *Echinops gigateus, T. tetraptera* and *X. aethiopica*; - Leucine meanwhile, was the first limiting essential amino acide for: *A. daniellii, D. psilurus, Echinops gigateus* and *P. brazzeana* and it was the second one in the: *D. glomerata, H. gabonii* (bark), *P. umbellatum, S. zenkeri* (bark) and *T. tetraptera*; - while In spices *A. daniellii, Echinops gigateus, F. leprieuri, P. brazzeana, S. zenkeri* (bark) and *T. tetraptera,* Phe+Tyr were the first limiting essential amino acid and the second limiting amino acid of *S. zenkeri* (fruit) and *X. aethiopica*; - About the lysine, it was the first limiting amino acid in *A. daniellii* and the second for: *C. frutescens* and *S. Striatinux*.

Based on these results, it concluded that certain of spices used in this study could be qualified as a good source of protein because the quality of protein in food was determined by the content of amino acids especially the essential amino acids [20]. However, to be effective consumption of the mixture of these spices is recommended to cover the needs for essential amino acid.

Table 4. Chemical score of spices (%)

Spices	Histidine	Valine	Meth + Cys	Threonine	Isoleucine	Leucine	Phe+ Tyr	Lysine
A. daniellii	ND	239.27	827.6	71.85	155.81	ND	ND	1.54
C. frutescens	114.44	54.39	175.2	236.66	220	121.90	122.39	97.69
D. glomerata	ND	ND	ND	ND	313.87	61.75	248.91	294.62
D. psilurus	47.77	193.17	ND	220.37	283.87	ND	85.22	87.5
E. giganteus	ND	ND	268.8	87.04	ND	ND	ND	168.85
F. leprieuri	40	ND	287.2	ND	416.13	76.66	ND	188.46
F. xanthoxyloïdes	89.44	20.24	308.8	116.30	215.81	187.77	115.87	105.19
H. gabonii (fruit)	ND	62.68	29.2	ND	195.81	124.44	125.43	152.12
H. gabonii (bark)	252.22	15.85	132.4	354.07	248.39	51.90	110.87	136.15
M. myristica	530.55	52.44	6	ND	210	125.24	193.48	124.81
M. whitei	87.77	5.36	8.8	238.52	130.97	101.59	313.91	328.85
P. brazzeana	419.44	10	1448.8	ND	337.10	ND	53.48	16.35
P. gunineense	4.44	148.54	39.2	154.07	208.06	128.25	ND	103.85
P. umbellatum	ND	ND	291.2	251.48	4.52	1.35	376.30	73.27
Scleria striatinux	135	3.17	813.2	85.55	126.45	45.24	161.95	40.19
S. melongena	207.22	2.19	ND	ND	233.55	153.65	11.52	111.15
S. zenkeri (fruit)	594.44	ND	ND	ND	282.58	155.39	87.83	237.12
S. zenkeri (bark)	176.66	ND	444.4	1459.63	438.39	105.39	ND	113.08
T. tetraptera	ND	ND	ND	ND	ND	114.76	ND	165.77
X. aethiopica	176.11	177.32	ND	ND	ND	144.44	96.74	134.04
FAO/WHO/UNU [2007] reference pattern	1.8	4.1	2.5	2.7	3.1	6.3	4.6	5.2

Mean values that have the same superscripts letters in the same column are not significantly different at the 5% level; Met = methionine; Cys = cysteine. Phe = phenylalanine; Tyr = tyrosine; Mean ± standard deviation; n = 3.

3.4. Fat and Fatty Acid Composition

The characterization and levels of fatty acids of oils of 20 studied spices recorded in Table 5 and Table 5 continued. This tables shown that these oils formed by the chain of fatty acids ranged from 8 to 24 carbons. Saturated fatty acids varied rare in our spices (C8: 0, C12: 0, C13: 0, C16: 0, C17: 0, C18: 0, C22:0). Caprylic acid (C8), described as partially volatile, presented in only two spices with 18.74% for *E. giganteus* and 6.56% for *S. melongena*. On the other hand, the C10 (capric acid) was much in *H. gabonii* (fruit) with a content of 49.9%. The presence of these fatty acids with short chain (C6 to C10) in relatively high quantities and the development of fairly strong odors have been demonstrated on *Morinda citrifolia* [28]. Indeed, these short-chain acids were highly volatile and could be released by intracellular lipase during technological treatments.

Table 5. Fats content (g/100g DM) and fatty acid composition of spices as g/100g DM of total fatty acids

Spices	Fats content*	C8	C10	C12	C13	C14: 1n9	C16	C16 : 1n9C	C17	C18
A. daniellii	23.1 ± 0.43^{l}	ND	ND	ND	2.73±0.10^{b}	2.42±0.04^{a}	ND	17.52±0.12^{f}	ND	ND
C. frutescens	11.09 ± 0.16^{e}	ND	ND	ND	ND	ND	ND	20.94±0.02^{i}	ND	ND
D. glomerata	11.25 ± 0.35^{e}	ND	ND	ND	ND	ND	ND	17±0.08ef	ND	ND
D. psilurus	15.33±0.63^{h}	ND	ND	4.57±0.99^{b}	ND	5.37±0.46^{b}	ND	16.14±0.48^{cd}	ND	ND
E. giganteus	12.24±0.28^{f}	18.74±0.28^{b}	1.39±0.28^{a}	ND	ND	ND	ND	12.34±0.25^{b}	ND	ND
F. leprieuri	32.14±0.14^{m}	ND	ND	ND	ND	ND	ND	31.58±0.56^{m}	ND	ND
F. xanthoxyloïdes	19.21 ± 0.76^{j}	ND	ND	ND	ND	ND	ND	19.86±0.28^{h}	ND	ND
H. gabonii (fruit)	16.96 ± 0.23^{i}	ND	49.9±0.13^{b}	ND	ND	ND	ND	13±0.28^{b}	ND	3.83±0.14
H. gabonii (bark)	2.19 ± 0.21^{a}	ND	ND	ND	ND	ND	ND	26.21±0.28^{l}	ND	ND
M. myristica	53.36±0.14^{o}	ND	ND	2.21±0.15^{a}	ND	ND	ND	ND	ND	ND
M. whitei	4.16 ± 0.14^{b}	ND	ND	ND	ND	ND	7.62±0.44^{b}	ND	ND	ND
P. brazzeana	6.15 ± 0.16^{c}	ND	ND	ND	ND	ND	ND	26.7±0.21^{l}	ND	ND
P. gunineense	20.31 ± 0.46^{k}	ND	ND	ND	ND	ND	ND	24.74±0.34^{k}	ND	ND
P. umbellatum	5.98 ± 0.26^{c}	ND	ND	5.89±0.41^{b}	ND	ND	ND	15.51±0.15^{c}	ND	ND
Scleria striatinux	17.39±0.79^{i}	6.56±0.36^{a}	5.39±0.01^{b}	ND	ND	ND	ND	16.74±0.42^{de}	ND	ND
S. melongena	7.14±0.33^{d}	ND	7.24±0.28^{c}	ND	1.28±0.14^{a}	ND	ND	26.93±0.56^{l}	6.4±0.20	ND
S. zenkeri (fruit)	1.95 ± 0.24^{a}	ND	ND	ND	ND	ND	ND	18.57±0.28^{g}	ND	ND
S. zenkeri (bark)	3.65 ± 0.21^{b}	ND	ND	ND	ND	5.61±0.28^{b}	5.89±0.14^{a}	23.41±0.14j	ND	ND
T. tetraptera	5.64 ± 0.32^{c}	ND	ND	ND	ND	ND	ND	22.77±0.85^{j}	ND	ND
X. aethiopica	33.69±0.72^{n}	ND	ND	ND	ND	ND	ND	10.13±0.14^{a}	ND	ND

*Source: Abdou Bouba et al. (2012). Mean values that have the same superscripts letters in the same column are significantly different at 5%; Mean ± standard deviation; n = 3. ND = Non Detected

Table 5. Continued (g/100g DM)

Spices	C18: 1n9t	C18 : 2n9C	C18 : 2n6C	C18 : 3n3	C20	C20 : 1	C22	C22 : 1n9	C22: 6n3	C24: 1n9
A. daniellii	ND	3.47±0.01^{bc}	25.69±0.65^{e}	9.37±0.26^{b}	ND	ND	ND	ND	8.29±0.14^{b}	ND
C. frutescens	ND	4.94±0.04^{ef}	31.76±0.08^{h}	31.98±0.05^{g}	8.54±0.11^{gh}	ND	ND	ND	ND	ND
D. glomerata	ND	8.62±0.17^{h}	23.15±0.20^{d}	26.09±0.69^{f}	9.87±0.56^{i}	ND	6.23±0.17^{c}	6.68±0.66^{c}	ND	ND
D. psilurus	ND	6.74±0.11^{g}	49.29±1.42^{n}	ND	5.95±1.27^{cd}	ND	ND	ND	13.5±0.42^{d}	ND
E. giganteus	ND	3.62±0.28^{bc}	36.88±0.45^{k}	23.6±0.19^{e}	3.91±0.14^{b}	ND	ND	ND	ND	ND
F. leprieuri	ND	4.83±0.80^{e}	18.97±0.56^{c}	40.07±0.60^{k}	4.77±0.28^{bc}	ND	ND	ND	ND	ND
F. xanthoxyloïdes	ND	3.27±0.08^{b}	43.79±0.48^{l}	24.08±0.08^{e}	8.96±0.56^{hi}	ND	ND	ND	ND	ND
H. gabonii (fruit)	ND	5.59±0.28^{f}	17.93±0.07^{b}	34.84±0.76^{ij}	ND	ND	4.03±0.04^{a}	3.95±0.42^{a}	13.17±0.49^{d}	ND
H. gabonii (bark)	3.65±0.28^{b}	ND	47.91±0.28^{m}	16.42±0.14^{d}	6.26±0.14^{d}	ND	ND	ND	ND	ND
M. myristica	ND	ND	ND	ND	2.45±0.14^{a}	ND	ND	ND	ND	ND
M. whitei	ND	ND	32.9±0.56^{i}	54.77±0.99^{l}	ND	ND	ND	ND	ND	ND
P. brazzeana	ND	3.98±0.28^{d}	35.97±0.1^{j}	26.96±0.28^{f}	6.39±0.28^{ef}	ND	ND	ND	ND	ND
P. gunineense	ND	2.45±0.35^{a}	39.5±0.70^{k}	14.74±0.08^{c}	5.8±0.28^{cd}	ND	ND	12.76±0.08^{d}	ND	ND
P. umbellatum	ND	3.84±0.05^{bc}	30.08±0.11^{g}	39.25±0.21^{k}	5.43±0.14cd	ND	ND	ND	ND	ND
Scleria striatinux	ND	3.68±0.37^{bc}	28.26±0.28^{f}	34.23±0.14^{hi}	5.19±0.14^{cd}	ND	ND	ND	ND	ND
S. melongena	5.25± 0.28^{c}	8.12±0.14^{h}	28.82±0.28^{f}	ND	7.59±0.28^{fg}	ND	4.85±0.28^{b}	ND	5.59±0.13^{a}	ND
S. zenkeri (fruit)	6.57± 0.14^{d}	ND	18.18±0.14^{bc}	35.68±0.28^{j}	2.16±0.14^{a}	3.48±0.28	ND	5.19±0.11^{b}	11.29±0.15^{c}	ND
S. zenkeri (bark)	2.27± 0.07^{a}	ND	13.03±0.04^{a}	26.69±0.05^{f}	4.91±0.07^{bc}	ND	ND	ND	20.45±0.07^{e}	ND
T. tetraptera	ND	ND	ND	33.64±0.42^{h}	5.96±1.55^{cd}	ND	ND	ND	11.9±0.28^{c}	ND
X. aethiopica	ND	ND	17.55±0.42^{bc}	6.72±0.11^{a}	ND	ND	3.86±0.08^{a}	3.87±0.03^{a}	30.1±0.07^{f}	22 .81±0.04

Mean values that have the same superscripts letters in the same column are significantly different at 5%; Mean ± standard deviation; n = 3. ND = Non Detected.

N-heptadecanoic acid (C17: 0) was not naturally present in vegetable oils even if amounts were observed in *S. striatinux*. The lauric (C12: 0) which is found mainly in seeds with oils are called lauric as oil of coconut or palm kernel. It also varied at a few levels of our spices (*D. psilurus, M. myristica, P. umbellatum*).

Nevertheless, the high levels of unsaturated fatty acids in long-chain found at least in all the spices. It's the case of: C18:2n9C, C18:2n6C, C18:3n3 which these levels varied significantly ($p<0.05$). Contrary to the other unsaturated fatty acids, fatty acid C18:1n9t, C22:1n9, C22: 6n3 and C24:1n9 varied and were not found in all the spices. In general, the unsaturated fatty acids belong on the diet plan in three main series: the ω-3 and ω-6 and ω-9. The ω-3 founded primarily in fish oils and certain vegetable oils while the ω-6 is founded primarily in vegetable oils. They play a significant role in the development and functioning of the brain [29].

Palmitic acid (C16: 1n9) which is ω-9 varied considerably between the spices. The highest content was encountered in *F. leprieuri* (31.58%) while the spices such as *M. myristica* and *M. whitei* were not detectable. However these values have proved to be lower than those found in olive oil which characterized by a high content fatty acid C18: 1n9C (72%). Monounsaturated fatty acids allow the reduction of LDL cholesterol in serum and as such plays a protective role in several cardiovascular diseases [23]. C18:2n9C fatty acid was presented in all the spices except *T. tetraptera* and *D. psilurus* which presented the highest content (49.29%) followed by *H. gabonii* "bark" (47.91%), *F. xanthoxyloides* (43.79%) and *P. guineense* (39.5%), then the spice having average grades with *E. giganteus* (36.38%).

Fatty acid C24: 1n9 is not common. It was located only in *X. aethiopica* (22.81%). It could be considered not only as a nutritional fatty acid but also share nutraceutical polyunsaturated long-chain according to its structure. Consumption of fatty acid to long strings such as arachidonic acid, precursor of prostaglandins has an antiplatelet effect [30].

In general, the ω-3 fatty acids play an important role in the development and maintenance of the human bodies, especially of the brain and which involved in the prevention of various diseases such as cardiovascular, psychiatric, neurological, dermatological diseases and rheumatic disorders [7,29,31,32]. The ω -3 fatty acids are essential nutrients for humans because it cannot be manufactured and then recommended daily, doses which range between 0.18 and 10.15 g/day. In our spices, the ω-3 fatty acids were fatty acid C18:3n3 and C22: 6n3. The content of fatty acid C18:3n3 varied significantly ($p < 0.05$) between the spices. Particularly rich in ω - 3 spices were: *M. whitei* (54.77%), *F. leprieuri* (40.07%), *P. umbellatum* (39.25%), *S. zenkeri* "fruit" (35.65%), *H. gabonii* (fruit) (34.84%), *T. tetraptera* (33.64%), *C. frutescens* (31.98%). While the fatty acid C22:6n3 were presented of eight of the studied spices. The higher content was *X. aethiopica* (30.1%) and the smallest content is the *S. striatinux* (5.59%). It is important to recall that the main dietary sources of these fatty acids are usually fish and marine products. Especially, the consumption of fatty acids (Eicosapentaenoic Acid: C20: 5ω - 3 and acid docohexaenoique: C22: 6ω-3) induced a decrease in triglycerides due to a reduction in the production of LDL, and sometimes, an increase of HDL [29,33]. For this reason, the spices rich in ω - 3 spices could be considered as an alternative source in ω - 3 such as *M. whitei, X. aethiopica, F. leprieuri, P. umbellatum, T. tetraptera, S. zenkeri* (fruit), *H. gabonii* (fruit) to complement the consumption of ω-3 from fish and marine products. It was a good new source of ω - 3 in our diet [34]. In addition it has been shown that there is a positive correlation between unsaturated fatty acids and lower cholesterol level in the plasma [12].

Linoleic acid (C18:2n6) still called ω - 6 is an essential fatty acid, and its content in our spices varies significantly ($p < 0.05$). Almost all spices have this fatty acid. The highest levels were found in *D. psilurus* (49.29%), *H. gabonii* (bark) (47.91%), *F. xanthoxyloides* (43.79%) and the lowest value was found in the *S. zenkeri* (bark) (13.03%). These values are similar to those founded in olive oil with 53.0% [28] and in the goat milk [8]. Oils rich in linoleic acid are important for human health. These oils reduce atherosclerosis by interaction with HDL in blood [Necmettin *et al.*, 2006] and the cholesterol levels of 10% with a maximum effect for a ratio of PUFA/SFA close to [35]. Some polyunsaturated fatty acids such as linoleic acid and arachidonic acid known as vitamin F are required for the growth and protection of the skin. These essential fatty acids deficiency would result in a skin condition such as alopecia and edema. This vitamin F is incorporated, about 5%, in some dermatological creams, especially in tanning lotions. This shows that certain spices such as *M. whitei, F. leprieuri, D. psilurus, H. gabonii* (bark), *F. xanthoxyloides* could be used in cosmetics [36].

4. Conclusions

With a view to improve the nutritional status and enhance natural resources in use, only the analysis of some nutrients of spices allowed retaining in a general way that, almost, all the spices powder analyzed are considered as nutritional resources. Nevertheless, *S. zenkeri* (bark), *P. brazzeana* and *M. whitei* are sources of essential amino acids whose levels are high. While spices *P. brazzeana*, *A. daniellii*, and *S. striatinux*, with a large percentage of sulphur amino acids therefore, could be spices to antioxidant potential, while spices *M. whitei*, *D. glomerata* and *S. zenkeri* (fruit) are shown to be rich in lysine which is an essential amino acid in the younger ones. Consumption of the spice mixtures could be a source of protein and essential amino acid supplement. Spices also can be a supplementary source of essential fatty acids (*D. psilurus*, *H. gabonii* "bark", *F. xanthoxyloides*) and especially in ω-3 other than fish oil and marine products (*M. whitei, F. leprieuri, P. umbellatum*). *X. aethiopica* which its high content in oil and long-chain polyunsaturated fatty acids which may have a role in addition to their nutritional importance in reducing the risk of cardiovascular disease contraction. Special attention must be given to spices characterized by a high content ω-3, ω-6 and methionine plus cysteine antioxidant role it plays in order to reduce and for their interest in reducing the risk of chronic diseases. It can be concluded in the light of these results that tropical edible plants found in Cameroon could contribute significantly to the diets of indigenous population in sub-Saharan Africa and thereby should provide their public health.

References

[1] Jiang, A. and Wang C., Antioxidant properties of natural components from Salvia plebeia on oxidative stability of ascidian oil. *Process Biochemistry*, 2, 41, 1111-1116, 2006.

[2] Dashti, Al-Awadi, Sawaya, Al-Otaibi and Al-Sayegh., Fatty acid profile and cholesterol content of 32 selected dishes in the state of Kuwait. *Food Chemistry*, 80, 377-386, 2003.

[3] De Campos, R.M.L., Hierro, E., Ordoñez, J.A., Bertol T.M., Terra N.N. and De la Hoz L., Fatty acid and volatile compounds from salami manufactured with yerba mate (Ilex paraguariensis) extract and pork back fat and meat from pigs fed on diets with partial replacement of maize with rice bran. *Food Chemistry*, 103, 1159-1167, 2007.

[4] Yoshinori ,O. and Mizue, O. Scavenging Effect of Water Soluble Proteins in Broad Beans on Free Radicals and Active Oxygen Species. *Journal of Agricultural and Food Chemistry*, 46, 401-406, 1998.

[5] Bera, D., Lahiri, D. and Nag, A., Studies on a natural antioxidant for stabilization of edible oil and comparison with synthetic antioxidants. *Journal of Food Engineering*, 74,542-545, 2006.

[6] Necmettin, Y., Mehtap, S., İbrahim, T. and Mahfuz, E., Fatty acid composition in some wild edible mushrooms growing in the middle Black Sea region of Turkey. *Food Chemistry*, 99, 168-174, 2006.

[7] Marianna, M., Massimiliano, P., Luigi, J., Pietro B. and Salvatore M., Omega-3 fatty acids and antioxidants in neurological and psychiatric diseases: An overview. *Progress in Neuro-Psychopharmacology & Biological Psychiatry*, 31, 12-26, 2007.

[8] Lina, C., Laura, G., Eleonora, U., Maria, S., Simonetti and Francesca, B., Fatty acid composition and CLA content in goat milk and cheese samples from Umbrian market. *European Food Research and Technology*, 239, 905-911, 2014.

[9] Avallone, S., Brault, S., Mouquet, C. and Treche S., Home processing of the dishes contributing the main sources of micronutrients in the diet of preschool children in rural Burkina-Faso. *International Journal of Food Sciences and Nutrition*, 58,108-115, 2007.

[10] Chavan, U.D., Mc Kenzie, D.B. and Shahidi F., Functional properties of proteins isolates from beach pea (*Lathyrus maritimus* L). *Food Chemistry*, 74,177-18, 2001.

[11] FAO/WHO/UNU. Technical Report Series 935. Protein and Amino acid. Requirements in Human Nutrition. Report of a Joint FAO/WHO/UNU Expert Consultation, 2007.

[12] Ackman, R.G., Remarks on official methods employing boron trifluoride in the preparation of methyl esters of fatty acids of fish oils. *Journal of the American Oil Chemists' Society*, 75, 541-546, 1998.

[13] Statgraphics. Statgraphics plus for windows 3.0. Rockville, ,MD: *Manugistics Inc.* 1997.

[14] Abdou Bouba, A., Yanou, N.N., Foyet, H.S., Scher, J., Montet, D. and Mbofung C.M.F., Proximate Composition, Mineral and Vitamin Content of Some Wild Plants Used as Spices in Cameroon. *Food and Nutrition Sciences*, 3, 423-432, 2012.

[15] Gülsün, Ö. and Abdurrahman, P., Amino acid and fatty acid composition of wild sea bass (*Dicentrarchus labrax*): a seasonal differentiation. *European Food Research and Technology* . 222, 316-320, 2006.

[16] Enza, T., Andrea, S., Antimo, D., Salvatore, R. and Gemma D. The role and molecular mechanism of D-aspartic acid in the release and synthesis of LH and testosterone in humans and rats. *Reproductive Biology and Endocrinology*, 7, 120, 2009.

[17] Peng, L., Yu-Long, Y., Defa, L., Sung Woo, K. and Guoyao W., Amino acids and immune function. *British Journal of Nutrition*, 98, 237-252, 2007.

[18] Bei-Zhong, H., Frans, M.R. and Robert, N.M.J., Amino acid profiles of sufu, a Chinese fermented soybean food. *Journal of Food Composition and Analysis*, 17, 689-698, 2004.

[19] Judith, K.S., Charmaine, W., Janet, M.L.and Douglas, W.W., Glutamine-Antioxidant Supplementation Increases Body Cell Mass in AIDS Patients with Weight Loss: A Randomized, Double-Blind Controlled Trial. *Nutrition*, 15, 860-864, 1999.

[20] Onuegbu, N.C., Adedokun II, Kabuo N.O. and Nwosu J.N., Amino acid profile and micronutrient composition of the African Pear (*Dacryodes edulis*) pulp. *Pakistan Journal of Nutrition*, 10 , 555-557, 2011.

[21] Jiyu, C., Xingqian, Y., Jianchu, C., Donghong, L., Shaohuan, Z., Nutritional composition of underutilized bayberry (Myrica rubra Sieb. and Zucc.) kernels. *Food Chemistry*, 107, 1674-1680, 2008.

[22] Tchiegang, C., Mezajoug, K.L.B, Tenin, D., Ndjouenkeu, R., Physicochemical and functional properties of defatted caked from two euphorbiaceae from Cameroon: *Riconodendron heudelotii* (Bail) and *Tetracarpium conophrorum* (Müll. Arg). *Journal of Food Technology*, 4, 96-100, 2006.

[23] Abdalbasit, A.M., Sara, Y.A., Siddig, I.A., Sit, F.C., Ahmed, M.E., Samia, O.Y. and Shiou W.G., Effects of roasting and boiling on the chemical composition, amino acids and oil stability of safflower seeds. *International Journal of Food Science and Technology*, 47, 1737-1743, 2012.

[24] Michael, A.W. and Hugh, N.T., Antioxidant characteristics of L-histidine. *The Journal of Nutritional Biochemistry*, 9, 308-315, 1998.

[25] Williamson, D.J., McTavish, S.F., Park S.B. and Cowen P.J., Effect of valine on 5-HT-mediated prolactin release in healthy

volunteers, and on mood in remitted depressed patients. *British Journal of Psychiatry*, 167, 238-242, 1995.

[26] John D., Fernstrom, Madelyn H. and Fernstrom, Tyrosine, Phenylalanine, and Catecholamine Synthesis and Function in the Brain. *Journal of Nutrition* 137, 1539-1547, 2007.

[27] Ronny, H., Navam, H., Kannan, A.and Pengyin C., Proximate composition and amino acid and mineral contents of *Mormordica charantia* L. pericarp and seeds at different maturity stages. *Food chemistry*, 122, 1111-1115, 2010.

[28] Chunhieng, T., HAY, L. and Montet, D., Detailed study of the juice composition of noni (*Morinda citrifolia*) fruits from Cambodia. *Fruits*, 60, 13-24, 2005.

[29] Ewelina, M.G., Joachim, H., Thomas, A. S. and Klaus, W.L., The effects of dietary omega-3 polyunsaturated fatty acid supplementation on attention and impulsivity in an animal model of attention deficit/hyperactivity disorder (ADHD). *Functional Foods in Health and Disease*, 4, 285-298, 2014.

[30] Schlienger, J.L. Nutrition et dyslipoprotéiémie. Revue Française des Laboratoires, 334, 51-54, 2001.

[31] Joseph, C.M.M.D. and Jeffrey, W.B.P.A.C. ω-3 Fatty acids (fish oil) as an anti-inflammatory: an alternative to nonsteroidal anti-inflammatory drugs for discogenic painy. *Surgical Neurology*, 65, 326-331, 2006.

[32] Maria, A.P., Vittorio, M.M., Giuseppe, B.and Fabio, C., Chemical parameters, fatty acids and volatile compounds of salted and ripened goat thigh. *Small Ruminant Research*, 74, 140-148, 2008.

[33] Kratz, M.C.P.and Wahrburg, U., The impact of dietary mono and poly-unsaturated fatty acids on risk factors for atheiosclerosis in humans. *European Journal of Lipid Science and Technology*, 104, 300-311, 2002.

[34] Kimberly, H., Nancy, L., Sharon E., Survey of omega-3 fatty acid intakes and omega-3 food selections in cardiac patients living in a section of the Midwestern United States. *Nutrition Research*, 24,741-747, 2004.

[35] Grungy, S.M. and Denke, A.M., Dietary influence on serum lipids and lipoproteins. *Journal of Lipid Research*, 31, 1149-1172, 1990.

[36] Ahmet, C., Essential oil and fatty acid composition of the fruits of *Hippophae rhamnoides* L. (Sea Buckthorn) and *Myrtus communis* L. from Turkey. *Biochemical Systematics and Ecology*, 32, 809-816, 2004.

13

Effect of Some Thermal Processing Techniques on the Anti- nutritional Factors of *Canavalia plagiosperma* Piper Seeds

Serah O. Alagbaoso*, Justina N. Nwosu, Ijeoma .M. Agunwa, Njideka E. Njoku, Linda O. Akajiaku, Chioma N. Eluchie

Department of Food Science and Technology, Federal University of Technology, P.M.B 1526, Owerri, Imo State, Nigeria
*Corresponding author: alagbaososerah@gmail.com

Abstract The effect of some thermal processing techniques on the anti-nutritional factors of *Canavalia plagiosperma* Piper seeds was studied. Raw seeds of *Canavalia plagiosperma* Piper (Oblique- seeded Jack bean) were autoclaved (at 121°C, 15 psi for 25, 30 and 35 minutes), cooked (at 100°C for 30, 40 and 50 minutes) and roasted (at 120°C for 40, 60 and 80 minutes). The anti-nutritional factors of the processed and raw samples were determined. Six (6) anti-nutrients (alkaloid: 1.17%, haemagglutinin: 20.31 Hu/mg, hydrogen cyanide: 4.70 mg/kg, raffinose: 0.64%, stachyose: 1.09%, steroid: 0.38%) were determined in the raw seed. The three processing treatments given to the seed showed a general reduction trend on these six anti-nutrients at different rates and levels. From the results, autoclaving was the best processing method for haemagglutinin (0.05 Hu/mg), raffinose (0.06%), stachyose (0.22%) and steroid (0.02%) while cooking was the best processing treatment for alkaloid (0.04%) and hydrogen cyanide (0.83 mg/kg). Roasting had the least reduction effect on the six anti- nutrients investigated. The results, therefore, suggest that moist heat treatments (autoclaving and cooking) significantly lowered the levels of antinutrients in *Canavalia plagiosperma* Piper seeds, thereby making it safer for consumption.

Keywords: *Canavalia plagiosperma piper, autoclaving, cooking, roasting, anti-nutrients*

1. Introduction

Legumes are good sources of cheap and widely available proteins for human consumption. They are staple foods for many people in different parts of the world [28]. Legume seeds have an average of twice as much protein as cereals and the nutritive value of the proteins are usually high [26]. Studies have shown that the lesser known legumes together with other conventional legumes can be used for combating protein malnutrition prevalent in the third world. The Protein Energy Malnutrition (PEM) problem can be alleviated by finding alternative cost effective sources of proteins. This can be achieved by the consumption of the legumes whole and in various processed forms (condiments) [4].

Oblique- seeded jack bean or giant bean (*Canavalia plagiosperma* piper) is also called *Riesenbohene* in German, *Pellarde losgentills* or *Promo gigante* in Spanish, and "ukpo Ghana" in the Eastern part of Nigeria. *Canavalia plagiosperma* belongs to the family of *leguminosae* known as *fabaceae* and subfamily of *papilionoideae*. The genus comprises of approximately 70-75 species of tropical origin. *Canavalia plagiosperma* piper is a hybrid of *Canavalia ensiformis* (Jack bean) and *Canavalia gliadiata* (sword bean) [3]. *Canavalia plagiosperma* is a high forage and seed yielding tropical legume with high energy and protein content and quality for Nigerian livestock industry [17]. Initial effort at determining the nutritional composition of the raw unprocessed dry seed of C. *plagiosperma* suggests that it has crude protein content of 35.53% which could be used to fortify cereal based diets. Recent research by Esonu *et al.* [11] showed that raw seed of *Canavalia plagiosperma* contains crude protein 36.11%, moisture 19.35%, crude fibre 5.12%, ash 1.80%, carbohydrate 51.46%, crude fat 5.48% and dry matter 80.65%. Also the nutritive and protein quality of the seed as shown by some studies seems to be similar to that of most of the edible legume grains and hence, they are advocated to be a good source for extending protein sources [15]. The use of *Canavalia plagiosperma* (raw) as a high protein food and forage crop for some countries like Mexico, Brazil and Central America has been documented [17]. In "Ikwuano" Umuahia in Abia State of Nigeria, the seed is used as soup thickener [10]. Also, the young pods and immature seeds of *Canavalia plagiosperma* are used generally as vegetables. The immature pods are made into a dish directly or often boiled with water; also the immature seeds are often consumed as curries and as a substitute for mashed potatoes [10]. Despite the desirable nutritive features of C. *plagiosperma* seed, it is not extensively utilized as food for man and /or feed for ruminants mainly due to the presence of certain anti-nutritional substances which may have adverse effects on human or animal nutrition. These antinutrients may include alkaloid, hydrogen cyanide, lectins and flatus-producing oligosaccharides [7,10,24].

Some of the problems associated with *Canavalia plagiosperma* is unavailability due to poor utilization and unfamiliarity. There are also uncertainties concerning the appropriateness of heat treatment and the degree/timing of such treatments. *Canavalia plagiosperma* has been noted by Esonu [10] to contain some anti- nutrients. Also, Odoemelam [17] and Moreina *et al.* [15] documented that the raw seed contains 35.53% of crude protein. Despite the high protein content of the seed, yet it is not generally utilized as food or in food formulations for human consumption. This could be as a result of anti-nutrients contained by the seed [15].

The aim of this work was to evaluate the efficacy of some processing techniques (autoclaving, cooking and roasting) on the removal or manipulation of antinutritional factors in oblique-seeded jack bean seeds (*Canavalia plagiosperma* piper).

2. Materials and Method

2.1. Materials

2.1.1. Source of Materials

The fresh seeds of *Canavalia plagiosperma* for this study were obtained from Ikwuano Umuahia in Abia State of Nigeria.

All equipment and chemicals used were available at reliable research laboratory services Umuahia, Abia state; National Root Crops Research Institute (NRCRI) Umudike; and Federal University of Technology (FUTO), Owerri, Imo State, Nigeria. All the chemicals used were of analytical grade.

2.2. Methods

2.2.1. Sample Preparation

The healthy seeds of *C. plagiosperma were* dehulled by parboiling at the temperature of 100°C for 5 min to ease decoating of the seeds, after which the coats were removed manually. Then, the dehulled /decoated seeds were dried in a hot air oven (IPro75.xx1.5, Gallenkamp, USA) at 60°C for 3 h.

The decoated seeds were divided into 4 batches:

Cooking

One batch was further divided into 3 portions, the 3 portions were cooked in distilled water (100°C) in a bean: water ratio of 1:10 (w/v) for different times of 30, 40, and 50 min. The cooked seeds were rinsed with distilled water and dried at 60°C for 3 h in a hot air oven (IPro75.xx1.5, Gallenkamp, USA).

Autoclaving

The second batch was also divided into 3 portions. The 3 portions were autoclaved at 15 psi (121°C) in distilled water in the bean: water ratio of 1:10 (w/v) for different time intervals of 25, 30 and 35 min, respectively. After treatment, the seeds were rinsed with distilled water and dried at 60°C for 3 h in a hot air oven (IPro75.xx1.5, Gallenkamp, USA).

Roasting

The third batch was divided into 3 portions. The 3 portions were roasted at 120°C for 40, 60 and 80 min, respectively.

The Fourth batch was left raw as control.

The processed dried seeds were milled and stored in airtight containers. Samples were taken from the airtight containers for anti-nutritional factors determination.

Plate 1. UNDEHULLED, RAW *CANAVALIA PLAGIOSPERMA* SEEDS

Plate 2. DEHULLED, RAW *CANAVALIA PLAGIOSPERMA* SEEDS

2.2.2. Determination of Anti-Nutritional Factors

Determination of alkaloids

The gravimetric method [14] was used.

A measured weight (5.0 g) of each sample was dispersed in 50 ml of 10% acetic acid solution in ethanol. The mixture was shaken and allowed to stand for 4 h at room temperature. At the end of this period, the mixture was filtered through Whatman No. 42. The filtrate was evaporated to one-quarter (1/4) its original volume. The extract was treated with dropwise addition of concentrated NH_4OH solution to precipitate the alkaloid. The alkaloid precipitate was filtered off and washed with 1% NH_4OH solution. The filtering was done with a weighed filter paper. The precipitate in filter paper was dried at 60°C in an oven (IPro75.xx1.5, Gallenkamp, USA) and weighed after cooling in a desiccators and reweighed.

The alkaloid content was calculated as shown below:

$$\% \text{ Alkaloids} = \frac{W_2 - W_1}{W} \times \frac{100}{1}$$

Given that;

W = weight of sample

W_1 = weight of empty filter paper

W_2 = weight of paper + alkaloid precipitate

Determination of Hemagglutinin

Hemagglutinin was determined using the procedure of Onwuka [20]. One gram (1 g) of the sample was weighed out and dispersed in a 10 ml normal saline solution buffered at pH 6.4 with a 0.01M phosphate buffer solution. The mixture was allowed to stand at room temperature for 30 min and was centrifuged afterwards to obtain the extract. 1 ml of 2% (v/v) trypsinized rabbit blood erythrocyte suspension in saline phosphate buffer (pH 7.0) was added to 0.1 ml of the extract diluents in a test tube. A control sample was also prepared containing only the blood cells. The test tubes containing the sample mixture and the ones containing the control sample were allowed to stand for 4 h at room temperature. 1ml of normal saline was added to all the test tubes and they were allowed to stand for 10 min, after which their absorbance were read in a colorimeter (Model 6051, Jenway Company, UK) at 600 nm. The hemagglutinin units per milligram of the sample was thus calculated:

$$\text{Hemagglutinin unit / mg} = (b - a) \times \text{F}$$

Where b = Absorbance of test sample solution

a = Absorbance of the blank control

F = Experimental factor given by:

$$F = \frac{1}{W} \times \frac{V_f}{V_a} \times D$$

Where V_f = Total volume of extract

V_a = Volume of extract used in the essay

W = Weight of the sample used

D = Dilution factor

Determination of hydrogen cyanide (HCN)

The method described by Ezegbe [12] was used. One gram (1 g) of the sample was dispersed in 50 ml of distilled water in a 250 ml conical flask. An alkaline picrate paper was hung over the sample mixture and the blank in their respective flasks.

The set up were incubated overnight at room temperature and each picrate paper was eluted (or dipped) into 60 ml distilled water. A standard cyanide solution was prepared and diluted to a required concentration. The absorbance of the eluted sample solution and of the standard were measured colourimetrically at 540 nm against a blank reagent at zero.

The hydrogen cyanide content was determined by the formula shown below:

$$\text{HCN(mg / kg)} = \frac{1000}{W} \times \frac{au}{as} \times C \times D$$

Where W = weight of sample analyzed

au = absorbance of test sample

as = absorbance of standard HCN solution

C = Concentration of standard in mg/dl

D = Dilution factor where applicable

Determination of oligosaccharides

The method of Ezegbe [12] was used. One gram (1 g) of sample was boiled in 100ml of 2M HCl solution until it was negative to iodine starch test. It was centrifuged and the hydrolysate (supernatant) used for the analysis. 1ml of the hydrolysate was mixed with 4 ml of anthrone reagent in a test tube and boiled for 10 min in a water bath while covering the test tubes. After boiling, the mixture was filtered and diluted with distilled water. Similarly, a standard sugar solution (glucose) was prepared and treated as described above and the absorbance of both the sample and sugar solutions were read with (Model 6051, Jenway Company, UK) at 432 nm against a blank reagent at zero.

The oligosaccharides content was calculated thus:

$$\text{Percentage sugar} = \frac{Au}{As} \times C \times F$$

Where Au = Absorbance of test sample

As =Absorbance of standard sugar solution

C = Concentration of sugar solution

F = Experimental factor given by $\frac{V_f}{V_a} \times D \times \frac{100}{W}$

Where V_f =Total filtrate volume

V_a =Volume of aliquot analyzed

D = Dilution factor (where applicable)

W = Weight of sample used

$$\text{Raffinose} = \frac{1}{4} \times \%Sugar$$

$$\text{Stachyose} = \frac{3}{4} \times \%Sugar$$

Determination of total steroid

The total steroid was determined colorimetrically with reference to the saponin content [16]. The saponin crystals were dissolved in a 50 ml formaldehyde- Conc. H_2S0_4 mixture (1:1 v/v) and the absorbance was measured at 500 nm with a colorimeter (Model 6051, Jenway Company, UK).

The steroid content was calculated as follows:

$$\%\text{Steroid} = \frac{Absorbance}{Weight\ of\ sample} \times \frac{100}{1}$$

2.3. Statistical Analysis

All analyses were carried out in triplicate and results were expressed as mean ± SD. Analysis of variance (ANOVA) using general linear model procedure of the Statistical Analysis System [22]. Means comparisons were performed using the Duncan's Multiple Range Test.

3. Results and Discussion

3.1. Effect of autoclaving on the anti-nutritional factors in *Canavalia plagiosperma* piper seeds

The effect of autoclaving on the anti-nutritional factors in *Canavalia plagiosperma* seed was presented in Table 1. Autoclaving significantly reduced (P≤0.05) the alkaloid content of *Canavalia plagiosperma* piper seed as processing time increased (Table 1). The autoclaving for 35 min had the highest reduction effect on alkaloid, followed by autoclaved at 30 min and lastly autoclaving for 25 min. Autoclaving reduced alkaloid by 95.73%. This finding agrees with the work of Odoemelam [17] who reported that moist heat treatment (autoclaving and boiling) was the best processing method for *Canavalia plagiosperma* seed. Alkaloid causes gastro-intestinal and neurological disorders [23].

Table 1. Effect of autoclaving on the anti-nutritional factors in *Canavalia Plagiosperma* seed

Sample	Alkaloid (%)	Haemagglutinnin(Hu/mg)	Hydrogen cyanide (mg/kg)	Raffinose (%)	Stachyose(%)	Steroid(%)
Raw Seed	1.17a ±0.01	20.31a ±0.01	4.70a ±0.06	0.64a ±0.00	1.09a ±0.05	0.38a ±0.00
Autoclaved, 25min	0.12e ±0.01	14.38g ±0.72	1.25e ±0.02	0.23f ±0.01	0.52c ±0.01	0.10cd ±0.01
Autoclaved, 30 min	0.13e ±0.06	10.34h ±0.02	1.24e ±0.01	0.11h ±0.00	0.33e ±0.02	0.05gh ± 0.03
Autoclaved, 35 min	0.05f ±0.01	4.11i ±0.00	1.05g ±0.01	0.06I ±0.01	0.22f ±0.02	0.02g ±0.00

Values are means ± SD of triplicate determinations; Values followed by different superscripts in the same row are significantly *(p <0.05)* different

The raw seed of *Canavalia plagiosperma* was found to contain haemagglutinin level of 20.31 Hu/mg (Table 1) which is in line with the work of Esonu *et al.* [11] which said that raw C. *plagiosperma* piper has haemagglutinin content of 20.10 Hu/mg. Autoclaving significantly reduced the haemagglutinin level with the processing time increased. Autoclaving for 35 min had the highest reduction effect on haemagglutinin content of *Canavalia plagiosperma* piper seed. Autoclaving reduced haemagglutinin content of the seed by 79.76% (Table 1). Among the various processing techniques employed, the autoclaving was more effective in reducing the maximum level of haemagglutinin (79.76%) (Table 1). Similarly, significant reduction in level of haemagglutinin by autoclaving was reported for *Abrus precatorius*, *Mucuna pruriens* var. utilis, *Entada scandens*, *Vigna aconitifolia*, V. *sinensis, C. plagiosperma* piper and *Bauhinia purpurea* [11,21,27]. There was a significant different (p≤0.05) on the haemagglutinin content of all the autoclaved samples.

Hydrogen cyanide (HCN) level of *Canavalia plagiosperma* reduced as the processing times increased (Table 1). Autoclaving for 35 min showed the highest reduction effect on hydrogen cyanide. Autoclaving reduced the HCN content of C. *plagiosperma* by 77.66% (Table 1). These results agree with the work of Odoemelam [17] who reported that moist heat treatments (cooking and autoclaving) were the best processing methods for *C. plagiosperma* seed. Hydrogen cyanide causes gasping, convulsion and staggering [16]. The increase in the autoclaving time caused a significant reduction effect (p≤0.05) on the content of raffinose of *C. plagiosperma* seed (Table 1). Autoclaving for 35 min reduced raffinose content of *C. plagiosperma* piper by 90.63%. Similar results were reported by Ibrahim *et al.* [13] found that cooking caused complete removal of raffinose and about 60-80% of stachyose depending on cooking method. They also attributed this loss to the thermal hydrolysis of oligosaccharides to simple disaccharides and monosaccharides.

The stachyose content of *C. plagiosperma* piper seed increased as the durations of autoclaving increased (Table 1). Autoclaving caused significant reduction effects (p≤0.05) in the stachyose content of *C. plagiosperma* (Table 1). Autoclaving for 35 min had the highest reduction effect, followed by 30 min (Table 1). Autoclaving reduced stachyose by 79.82 % (Table 4). Autoclaving is the best processing method to eliminate stachyose (Table 1) when compared with cooking and roasting (Table 2 and Table 3). These decreases in levels of oligosaccharides (stachyose) due to autoclaving might be attributed to heat hydrolysis of the oligosaccharides with the formation of simple sugars [13,19]. These results coincide with the findings of Ugwu and Oranye [25] where autoclaving for 60 min totally eliminated stachyose in African breadfruit. Oligosaccharides (stachyose and raffinose) are the major contributory factors of flatulence [18]. The steroid level of *C. plagiosperma* piper reduced significantly (P≤0.05) as the processing time increased (Table 1). Autoclaving 35 min had the highest reduction effect on steroid content of *C. plagiosperma* piper seed (Table 1). Autoclaving reduced the steroid levels by 94.74% (Table 4). This finding agrees with the work of Odoemelam [17] who reported that moist heat treatment (autoclaving and boiling) was the best processing method for *C. plagiosperma* seed.

3.2. Effect of cooking on the anti-nutritional factors in *Canavalia plagiosperma* piper seeds

The effect of cooking on the anti-nutritional factors in *C. plagiosperma* piper seeds was shown in Table 2. Cooking had significant decreasing effect (P≤0.05) on alkaloid content of *C. plagiosperma* piper seeds (Table 2). The alkaloid levels reduced significantly (P≤0.05) as the processing (cooking) time increased (Table 2). Cooking for 50 min showed the highest reduction effect on alkaloid content of the seed. Cooking reduced the alkaloid level by 96.58% (Table 4). From (Table 2) cooking had the highest reducing effect on alkaloid when compared with autoclaving and roasting. This finding agrees with the work of Odoemelam [17] who reported that moist heat treatment (autoclaving and boiling) was the best processing method for *C. plagiosperma* seed.

Table 2. Effect of cooking on the anti-nutritional factors in *Canavalia plagiosperma* seed

Sample	alkaloid (%)	Haemaglutinin (Hu/mg)	Hydrogen cyanide (mg/kg)	Raffinose (%)	Stachyose (%)	Steroid (%)
Raw Seed	$1.17^{a}\pm0.01$	$20.31^{a}\pm0.01$	$4.70^{a}\pm0.06$	$0.64^{a}\pm0.00$	$1.09^{a}\pm0.05$	$0.38^{a}\pm0.00$
Cooked, 30 min	$0.10^{e}\pm0.01$	$18.01^{d}\pm0.00$	$1.24^{e}\pm0.01$	$0.27^{e}\pm0.01$	$0.47^{c}\pm0.02$	$0.14^{d}\pm0.01$
Cooked, 40 min	$0.06^{f}\pm0.01$	$17.13^{e}\pm0.06$	$0.18^{f}\pm0.02$	$0.22^{f}\pm0.02$	$0.32^{e}\pm0.01$	$0.14^{d}\pm0.03$
Cooked, 50 min	$0.04^{f}\pm0.01$	$15.79^{f}\pm0.11$	$0.83^{h}\pm0.01$	$0.13^{g}\pm0.00$	$0.23^{F}\pm0.01$	$0.08^{ef}\pm0.03$

Values are means ± SD of triplicate determinations; Values followed by different superscripts in the same column are significantly *($p < 0.05$)* different

The data in Table 2 shows the effect of cooking on haemagglutinin content of raw seeds. The reduction in haemagglutinin level increases as the cooking time increases. Cooking for 50 min significantly reduced ($P \le 0.05$) the haemagglutinin concentration of raw seed to 15.79 Hu/mg as shown in Table 2. This coincides with the work of Enwere [8], who noted that traditional method of household cooking and industrial autoclaving or retorting are capable of detoxifying haemagglutinin using moist heat.

Cooking remarkably reduced the hydrogen cyanide (HCN) level of *Canavalia plagiosperma* piper seeds as the processing time increased (Table 3). The initial level value of HCN (4.70 mg/kg) was reduced to 0.83 mg/kg after 50 min cooking which indicates very significant ($P \le 0.05$) reduction of the HCN content. This indicates that boiling could therefore be employed for total destruction of the toxins without destroying the nutrient. Odoemelam [17] who reported that moist heat treatments (cooking and autoclaving) were the best processing methods for *C. plagiosperma* seed. Cooking for 50 min gave hydrogen cyanide value of 0.83 mg/kg which is far lower than the fatal dose level of 50 mg/100g which is equivalent to 500 mg/kg [16]. Cooking reduced the HCN level by 82.34% (Table 4). This result agrees with the work and staggering [16] and also inhibits the cytochrome oxidase through combination with their copper and iron ions, respectively [20].

The oligosaccharides (stachyose and raffinose) content in raw *C. plagiosperma* piper and the effect of cooking on them were presented in Table 2. The increase in the durations cooking caused a significant reduction ($p \le 0.05$) in the content of stachyose and raffinose of *C. plagiosperma* (Table 2). The cooking reduced them by 78.90% and 79.69% (Table 4), respectively. Cooking for 50 min showed the highest reduction effect in stachyose and raffinose. This results agree with Ibrahim *et al.* [13] who reported that cooking caused complete removal of raffinose and about 60-80% of stachyose depending on cooking method. Ordinary cooking of pre-soaked seeds did not increase the removal percentage of stachyose, whereas pressure cooking of presoaked seeds relatively increased stachyose loss from 59.8 to 66.17%. Steroid levels reduced significantly ($P \le 0.05$) as the processing time increased (Table 2). Cooking for 50 min reduced steroid content of the seed by 78.95% (Table 4), this finding agrees with the work of Odoemelam [17].

3.3. Effect of roasting on the anti-nutritional factors of *Canavalia plagiosperma* piper seeds

The effect of roasting on the anti-nutritional factors in *C. plagiosperma* piper seeds was shown in Table 3. Roasting significantly reduced the alkaloid content of *C. plagiosperma* piper. The level alkaloid increased significantly ($P \le 0.05$) as the roasting time increased (Table 3). Roasting for 80 min showed the highest reduction effect (70.09%) (Table 4). When compared the level of reduction in alkaloid in Table 3 with those of Tables 1 and 2, it is established that moist heat treatment is better in reduction of alkaloid, which is in accordance with the work of Odoemelam [17]. Also, studies by Esonu *et al.* [9] have demonstrated that dry heat treatment alone as a method of processing jack bean seeds did not appreciably reduce the level of toxic factors.

The haemagglutinin content of raw seed of *C. plagiosperma* was found to be 20.31 Hu/mg (Table 3) which is in line with the work of Esonu *et al.* [11] which said that raw *C. plagiosperma* piper has haemagglutinin content of 20.10 Hu/mg. Roasting significantly reduced the haemagglutinin level. The level of haemagglutinin reduced as the processing time increased. Roasting for 80 min reduced the haemagglutinin level by 15.46%. The percentage reduction is low when compared with moist heat treatments: autoclaving (79.76%) and cooking (22.25%) (Table 4). The reduced percentage decrease in haemagglutinin by roasting agreed with the work of Esonu *et al.* [9].

Table 3. Effect of roasting on the anti-nutritional factors in *Canavalia plagiosperma* piper seeds

Sample	alkaloid (%)	Haemaglutinin (Hu/mg)	Hydrogen cyanide (mg/kg)	Raffinose (%)	Stachyose (%)	Steroid (%)
Raw Seed	$1.17^{a}\pm0.01$	$20.31^{a}\pm0.01$	$4.70^{a}\pm0.06$	$0.64^{a}\pm0.00$	$1.09^{a}\pm0.05$	$0.38^{a}\pm0.00$
Roasted, 40 min	$0.48^{b}\pm0.00$	$19.21^{b}\pm0.02$	$2.46^{b}\pm0.01$	$0.52^{b}\pm0.00$	$0.85^{b}\pm0.00$	$0.33^{b}\pm0.01$
Roasted, 60 min	$0.44^{c}\pm0.01$	$18.39^{c}\pm0.36$	$2.17^{c}\pm0.01$	$0.45^{c}\pm0.00$	$0.82^{b}\pm0.02$	$0.25^{c}\pm0.01$
Roasted, 80 min	$0.35^{d}\pm0.00$	$17.17^{e}\pm0.07$	$1.84^{d}\pm0.01$	$0.37^{d}\pm0.01$	$0.40^{d}\pm0.05$	$0.24^{c}\pm0.03$

Values are means ± SD of triplicate determinations; Values followed by different superscripts in the same column are significantly *($p < 0.05$)* different

Table 4. Reduction (%) of some anti-nutritional factors in *Canavalia plagiosperma* piper for the best processing treatments

Anti- nutritional factors	Autoclaving (35 min)	Cooking (50 min)	Roasting (80 min)
Alkaloid (%)	95.73	96.58	70.09
Haemagglutinin (Hu/mg)	79.76	22.25	15.46
Hydrogen cyanide (mg/kg)	77.66	82.34	60.85
Raffinose (%)	90.63	79.69	63.70
Stachyose (%)	79.82	78.90	42.19
Steroid (%)	94.74	78.95	36.84

The oligosaccharides (raffinose and stachyose) content in *C. plagiosperma* piper increased decreased as the processing (roasting) time increased. The increase in the durations roasting caused a significant reduction ($p \le 0.05$) in the content of stachyose and raffinose of *C. plagiosperma* (Table 3). Roasting reduced stachyose and raffinose by 63.70% and 42.19% (Table 4), respectively. Roasting for 80 min showed the highest reduction effect on the two oligosaccharides (Table 3), but they are low when compared to the reductions given by autoclaving (Table 1) and cooking (Table 2). This coincides with the work of Akande and Fabiyi [2] who reported superiority of cooking (moist heat) to dry heat for improvement of growths promoting action in soybeans and in Jack beans [5,6].

The hydrogen cyanide content of *C. plagiosperma* piper reduced significantly (p≤0.05) as the processing (roasting) time increased (Table 3). Roasting for 80 min reduced hydrogen cyanide level by 60.85 % (Table 4). The reduction effect of roasting was lower when compared with that of cooking and autoclaving. This result agrees with the work of Odoemelam [17].

The steroid level reduced significantly (P≤0.05) as the processing (roasting) time increased (Table 3). Roasting for 80 min reduced steroid content of raw seed by 36.84%. This reduction on steroid caused by roasting is low when compared with the level of reduction caused by autoclaving (Table 1) and cooking (Table 2). This result agreed with the work of Ahamefule *et al.* [1] who reported that boiling, roasting and soaking are some of the processing methods employed but they do not completely eliminate anti-nutritional factors but inactivate them.

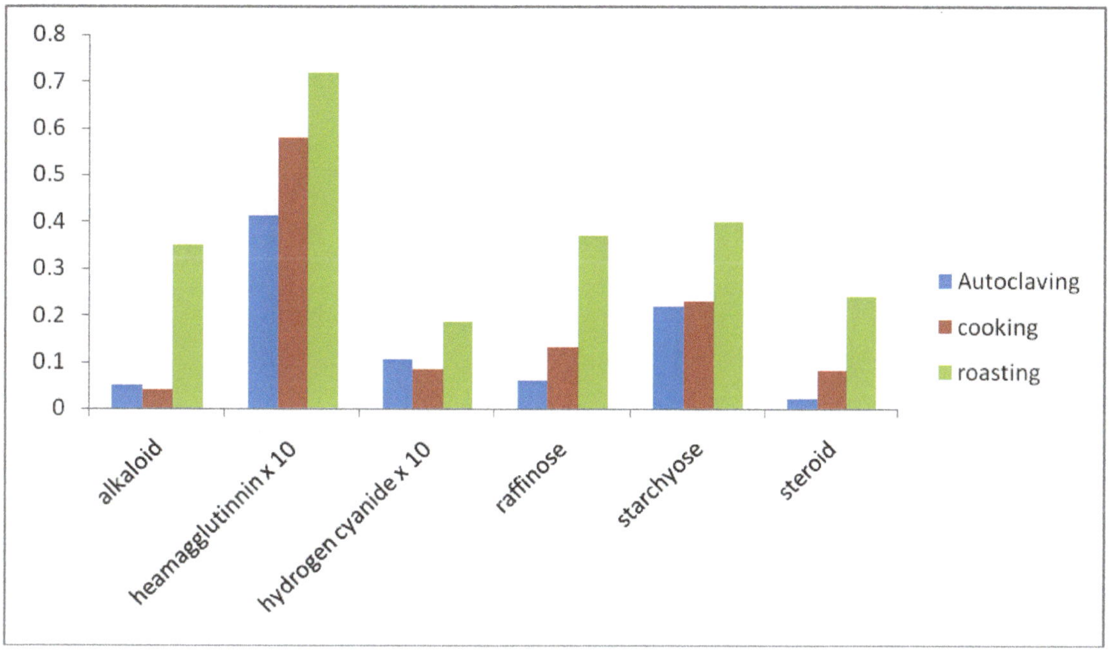

Figure 1. The effect of some processing techniques on the anti-nutritional composition of *Canavalia plagiosperma* piper seed

Table 4 presents the percentage reduction of some anti-nutritional factors in *C. plagiosperma* piper seeds after processing. Generally, the three processing treatments reduced the anti-nutrients levels at different rates. Autoclaving was the best processing method for stachyose, raffinose, haemagglutinin and steroid while cooking was the best processing treatment for hydrogen cyanide and alkaloid. It could be deduced therefore that moist heat treatments (autoclaving and cooking) were the best in reducing anti-nutritional factors when compared with dry heat treatment (roasting). Figure 2 clearly shows the highest reduction caused by each of the three processing methods on the six different anti-nutritional factors investigated.

4. Conclusion and Recommendations

The results revealed that moist heat treatments (autoclaving and cooking) are good in reducing anti-nutrients. The three treatments drastically reduced the level of anti-nutritional factors; autoclaving and cooking reduced it to the level that is safe for human consumption. The seed when processed can be used as a good plant protein source for this teaming population in the world since animal protein is very expensive. It could also be used to fortify cereal based diets such as pap. The processing treatments used in this work caused reduction at different rates in the level of anti-nutrients present in the raw seed of *C. plagiosperma* to tolerable level safe for consumption; it is therefore recommended that slightly increasing the processing time may totally eliminate the anti-nutrients in the seed. It is also recommended that increasing the processing time of roasting will be more efficient in reducing the anti nutrients to tolerable levels. Also, other methods of reduction of anti-nutrients in foods other than heat (such as malting, fermentation etc) should be carried out on *C. plagiosperma* seed so as to know their own suitability and rate of reduction of the anti-nutrients in the seed.

References

[1] Ahamefule FO, Obua BE, Ukweni IA, Oguike MA, Amaka RA (2008). Haematological and biochemical profile of weaner rabbit fed raw or processed pigeon pea seed meal based diets. Afr. J. Agric. Res. 3(4): 315-319.

[2] Akande KE, Fabiyi EF (2010). Effect of processing methods on some anti-nutritional factors in legume seed for poultry feeding. Int. J. Poultry Sci. 9(10): 996-1001.

[3] Akinmutimi AH (2004). Evaluation of sword bean (Canavalia gladiata) as an alternative feed resources for broiler chickens. PhD

Thesis, Department of Non-ruminant Animal Production, Micheal Okpara University of Agriculture Umudike, Nigeria.

[4] Arisa NU, Aworh OC (2007). Production, quality assessment and acceptability of African yam bean Sphenostylis stenoscarpa sauce. J. Food Proc. Pres. 31: 771-778.

[5] Bressani R, Sosa JL (1990). Effect of processing on the nutritive value of Jack bean (Canavalia ensiformis). Plant Food for Human Nutr. 40: 207-214.

[6] Carlini CR, Udedibie ABI (1997). Comparative effect of processing methods of haemagglutinating and anti-trypsin activities of Canavalia braziliensis seeds. J. Agric. Food Chem. 45: 4372-4377.

[7] Doss A, Pugalenthi M, Vadivel V G, Subhashini G, Anitha Subash R (2011). Effects of processing technique on the nutritional composition and antinutrients content of under –utilized food legume Canavalia ensiformis L. DC. Int. Food Res. J. 18(3): 965-970.

[8] Enwere NJ (1998). Foods of plant origin. Afro-Orbis Publications Ltd. Pp. 194-199.

[9] Esonu BO, Udedebie ABI, Carlini CR (1998). The effect of roasting, dry urea treatment and sprouting on some thermostable toxic Factors in the Jack bean seed. Nig. Animal Prod. 25(1):36-39.

[10] Esonu BO (2009). Unconventional feed resource for livestock development and food security, paradigms from Nigerian livestock industry. 14th Inaugural Lecture of Federal University of Technology, Owerri, Imo State, Nigeria. Delivered on Wednesday 24th of June, 2009.

[11] Esonu BO, Anumni PE, Udedibie ABI, Emenalom OO, Etuk B, Odoemelam V, Okorie KC (2013). Evaluation of two-stage cooked Canavalia plagiosperma (piper) seed meal as feed ingredient in layer diets. Nig. J. Ani. Sci. 15: 95-103.

[12] Ezegbe CC (2012). Nutritional quality and physio-chemical properties of the seed and oil of Chinese fan palm. M.Sc. Thesis, Department of Food Science and Technology, Federal University of Technology Owerri, Nigeria.

[13] Ibrahim SS, Habiba RA, Shatta AA, Embaby HE (2002). Effect of soaking, germination, cooking and fermentation on anti-nutritional factors in cowpea. Nahrung 46 (2): 92-95.

[14] Inuwa HM, Aina VO, Gabi B, Aimola I, Toyin A (2011). Comparative determination of anti-nutritional factors in ground oil and palm oil. Adv. J. Food Sci. Technol. 3(4): 275-279.

[15] Moreina RA, Cordeiro EF, Cavada BS, Nunes ED, Fernerndes AG, Oliveine JTA (1993). Plant seed lectins. A possible marker for chemotaxonomy of the genus Canavalia. R. Bras. Veg. Brazil Pp. 127-132.

[16] Nwosu JN (2011). The effects of processing on the anti-nutritional properties of 'Oze' (Bosqueia angolensis) seed. J. Am. Sci. 7(1): 1-6.

[17] Odoemelam VU (2007). Proximate composition, anti-nutritional properties and mineral assay of raw and processed Canavalia plagiosperma seed. M.Sc. Thesis, Department of Animal Production and Management, Micheal Okpara University of Agriculture, Umudike, Abia State, Nigeria.

[18] Okaka JC, Akobundu ENT, Okaka ANC (1992). Human nutrition: An Integrated approach (2nd ed.) Academic publishers, Enugu. Pp. 355-375.

[19] Onigbinde AO, Akinyele IO (1983). Oligosaccharide content of 20 varieties of cowpea in Nigeria. J. Food Sci. 48: 1250-1254.

[20] Onwuka GI (2005). Food analysis and instrumentation: Theory and practice. Naphthali prints, Lagos. Pp 129-144.

[21] Pugalenthi M, Vadivel V, Janak P (2007). Comparative evaluation of protein quality of raw and differentially processed seeds of an underutilized food legume, (Abrus precatorius L). Livestock Res. Rural Devel. Article No. 168. Available in the online website. http://www.lrrd.Org/lrrd19/11/pugal9168.htm.

[22] S.A.S. (1999). Statistical Analysis System User's Guide, A.A. Ray, (Ed.), SAS Institute, Inc. Cary, NC.

[23] Soetan KO, Oyewole OE (2009). The need for adequate processing to reduce the anti-nutritional factors in plants used as human foods and animal feeds. A review. Afr. J. Food Sci. 3(9): 223-231.

[24] Udedibie ABI, Esonu BO, Agbabiaka LA (2000). Comparative performance of broilers fed diets containing differentially processed Jack bean meals. N.P.S.J. I: 32-41.

[25] Ugwu FM, Oranye NA (2006). Effects of some processing methods on the toxic components of African bread fruit (Treculia africana). Afr. J. Biotechnol. 5 (22): 2329-2333.

[26] Vijayakumari K, Siddhuraji P, Janardhanan K (1997). Effect of domestic processing the levels of certain antinutrients in Prosopis chilensis (Molina) stunz. Seeds. Food Chem. 59(3): 367-371.

[27] Vijayakumari K, Pugalenthi M, Vadivel V (2007). Effect of soaking and hydrothermal processing methods on levels on anti-nutrients and in vitro protein digestibility of Burhia purpunea. L. seeds. Food Chem. 103: 968-975.

[28] Youseff MM, Abdal MA, Shekibs LAE, Ziena HM (1989). Effects of dehulling, soaking and germination of chemical composition, mineral elements and protein patterns of feba beans (Vicia feba L.). Food Chem. 23: 129-136.

14

Effect of Phytol on Dehydrogenase Activity of Bacterial Isolates from Grilled Meat

ELUCHIE C.N.[1], ORANUSI S.[2], AKUJOBI C. O.[2], ALAGBAOSO, S.O.[1,*]

[1]Department of food Science and Technology, Federal University of Technology P.M.B 1526, Owerri Imo State
[2]Department of Industrial Microbiology, Federal University of Technology P.M.B 1526, Owerri Imo State
*Corresponding author: alagbaososerah@gmail.com

Abstract Effect of phytol on the dehydrogenase activity of *Staphylococcus aureus Bacillus cereus*, *Pseudomonas aeruginosa* and *Enterococcus faecalis* isolated from grilled meat was examined via dehydrogenase assay using 2,3,5 – tripheny tetrazolium chloride (TTC) as the electron acceptor. The bacteria isolates were exposed to various concentrations of phytol (0-2000 μg) in a nutrient broth. The minimum inhibitory concentration (MIC) and minium bactericidal concentration (MBC) values of phytol against the bacterial isolates were found in the range of 31.25 μg /ml to 250 μg /ml and 125 μg /ml to 250 μg /ml respectively. Effect of phytol on the dehydrogenase activity of the bacterial isolates showed that gram negative *Pseudomonas aeruginosa* had higher dehydrogenase activity than the gram positive *Bacillus cereus*, *Enterococcus faecalis* and *Stpahylococcus aureus*. The effect of phytol on the dehydrogenase activity varied significantly among the bacterial isolates. This result indicates that phytol could be used as an antibacterial agent in food industry to inhibit the growth of certain food borne pathogens.

Keywords: *Phytol, dehydrogenase activity, grilled meat*

1. Introduction

For years, food borne illness resulting from consumption of food contaminated with pathogenic bacteria and/or their toxins has been of vital concern to public health. Among food borne outbreaks in the United States during 1983-1987 in which the etiology was determined, bacterial pathogens caused the largest number of out breaks (66%) and cases (92%) [15]. Thus, controlling pathogens could reduce food borne outbreaks and assure consumers a safe, wholesome and nutritious food supply.

Plants contain a great number of secondary metabolites, many of which display biological activity as "natural products" with a role in plant defense against bacteria, fungi and other microorganisms. There has been a growing interest in new and effective antibacterial substances from natural sources like plants to reduce cases of bacterial diseases. In many cases, antibacterial compounds from plants, herbs and spices in low concentrations have been used effectively in the food industry. Biologically active secondary metabolites of spices, herbs and medicinal plants, due to their diverse range of antibacterial activities, are becoming increasingly important in food industry as potential food preservatives to control food borne pathogenic bacteria [1].

Food borne pathogenic bacteria cause several food borne diseases [15]. The indiscriminate and excessive use of a wide range of chemical bactericides has led to extended environmental pollution and the production of resistant pathogen populations. During the last few years, research on plant-based antibacterial agents has produced a diverse range of products with novel modes of action, which are expected to have a significant impact on the control of food borne diseases in coming decade. Therefore the demand for plant based antimicrobials as novel bactericide sources for controlling food borne pathogens is rapidly increasing (Ulubelen *et al.*, 2000; Kubo *et al.*, 2004; Giang *et al.*, 2006).

Some synthetic chemicals have been used to control microbial growth and reduce the incidence of food borne illness. Although these synthetic preservatives are effective, they might be detrimental to human health. Consumers are concerned about the safety of food products that contain artificial preservatives. The ban of chemical additives by some consumers has driven the food industry and food research towards the search for natural antibacterial compounds from plant origin. Many natural compounds (e.g. terpenoids) found in dietary plants possess antibacterial activities against food borne pathogens (Deller *et al.*, 1996; Singh and Singh 2003; Aiyelaagbe *et al.*, 2007).

Phytol is an acyclic diterpene alcohol that can be used as a precursor for the manufacture of synthetic forms of vitamin E [7] and vitamin K1 [5]. Phytol is likely the most abundant acyclic isoprenoid compound present in the biosphere and its degradation products have been used as biogeochemical tracers in aquatic environments [14].

Dehydrogenase is an enzyme that catalyzes the removal and transfer of hydrogen from a substrate in an oxidation-reduction reaction. It is an enzyme that mobilizes the hydrogen of a substrate so that it can pass to a hydrogen acceptor [10]. Measurement of microbial enzyme activity has been used in the assessment of ecotoxicological impacts of environmental substrates. In this regard, dehydrogenase activity has been widely used. The dehydrogenase assay is an effective primary test for assessing the potential toxicity of metals to soil microbial activities [9], toxicity of metals to planktonic [9] and heterotrophic [8] bacteria from tropical river sediments. Toxicity of antimicrobial agents to pathogenic bacteria has been accessed using the dehydrogenase assay [10,11]. The aim of this work is to determine the effect of phytol on the bacterial isolates from grilled meat.

2. Materials and Methods

2.1. Materials

The materials used in this research work include; phytol from sigma chemical industry, USA, grilled meat (suya and pork),sterile bottles, Petri dishes, conical flask, test tubes, measuring cylinder, wire loop, glass rod, aluminum foil, cotton wool, pipettes, forceps, cork borer, autoclave, microscope, weighing balance, Bunsen burner ,incubator, spectrophotometer, rotary incubator. Reagents used include; dimethyisulphoxide (DMSO) obtained from chevron Philips chemical company USA, tannic acid from powder park chemical, India, casein-Universal buffer, amyl alcohol and triphenyl formazan were obtained from sigma chemical industry, USA. Trichloroacetic acid from hydrite chemical company, brook field,2,3,5-triphenyl tetrazolium chloride [TTC] from BDH England, methyl red from jigchem universal, mumbia and NaCL from cellmark , USA. All reagents were of analytical grade. The media used were nutrient agar, macConkey agar and nutrient broth-glucose obtained from becton Dickinson and company, USA respectively. Luria-bertani broth and yeast extract casein broth were obtained from sigma chemical industry, USA. Brain heart infusion broth and agar-solidified Mueller hinton broth were obtained from hardy diagnostic, USA.

2.2. Methods

2.2.1. Source of Sample

A total of thirty samples of grilled meat (suya and pork) were collected from five different locations namely; world bank housing estate in new Owerri, Ihiagwa in Owerri suburb, Ama Awusa in Owerri town, Mbieri in Mbaitolu L.G.A and Akabo in Ikeduru L.G.A, all in Imo state.

2.2.2. Media Preparation

Nutrient agar and MacConkey agar were used for the isolation of the bacterial organisms and they were prepared according to the manufacturers' instructions.

2.2.3. Isolation of Bacterial Strains

Food-associated bacteria were isolated from grilled meat products, Suya and pork meat, using Nutrient agar and MacConkey agar. Ten grams of suya and pork meat were weighed aseptically into 90ml sterile distilled water, agitated for 5mins and serially diluted ten fold. 0.1ml of (10^{-8} dilution) was spread on Nutrient agar and MacConkey agar and the plates incubated at 37°C for 24hrs and 48hrs respectively. After incubation, single morphologically well-formed colonies were picked and subcultured. The colony forming unit per gram (CFU/g) was calculated. Colonial morphology, microscopic characteristics, and biochemical and carbohydrate fermentation tests of the isolated organisms were used for characterization.

2.2.4. Purification of Isolates

Each of the sub-cultured isolates were further sub-cultured aseptically from the plates by streaking a loopful of the bacterial colony onto a freshly prepared Nutrient agar and MacConkey agar plates and incubated at 37°C for 24hrs and 48hrs respectively. After incubation, discrete colonies were picked and Gram stained. Microscopy was carried out inorder to confirm their purity. Each pure colony was picked and inoculated onto sterile Nutrient agar slants in screw- capped bottles. These were incubated at 37°C for 24hrs and stored in the refrigerator at 4°C as stock culture for further tests.

2.2.5. Characterization of Bacterial Isolates

Colonial morphology, microscopic characteristics, and biochemical and carbohydrate fermentation tests of the isolated organisms were used for characterization.

2.2.6. Sugar Assimilation Test

This was carried out using a broth medium composed of 10g of peptone, 5g NaCl, 0.03g methyl red and 1litre deionised water. After shaking, 9ml of the above mixture were dispensed into test tubes. Durham tubes were then introduced into the tubes in an inverted position to ensure that each durham tube was filled with the medium. The media was sterilized at 121°C for 15 mins.

A solution of each sugar (Glucose, Sucrose, Maltose, Lactose, Mannose and Xylose) was prepared by dissolving 1g of each sugar in 100ml of distilled water and sterilized at 121°C for 8 minutes. After cooling, 1ml of each sugar solution was aseptically added into the 9ml of the basal medium. Each tube was inoculated with the bacterial isolate. This was incubated for 48hrs.The tubes were checked for gas production in the durham tubes and change in colour of the medium from red to yellow indicated acid production. The tube which was not inoculated with the bacterial isolate served as the control.

2.3. Determination of Minimum Inhibitory and Bactericidal Concentrations

The minimum inhibitory concentration (MIC) of the compound (phytol) was tested by a two-fold serial dilution method [4]. The test compound was first dissolved in dimethylsulphoxide (DMSO), incorporated into Luria-Bertani medium for bacterial pathogens to obtain a concentration of 2000μg/ml and serially diluted to achieve 1000, 500, 250, 125, 62.5, and 31.25μg/ml. The final concentration of DMSO in the culture medium was maintained at 0.5% (v/v). A 10 μl standardized suspension of each tested organism (1.5×10^{8} CFU/ml) was transferred to each tube. The control tubes containing only bacterial suspensions were incubated at 37°C for 24h. The lowest concentration of the compound, which did not show any growth of test organisms after macroscopic evaluation was determined as MIC, and was expressed in μg/ml. Further, the concentrations showing complete inhibition of visual growth of bacterial pathogens were identified, and 50µl of each culture broth was transferred onto the agar plates and incubated at 37°C for 24hrs. The complete absence of growth of bacterial colonies on agar surface in the lowest concentration sample was defined as the minimum bactericidal concentration (MBC). Each assay in this experiment was replicated three times.

2.4. Determination of Antibacterial Activity Using the BDS Method (Broth Dilution with Shaking)

The method as described by Arai *et al.* [3], was adopted. The compound to be tested was added at the indicated concentrations, to 10ml aliquots of brain heart infusion (BHI) broth in test tubes without any solubilizing agent or surfactant. An aliquot of an overnight culture of bacterial isolates was added to each sample to give approximately 10^5 CFU of the different bacterial isolates per ml. Each culture was incubated, with shaking at 40rpm, for 24hrs at 37°C. The inhibitory activity of the compound being tested was monitored turbidimetrically. The optical density at 660nm was determined with a spectrophotometer.

2.5. Effect of Phytol Concentration on Dehydrogenase Activity of Bacterial Isolates

Dehydrogenase assay method as described by Nweke *et al.* [8] was employed.

Dehydrogenase activity was determined using 2, 3, 5-triphenyltetrazolium chloride (TTC) as the artificial electron acceptor, which is reduced to the red-coloured triphenylformazan (TPF). The dehydrogenase activity of the bacterial isolates was determined in 3.4ml volume of nutrient broth-glucose-TTC medium supplemented with varying concentrations of phytol (0 -2000µg/ml) in separate screw-capped test tubes. A 0.3ml of suspension of the bacterial isolate was inoculated into triplicate 20ml screw-capped tubes containing 3.4ml of phosphate buffered (pH 7.2) nutrient broth-glucose medium amended with toxicant (phytol 0-2000 µg/ml) and preincubated on a rotary incubator (150rpm) at room temperature (28± 2°C) for 24 hrs.

The TPF produced was extracted in 4ml of amyl alcohol and determined spectrophotometrically.

The amount of formazan produced was determined from a standard dose-response curve (0-50mg/l TPF in amyl alcohol) at a reasonable R^2 value. Dehydrogenase activity was expressed as micrograms of TPF formed per mg dry weight of cell biomass per hour.

2.6. Data Analysis

The results obtained from the data were subjected to Analysis of variance (ANOVA) and values of $P \leq 0.05$ were considered statistically significant according to Onuh and Igwemma [12].

3. Result and Discussion

3.1. Characterization of Bacterial Isolates

Out of a total of 30 samples of grilled meat analyzed, 8 pathogenic organisms were isolated. The most common isolates were *B. cereus*, *S. aureus*, *E. faecalis* and *P. aeruginosa* as shown in Table 1 and Table 2.

Table 1. Colonial morphology and microscopic characteristics of bacteria isolated from grilled meat

Sample code	Colony code	Colony count	Cell morphology	Colonial morphology	Motility	Gram stain	Spore	Flagellum	Capsule	Probable organism
Suya And Pork Meat	X1	$1.6x10^{10}$	Spherical cells in pairs and short chains	Small smooth and shiny round colonies	-	+S	-	-	-	*Enterococcus sp*
	X2	$5.0x10^9$	Straight rods in short chains with oval central spores	Dull and dry irregular flat cream colonies	+	+R	+	+	-	*Bacillus sp*
	X3	$3.0x10^9$	Straight or slightly curved rods with tapered or clubbed end	Circular umbonate cream colonies	-	+R	-	-	-	*Corynebacterium sp*
	X4	$2.0x10^9$	Slightly curved rods in singles	Bluish-green colonies spreading on plates	+	-R	-	+	-	*Pseudomonas sp*
	X5	$1.6x10^{10}$	Spherical cells in irregular clusters. Few in pairs.	Golden yellow smooth and circular colonies	-	+S	-	-	-	*Staphylococcus sp*
	X6	$5.0x10^9$	Cocci in pairs, tetrads.	Small round yellow colonies	-	+S	-	-	-	*Micrococcus sp*
	X7	$3.0x10^9$	Straight rods with oval and central spores	Mucoid and slimy irregular raised colonies	+	+R	+	+	-	*Bacillus sp*
	X8	2.0x10	Cocci predominantly in tetrads. Few in pairs and clusters	Small round convexed orange colonies	-	+S	-	-	-	*Micrococcus sp*

R – Rods
S – Spherical cocci

Table 2. Biochemical and carbohydrate fermentation test of bacteria isolated from samples

Colony	Cat	Oxi	Coag	In	MR	VP	Cit	H_2S	$N0_3$	Urease	TSI	Carbohydrate fermentation						Identity of isolates
												Glu	Suc	Mal	Lac	Mann	Xyl	
X1	-	-	-	-	+	-	+	nd	-	+	Nd	+	+	-	+	+	nd	***E. faecalis***
X2	+	-	-	-	-	+	+	nd	+	-	Nd	+	-	-	$+^S$	+	nd	***B. subtilis***
X3	+	-	-	-	-	+	+	nd	+	-	Nd	+	-	+	$+^S$	-	nd	***Corynebacterium sp***
X4	+	+	-	-	+	-	+	nd	+	+	Nd	+	-	-	-	+	nd	***P.aeruginosa***
X5	+	-	+	-	-	+	-	nd	+	+	Nd	+	+	+	+	+	nd	***S. aureus***
X6	+	-	-	-	+	-	+	nd	-	+	Nd	-	-	-	-	-	nd	***M.cuteus***
X7	+	-	-	-	-	+	+	nd	+	-	Nd	+	-	-	$+^S$	-	nd	***B.cereus***
X8	+	-	-	-	+	-	+	nd	+	-	Nd	+	+	-	-	-	**nd**	***M.roseus***

Cat, Catalase; Oxi, Oxidase; Coag, coagulase; In, indole; MR, methyl red; VP, Voges Proskaeur; Cit, citrate; H_2S, Hydrogen sulphide utilization; NO_3, Nitrate reduction; Glu, glucose; Suc, sucrose; Mal, maltose; Lac, lactose; Mann, mannose; Xyl, xylose, $+^s$, slow reaction

3.2. Determination of Minimum Inhibitory and Bacteriocidal Concentrations

The response of these bacterial isolates to phytol showed that phytol has potent inhibitory effect in terms of MIC and MBC values against all the tested bacterial pathogens. As shown in Table 3, the MIC and MBC values of phytol against the tested bacterial pathogens were found in the range of 31.25µg/ml to 62.5µg/ml and 125µg/ml to 250 µg/ml, respectively. *Pseudomonas aeruginosa* and *Staphylococcus aureus* were found highly susceptible bacterial pathogens to the phytol having MIC values of 62.5µg/ml and 31.25µg/ml respectively and MBC values of 125µg/ml respectively, for each bacterial pathogen. Phytol also had potential antibacterial effect against *B.cereus* and *E. faecalis.*

Similarly, Mathabe *et al.* (2008), reported the antibacterial activity of plant based terpenoid compounds against a panel of food borne pathogenic bacteria,showing their potential antibacterial effect.

Table 3. Determination of MIC and MBC of phytol against bacterial pathogens isolated from grilled meat

Bacterial pathogen	Gram reaction	PHYTOL	
		MIC(µg/ml)	MBC(µg/ml)
Bacillus cereus	+	31.25	250
Pseudomonas aeruginosa	-	62.5	125
Staphylococcus aureus	+	31.25	125
Enterococcus faecalis	+	31.25	250

MIC) Minimum inhibitory concentration (value in µg/ml)
MBC) Minimum bactericidal concentration (value in µg/ml)

3.3. Determination of Inhibitory Effect of Phytol on the Growth of the Bacterial Isolates

Broth dilution with shaking (BDS) method was used to investigate the inhibitory effect of phytol on the growth of *S. aureus*, *B. cereus*, *P. aeruginosa* and *E. faecalis*. Figure 1 shows that in comparison to control absorbance; phytol was stimulatory to the growth of *B.cereus*; *P. aeruginosa* and *E. faecalis* at all concentrations but inhibitory to *S.aureus* at concentrations between 250µg/ml and 2000µg/ml.

The time kill assay demonstrated that the growth of the different bacterial isolates decreased with increase in concentration except in *S. aureus* and *P. aeruginosa*. Inhibitory effects of cell growth decreased at elevated concentrations even though a certain fraction of the bacteria in each bacterial suspension was killed. Thus, surviving bacteria grew faster than did cells under control conditions. It appears that phytol has both growth-inhibitory and growth accelerating activities. The net effects depended on the final concentration, with inhibitory activity predominating at lower concentration.

3.4. Effect of Phytol on Dehydrogenase Activity of the Bacterial Isolates

Results obtained from the control samples showed that these organisms were able to reduce TTC to red formazan at variable rates and extent. The gram-negative *P. aeruginosa* had higher rates of dehydrogenase activity than the gram + positive *B. cereus*, *E. fecalis* and *S. aureus*. This variation may be due to differences in cell wall components or dehydrogenase systems, since different micro-organisms have been reported to have different dehydrogenase systems [13]. This result is also in consonance with the work of Nweke *et al.*, [9].

The effects of the different concentrations of phytol on the bacterial isolates with respect to the dehydrogenase activity and its inhibition are shown in Figure 2. The responses of the bacterial dehydrogenase activities to phytol are concentration – dependent and vary between the organisms. The dehydrogenase activity inhibition observed in this study is consistent with those observed by other workers using plant extracts [2,10,11]. Result presented in Figure 3 showed that *P. aeruginosa* had higher percentage inhibition than the other bacterial isolates at all concentrations of phytol. This implies that *P. aeruginosa* was more sensitive to the deleterious effect of phytol than the other bacterial isolates.

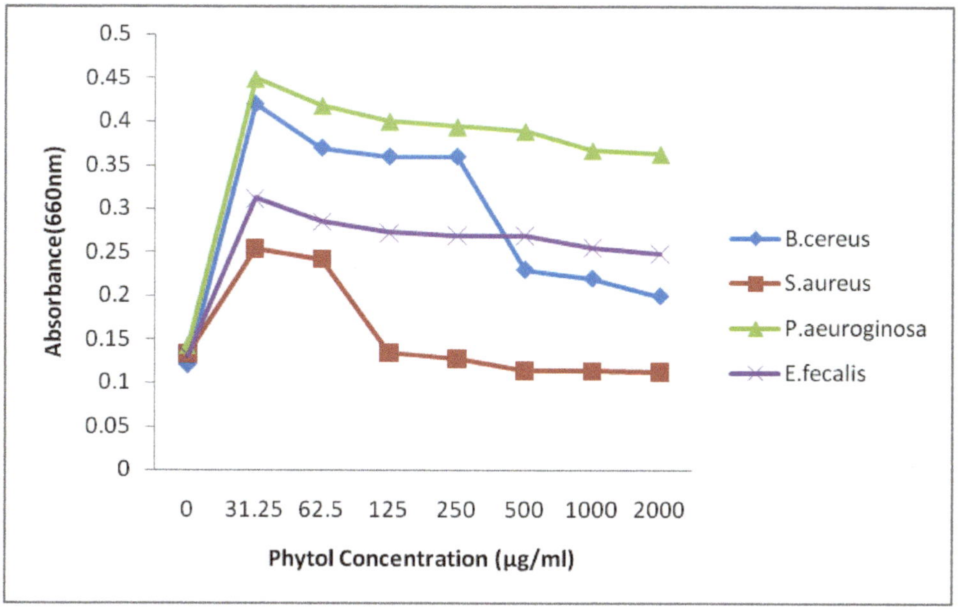

Figure 1. Effect of Phytol concentration on the growth of the bacterial isolates

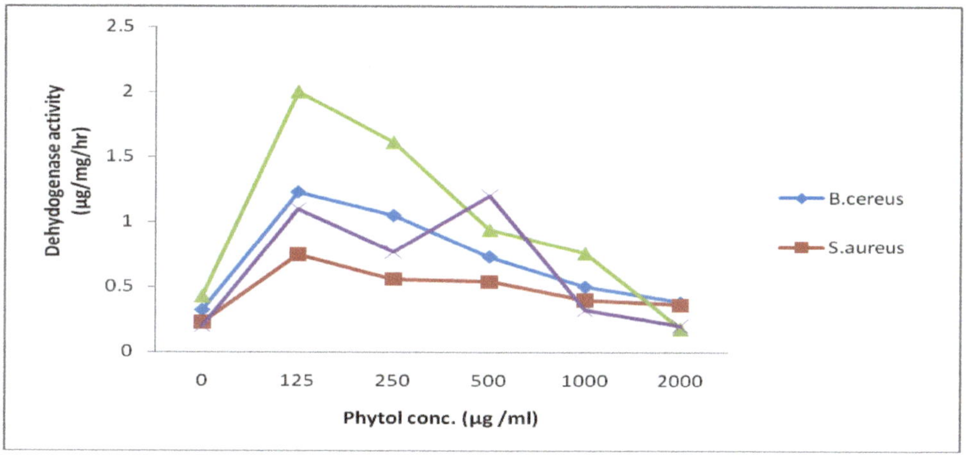

Figure 2. Effect of phytol concentration on dehydrogenase activity of bacterial isolates

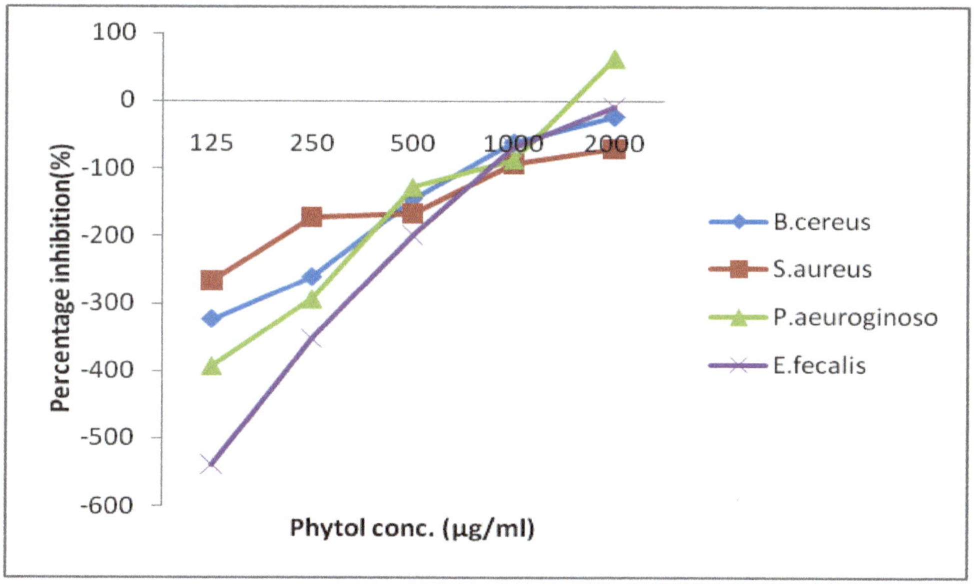

Figure 3. Inhibition of Dehydrogenase Activities of the Bacterial isolates by phytol

4. Conclusion

Food safety is a fundamental concern of both consumers and food industry, especially as the number of reported cases of food – associated infections continues to increase and is rapidly changing (Alzoreky and Nakahara, 2003). Recurring out breaks of food borne illness caused by food borne pathogenic bacteria have sustained the demand for preservation systems that limit the proliferation of food borne pathogens in refrigerated, minimally processed and ready to eat foods. In this regard, plant – based antimicrobials could be potential alternatives in the food industry to control food borne pathogens.

Utilization of bioactive secondary metabolites as natural antibacterial agents may offer many new applications for food industry. Natural food preservatives targeted at food and food products that are easily contaminated by bacteria are highly desired.

The availability of natural compounds developed by exploring more chemistry, activation of plants natural resistance mechanisms and natural products will contribute to sustainable food and food products.

The results described in this study clearly indicate that phytol, a diterpenoid possesses the potential to control food borne pathogens, and these finding are in strong agreement with previous reports (Singh and Singh 2003; Murthy *et al.*, 2005).

References

[1] Aiyelaagbe, O.O., Adesoganm, k., Ekundayo, O., and Gloer, J.B. (2007). Antibacterial Diterpenoids from *Jatropha podagrica* Hook; *Phytochemistry* 68: 2420-2425.

[2] Alisi, C.S., Nwanyanwu C.E, Akujobi, C.O and Ibegbulem, C.O. (2008). Inhibition of Dehydrogenase Activity in Pathogenic Bacteria Isolates by Aqueous Extracts *Musa paradisiaca var sapientum*. *African Journal of Biotechnology* 7(12): 1821-1825.

[3] Arai, T., Hamashima, H., and Sasatsu, M. (1996). Inhibiting Effects of Fatty Acids, Purified Camellia Oil and Olive Oil on the Growth of *Staphylococcus aureus*. Jpn. J. *Chemother*. 44: 786-791.

[4] Bajpai, V.K., Al-Reza, S.M., Lee, J.H., and kang, S.C. (2009a). Chemical Composition, Antibacterial and Anxioxidant Activities of Leaf Essential Oil and Extracts of *Motasequoia glytostroboides*. Miki ex Hu; *Food Chem*. Toxicol. 47: 1876-1883.

[5] Daines, A.M. (2003). The Synthesis of Naturally Occurring Vitamin K and Vitamin K Analogues. *Current Org. Chem*. 7: 1625-1634.

[6] Dellar, J.E., Cole, M.D and Waterman, P.G. (1996). Antimicrobial Abietane Diterpenoids from *Plectranthus elegan*. *Phytochemistry* 41.735-738.

[7] Netscher, T. (2007). Synthesis of Vitamin E. *Vit. and Hormones*. 76: 155- 2002.

[8] Nweke, C.O., Alisi, C.S., Okolo, J.C. and Nwanyanwu, C.E. (2007). Toxicity of Zinc to Heterotrophic Bacteria of Tropical River Sediments. Appl. *Eco. Envron. Res*. 5(1): 123-132.

[9] Nweke, C.O., Okolo, J.C., Nwanyanwu, C.E and Alisi, C.S. (2006). Response of Planktonic bacteria of New Calabar River to Zinc Stress. *Afr. J. Biotechnol*. 5(8): 653-658.

[10] Nwogu, L.A., Alisi, C.S, Ibegbulem, C.O, Igwe, C.U. (2007). Phytochemical and antimicrobial activity of ethanolic extract of *Landolphia owariensis* leaf. *Afr. J. Biotechnol*. 6(7):890-893.

[11] Nwogu, L.A., Alisi, C.S, Igwe, C.U., Ujowundu, C.O. (2008). A Comparative Study of the antimicrobial properties of the ethanolic extracts of *Landolphia owariensis* leaf and root. *Afri. J. Biotechnol*. 7(4): 368-372.

[12] Onuh, M.O. and Igwemma, A.A. (2000). Applied Statistical Techniques for Business and Basic Sciences. Skillmark Media Ltd. Owerri, Imo State. Pp. 176-190, 205-215, 322-338.

[13] Praveen-Kumar, J.C. (2003). 2,3,5 – Triphenytetrazolium Chloride (TTC) and Electron Acceptor of Culturable Cell Bacteria, Fungi and *Antinomycetes*. *Biol. Fert. Soil* 38:186-189.

[14] Rotani, J.F and Volkman, J.K (2003). Organic Geochemistry. Phytol Degradation Products as Biogeochemical Tracers in Aquatic Environments. 34, 1-35.

[15] Tauxe, V.R. (2002), Emerging Food Borne Pathogens. *Int. J. Food Microbial*. 78: 31-34.

[16] Ulubelen, A., Oksuz, S., Kolak, U., Bozok-Johansson, C., Celik, C., and Voelter, W. (2000). Antibacterial diterpenes from the Roots of *Salvia viridis; Planta Med*.66: 458- 462.

Effect of Wrapping Materials on Physico-chemical and Microbiological Qualities of Fermented Melon Seed (*Citrullus colocynthis* L.) Used as Condiment

Anne I. Peter-Ikechukwu*, Ngozi O. Kabuo, Serah O. Alagbaoso, Njideka E. Njoku, Chioma N.Eluchie., Williams O. Momoh

Department of Food Science and Technology, Federal University of Technology, Owerri, Owerri, Imo State, Nigeria
*Corresponding author: alagbaososerah@gmail.com

Abstract The effect of six different wrapping materials on physico-chemical and microbiological qualities of fermented melon seed (*Citrullus colocynthis* L) was studied. Melon seeds were sorted, washed, boiled (48 hrs), dehulled and wrapped in blanched plantain leaf, then it was boiled again for 2 hours, drained, cooled and allowed to ferment naturally for 72 hours. The fermented melon seeds were then mashed, wrapped with six different wrapping materials include: fresh, blanched and dried of *thaumatococcus danielli* leaf, aluminum foil, black polyethylene and transparent polyethylene, placed in a warm environment for 72 hours (maturation period). The products which are known as *Ogiri* condiment were subjected to microbiological and physico-chemical evaluation. The results of the microbiological evaluation showed that the predominant bacteria involved were *Bacillus* spp, *Enterococcus* spp and *Corynebacterium* spp. While *Saccharomyces cerevisae* was isolated. The results of the physico-chemical evaluation showed increase in pH from 5.5 to 6.3, Titratable acidity (TTA) increased from 0.2% to 0.49% during fermentation and varied during maturation. Aluminінum foil had the highest TTA of 0.43% while transparent polyethylene had the lowest TTA of 0.23%. During fermentation, the amino nitrogen increased from 2% to 5.5%. The amino nitrogen also increased appreciably during maturation. Black polyethylene had the highest amino nitrogen value (21.15%) while aluminum foil had the least value of amino nitrogen (15.65%) at 72 hours.

Keywords: *melon seed, Ogiri, fermentation, packaging, microorganisms*

1. Introduction

The origin of food packaging can be said to be as old as civilization. It is an integral part of food processing and entails the use of some materials in the packaging of foods [5]. Packaging is an important descriptive area of food technology, as it concerns the preservation and protection of all types of food materials from microbial spoilage and oxidation. It also extends the shelf-life characteristics of products [16]. Packaging protects a food product from dirt and other contaminants prevent damage that could arise while handling, transporting and distributing the products. It protects food from light, moisture re-absorption, rodents etc and also facilitates distributions and marketing of the food product and raw materials [5]. Traditionally, all sorts of materials are used to package food. In Africa, corn sheet both dry and fresh are used for wrapping various types of foods. Plantain or banana leaves, other broad leaves, calabash gourds, cane and raffia leaves are all used as packaging materials [16,17]. All these leaves are used as packaging material because of the natural flavor they impacted on the food material [12,33]. The primary purpose of a manufacturer is to protect the food product, to keep it in good condition, and to preserve the flavor until it reaches the consumer [17]. The journey from the processor to the ultimate consumer may be a long one, but the product must still reach the consumers with its freshness, flavor, attractiveness and appearance unimpaired [17].

In Nigeria, traditional packaging materials such as *thaumatococcus Danielli*, banana leaves, cocoyam leaves and plantain leaves have gained ground for long [31]. The use of synthetic material such as tin can, aluminum foil, cellophane paper, glass, plastics etc. were invented with civilization to make packaging easier and presentable [28]. Leaves in general, only have limited protection against contamination of the food product packed in them, but have no resistance to impact injury and are easily tampered with [4]. The early man has practiced the preservation of food through various ways which includes the fermentation of food, drying of food over a fire or with the aid of sunlight and salting, and thereafter, packaging with various materials [31].

Egusi Melon (*Citrullus colocynthis L.*) belongs to the species of the genus *Citrullus* of *Cucurbitaceae* family, which usually consists of a large number of varieties that are generally known as melons. Egusi (*Citrullus colocynthis L.*) is among the 300 species of melon found in tropical Africa and it is cultivated for its seeds, which have been reported to be rich in oil and protein. The regions of its cultivation are Middle East, West African (Nigeria, Ghana, Togo, Benin) and other African countries for the food in the seeds and as a crop inter-planted with maize, cassava and yam. As reported by Giwa *et al.*, 2010, in Nigeria only, ''egusi'' is cultivated over an area of 361,000 ha with a production figure of 347,000 tonnes (as seeds) in 2002. It is used both as condiment and thickener in Nigerian local soup, and the industrial scale production of the oil is yet to be utilized despite the huge potential.

'Ogiri' generally refers to as an oily paste made from oil seeds in West Africa. They are also used as soup condiments with strong smell. It is a product of fermentation of melon seeds (*Citrullus vulgaris*) [2] consumed by the Ijebu and Ondo tribes in the forest zone of South Western Nigeria. *Ogiri egusi* is a food flavouring condiment prepared by traditional methods of uncontrolled solid state fermentation of melon seeds involving the use of chance fermentation. In ogiri preparation, melon seeds are boiled until they are very soft and mashed. The mashed melon seeds are then wrapped tightly in banana leaves and left to ferment for five to seven days. Thereafter, the fermented mashed melon is placed in earthen well pot and covered with jute bags which provide low oxygen tension [23,26]. The fermenting mashed melon is still wrapped in leaves, placed on a wire mesh, smoked over charcoal heat at a distance for about two hour and pulverished before it can be used in cooking [1,23].

In traditional fermentation processes, natural microorganisms are employed in the preparation and preservation of different types of food. These processes add to the nutritive value of foods as well as enhancing flavor and other desirable qualities associated with digestibility and edibility. The fermentation techniques are often characterized by the use of simple, non-sterile equipment, chance or natural inoculum, unregulated conditions, sensory fluctuations, poor digestibility, and unattractive packaging of the processed product [20]. Ogiri is a fermented food condiment of flavoring agent, whose character and organoleptic properties depends on microbial activities [21]. The production process is still a traditional family art and the fermentation is by chance inoculation [37]. It is consumed mainly in Southern Nigeria especially by the Igbos [36].

Nigeria is endowed with a wide range of indigenously fermented foods and condiment [1], which are traditionally packaged with leaves [22]. The traditional food condiments are often stigmatized partly due to poor packaging practices and are such regarded as food for the poor. However, inadequate packaging of this product for improved patronage and effective commercialization has been identified as a major problem as it is still being packaged with leaves and sometimes covered with cement papers which are unhygienic and unsafe for consumption, therefore, this research is aimed at determining the effects of both treated and untreated (fresh, blanched and dried) *thaumatococcus Danielli* leaves and modern (aluminum foil, polyethylene transparent and black) wrapping materials on microbial and physico-chemical qualities of fermented melon seeds as condiment.

2. Materials and Method

2.1. Sample Collection

Melon seed (*Citrullus colocynthns. L*) used for this work were purchased from, Ekeonunwa market in Owerri, the capital city of Imo State, Nigeria. Aluminum foil, (Raynolds, USA) and polyethylene were purchased from Maris Super Market, Ikenegbu in Owerri Municipal L.G.A Imo State Nigeria. The *thaumatococcus Danielli* leaves used were also sourced from Eziobodo village, Owerri in Imo State, Nigeria. All the equipment used were available at food science and Technology Department, Federal University of Technology Owerri, in Imo State of Nigeria and all the chemicals used are of analytical grade.

2.2. Methods

2.2.1. Sample Preparation

2.2.2. Traditional Production of Fermented Melon Seed (Ogiri)

Undehulled melon seeds were properly washed, boiled for 3 hours, cooled and dehulled. The dehulled seeds were wrapped tightly in layers of blanched plantain leaves and pierced with fork. The wrapped cotyledons were thereafter boiled for 2 hours, removed from water and placed on a wire mesh to drain for 1 hour. The wrapped cotyledon was then left to ferment at the prevailing ambient temperature (28°C) for 72 hours. At the end of the fermentation period, the seeds were pounded into paste. The paste was subsequently distributed into the various wrapping material which comprised blanched, unblanched and dried leaves of *thaumatococcus danielli*, and aluminum foil, transparent polyethylene and black polyethylene bags and were left to ferment for 72 hours over a fire place for maturation. Figure 1 shows the flow diagram of ogiri egusi condiment production.

Determination of Microbiological Properties of the Samples

Sample preparation and enumeration of microbial isolates: Microbial isolation and identification was by standard microbiological techniques [25]. Ten grams of each sample for microbiological evaluation was aseptically homogenized in 90 ml of 0.1% sterile peptone water, shaken thoroughly, and appropriate dilutions (up to 10^{-9}) were prepared for microbiological studies. Aliquot 0.1 mL of appropriate dilutions was spread inoculated in duplicate onto Nutrient agar (Oxoid, Hampshire, U.K.), and Potato Dextrose Agar (PDA) (PDA, Oxoid). The inocula were spread with sterile spreader to ensure even distributions before incubating the plates. Nutrient Agar was incubated at 37± 2°C for 24 – 48h for the growth of heterotrophic bacteria, while PDA plates were incubated at 25 ± 2°C for 3 days [3]. Colonies were enumerated at the end of incubation period using digital colony counter (Gallenkamp, England). The isolates were characterized on the bases of colonial morphology, microscopic and biochemical characteristics [7,8] to include: indole production, methyl red, Voges-Proskauer, citrate utilization, motility, spore stain, urease production, catalase, oxidase, coagulase, starch hydrolysis, gelatin liquefaction, fermentation of glucose, lactose, sucrose, maltose, mannitol, xylose, raffinose, arabinose, temperature and salt tolerance tests.

Further identification of bacteria isolates was by Biomerieux® sa API kit and with reference to standard identification manuals [8,15,19].

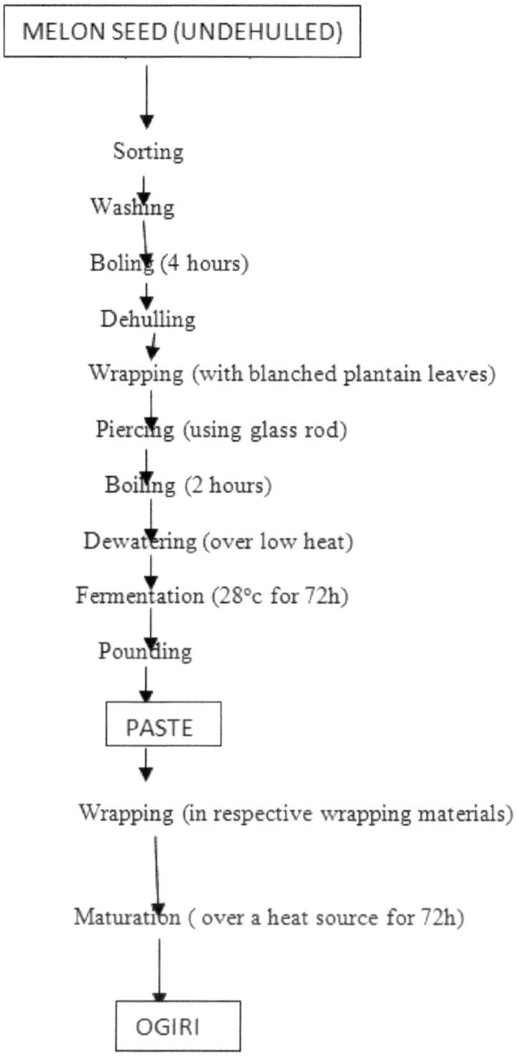

Figure 1. The Flow diagram of Production of Ogiri

Physico-chemical Analyses

Physico-chemical analysis was carried out as described by Ogueke *et al.*, [24], Onwuka, [29] and Han *et al.*, [11].

Amino Nitrogen Determination

The amino nitrogen of the mashed seeds for each fermentation process was measured at 24 h interval, for fermentation and maturation periods. Two grams of the mashed sample were placed in a conical flask, 0.5 ml of phenolphthalein (0.5%) and 0.4 ml of neutral saturated potassium oxalate were added. The mixture was allowed to stand for few minutes and this was neutralized with NaOH (0.1N) to a standard pink colour. Two (2.0 ml) of 40% formaldehyde solution was added and allowed to stand for few minutes (until mixture was colourless), This was titrated with NaOH (0.1N) to pink colour. The percentage amino nitrogen was calculated using the titre value (V) as shown below;

$$\%\ \text{amino nitrogen} = \frac{NNaOH \times VNaOH \times 100}{Mass\ of\ sample}$$

Where N = Normality; V = Titre value.

Titratable Acidity Determination

The titratable acidity of the inoculated seeds (expressed as percentage lactic acid) for each fermentation process was measured on inoculation and at 24 hours interval, for fermentation maturation periods Five grams of mashed seeds were dissolved in 10.0ml distilled water and 2-5 drops of phenolphthalein was added. The solution was carefully neutralized with 0.1N sodium hydroxide (NaOH) and titrated with aqueous 0.1N sodium hydroxide (NaOH), constantly swirled until a pink colour persisted for 15 seconds. The titre value was recorded and the titratable acidity was calculated as % lactic acid thus;

$$\%\ \text{TTA}\ (\%\ \text{Lactic Acid}) = \frac{Titre\ value \times 0.09 \times N \times 100}{Wt}$$

Where N = Normality of NaOH; 0.09 = Equivalent weight of predominant acid, wt = Weight of sample used.

Temperature

Temperature was measured with a thermometer

pH Value

The pH value of the samples was determined using a standard pH meter (P3935100, PHYWE, GERMANY)

DATA ANALYSIS

The results obtained from the data were subjected to Analysis of variance (ANOVA) according to Onuh and Igwemma [27] and SAS [35]. Significant means at $p \leq 0.05$ were separated using Fisher's least significant difference (LSD) test [27].

3. Results and Discussion

3.1. Microbial Assessments

3.1.1. Total Viable Count for Fermentation and Maturation Periods

The result of the total viable count on fermented melon seed is shown in Table 1. The total viable count increased exponentially from zero time to72 hrs both for bacteria and fungi. This indicates that the organisms were at their exponential phase of growth [6] and that the fermenting melon seed was a suitable substrate for the microorganism to grow. At zero time, the melon substrate had 8.0 x10^9cfu /g of total viable bacterial count and increased to 2.32 x10^{10}cfu /g while yeast and fungi count was 5.3 x10^7cfu /g which increased to uncontrollable number.

Table 1. Total viable count of fermented melon seeds (*Citrullus colocynthns* L.) during fermentation at 28°C

Fermentation period (hrs)	Total viable bacteria count	Yeast and mold count
	cfu g^{-1}	
Zero time	8×10^9	5.3×10^7
24	7.5×10^9	2.05×10^8
48	1.95×10^{10}	2.89×10^8
72	2.32×10^{10}	TNC

Note; TNC –TOO NUMEROUS TO COUNT.

The result of the total viable bacterial count on fermented melon seeds wrapped in various wrapping materials during maturation at 24 hrs and 72 hrs is shown in Table 2. An increase was observed in the total viable count bacteria, yeast and fungi. The sample wrapped in Aluminum foil had the least bacteria count of 1.28 $x10^{12}$cfu /g after 72 hrs while fresh leaf was the highest in bacteria count ($2.37x10^{12}$cfu/g). Sample wrapped in dried leaf had the least yeast and fungi count of 9.5 $x10^{9}$cfu /g at 72 hrs, followed by aluminum foil, while fresh leaf had the highest ($2.12x10^{10}$cfu/g). The least fungi count observed in dried leaf may be due to drying which reduced the moisture level of the leaf making it unavailable to the substrate for microbial activity [31]. On the other hand, the least bacteria count observed in aluminum foil could be attributed to the non-porous nature of aluminum foil which prevented the entrance of external growth factors such as moisture, air for microbial growth and oxygen [32].

Table 2. Total viable count of fermented melon seeds (*Citrullus colocynthns* L.) during maturation

Packaging materials	Total viable bacteria count		Yeast and mold count	
	cfu g^{-1}			
	24 hrs	**72 hrs**	**24 hrs**	**72 hrs**
Unblanched leaf	1.84×10^{12}	2.37×10^{12}	1.74×10^{10}	2.12×10^{10}
Blanched leaf	1.41×10^{12}	2.07×10^{12}	1.85×10^{10}	1.95×10^{10}
Dried leaf	8.3×10^{11}	1.62×10^{12}	1.32×10^{10}	9.5×10^{9}
Aluminum foils	9.3×10^{11}	1.28×10^{12}	1.47×10^{10}	1.22×10^{10}
Polyethylene (Transparent)	1.76×10^{12}	1.86×10^{12}	TNC	1.62×10^{10}
Polyethylene (Black)	1.22×10^{12}	2.15×10^{12}	2.05×10^{10}	1.42×10^{10}

Note; TNC – TOO NUMEROUS TO COUNT.

3.1.2. Cultural, Morphological and Microscopic Characteristics of Bacteria and Fungi Detected

The result of the cultural and morphological characteristics of bacteria detected during fermentation is shown in Table 3. From the result, three types of colonies were observed on the nutrient agar plate. The length of the colonies observed ranged from 1-5 mm. Azu [6] reported that the usefulness of using size as a distinguishing feature is however limited by the fact that the sizes change with change in prevailing condition. Colonies grown in conditions of over –crowding tend to be considerably smaller than those of the same organism growing on plates containing only a few colonies. So, it could be assumed that the length of the organism in the agar plate was small due to the overcrowdiness in the agar plate. The organisms on the agar plate showed regular to irregular shape, they were yellow creamy in colour with entire or serrated margin. The bacteria have low convex, flat and unbonate elevation with surface appearances which were moist and shiny or dull and dry. The characteristics showed a continuous and consistent succession of microorganisms with similarities.

Table 3. CULTURAL AND MORPHOLOGICAL CHARACTERISTICS OF BACTERIA ORGANISMS

Sample code	Colony code	Size (mm)	Shape	Colour	Elevation	Margin	Appearance
0 day	Aa	2	Regular	Yellow	Low convex	Entire	Moist and Shiny
	Ab	1	Regular	Creamy	Low convex	Entire	Moist and Shiny
	Ac	4-5	Irregular	Creamy	Flat	Serrated	Dull and dry
24 hours	Aa	1-2	Irregular	Creamy	Flat	Serrated	Dull and dry
	Ab	2-4	Regular	Creamy	Umbonate	Entire	Moist and Shiny
	Ac	1	Regular	Creamy	Low convex	Entire	Moist and Shiy
48 hours	Aa	2-3	Irregular	Creamy	Flat	Serrated	Dull and dry
	Ab	1-2	Regular	Creamy	Low convex	Entire	moist and shiny
	Ac	3-1	Regular	Creamy	Umbonate	Entire	Moist and Shiny

The result in Table 4 showed the cultural and microscopic characteristics of fungi isolated during fermentation of the melon seeds. No growth was observed on the potato dextrose agar plate on the zero time. Furthermore, after 24 hrs of fermentation, *sacharomyces cerevisae*) was identified on the potato dextrose agar plate, which is an organisms known to ferment carbohydrates, breaking them down to sugar in cereal based product mainly. Therefore, it is present due to the carbohydrate content of the melon seed. The presence of fungi during fermentation was due to contamination [31] possibly from the processing environment.

Microscopic and biochemical characteristics of bacteria isolated.

The result of the biochemical test conducted on the bacteria colonies during fermentation period is shown in Table 5. The most predominant organisms identified from the biochemical test were *Enterococcus faecalis*, *Corynebacterium* species and *Bacillus* species, Osho *et al* [30], reported that quite a number of *Bacillus* species have been isolated from various fermented food condiments, although yeast and other bacteria were also seen, only part of them can be considered to play a substantial role in fermentation process. Iwuoha and Eke [14] also reported that the combination of *Bacillus* species and Alkaligence species in Ogiri-egusi was capable of producing the quality characteristics of a good Ogiri which was confirmed from the result of biochemical test in conformity with the above report, It could be deduced that Bacillus species were the main fermenting organisms in melon seed (*citrullus colocynthns L*). However, from the above result, it could be detected that *Bacillus* spp were not the only microorganism that effected the fermentation of melon seed especially with the dependence on natural innoculium. There is evidence that a combination of *Bacillus* species, *micrococcus leteus*, *enterococcus feacalis*, and *corynebacterium* species which were seen in the study effected natural fermentation of the melon seed.

Table 4. Microscopic characteristics of yeast isolates

Hours	Colony Code	Colonial characteristics	Microscopic appearance	Identity of isolates
0	Ax	None	No definite appearance	*No growth*
	Ay	Tiny Creamy circular colonies	Gram- positive large spherical Budding cells	*Saccharomyces cerevisiae*
24	Az	Creamy circular mucoid Butyrous colonies	Gram –positive large ellipsoidal and spherical budding cells.	Saccharomyces cerevisiae *var. ellipsoideus*
	Ay	Tiny creamy circular colonies	Gram- positive large spherical Budding cells.	*Saccharomyces cerevisiae*
48	Az	Creamy circular mucoid Butyrous colonies	Gram positive large ellipsoidal spherical budding cells.	*Saccharomyces cerevisiae var.ellipsoideus*
	Ay	Tiny creamy circular colonies	Gram positive large spherical Budding cells	*Saccharomyces cerevisiae*

Table 5. Microscopic and biochemical characteristics of bacteria isolates

Hour	Type of colony	Microscopic characteristics	Catalase	oxidase	Coagulase	Indole	Citrate	Motility	Sugar fermentation					Most probable Organism Identified
									Glue	Sucr	Lact	Malt	Mann	
Zero time	1	Gram +ve spherical chain	+	-	-	-	+	-	+	+	+	+	+	*Enterococcus faecalis*
	2	Gram +ve rod short chain	+	-	-	-	+	+	+	+	+	+	+	*Bacillus* spp
24	1	Gram +ve spherical chain	+	-	-	-	+	-	+	+	+	+	+	*Enterococcus faecalis*
	2	Gram +ve rod Short chain	+	-	-	-	+	+	+	+	+	+	+	*Bacillus* spp
48	1	Gram +ve rod Pleamorphic	+	-	-	-	-	-	-	-	-	-	-	*Corynebacterium* spp
	2	Gram +ve rod Short chain	+	-	-	-	+	+	+	+	+	+	+	*Bacillus* spp

Where +ve =positive, Gluc= Glucose, Sucr =Sucrose, Lact =Lactose, Malt= Maltose, Mann= Manitol.

Physico-chemical Assessment

The result of the chemical parameters that were monitored during fermentation of melon seed (*citrullus colocynthis* L.) is shown in Table 6. Temperature was fairly constant (33°C) at 24 hours to 48 hrs, which reduced to 29°C at 72 hrs of fermentation. The initial temperature (33°C) was as a result of the drop in the cooking temperature of the melon seeds which finally reduced to the prevailing ambient temperature of 29°C as at the time of fermentation. The pH of the fermented seed was observed to increase during fermentation from 5.5 to 6.3; similar report was made on *Citrullus vulgaris* L. series [31]. Also Odunfa [22] reported the pH of Ogiri-egusi increase from 6.5 to 8.1 and fermentation temperature from 29°C – 30°C. The rise in pH during production of alkaline fermented food is due to the ability of the dominant microorganisms *Bacillus* species to hydrolyze protein into amino acid and ammonia. Sanni *el al* [34] reported that the higher pH values of fermented legumes compared to other materials under similar condition were as a result of their higher protein content. Therefore, the increase in the pH of fermented melon (*citrulls colocynthis L.*) could be attributed to its high protein content, which was hydrolyzed to amino acid and ammonia. There was an increase in the amino nitrogen produced during fermentation of melon seed (Table 6) as the period of fermentation increased. Amino nitrogen can be used as an indicator for autolysis and microbial degradation of food protein [13] as well as a good index of yeast growth [18]. It is also used as a good indicator for increase in pH value.

Table 6. Physico-chemical properties of melon seeds during fermentation

Parameters	24 hours	48 hours	72 hours
Temperature (°C)	33	33	29
p^H	5.5	6.0	6.3
Amino Nitrogen(%)	2 %	4%	5.5%
Total Titratable Acidity(%)	0.2%	0.45%	0.49%

Table 6 reveals an increase in total titratable acidity (TTA) from 0.21% to 0.49%. Titratable acidity is calculated as lactic acid produced by bacteria cells or enzyme from the raw material [38]. The result of the total titratable acidity obtained showed the percentage of lactic acid that was produced during fermentation of melon seeds. In alkaline fermented foods, lactic acid production would not be much beneficial due to the desired alkaline nature of the product expected at the end.

Table 7. Physico- chemical properties of melon seeds during maturation

Packaging Material	p^H		TTA		Amino Nitrogen	
	24 hours	72 hours	24 hours	72 hours	24 hours	72 hours
Unblanched leaf	6.8	7	0.66%	0.36%	13.65%	19.50%
Blanched leaf	6.5	7.5	0.64 %	0.25%	11.25%	18.15%
Dried leaf	6.4	7	0.54%	0.32%	9.35%	17.65%
Aluminum foil	6.8	7.5	0.66%	0.43%	10.50%	15.65%
Polyethylene (Transparent)	6.8	7.8	0.66%	0.23%	16.58%	20.00%
Polyethylene (Black)	6.5	7.7	0.57%	0.36%	17.15%	21.15%

Physico-chemical assessment of melon seeds during maturation

The result of the physico-chemical parameters that were examined during maturation period of fermented melon seed paste was wrapped into various wrapping materials (aluminum foil, leaves, transparent and black polyethylene) are shown in Table 7. The value of pH of the fermented melon paste increased from 6.4 to 7.8 as the maturation periods increased from 24 to 72 hrs. The total titratable acidity at this stage varied among the packaging materials.

The total titratable acidity (TTA) result showed that the samples packaged in transparent polyethylene, blanched leaf and dried leaf had low TTA of 0.23%, 0.25% and 0.32% respectively at 72 hrs. While the aluminum foil has the highest TTA (0.43%).

The amino nitrogen increased appreciably during the maturation. Black polyethylene had the highest amino nitrogen value (21.15%) while aluminum foil had the least value of amino nitrogen (15.65%) at 72 hours. According to Ogueke *et al.*, [24], increased amino nitrogen is due to the breakdown of proteins to amino acids by proteases. The putrid aroma which is a characteristic of fermented melon seed (Ogiri) is due to the ammonium ion produced from the amino nitrogen content of the melon.

4. Conclusion

Modern packaging materials maintained and improved the chemical composition of fermented melon seed such as pH and amino nitrogen, which are greatly responsible for the characteristic flavor of the condiment. Therefore, packaging this condiment with modern packaging materials like aluminum foil and polyethylene will enhance the chemical quality, aesthetic value and marketability in the world.

References

[1] Achi, O.K. (2005b) Traditional Fermented Condiments in Nigeria. *Afr.* J. *Biotechno*. 4(13): 1612-1621.

[2] Achi OK (2005a). The Upgrading of traditional fermented foods through biotechnology. Afr. J. Biotechnol. 4: 375-380.

[3] Akinyele, B. J., & Oloruntoba, O. S. (2013). Comparative Studies on *Citrullus vulgaris*, *Citrullus colocynthis* and *Cucumer opsismannii* for Ogiri production. Brit. Micro. Res. J. 3(1) :1-18.

[4] Awan, J.A. and Okara, J.C.(2015). Food packaging: In; Element of food spoilage and preservation.Institute of Management and Technology, Enugu, Nigeria.(2nd edn). Pp 236-237.

[5] Ayo, J. A. (2003) Food packaging: Origin, Trend, Principles and Applications. AMANA Publishers.Kaduna Nigeria. pp1.

[6] Azu, O. (2004). Laboratory Guide for Microbiology, Crystal Publishers, Okigwe, Imo State, Nigeria.

[7] Barnnett JA, Payne RE, Yarrow D (Eds) (2011). Yeasts: Characteristics and Identification. 3rd Eds.

[8] Cheesbrough M. (Eds) (2002). Biochemical tests to identify bacteria. In: Laboratory practice in tropical countries, Cambridge edn. pp. 36-70.

[9] Cheesbrough, M. (Ed.) (2006). District Laboratory Practice in Tropical Countries. Part 2, Cambridge.

[10] Giwa, S., Abdullah, L. C., Adam, N. M. 2010. Investigating "Egusi" (*Citrullus colocynthis* L.) Seed Oil as Potential Biodiesel Feedstock. Energies, 3 (4): 607-618.

[11] Han, B.Z., Beumer, R.R., Rombouts, F.M. and Nout, M.J.R. (2001). Microbiological safety and quality of Commercial Sufu- a Chinese fermented soybean food. Food Control. 12: 541-547.

[12] Hotz, C. and Gibson, R.S. (2007) Traditional food processing and preparation practice to enhance the bio-availability of micronutrient in plant based diet. J. *Nutr.* 134(4): 77-100.

[13] Ijong, F.G. and Olita, Y. (1995). Microflora and Chemical Assessment of an Indonesia Traditional Fermented Fish Sauce. Bakasangg. Fac. Appl. Bio.Sci. 34: 95-100.

[14] Iwuoha, C.I. and Eke, S.O. (1996). Nigerian Indigenous Fermented Foods, Their traditional process operation, inherent problems, Improvements and current status. Food Res. Int. 29: 527 -540.

[15] Jolt, J. G.; Krieg, N. R., Sneath, P. H. A., Stanley, J. T. and Williams, S. T. (1994). Bergey's manual of Systematic bacteriology, 9th edn., Williams and Wilkins Co. Baltimore, Maryland.

[16] Komolafe, E.A. (2005) Food Packaging. Biotidara Publisher Eringbo, Akure. pp 47- 51.

[17] Kordylas, J. M. (1991) Processing and preservation of tropical and sub- tropical foods. Macmillan Education Ltd. London. pp. 49-73.

[18] Lekkas, C., Stewart., A.A., Hill, A.E., Taid, B. and Hodgson, J. (2007). Elucidation of the role Nitrogenous wort in Yeast fermented. J. Inst. Brew. 113: 3-8.

[19] MacFaddin, J. F. (2000). Biochemical Tests for Identification of Medical Bacteria, 3rd edn., Williams and Wilkins, Philadelphia, P. A.

[20] Nout, M.J.R and Rombout, F.M (1990). Recent development in Tempe research. J. Appl. Bacteriol. 69: 609-633.

[21] Nwosu, C. D., & Ojimelukwe, P. C. (1993). Improvement of the traditional method of Ogiri production and identification of the micro-organisms associated with the fermentation process. *Plant Foods for Human Nutrition*, *43*(3), 267-272.

[22] Odunfa, S.A. (1981). A note on the microorganism Association with the fermention of African Locust Bean (*Parkia filicoideea*) During Iru production. J. Plant Food, 3: 245-250.

[23] Odunfa SA. (1985). African Fermented Foods, In: Woods, B.J.B (eds.) Microbiology of Foods, Vol 11, Amsterdam, Elsevier, Applied Science Publishers. pp. 155 -191.

[24] Ogueke, C.C., Onwuamanam, C.I., Peter-Ikechukwu, A.I., Iwouno, J.O. and Ahaotu, A. (2013). Castor seed fermentation using free and attached cells of Bacillus subtilis for production of a soup flavouring. Int. J. life Sci. 2(3): 150-155.

[25] Olutiola PO, Famurewa O, Sontag HG. (2nd Eds) (2000). An introduction to general microbiology, A practical Approach: Heldelberg Publishers.

[26] Omafuvbe BO, Olumuyiwa SE, Bolanle AO, Steve RA. Chemical and biochemical changes in African locust bean (*Parkia biglobosa*) and melon (*Citrullus vulgaris*) seeds during fermentation of condiments. Pakistan Journal of Nutrition. 2004;3:140-145.

[27] Onuh, M.O. and Igwemma, A.A. (2000) Applied Statistical techniques for Business and Basic Sciences. Skillmark Media Ltd. Owerri, Imo State. Pp. 176-190, 205-215, 322-338.

[28] Onweluzo, J.C. and Eillita, M. (2003) Surveying Mucuna utilization as a food in Enugu and Kogi State of Nigeria. Tropical and Sub-tropical Agroeco. 1.1: 213-225.

[29] Onwuka, G.I. (2005). Food Analysis: Theory and Practice. Naphthali Print Lagos. pp 103-104.

[30] Osho, A., Mabekoje, O.O. and Bello, O.O. (2009) Comparative study on the microbial load of Garri, Elusbo- Isu and Iru in Nigeria. Afr. J. Food Sci. 4(10): 646-649.

[31] Peter- Ikechukwu, A.I.(2015) Effect of wrapping material on chemical and microbiological qualities of Fermented melon seed (Citrullus vulgaris L. Series); J. of Analytical Chemistry; 15(2) 74-82.

[32] Potter, N.N. and Hotchkiss, J.H. (1995). Food Sciences, (5th edn.); Chapman and Hall, New York, U.S.A.

[33] Rabash, T.M., Feng, H., Yang, W., Khahil, E. and Mohammad, O. (2011) Effect of type of packaging material On the physicochemical and sensory properties of olive oil. Int. Agric. Biol. 24 (4): 66-72.

[34] Sannni, A.I., Agernor, G.S., Sakyi-Davion, E. and Sefa-Dehah, S. (2000). Aerobic spore forming bacteria and Chemical composition of some Nigerian fermented soup condiments. Plant Foods Human Nutr. 55: 111-118.

[35] SAS, (1999) Statistical Analysis System User' Guide Statistics. SAS Institute Inc. Cary NC 27513 USA.

[36] Steinkraus, K. (1995). *Handbook of Indigenous Fermented Foods, revised and expanded*. CRC Press.

[37] Steinkraus, K. H. (2002). Fermentations in world food processing. *Comprehensive Reviews in Food Science and Food Safety*, *1*(1), 23-32.

[38] Vugst, L.D. (2002). Technology aspect related to the application of functional starter cultures, Food Techno. Biotechnol. 38: 105-112.

16

Development and Validation of Multi-Residue Analysis of 82 Pesticides in Grapes and Pomegranate as per the Requirements of the European Union (EU) and Codex Alimentarius Using GC-MS/MS with Compound Based Screening

Gouri Satpathy[1], Yogesh Kumar Tyagi[1,*], Rajinder Kumar Gupta[2]

[1]University School of Basic and Applied Sciences, GGS Indraprastha University, Delhi, India
[2]University School of Biotechnology, GGS Indraprastha University, Delhi, India
*Corresponding author: drytyagi@gmail.com

Abstract The Quick Easy Cheap Effective Rugged and Safe (QuEChERS) multiresidue method has been validated for the extraction of 82 pesticides belonging to various chemical classes from grapes and pomegranate (commodities with high sugar and low lipid contents). A mixture of 82 pesticides amenable to gas chromatography (GC) was quantitatively recovered from spiked grapes and pomegranate and determined using gas chromatography-tandem mass spectrometry (GC-MS/MS). The method employed involved initial extraction in a water/ethyl acetate system, an extraction/partitioning step after the addition of salt, and a cleanup step utilizing dispersive solid-phase extraction (d-SPE); this combination ensured that it was a rapid, simple and cost-effective procedure. The method setup is streamlined with the new software approach of Compound Based Scanning (CBS). The matrix-matched calibration results have demonstrated good reproducibility, robustness and linearity. The spiking levels for the recovery experiments were 0.005, 0.01 and 0.1 mg kg^{-1} for GC-MS/MS analyses. Adequate pesticide quantification and identity confirmation were attained, even at the lowest concentration levels, considering the high signal-to-noise ratios, the very good accuracies and precisions, as well as the good matches between the observed ion ratios. Mean recoveries mostly ranged between 70 and 110 % (91% on average), and RSD were generally below 12% (7.3% on average). The use of analyte protectants during GC analysis was demonstrated to provide a good alternative to the use of matrix-matched standards to minimize matrix-effect-related errors. For all compounds LODs were 0.001 to 0.005 $mgkg^{-1}$ and LOQs were 0.005 to 0.020 $mgkg^{-1}$. Correlation coefficients of the calibration curves were >0.991. Based on these results, the methodology has been proven to be highly efficient and robust and thus suitable for monitoring the Maximum Residual Limit (MRL) compliance of a wide range of commodity/pesticide combinations.

Keywords: *grapes, pomegranate, pesticides, QuEChERS, method validation, GC-MS/MS*

1. Introduction

The term "pesticides" is commonly used to describe a very broad class of crop-protecting chemicals, including insecticides, herbicides and fungicides. Together, these compounds have become a primary tool in modern agriculture and have contributed to a dramatic increase in crop yields in recent decades for most field, fruit, and vegetable crops. Along with the obvious benefits, concerns have also surfaced regarding the effect of these chemicals on human health. Considering the lethal effects of pesticides on human health, their residues in food commodities are issues of public concern and are regulated by several legislations. In order to address these concerns, governments have initiated monitoring programs to enforce regulatory compliance and ensure food safety. Given that pesticides are polluting the earth and causing problems in human beings and wildlife, the quantity of pesticide being consumed becomes a necessary knowledge. Pesticide residue analysis is an important process in determining the safety inusing certain pesticides.

Grapes and pomegranates are important fruit crops in India, the commercial cultivation of which receives frequent application of a large volume of pesticides throughout the cropping season to control a variety of pests and diseases. Pesticide residue is a major concern for the stakeholders of the grape industry, since quality regulations and food safety standards are becoming more stringent in most countries. The management of pesticide residues in grapes is challenging because, besides direct application, pesticide residues may also appear in grape berries from indirect sources like soil, contaminated agro-inputs (e.g. manures, fertilizers, growth regulators, irrigation water, etc.), drift from adjoining fields of other agricultural crops, etc. Hencehundreds of chemicals are regularly monitored in the grape and pomegranate samples of Indian origin for export, although the number of recommended chemicals for direct use in grape cultivation is only 45 [1]. Considering the importance of food safety, the Government of India has made a satisfactory certificate of residue analysis a mandatory prerequisite for issuance of a phytosanitary certificate for export [2]. Therefore analysis of pesticide residues is an important process to determine the safety of using certain pesticides. Consequently, governments, food producers and food retailers have a duty to ensure that any residues occurring in the foods for human consumption are at or below statutory MRLs. Europe and US, among others, look for certified quality produce for import and have stringent agricultural practices and are particular about residue content. Commission decision 90/642/EEC adopted in the European Union sets MRLs for more than 500 different pesticides in over 300 different food commodities. Many of these MRLs are set at a default value of 0.01 $mgkg^{-1}$, the typical limit of determination of routine analytical methods. Thus there is requirement to have an analytical test method to analyze grapes and pomegranates for large numbers of pesticides at concentrations at or below 0.01 $mgkg^{-1}$, preferably with low costs and rapid analysis (rapid turnaround times < 24 h). This is not possible with single Quadrupole MS and is often achieved using MRM based on the use of a combination of LC-MS/MS and GC-MS/MS due to the most selective and sensitive nature of the detectors [3,4,5,6,7,8].

Every good detection technique requires purified sample that can meet quality requirements for any particular analysis. It also requires a good sample preparation technique (high sample throughput, ruggedness, ease-of-use, low cost and labor, minimal solvent usage and waste generation, occupational and environmental friendliness, small space requirements and minimal material and glassware needs) and needs to be suitable for both single processing as well as batch processing. A number of methods have been reported for extraction of pesticide residues in fruits and vegetables, which involve an aqueous acetonitrile extraction with d-SPE cleanup and GC-MS/MS detection. However, they were not in total compliance with the criteria established by the EU [9]. Although acetonitrile based extraction is readily compatible with LC-MS/MS, certain issues are encountered when less polar semi-volatile pesticides are analyzed by GC-MS/MS. Direct analysis of acetonitrile by GC is not compatible and can degrade the GC column phase. It also leads to poor focusing of chromatographic peaks because the vapour created by the sample overloads the insert liner dimension due to a high thermal expansion coefficient [10,11]. Moreover, evaporation and reconstitution of acetonitrile can cause loss of concentration of some of organophosphorous analytes such as monocrotophos, phorate, etc. Another problem is the high cost and availability of acetonitrile in the market.

Hence, the main objective of the study was to develop an improved method for reliable identification and accurate quantification of pesticide residues in grapes and pomegranates. The methodology was based on prior work with enhancement of a few analytical properties. Some of the important aspects considered during the present work were: (i) extension of the scope of previous methods so as to include as many target analytes as possible, (ii) design and optimization of MRM for the GC-MS/MS acquisition method, and (iii) testing the effectiveness of alternate cheaper solvent (ethyl acetate) extraction followed by PSA and GCB clean-up as means of reducing adverse influence of co-extractants on method and instrument performance. In addition, an important part of this work was demonstrating the performance of the developed method as per section 1.8 and for regular analysis.

2. Experimentals

2.1. Materials

The composite samples, consisting of 1-2 kg of grapes and pomegranate were obtained from a local market, Delhi. The standard stock solutions (1000 ppm) of all the 82 pesticides listed in Table 1 were prepared in ethyl acetate and stored at 4°C. Ethyl acetate was procured from RFCL, Delhi, India. Other chemicals: anhydrous magnesium sulfate, Primary Secondary Amine (PSA) and Graphatized Carbon Black (GCB)were from Agilent Technologies.

2.2.Sample Preparation

A modified QuEChERS method using ethyl acetate was used for the preparation of samples [9,11,12]. According to this method, 10 g of finely ground sub-sample was placed in a polypropylene centrifuge tube (50 mL). 10 mL ethyl acetate was added to it and the mixture was shaken vigorously on a laboratory shaker for 2 min and then centrifuged at 3000 rpm for 2 min. For clean-up, an aliquot (2 mL) of the supernatant was transferred into a polypropylene centrifuge tube containing 100 mg anhydrous magnesium sulphate, 40 mg PSA and 10 mg GCB. The tube was then vortexed for 1 min and centrifuged at 3000 rpm for 2 min. An aliquot of the supernatant was transferred into a glass vial for GC-MS/MS analysis. The sample concentration in the final extracts was 1 gmL^{-1}.

2.3. Instrumental Conditions

GC analyses were performed using a Scion TQ-MS/MS gas chromatograph, equipped with electronic flow control (EFC), a 1079 universal capillary injector and a CP-8400 auto sampler. The injector temperature was maintained at 260°C. 2 μl injections were performed by split-splitless injections with the split ratio as follows: initially set at 20:1; then, the split vent was closed from 0.01 min to 1.5 min, after which the split ratio was held at 100:1 till the 20th min, and finally reduced to 20:1. A BR-5 ms (30m × 0.25mm, 0.25 μm) column was used for separation. Helium (of 99.999 % purity) was used as carrier gas at a flow rate of 1.5 mLmin^{-1}. The column was held at 90°C for 3 min after injection, then programmed at 20°C to 150°C min^{-1} and then to 300°C at 1°C min^{-1} and was held for 4 min (total runtime 37 min). In this study, GC-MS/MS with an Axial Ion Source, Active-focusing q0, curved collision cell (for multiple noise cancellation), Extended Dynamic Range (EDR) detector (for analyzing samples at ppb to ppm level) and a new compound-based screening (CBS) approach was used. MS was operated in the EI mode at -70 eV. A collect delay time of 5.95 min was set to prevent instrument damage. The temperatures of the transfer line, ion source, manifold, active focuses were set at 300, 230, 40 and 135°C, respectively. The MS was calibrated with perfluorotributylamine (PFTBA). For the MS/MS experiments, argon (99.9998 % purity) was used as the collision gas with the collision cell pressure set at 1.7 mTorr. Data acquisition and processing were performed using MS Workstation, version 8.0.

2.4. Method Performance

Method validation and uncertainty of measurement were estimated using international guidelines [15,16]. The linearity of the pesticide mixtures were prepared containing 82 compounds and the calibration curves were studied at a concentration range between 0.005 and 0.1 mgkg^{-1} with duplicate injections prepared in both blank matrix extract and in ethyl acetate. The accuracy and precision of the method was assessed using spiked samples of grapes and pomegranate for six replicates at three spiking levels of 0.01, 0.05, and 0.1 mgkg^{-1} to evaluate efficiency of the method. The results were always calculated using matrix-matched multi-level calibration standards bracketing the samples. LODs of all tested analytes were taken where S/N $\geq$ 3 and LOQs were taken as the lowest spiking level, at which the validation was achieved (i.e. the average recovery was in the range 70 - 120 % with RSD $\leq$ 20 %), and was determined based on the accuracy and precision data obtained through the recovery studies.

3. Results and Discussion

3.1. Optimization of Sample Preparation and GC-MS/MS Acquisition Method

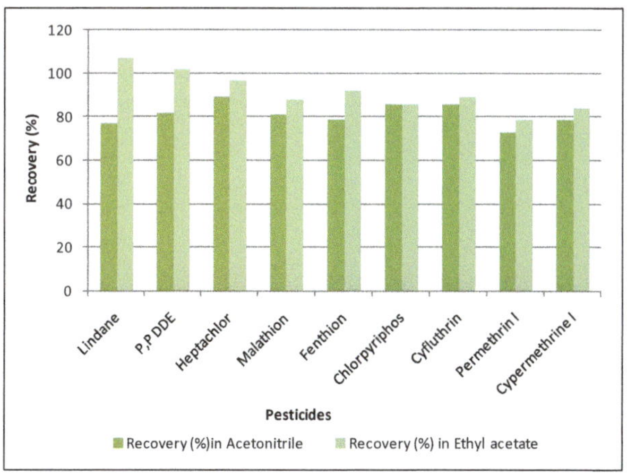

Figure 1. Comparison of recovery (%) obtained from different solvent extract of grapes

In this method, a single and quick step of extraction using ethyl acetate was used because it is equally acceptable for extraction of non-fatty food, and especially suitable for the extraction of high-sugar commodities like grapes, since sugar has limited solubility in ethyl acetate [13,14]. Furthermore, substitution of acetonitrile by ethyl acetate increased the recovery of OC, OP and SP and it substantially reduces the input cost of analysis (Figure 1).

The purification step was optimized by using different weight of dSPE materials (PSA and GCB) to remove fatty acids and pigments from extracts of grapes and pomegranate. As a comparison of the results obtained from different weights of dSPE material in grapes extract, 40 mg PSA and GCB of 10 mg were found sufficient and gave equivalent response as of 100 mg each (Figure 2). Therefore the above weights of PSA and GCB were selected for this experiment to remove interference from the matrix successfully and to reduce the cost of analysis.

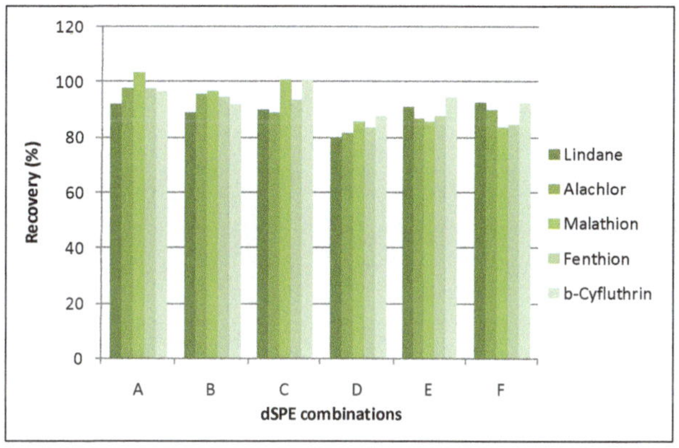

Figure 2. Comparison of recovery (%) obtained from different weight of dSE of grapes ((A)100 mg of each PSA and GCB(B) 50 mg of each PSA and GCB,(C) 40 mg of each PSA and GCB (D) 30 mg of each PSA and GCB (E) 40 mg PSA and 20 mg GCB (F) 40 mg PSA and 10 mg GCB

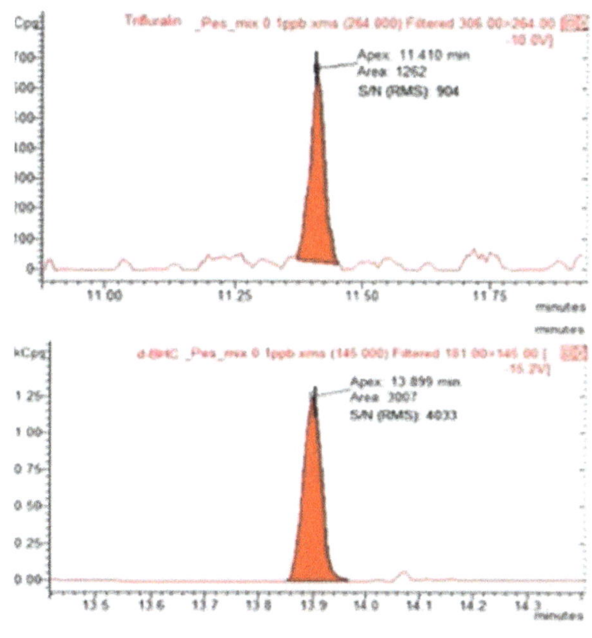

Figure 3. Peak of trifluralin and d-BHC at 0.001 mg kg^{-1} with RT (min), area and signal/noise ratio (both in counts per second)

The application of GC connected to tandem mass spectrometry (GC-MS/MS) in multi-class pesticide residue analysis has made the multi-residue analysis more rugged and convenient by offering the possibility of simultaneous determination of a large number of pesticides with varied physico-chemical properties without the need for chromatographic baseline separation. It has minimized the requirement for extensive sample cleanup, which was otherwise essential in earlier methods of analysis. The use of MRM mode removed the interfering peaks in the data acquisition and integration process. As a result, background noise was reduced and no peak was found missing due to interactions with matrix components. The use of this ethyl acetate extraction followed by a single-step clean-up and GC-MS/MS detection procedure could reliably identify pesticides at 0.001 $mgkg^{-1}$ (Figure 3).

Table 1. Different optimized GC-MS/MS parameters used for analysis of grapes and pomegranate and their validation results

Pesticide	RT (min)	MRM/CE (V)	Scan time (ms)	r^2	LOD (mg kg^{-1})	LOQ (mg kg^{-1})	% recovery	Inter-day precision(%)	±UM (%)
Hexaflumuron	8.08	202>110 (-25) 202>174 (-10)	30	0.992	0.0015	0.005	84-103	12	18
Trifluralin	11.416	306>160/(-25) 306>264/(-10)	45	0.996	0.0015	0.005	89-102	11.3	19
Phorate	11.835	121>93/ (-5) 260>75/ (-10)	45	0.994	0.0015	0.005	87-105	9	16
α-HCH	11.951	181>145/(-15) 219>183/(-10)	30	0.993	0.0005	0.005	86-103	9.6	17
Thiometon	12.204	125>47/(-15) 125>79/(-10)	75	0.996	0.0015	0.005	85-106	8.9	14
Dimethoate	12.347	125>79/(-10) 229>87/(-10)	30	0.997	0.0015	0.005	87-107	10.5	13
Carbofuran	12.575	164>103/(-25) 164>149/(-10)	30	0.995	0.0015	0.005	88-108	11.2	19
Simazine	12.626	201>173/(-5) 201>186/(-5)	45	0.992	0.0015	0.005	94-103	10.2	13
β-HCH	12.712	181>145/(15.1) 219>183/(10.1)	45	0.997	0.0005	0.005	94-106	11.5	15
γ-HCH	13.008	181>145/(-15) 219>183/(-10)	60	0.999	0.0001	0.005	85-107	11.7	16
Phosphamidon	13.471	264>127/(-15) 264>193/(-10)	45	0.999	0.0015	0.005	88-106	10.7	15
Diazinon	13.486	179>137/(-15) 304>179/(-10)	60	0.992	0.0015	0.005	88-104	9.8	14
Chlorothalonil	13.531	266>133/(-35) 266>231/(-15)	45	0.993	0.0015	0.005	87-105	9.4	15
δ-HCH	13.902	181>145/(15.2) 219>183/(10.2)	45	0.994	0.0005	0.005	87-108	10.3	14
Formothion	14.433	170>93/(-10) 125>93/(-10)	30	0.995	0.0015	0.005	86-106	8.9	12
Metribuzin	15.003	198>82/(-15) 198>110/(-10)	75	0.997	0.0015	0.005	85-109	8.8	12
Methyl parathion	15.195	263>109/(-15) 263>246/(-5)	30	0.995	0.0015	0.005	88-107	9.6	13
Alachlor	15.28	160>132/(-10) 188>160/(-10)	45	0.999	0.0005	0.005	84-108	7.6	11
Paraxon	15.664	149>91(-10) 149>119(-5)	90	0.994	0.0015	0.005	88-107	5.7	10
Pirimiphos methyl	16.07	290>125/(-20) 290>151/(-20)	30	0.993	0.0015	0.005	85-107	8.9	13
Fenarimol	16.088	277>109/(-20) 277>260/(-10)	45	0.992	0.0015	0.005	89-108	6.8	11
Linuron	16.295	160>124/(-10) 248>61/(-10)	75	0.994	0.0015	0.005	91-105	10.3	18
Malathion	16.477	173>99/(-15) 173>127/(-5)	30	0.992	0.0015	0.005	88-103	12	17
Metolachlor	16.59	162>133/(-15) 238>162/(-15)	45	0.995	0.0015	0.005	87-108	9.7	13
Aldrine	16.673	263>193/(-35) 263>228/(-10)	30	0.994	0.0015	0.005	86-108	9.7	14
Chlorpyrifos	16.673	314>258/(-15) 314>286/(-10)	45	0.995	0.0015	0.005	89-109	10.2	15
Fenthion	16.8	278>109/(-20) 278>125/(-20)	30	0.996	0.0015	0.005	85-106	7.7	14
Parathion	16.927	291>109/(-15) 291>137/(-10)	30	0.993	0.0015	0.005	86-108	7.8	15
Triadimefon	17.053	208>111/(-20) 208>181/(-10)	45	0.995	0.0015	0.005	87-102	8.5	14
Pendimethalin	17.839	252>162/(-10) 252>191/(-10)	45	0.992	0.0015	0.005	84-103	8.7	13

Pesticide	RT (min)	MRM/CE (V)	Scan time (ms)	r^2	LOD (mg kg^{-1})	LOQ (mg kg^{-1})	% recovery	Inter-day precision(%)	±UM (%)
Penconazole	18.033	248>157/(-25) 248>192/(-15)	30	0.994	0.0015	0.005	85-105	14.8	19
Heptachlor epoxide	18.049	353>263/(-15) 353>282/(-15)	45	0.993	0.001	0.005	88-102	11.4	16
Chlorienvinphos - α	18.233	267>159(-20) 323>267(-15)	30	0.994	0.0015	0.005	94-103	10.7	15
Captan	18.355	149>79/(-10.1) 149>105/(-10)	30	0.996	0.0015	0.005	88-105	10.4	14
Phenthoate	18.39	274>121/(-10) 274>246/(-10)	30	0.994	0.0015	0.005	89-104	11.9	16
Quinalphos	18.413	146>118/(-20) 157>129/(-15)	30	0.994	0.0015	0.005	94-102	11.3	17
Dieldrin	18.888	263>193/(-25) 263>228/(-20)	60	0.995	0.001	0.005	86-101	9.8	14
Chlordane - II	18.888	373>266/(-20) 373>301/(-10)	30	0.992	0.001	0.005	86-107	11.5	17
O, P DDE	19.048	246>176/(-30) 318>246/(-25)	75	0.996	0.001	0.005	88-107	11.3	16
Butachlor	19.325	176>147/(-15) 237>160/(-10)	75	0.992	0.001	0.005	89-106	9.8	13
Endosulfan - α	19.333	241>170/(-25) 241>206/(-15)	30	0.992	0.001	0.005	92-108	9.6	15
Chlordane - I	19.351	373>266/(-20) 373>301/(-10)	60	0.994	0.001	0.005	84-106	8.9	14
Hexaconazole	19.858	214>152/(-20) 214>159/(-20)	30	0.996	0.0015	0.005	90-105	8.9	12
Tricyclazole	19.88	162>134/(-20) 189>161/(-15)	90	0.992	0.0015	0.005	91-108	7.8	11
Pretilachlor	20.066	162>147/(-15) 176>134/(-15)	75	0.994	0.0015	0.005	89-107	6.7	11
Profenofos	20.105	139>97/(-10) 339>269/(-15)	30	0.993	0.0015	0.005	87-105	7.5	12
P, P DDE	20.229	246>176/(-20) 246>211/(-20)	60	0.993	0.001	0.005	89-109	9.4	14
Fenitrothion	20.238	383>255/(-20) 383>335/(-10)	60	0.997	0.0015	0.005	85-107	5.4	11
Oxadiazon	20.357	258>112/(-25) 258>175/(-10)	30	0.995	0.0015	0.005	84-109	4.4	10
O, P DDD	20.433	235>165/(-25) 235>199/(-15)	60	0.995	0.001	0.005	84-106	8.9	13
Buprofezin	20.541	172>115/(-10) 249>193/(-10)	60	0.993	0.0015	0.005	88-108	7.9	13
Oxyfluorfen	20.57	252>146/(-25) 302>274/(-15)	30	0.996	0.0015	0.005	86-103	9.8	15
Endosulfan - β	21.399	195>159/(-10) 241>206/(-15)	75	0.998	0.0015	0.005	88-98	10.5	15
P, P DDD	21.705	235>165/(-15) 235>199/(-15)	30	0.994	0.001	0.005	94-101	10.8	16
Ethion	21.776	231>129/(-25) 231>175/(-10)	75	0.993	0.0015	0.005	85-101	10.2	16
DDT - OP	21.776	235>165/(15.1) 235>199/(15.1)	45	0.992	0.001	0.005	93-107	11.2	17
Endosulfan sulfate	22.783	272>237/(-15) 387>253/(-10)	30	0.996	0.0015	0.005	84-103	9.4	13
Propiconazole	23.041	173>145/(-15) 259>69/(-10)	75	0.994	0.0015	0.005	87-102	13.8	18
DDT - PP	23.049	235>165/(-15) 235>200/(-10)	45	0.997	0.001	0.005	88-106	7.8	15
Heptachlor	23.225	171>71/(-15) 171>85/(-15)	45	0.992	0.001	0.005	89-106	9.8	14
Tebuconazole	23.532	250>125/(-10) 250>163/(-10)	75	0.993	0.0015	0.005	85-108	13.6	19
Diclofop-methyl	23.681	253>162/(-15) 340>253/(-10)	90	0.995	0.0015	0.005	84-105	9.8	14
Iprodione	24.565	314>245/(-15) 314>271/(-10)	45	0.992	0.0015	0.005	88-102	10.9	13
Bifenthrin	24.968	181>141/(-20) 181>166/(-15)	60	0.993	0.0015	0.005	92-101	11.2	15
Fenpropathrin	25.264	181>152/(-20) 265>210/(-15)	75	0.996	0.0015	0.005	86-103	10.6	14

Pesticide	RT (min)	MRM/CE (V)	Scan time (ms)	r^2	LOD (mg kg^{-1})	LOQ (mg kg^{-1})	% recovery	Inter-day precision(%)	±UM (%)
Phosalone	25.987	182>111/(-15) 182>138/(-10)	30	0.995	0.0015	0.005	86-104	9.6	14
Cyhalothrin λ	26.893	181>152/(-20) 197>141/(-15)	30	0.997	0.0015	0.005	89-98	8.5	14
Edifenphos	26.985	139>111/(-15) 251>139/(-15)	60	0.999	0.0015	0.005	88-106	8.7	13
Bitertanol	28.137	170>115/(-30) 170>141/(-20)	30	0.994	0.0015	0.005	92-107	6.7	12
Permethrin I	28.349	163>127/(-5) 183>153/(-15)	30	0.992	0.0015	0.005	94-108	9.7	15
Permethrin II	28.60	163>127/(-5.1) 183>153/(15.1)	60	0.993	0.0015	0.005	85-109	10.4	16
b-Cyfluthrin	29.642	163>127/(-5.1) 206>151/(20.1)	75	0.997	0.0015	0.005	92-103	5.6	11
Cypermethrin I	30.035	163>127/(-5) 181>152/(-20)	50	0.995	0.0015	0.005	89-101	9	14
Cypermethrin II	30.25	163>127/(-5.1) 181>152/(-20.1)	50	0.995	0.0015	0.005	88-103	10.8	13
Cypermethrin III	30.346	163>127/(-5.2) 181>152/(-20.2)	30	0.993	0.0015	0.005	89-108	11.2	15
Cypermethrin IV	30.439	163>127/(-5.3) 181>152/(-20.3)	30	0.996	0.0015	0.005	84-101	11.7	16
Fenvalerate I	31.718	125>89/(-20) 167>125/(-10)	30	0.994	0.0015	0.005	85-104	9.7	14
Fluvalinate	32.029	250>55/(-10) 250>200/(-15)	45	0.993	0.0015	0.005	86-92	8.9	13
Esfenvalerate	32.122	125>89/(-20) 167>125/(-10)	45	0.995	0.0015	0.005	91-109	11	15
Fenvalerate II	32.127	125>89/(-20.1) 167>125/(-10.1)	45	0.992	0.0015	0.005	89-106	10.7	15
Difenoconazole	32.626	323>265/(-15) 323>202/(-25)	45	0.995	0.0015	0.005	88-101	14.6	19
Deltamethrin	33.158	253>93/(-20) 253>172/(-10)	45	0.992	0.0015	0.005	90-102	12	16

Compound name, Retention time (min), MRM transition with precursor and product ions and their applied Collision Energy (volt), Dwell time (ms), Limit of detection and quantification ((LOD and LOQ in $mgkg^{-1}$), coefficient of regression (r2), Recoveries (%) (Average range in 2 commodities), Repeatability Inter-day precision expressed as % RSD and overall uncertainties expressed as % (k=2) calculated at LOQ level.

GC-MS/MS was operated in the MRM mode due to its high specificity and capability of simultaneously monitoring numerous product ions of a large number of compounds in complex sample matrices. Therefore optimization began with the GC-MS/MS data acquisition parameters; optimization of precursor and product ions and collision energies (CE). The precursor ions, product ions and collision energies optimized for all 82 pesticides are tabulated in Table 1. Suitable MRM transitions for each compound were selected carefully to ensure specificity and sensitivity of the determination. Initially, the complete precursor and product ion scan for each pesticide was investigated and then the collision energies were optimized to obtain the best response for two selected product ions. Theoretically, the best option for the choice of precursor ion for MS/MS fragmentation is the base ion in the mass spectrum as it shows the highest intensity. But in practice, this ion does not necessarily represent the best choice, as usually even less intensive ions in the high mass range are usually more adequate in terms of specificity and higher signal-to-noise ratio. After considering all these factors, appropriate MS/MS MRM transitions were optimized, without a significant trade-off between sensitivity and selectivity. However, quantification was always done using one transition while the second transition served for unequivocal identification (confirmation) of detected pesticide residues. Results were regarded as positively identified when signal-to-noise ratio (S/N) was >3 for both transitions and RSD of the respective ion ratios fell within the tolerances specified in the EU legislation. Briefly, the EU Commission's decision in 2002 implements the concept of identification points (IPs) where 1 identification point is earned from a precursor ion, and 1.5 identification points are earned from a resulting product ion. For the unequivocal confirmation of a compounds' identity, at least 3 and 4 identification points are required for legal and banned substances, respectively. In this work, two product ions resulting from fragmentation of one precursor ion or two product ions each resulting from two different precursor ions were monitored. Hence, 4 or 5 identification points were acquired for the majority of compounds.

Secondly, for all studied analytes, the ion ratio for each MRM (ratios between 20 % and 50 % of the most intense transition) matches with the certified reference material within specified tolerances of ± 25 %.The use of CBS significantly simplifies development of MRM and processing of data by automatically linking data. This eliminates the need for separate method set-up, quantification and confirmation. Based on the elution profile of pesticides, a time-scheduled acquisition method was constructed, called "time segments" by CBS software.

The present scope of the method covers 82 compounds and a total of 184 MRM transitions (including crossed-over transitions) monitored with 109 RT-segments (2-3 transitions each). Figure 4 shows RT-windows (segments) of a spiked grape samples where some overlap of MRM transitions from one time-segment to another was divided automatically.

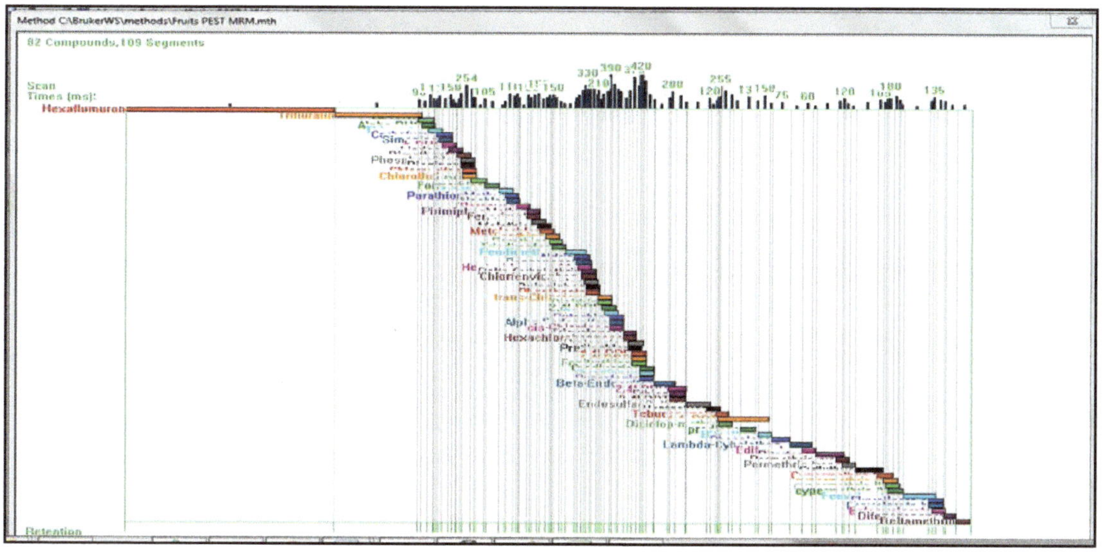

Figure 4. The RT-windows (segments) of spiked grapes (82 pesticides in 109 segments)

Thus, the transitions of critical compounds eluting close to the beginning or the end of adjacent time-segments were acquired in two segments to facilitate integration and prevent target compounds from being missed. The scan time for each compound (each with 2 MRMs) was optimized between 30 to 90 ms to satisfy sensitivity by considering sufficient data points (>15) across each peak (Figure 5).

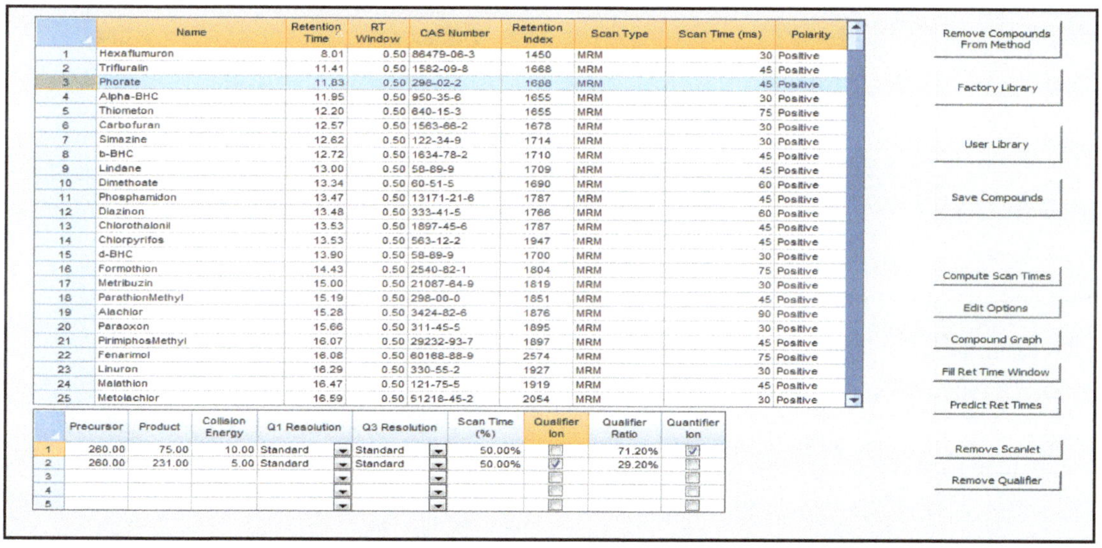

	Name	Retention Time	RT Window	CAS Number	Retention Index	Scan Type	Scan Time (ms)	Polarity
1	Hexaflumuron	8.01	0.50	86479-06-3	1450	MRM	30	Positive
2	Trifluralin	11.41	0.50	1582-09-8	1668	MRM	45	Positive
3	Phorate	11.83	0.50	298-02-2	1688	MRM	45	Positive
4	Alpha-BHC	11.95	0.50	950-35-6	1655	MRM	30	Positive
5	Thiometon	12.20	0.50	640-15-3	1655	MRM	75	Positive
6	Carbofuran	12.57	0.50	1563-66-2	1678	MRM	30	Positive
7	Simazine	12.62	0.50	122-34-9	1714	MRM	30	Positive
8	b-BHC	12.72	0.50	1634-78-2	1710	MRM	45	Positive
9	Lindane	13.00	0.50	58-89-9	1709	MRM	45	Positive
10	Dimethoate	13.34	0.50	60-51-5	1690	MRM	60	Positive
11	Phosphamidon	13.47	0.50	13171-21-6	1787	MRM	45	Positive
12	Diazinon	13.48	0.50	333-41-5	1766	MRM	60	Positive
13	Chlorothalonil	13.53	0.50	1897-45-6	1787	MRM	45	Positive
14	Chlorpyrifos	13.53	0.50	563-12-2	1947	MRM	45	Positive
15	d-BHC	13.90	0.50	58-89-9	1700	MRM	30	Positive
16	Formothion	14.43	0.50	2540-82-1	1804	MRM	75	Positive
17	Metribuzin	15.00	0.50	21087-64-9	1819	MRM	30	Positive
18	ParathionMethyl	15.19	0.50	298-00-0	1851	MRM	45	Positive
19	Alachlor	15.28	0.50	3424-82-6	1876	MRM	90	Positive
20	Paraoxon	15.66	0.50	311-45-5	1895	MRM	30	Positive
21	PirimiphosMethyl	16.07	0.50	29232-93-7	1897	MRM	45	Positive
22	Fenarimol	16.08	0.50	60168-88-9	2574	MRM	75	Positive
23	Linuron	16.29	0.50	330-55-2	1927	MRM	30	Positive
24	Malathion	16.47	0.50	121-75-5	1919	MRM	45	Positive
25	Metolachlor	16.59	0.50	51218-45-2	2054	MRM	30	Positive

	Precursor	Product	Collision Energy	Q1 Resolution	Q3 Resolution	Scan Time (%)	Qualifier Ion	Qualifier Ratio	Quantifier Ion
1	260.00	75.00	10.00	Standard	Standard	50.00%	☐	71.20%	☑
2	260.00	231.00	5.00	Standard	Standard	50.00%	☑	29.20%	☐
3							☐		☐
4							☐		☐
5							☐		☐

Figure 5. The CBS window for GC-MS/MS-MRM optimization (retention time, scan time, qualifier and quantifier ions and collision energy)

3.2. Method Performance

The combination of pesticides and foods which were tested, were determined by the potential of the analytical methods and the health risks to consumers due to pesticide contamination. Keeping this in mind, the method should be robust, give accurate results, meet detection limit needs, and cover the desired scope of matrices and analytes. Eventually, the developed method was validated in terms of linearity, recoveries and precision. All quantitative results were calculated using matrix-matched standards prepared by spiking extracts of the original grapes and pomegranate samples with different levels of pesticides as recommended by the European guidelines. Linearity was studied in matrix-matched standard solutions of grapes and pomegranate separately. All 82 pesticides belonging to various classes were separated and detected between 5.95 min and 33.2 min. Fig.6 shows the spiked grape sample with 28 pesticides at 0.005 $mgkg^{-1}$ level.

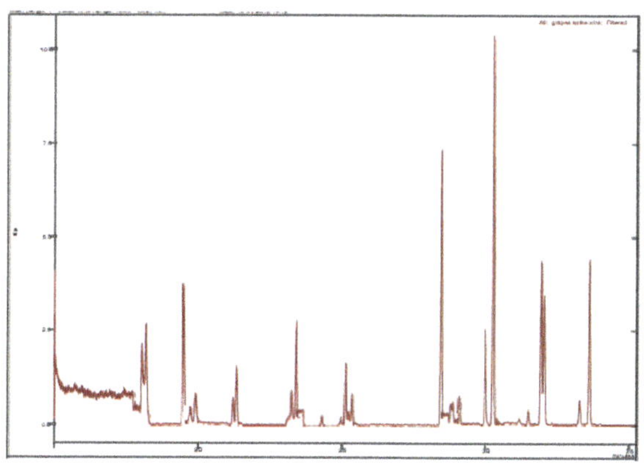

Figure 6. Spiked grapes sample with 28 pesticides at 0.005 $mgkg^{-1}$ level

Good linearity of the calibration curves was obtained for all compounds analysed by GC-MS/MS over the entire range of 0.005 - 0.1 $mgkg^{-1}$ evaluated by duplicate analyses at five different concentration levels. The target compounds showed coefficients of regression (r^2) more than 0.99 for all compounds (Figure 7).

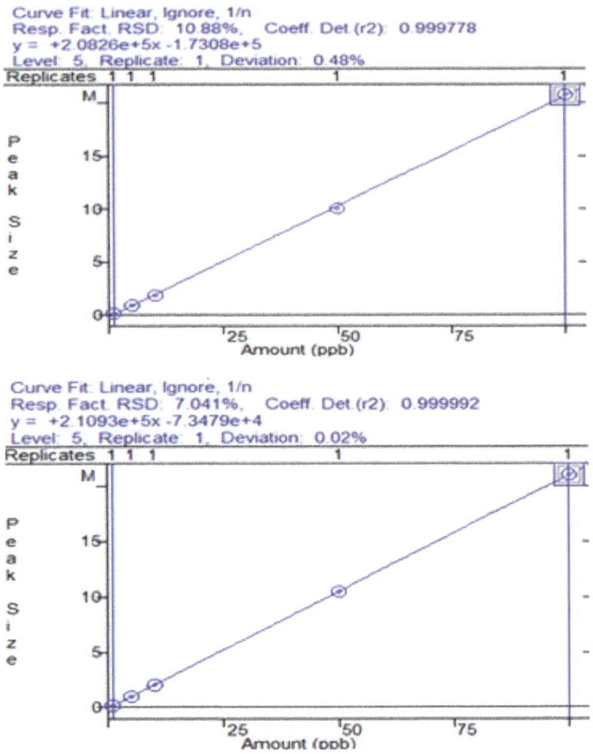

Figure 7. Calibration curve (0.005 to 0.1 mg kg^{-1}) of edifenphos (grapes) and alachlor (pomegranates) by GC-MS/MS

The analytical results obtained for the studied compounds are shown in Table 1.

Recovery experiments were carried out at three different spiking levels of 0.01, 0.05, and 0.1 mg kg^{-1}. The lowest spiking concentration of 0.01 mg $kg^{-1}$was selected to test the method performance at the restrictive residue level set by EU legislation for fruits. UM was calculated as per the ISO guide [15]. In the present work, the overall uncertainties were calculated at 0.01 mg kg^{-1}level (Table 1). These were calculated from the n = 20 results (each matrix) from the experiment performed under repeatable and reproducible conditions. The obtained results demonstrated that the method achieved satisfactory quantitative recoveries in the range of 84 - 109 % with RSD <12 % for repeatability and intermediate precision and UM of ± 12 - 19 %. The present study also included the analytes; penconazole, tricyclazole, difenoconazole, hexaconazole, tebuconazole, carbofuran and oxyfluorfen, which are nowadays analyzed by LC-MS/MS [17,18] to evaluate the usefulness of GC-MS/MS for their determination.

Analytical quality control was applied in order to assure the quality of the obtained results and identify limitations of the method. All sets of samples were accompanied by analysis of matrix blanks (if available for the particular matrix), spiked samples and matrix-matched calibration standards bracketing the samples. Method performance data was usually acquired by analysing spiking recovery at 0.01 $mgkg^{-1}$. For this purpose, randomly selected real samples (both grapes and pomegranate) were used; as a result, somewhat more recoveries within the range of 70 % and 120 % were obtained, because of some pesticide residues already present in the sample. In such cases the subsequent calibrations were all carried out with pesticides spiked in the fruit matrices. As a result, the RSD for average response of the pesticides over the entire calibration range were obtained within the limit (RSD ≤ 12 %), thus proving the excellent precision of the method for analysing these pesticides in a complex sample matrix (Figure 8).

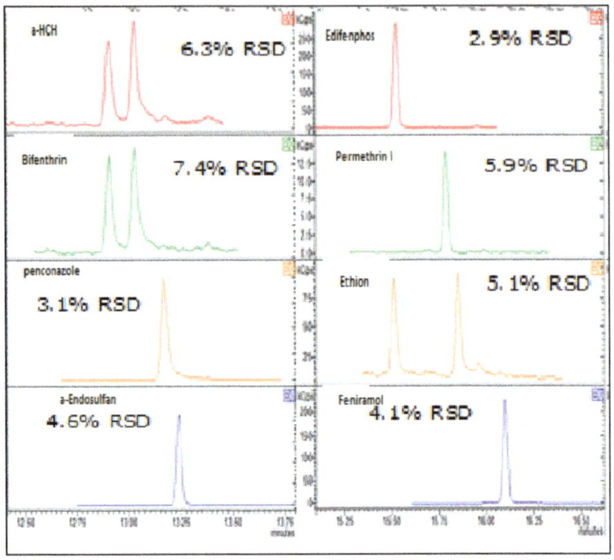

Figure 8. Precision of different group of pesticides in grapes (n = 5); α-HCH, edifenphos, bifenthrin, permethrine I, penconazole, ethion, α-endosulfan and feniramol

4. Conclusion

The proposed method using single step SLE with ethyl acetate followed by PSA and GCB clean-up has greatly simplified the method for multiresidue pesticide analysis and also improved the productivity for all non-polar GC grade pesticides. Contrary to single-stage MS technique (such as full scan and SIM), the MS/MS produces more unique spectra resulting from two consecutive steps of mass fragmentation and it helps to solve analytical problems where the single-stage MS techniques fail. This work has demonstrated that GC-MS/MS is a key tool for comprehensive multi-residue screening of pesticides belonging to various chemical classes in complex matrices such as grapes and pomegranate. The method has shown good sensitivity, linear calibration range, robustness in the matrix, and ability to meet the requirement of EU Commission decision and quantitate the pesticides at the level of 0.01 $mgkg^{-1}$.

References

[1] APEDA, Regulation of export of fresh grapes from India through monitoring of pesticide residues, Amendments in grape RMP - 2007, Amendment-5 (Revised Annexure – 7&11), 2006a. Available: http://www.apeda.com. [Accessed Oct. 30, 2011].

[2] APEDA,Regulation of export of fresh grapes from India through monitoring of pesticide residues, Annexure 1: Circular No.91-4/95 PQD dated 29th February 2000, 2006b. Available:http://www.apeda.com/apeda/GRAPES 05/ RMPGrapes 2007 17 oct06 3.pdf. [Accessed Oct. 30, 2011].

[3] Schachterle, S. andFeigel, C, "Pesticide residue analysis in fresh produce by gas chromatography-tandem mass spectrometry," *Journal of Chromatography A*, 754. 411-422.1996.

[4] Torres, C.M., Pico, Y. and Manes, J, "Determination of pesticide residues in fruit and vegetables – a review," *Journal of Chromatography A*, 754. 301-331. 1996.

[5] Herdman, R.C., Pahrham, W.E., and Shen, S,*Pesticide Residues in Food: Technologies for Detection,* Government Printing Office, Washington, 1988.

[6] Cajka, T.and Hajslová, J, "Gas chromatography–time-of-flight mass spectrometry in food analysis," *LC–GC Europe*, 20(1). 25-26. 2007.

[7] Lambropoulou, D.A. and Albanis, T.A, "Methods of sample preparation for determination of pesticide residues in food matrices by chromatography-mass spectrometry-based techniques: a review," *Analytical and Bioanalytical Chemistry*, 389 (6). 1663-83.2007.

[8] van der Lee, M.K., van derWeg, G., Traag, W.A. and Mol, H.G.J, "Qualitative screening and quantitative determination of pesticides and contaminants in animal feed using comprehensive two-dimensional gas chromatography with time-of-flight mass spectrometry," *Journal of Chromatography A*, 1186. 325.2008.

[9] M. Anastassiades, *Recovery Studies using the QuEChERS Method,* 2008. Available: http://www.quechers.com/. [Accessed Apr. 3, 2011].

[10] AOAC Official Method 2007.01. Pesticide Residues in Foods by Acetonitrile Extraction and Partitioning with Magnesium Sulfate.2007.

[11] Satpathy, G., Tyagi, Y.K. and Gupta, R.K, "A novel optimized and validated method for analysis of multi-residues of pesticides in fruits and vegetables by microwave-assisted extraction (MAE)–dispersive solid-phase extraction (d-SPE)–retention time locked (RTL)–gas chromatography–mass spectrometry with Deconvolution reporting software (DRS)", *Food Chemistry*, 127. 1300-1308. 2011.

[12] Walorczyk, S, "Development of a multi-residue screening method for the determination of pesticides in cereals and dry animal feed using gas chromatography-triple quadrupole tandem mass spectrometry," *Journal of Chromatography A*, 1165(1-2). 200-12. 2007.

[13] Mol, H.G.J., van Dam, R.C.J. and Steijger, O.M, "Determination of polar organophosphorus pesticides in vegetables and fruits using liquid chromatography with tandem mass spectrometry: selection of extraction solvent," *Journal of Chromatography A*, 1015. 119-127. 2003.

[14] Banerjee, K., Upadhyay, A.K., Adsule, P.G., Patil, S.H., Oulkar, D.P. and Jadhav, D.R, "Rate of degradation of lambda-cyhalothrin and methomyl in grapes (*Vitisvinifera* L.)," *Food Additives and Contaminates*, 23. 994-999.2006.

[15] GUM, ISO guide to the expression of uncertainty in measurement. BIPM, IEC, IFCC, ISO, IUPAC, IUPAP, OIML, Geneva. 1993.

[16] EURACHEM, Quantifying uncertainty in analytical measurements, EURACHEMSecretariat, P.O. Box 46, Teddington, Middlesex, TW110LY, UK, 1995.

[17] Hiemstra, M. and de Kok, A, "Comprehensive multi-residue method for the target analysis of pesticides in crops using liquid chromatography-tandem mass spectrometry," *Journal of Chromatography A*, 1154 (1-2). 3-25. 2007.

[18] Pizzutti, I.R., de Kok, A., Zanella, R., Adaime, M.B., Hiemstra, M., Wickert, C. and Prestes, O.D, "Method validation for the analysis of 169 pesticides in soya grain, without clean up, by liquid chromatography-tandem mass spectrometry using positive and negative electrospray ionization," *Journal of Chromatography A*, 1142(2). 123-36. 2007.

17

Utilization of Pulse Processing Waste (*Cajanus cajan* Husk) for Developing Metal Adsorbent: A Value-added Exploitation of Food Industry Waste

Vishal. R. Parate*, Mohammed. I. Talib

Department of Food Technology, UICT, North Maharashtra University, Jalgaon, 425001, India
*Corresponding author: vishal_parate@yahoo.com

Abstract India is the largest producer of pulses (edible legume) in the world. Pigeon pea or Tur (*Cajanus cajan*) is the second most major pulse produced in India and is mainly processed to convert into product known as "Tur dal". The manufacturing industries of Tur dal generate considerable amount of byproduct/waste in the form of husk. The present investigation explores the possibility of exploiting activated carbon prepared from the Tur dal husk for removing heavy metal. The objective of the work was to enhance the performance of prepared activated carbon by carrying out batch experiments study to optimize the condition of adsorption for complete removal of Cu ions from its 50 ml, 50 ppm solution. The char was produced by carbonizing the washed and dried Tur dal husk in air tight container at 500°C for 1 hr. The obtained char was then activated by soaking in concentrated sulfuric acid (1:1 ratio) for 24 hr. at room temperature to get activated carbon. The physicochemical characterization of achieved activated carbon was done for various parameters along with analysis of trace elements (ICP), crystal nature (XRD) and structural morphology (SEM). The optimum condition for adsorption was studied by altering pH (2-10), agitation speed (50-250 rpm), temperature (10-60°C), adsorbent dose (0.1- 2.25 g) and contact time (0.5-4 hr.). The work concludes activated carbon prepared from Tur dal husk possess considerable metal adsorption property and showing maximum activity in an optimum condition of 6 pH, 200 rpm agitation speed, 10°C temperature, 2.25 g adsorbent dose and 3 hr. contact time. The examination of thermodynamic data confirmed the adsorption by the developed activated carbon was exothermic (positive ΔH^0), changing from spontaneous to non-spontaneous with increase temperature (both negative and positive ΔG^0), making system ordered and feasible (positive ΔS^0) through adsorption. The work suggests way of giving value addition to pulse processing industry waste for their better utilization in eco-friendly and economical way.

Keywords: *adsorption, copper, food waste, activated carbon, tur dal husk*

1. Introduction

Food derived from plant source contain inedible portion in the form of stone, peels, seed coat and husk. One of the purposes of the processing of raw agricultural commodity is to make them consumable by remove this uneatable portion. The separated scraps are sometimes subject to cattle feed as by product of the industry and mostly thrown as waste. The agro food industry creates huge amount waste and is matter of concern due their potential of creating environmental pollution. Under the provision of various environmental acts the food processing industrial waste needs to be disposed safely and require significant investment on such waste disposal. The cost on such treatments and disposal however can be reduced if the produced leftover is utilized for some useful purpose instead of throwing [1].

The agro food industry has the tremendous potential to remove heavy metals from effluent by adsorption. The heavy metal adsorption properties of agro waste is natural and is due to the presence of various functional groups carried by cellulose, hemicelluloses, lignin, starches, sugar, protein, lipid present in agro based garbage. These functional groups have the capability of complexing heavy metals [2] and remove them by process of adsorption. The metal adsorption capacity of agro waste can be improved by their heating in controlled condition to produce its carbon (extremely porous). The adsorption performance of this agro based carbon can be further enhanced by their activation (alteration in surface functional groups and their charges) with various agents such as oxidizing agent (hydrogen peroxide), organic compounds (ethylenediamine, formaldehyde), organic acid (tartaric acid, citric acid, sulfuric acid, hydrochloric acid, nitric acid), base solutions (calcium or sodium hydroxide, sodium carbonate) [3].

The natural aquatic systems is getting polluted due to the introduction various toxic heavy metal ions such as Hg, As, Pb, Cr, Fe, Se, Co, Ni, Cd, Zn, Cu at a much higher concentration than the prescribed limits, coming from the waste water of mining, refining ores, fertilizer tanneries, batteries industries [4,5]. According to Indian Environment (Protection) Rule, 1986 [6], the permissible maximum limits of copper in industrial effluent for discharge in inland surface water, public sewers and in marine costal area is maximum 3.0 ppm. The available techniques to remove heavy metals like precipitation, ion exchange, reverse osmosis and electrolytic methods have limitations (high cost, production of sludge in huge quantity) and not much popular [7]. The removal of heavy metals by adsorption using activated carbon on the other handis more reliable method. The large surface area, chemical nature, micro-porous structure makes activated carbon ideal for the said purpose [8]. The activated carbon are always in high demand but prove expensive some times, if prepared from costly raw materials. The cost of activated carbon produced from agro food industrial waste is less as large quantities are readily and cheaply available [9]. The various agro based materials investigated for their metal adsorption property and reported to be successful. Husk or bran of rice, wheat flour, maize, corn, black gram husk, pearl millet, Shells and/or stones of coconut, almonds, apricots, peanuts, olive wastes, cherries, bael fruit, Brazil nut, hazelnut, wood apple, chestnut, Peels of different fruits such as, orange, banana, watermelon, garlic, cassava, mango, jack fruit, Wastes resulting from the production and processing of tea, coffee, as well as sugar cane bagasse, coir pith, grape bagasse, Agro waste from papaya seeds, sunflower, palm seed coat, pineapple stem, maize cob, pine bark, tamarind seed, barley straw already have been proved effective for metal adsorption purpose [10].

India is the largest consumer and producer of pulses (edible legume) in the world, which is around 15% of the total grain in country. Dal is a major form in which pulses are consumed in India. Dal is also popular in other Asian countries like Nepal, Pakistan, Sri Lanka and Bangladesh. Dal is a prepared by separating pulse (dried seeds) from their outer hulls and split. There are large number of dal mills scattered in the India which process more than 75% of the pulses produced. Chickpea (*Cicer arietinum*) also known as Bengal gram (chana) is the major pulse produced in India followed by Pigeon pea (*Cajanus cajan*) also known as Tur (arhar) or tropical green pea or yellow Pigeon pea or Red gram. Pigeon pea is mainly consumed in the form of split pulse as "Tur dal" in India. The processing of Pigeon pea to produce Tur dal involve two steps 1) Loosening of husk or seed coat and 2) Dehusking and splitting into cotyledon. The loosening of husk is facilitated by prolong sun drying as it is effective and economical. The loose husk then separated from the grain by imparting friction with grooved rollers in machine. Tur dal industry generate considerable amount of husks which is around 8% of this pulse [11]. Tur dal husk which is the low cost by product of this industry is generally considered for cattle feed only. During processing many times due to use of lime water, alkali and other chemicals the husk become unfit even for cattle feed. The Tur dal husk which is basically cellulosic material can be very well exploited for the preparation of activated carbon in economical way. Earlier investigators mainly studied *Cicerarientinum* husk for metal adsorption but in their raw form [12-17]. Their study confirms husk of *Cicerarientinum* to be an excellent material for biosorption of metals if present in low concentration in water. Husk of pulse *Lathyrussativus* (Indian pea) was studied by Panda, et al., 2006 [18]. Mungbean (*Phaseolus aureus/ Vignaradiata*) husk an agro waste material was evaluated by Saeed, A., et al., 2009 [19] as a new sorbent for the removal of cadmium (II) from aqueous solution. The hulls of *Vignaradiata* and *Cicer arietinum* have been exploited for the preparation activated carbon [20,21]. The referred literature review reveled that only Ahalya et al., (2007) carried out work on the raw husk of Tur dal as a new biosorbent for the removal of Ferric (III) and Chromium (VI) ions from aqueous solutions. Their study concludes Turdal husks in their raw form has potential to treat wastewaters containing low concentration of metal ions. The preparation of activated carbon from Tur dal husk is still not attempted.

The present investigation was carried out with the aim to prepare activated carbon from pulse processing industrial waste (Tur dal husk) and to study its adsorption potential for the effective removal of metal (Cu ion) from its aqueous solution.

2. Materials and Methods

Stock solution of copper (II) ion (1000 ppm) was made by dissolving 3.9295 g of copper sulfate ($CuSO_4.5H_2O$) in one liter of double distilled water. This prepared stock solution was used to prepare desired concentration of fresh Cu solution by appropriate dilution. The chemicals required for analysis were procured from Jinendra Scientific, Jalgaon (Maharashtra state, India) and of analytical grade from standard brands (MERCK, SDFCL). Niki Agro Products Pvt. Ltd., a local pulse processing industry located at M.I.D.C, Jalgaon, was contacted to obtain Tur dal husk.

2.1. Preparation of Activated Carbon

The activated carbon from Tur dal husk was prepared by slight modification and combining the methods described by El-Ashtoukhya et al., 2008 [22], Sivaraj et al., 2010 [23] and El-Sayed et al., 2011 [24]. The husk was first carbonized and the resulting char was then activated by treating with sulfuric acid as given in Figure 1.

2.2. Characterization of Activated Carbon

The prepared activated carbon was analysed for various physicochemical parameters, crystallography and morphological structure by standard methods using sophisticated instruments given in Table 1.

2.3. Initial Adsorption Study

Initially the batch adsorption experiments were carried out to find out roughly the adsorption potential of prepared activated carbon Ad_3 in adsorbing Cu ion from its solution. The whole sets of experiments were carried out changing Cu solution concentration and adsorbent carbon dose at natural pH. The other parameters for adsorption like volume of Cu ion solution (50 ml), particle size of adsorbent (0.15-0.25 mm), agitation speed (150 rpm) and temperature (30°C) were kept constant during preliminary study. First experiment carried out using Cu solution of 100 ppm (natural pH 5.63) and 0.1 g of adsorbent dose. Second experiment was performed keeping adsorbent dose same (0.1 g) but lowering the concentration of Cu solution to 50 ppm (natural pH 5.78). In third experiment concentration of Cu solution was kept same i.e 50 ppm but adsorbent dose increased to 0.25 g. Forth experiment was conducted in the same manner as third only changing adsorbent dose to 0.5 g.

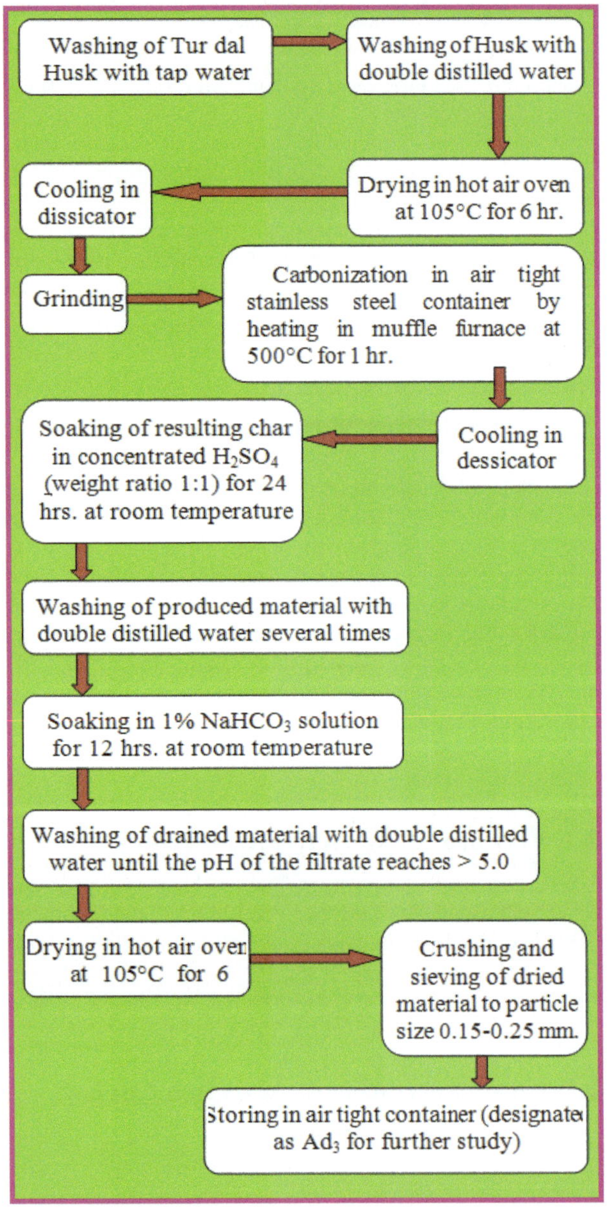

Figure 1. Method of preparation of adsorbent ad$_3$ from Tur dal husk

All the experiments were carried out in 100 ml capped conical flasks adding required amount of Cu solution and adsorbent dose. The flasks were agitated in orbital shaking incubator (REMI) for 0.5, 1, 1.5, 2, 2.5, 3, 3.5, 4, 4.5, 5, 5.5, 6, 6.5, 7, 7.5 and 8 hr. The flasks were agitated in orbital shaking incubator (REMI) for 0.5, 1, 1.5, 2, 2.5, 3, 3.5, 4, 4.5, 5, 5.5, 6, 6.5, 7, 7.5 and 8 hr. After every treatment period the suspension was filtered through Whatman No. 42 filter paper. The filtrate was then diluted to desired dilution with double distilled water to analyse residual Cu concentration using Atomic Absorption Spectrophotometer (ELICO Ltd., Model: SL 176) [28]. The% Adsorption/ Removal of Cu was calculated as per equation 1:

$$\%.\,Adsorption = \left\{ \frac{C_i - C_e}{C_i} \right\} x\ 100 \qquad (1)$$

Table 1. Analysis Methods and Instruments for Physicochemical Characterization of Ad$_3$

Parameter	Equipment/ instrument	Analysis Method
Moisture	Hot air oven	Bureau of Indian Standards method (IS 877: 1989) [25]
Water soluble content	Weighing balance (A/VD, Model HR-200)	
pH	pH Meter (Deluxe 101 (EI))	
Particle Size	IS Sieve & Sieve Shaker	
Conductivity	Conductivity Meter (Systronic, Model 304)	Bureau of Indian Standards method (IS 14767: 2000) [26]
Ash	Muffle furnace (Tempo Instruments and Equipment Pvt. Ltd.)	European Chemical Industry Council (CEFIC) methods, 1986 [27]
Bulk density	Bulk Density apparatus (DBK 5028-7)	
Yield	Weighing balance (A/VD, Model HR-200)	El-Ashtoukhya et al., 2008
Calorific value	Digital Bomb Calorimeter (Rajdhani Scientific, Model: RSB 6)	
Traces of elements	Coupled PlasmaOptical Emission Spectrometer (ICP-OES) (Model: VARIAN 720-ES)	
Adsorbent nature and shape of crystals	X-Ray Diffraction (XRD) (BRUKER, Germany, Model: D8 ADVANCE)	
Adsorbent morphology	Scanning Electron Microscope (SEM) (HITACHI, Model: S- 4800TypeII), Coupled with Energy Dispersive X-ray Spectroscopy (EDS) of BRUKER	

Where, C_i and Ce is initial and equilibrium concentration of Cu ions solution respectively in mg/L [29,30].

2.4. Study of Effect of Adsorption Conditions

The optimization study of adsorption parameters was planned on the basis of findings of preliminary batch adsorption trial. The effect of pH, agitation speed, temperature and adsorbent dose on adsorption capacity of Ad$_3$ was observed by keeping the other parameters of adsorption such as initial concentration of Cu solution (50 ppm), volume of Cu solution (50 ml), particle size of adsorbent (0.15-0.25 mm) and contact time (4 hr.) constant. The effect of pH was studied by adjusting the pH of solutions at 2, 3, 4, 5, 6, 7, 8, 9, 10 using 0.1-0.5 M solution of NaOH or HCl. The pH adjusted solutions were stirred at 150 rpm at 30°C with 0.5 g Ad$_3$ dose. The obtained optimum pH was then maintained in all the further study of adsorption. The effect of agitation on adsorption was examined by shaking the suspension of copper solution containing 0.5 g Ad$_3$ dose at various speed (50, 100, 150, 200 and 250 rpm) in an incubation temperature of 30°C. The temperature effect on Cu removal was observed by maintaining temperature at 10, 20, 30, 40, 50 and 60°C keeping pH and agitation speed at optimized level and Ad$_3$ dose 0.5 g. The effect of adsorbent dose was studied by shaking Cu solution with various dose of Ad$_3$ (0.1, 0.25, 0.5, 0.75, 1, 1.5, 2.0 and

2.25 g) keeping pH, agitation speed and temperature at optimum condition. The adsorption capacity q_e (mg/g) was determined by the equation 2:

$$q_e = V\frac{(C_i - C_e)}{W} \quad (2)$$

Where, C_i and Ce is initial and equilibrium concentration of Cu ions solution respectively in mg/L, V is the volume of the Cu solution in liter and W is the mass of the Ad_3 in g [31].

The study ended with the optimization of contact time, carrying out the adsorption at achieved optimum conditions.

2.5. Thermodynamic Study

The feasibility of the adsorption process was predicted by estimating thermodynamic parameters such as standard Gibb's free energy (ΔG^0), enthalpy change (ΔH^0) and entropy change (ΔS^0) from the adsorption data obtained during temperature optimization.

The ΔG^0 in J/mol was estimated using thermodynamic equilibrium constant (K_C) by equation 3:

$$\Delta G^0 = -RT\ lnK_C \quad (3)$$

Where, R is ideal gas constant in 8.314 J/mol K and T is temperature in K

Equation 4 is showing how the K_C is related to C_∂ and C_e

$$K_C = C_\partial / C_e \quad (4)$$

Where, C_∂ is mg of Cu adsorbed per liter (mg/L) and C_e is the equilibrium concentration of Cu solution in mg/L.

C_∂ was computed using equation 5:

$$C_\partial = C_i - C_e \quad (5)$$

Where, C_i is the initial concentration of Cu solution and C_e is is the equilibrium concentration of Cu solution both in mg/L.

Equation 6 represents Van't Hoff equation giving relation between ΔG^0 and ΔH^0 and ΔS^0.

$$\ln K_C = -\frac{\Delta G^0}{RT} = -\frac{\Delta H^0}{RT} + \frac{\Delta S^0}{R} \quad (6)$$

The values of ΔH^0 and ΔS^0 were calculated from the slope and intercept of the plot of $\ln K_C$ versus $1/T$ [32].

3. Results and Discussions

3.1. Characterization of Ad_3

The physicochemical characterization of developed adsorbent Ad_3 is given in Table 2. The moisture in Ad_3 was found to be 6.17%, ash 6.95% and water soluble content 0.91%. The calorific value of prepared adsorbent carbon was 7172.67 Cal/g. The observed pH of Ad_3 solution was 5.44 and may be due to treatment of prepared carbon with H_2SO_4. The recovery of Ad_3 from dried husk (yield) was 31.31%. The bulk density of Ad_3 was 0.5479% whereas conductivity 297.25 mS/m. The particles found in Ad_3 mixture was of size 210-250 µm (60.66%), 180-210 µm (9.12%) and 150-180 µm (30.22%).

Table 2. Physicochemical Characteristics of Adsorbent Ad_3

Parameter	Result
Moisture (%)	6.17 ± 0.06
Ash (%)	6.95±0.08
Water soluble content (%)	0.91 ±0.02
Calorific value (Cal/g)	7172.67 ±1.20
pH	5.44 ± 0.01
Yield (%)	31.31 ± 0.31
Bulk density (g/cm^3)	0.5479 ±3.32
Conductivity (mS/m)	297.25
Particle size	
210-250 µm	60.66%
180-210 µm	9.12%
150-180 µm	30.22%

Values are Mean (± SEM) of 3 determinations

Table 3 is showing the trace amounts of various elements present in Ad_3. The elemental analysis confirms the presence of Mg, Na, Ca, Fe, Zn, Cu, Mn and Pb in prepared activated carbon responsible for ash.

Table 3. Trace Element Analysis of Adsorbent Ad_3

Element	Result (ppm)	Element	Result (ppm)
Mg	3092.66	Na	3040.84
Ca	1441.89	Fe	145.70
Zn	46.11	Cu	18.26
Mn	13.17	Pb	5.16

Figure 2 is showing the standard XRD pattern of carbon whereas XRD of Ad_3 is given in Figure 3.

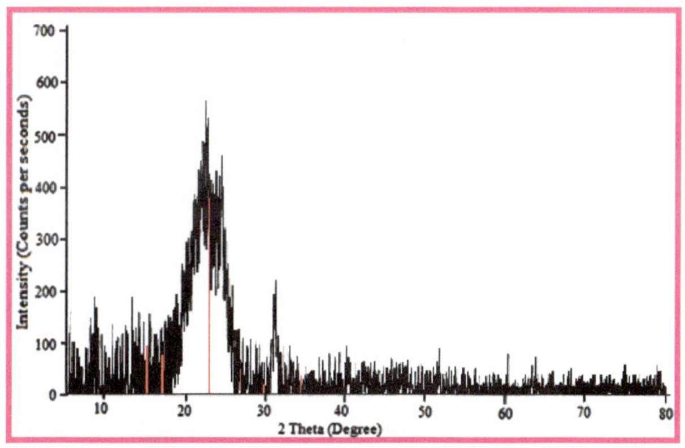

Figure 2. Standard X-ray diffraction of carbon

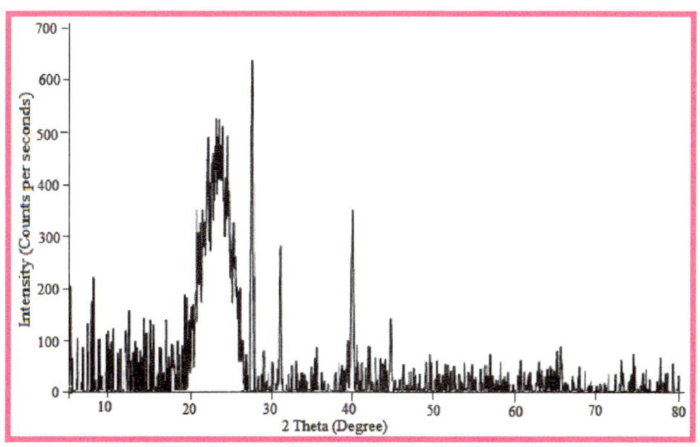

Figure 3. X-ray diffraction of Ad_3

The standard XRD peak for carbon was found to matching with XRD peak obtained for Ad_3 and verifies the produced material was carbon. The XRD result also confirmed the obtained activated carbon was a mixture of 60.6% amorphous and 39.4% hexagonal shaped crystalline crystals.

Figure 4 and Figure 5 is representing the porous structure in raw Tur dal husk and Ad_3 respectively when observed through SEM. It can be seen that raw husk surface had no or very few pores and thus acquiring less surface area whereas surface of Ad_3 was characterized by many pores and exhibiting large surface area.

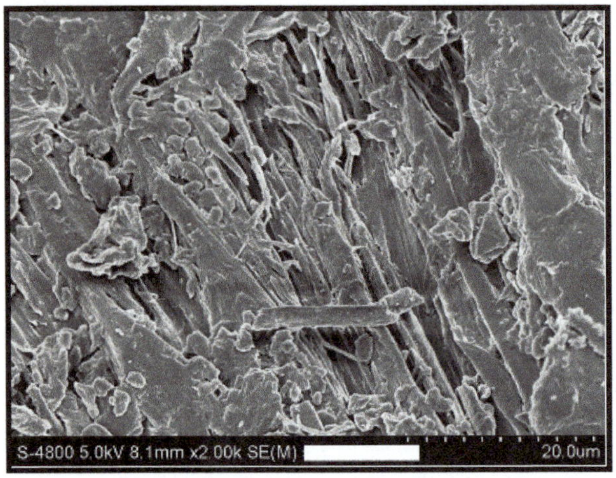

Figure 4. SEM image of raw tur dal husk

Figure 5. SEM image of Ad_3

Further treatment of this porous char with H_2SO_4 developed high porosity [33]. The developed activated carbon Ad_3 with its porous structure and large surface area was predicted to be effective metal adsorbent.

3.2. Initial Adsorption Trial

Table 4 illustrates the outcome of initial adsorption trial performed by varying the initial concentration of Cu solution and the dose of Ad_3. The adsorption of Cu ion from its 100 ppm solution by 0.1 g Ad_3 dose was low (10.5%) in first adsorption experiment. The second experiment conducted with same dose (0.1 g) but lowering initial Cu solution concentration to 50 ppm, shown improvement in Cu adsorption (18.8%). The third experiment carried out with 0.25 g dose and 50 ppm initial Cu solution concentration was also found to increase the adsorption performance (32.2%). Further to increase the adsorption, the fourth experiment carried out with 0.5 g dose and 50 ppm initial Cu solution concentration however shown no extreme progress in adsorption (39.1%). Based on the findings of preliminary adsorption study the optimization of adsorption condition was planned as investigation carried out by the previous researchers reported drastic change in the performance of adsorbent by the optimization of adsorption condition [34,35,36].

Table 4. Preliminary Adsorption Trials

Parameter	First Trial	Second Trial	Third Trial	Fourth Trial
Residual Cu ion concentration at equilibrium	88.4	40.1	33.5	30.1
% Adsorption of Cu ion	10.5	18.8	32.2	39.1
Equilibrium Time	3.0 hr	2.5 hr	3	3.5

3.3. Study of Effect of Adsorption Condition

3.3.1. Effect of pH on Adsorption

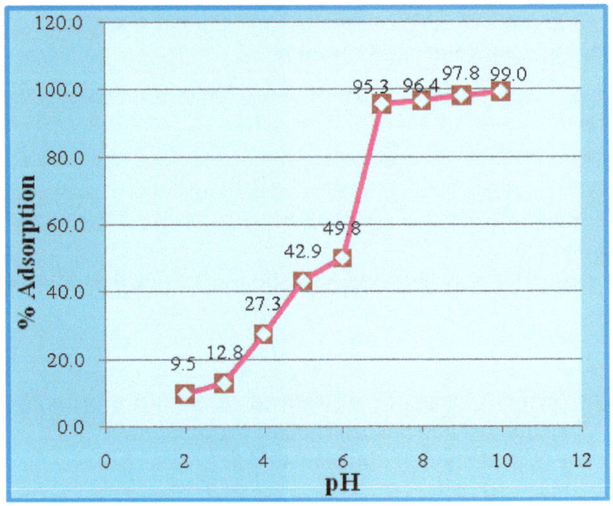

Figure 6. Effect of pH on Cu adsorption by Ad_3

The initial pH of metal solution is considered most important as it alter the charges on the adsorbent surface and affect adsorption. The effect of pH on adsorption is shown in Figure 6. The adsorption of Cu ion was found to increase linearly with increase in pH. The adsorption was observed to be low at pH 2 (9.5%) and 3 (12.8%). The moderate adsorption was found at pH 4, 5 and 6 (27.3%, 42.9% & 49.8% respectively). Very high adsorption was achieved at pH 7, 8, 9 and 10. The adsorption of metal ions by activated carbon is mainly due to electrostatic attraction between positively charged metal ions and negative charges on the surface of activated carbon. At low pH (2 & 3) the metal ion solution carried H^+ ions in high amount. As the Ad_3 added in the Cu solution the surface of Ad_3 completely got covered with H^+ ions. The shield of H^+ ions on the surface of Ad_3 was preventing the Cu ion (positively charged) from reaching to the adsorption sites of Ad_3 by repulsion, resulting less adsorption [37]. However the number of H^+ ions decreased as the pH increased (4, 5 & 6), due to which Cu ion became more accessible to negative charged sites on the Ad_3 and significant adsorption occurred. The obtained result was in accordance with the data found by Boonamnuayvitaya et al., 2004 [38] while removing heavy metals (Cd^{2+}, Cu^{2+}, Pb^{2+}, Zn^{2+} and Ni^{2+}) by coffee residues. The very high removal of Cu at pH 7 and above was not due to adsorption only but because of precipitation of Cu ion as copper hydroxide [39,40]. Practical point of view for the application of adsorption, 6 pH was selected as optimum without precipitation of Cu.

3.3.2. Effect of Agitation on Adsorption

Figure 7 is viewing the effect of agitation speed on Cu adsorption by Ad_3.

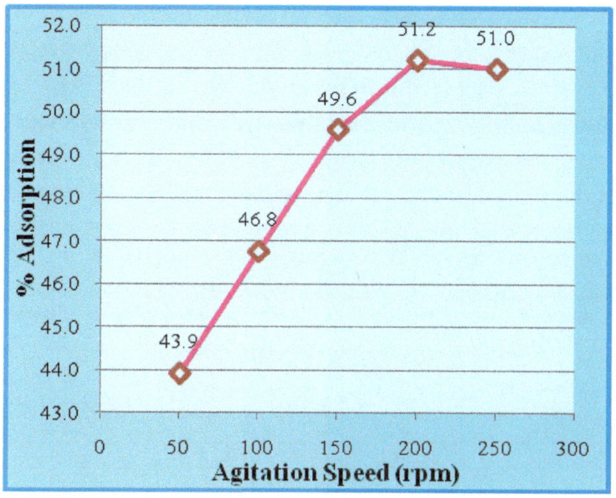

Figure 7. Effect of agitation speed on Cu adsorption by Ad_3

The adsorption found to increase with increase in agitation speed and after attending optimum speed start declining. The adsorption was 43.9%, 46.8%, 49.6%, 51.2% and 51.0% for agitation speed 50, 100, 150, 200 and 250 rpm respectively. The maximum adsorption was obtained at 200 rpm agitation speed and optimized for further study. The adsorption occur when adsorbate particle reach on the surface of adsorbent by crossing resistance offered by bulk of solution and boundary layer surrounding adsorbent surface. The adsorption further continues as Adsorbate molecules diffuse in the bulk of adsorbent [41]. The thickness of fluid film surrounding Ad_3 particle was greater when the agitation set to low rpm. The Cu ions were unable to cross this thick layer easily and diffusion of Cu ions within Ad_3 pores were also less, due to which low adsorption occurred [42]. As the agitation speed increased the thickness of fluid film surrounding Ad_3 reduced due to increase in turbulence and adsorbate particles (Cu ions) reaching easily to the Ad_3 surface as experiencing less resistance crossing the boundary layer. Also the diffusion of Cu ions wasat its extreme causing high adsorption [43]. The result agree with the data obtained by Zouboulis et al., 2007 [44] while removing toxic metals from an aqueous mixture containing zinc, copper and nickel exploiting microbes as biosorbent. The adsorption found to be almost same at 250 rpm as obtained at 200 rpm. This shows 200 rpm agitation speed was sufficient to make available all the binding sites of the Ad_3 for adsorbing Cu^{2+} ion in solution. Increasing agitation beyond 200 rpm was unable to establish the proper contact between Ad_3 and Cu^{2+} ion, also desorption at some sites occurred due to possible break-up of the Cu ions and Ad_3 bonding, resulting no further improvement in adsorption. The similar result i.e. increase in adsorption with increase in shaking speed up to certain rpm and then decreasing beyond optimum rpm, was also reported by other researchers [45,46].

3.3.3. Effect of Temperature on Adsorption

Figure 8 is showing the effect of temperature on adsorption of Cu by Ad_3. The adsorption was 54.9%, 53.2%, 51.0%, 49.6%, 46.6% and 44.7% at 10, 20, 30, 40, 50 and 60°C respectively. The Cu adsorption was fund to be decreasing with increasing temperature. This shows the process of adsorption of copper by Ad_3 was exothermic in nature. The possible cause for the observed effect may be due to the breaking of some internal bonds with increasing temperature, resulting lessening of few adsorption sites on Ad_3 surface and within pores [47]. Similar phenomenon was observed by other investigators while carrying out adsorption of metals or dyes by various biosorbents [48], [49]. Among all the studied temperature, low temperature (10°C) was optimized as most suitable temperature for the said adsorption.

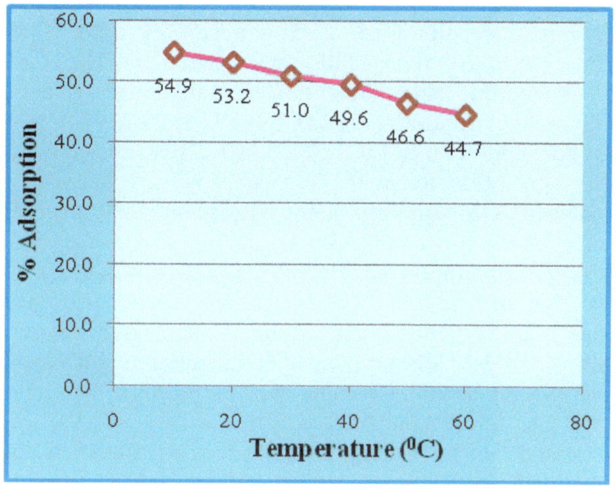

Figure 8. Effect of temperature on Cu adsorption by Ad_3

3.3.4. Effect of Adsorbent Dose on Adsorption

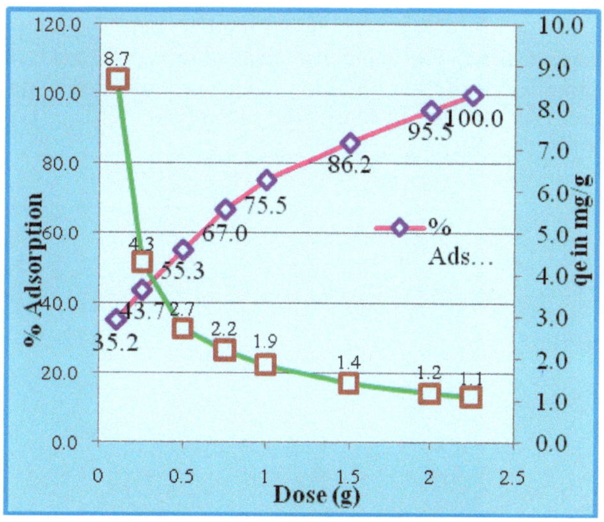

Figure 9. Effect of dose of Ad_3 on Cu adsorption

The effect of amount of Ad_3 for adsorbing Cu is depicted in Figure 9. The adsorption of Cu for Ad_3 quantity 0.1, 0.25, 0.5, 0.75, 1.0, 1.5, 2.0 and 2.25 g was 35.2, 43.7, 55.3, 67.0, 75.5, 86.2, 95.5 and 100%

respectively and adsorption capacity 8.7, 4.3, 2.7, 2.2, 1.9, 1.4, 1.2 and 1.1 mg/g respectively. The adsorption capacity q_e found to be decreasing and% adsorption increasing with increasing application of Ad_3 amount [50]. The corresponding increase in number of adsorption sites with the elevated doses of Ad_3 was the cause for improved adsorption. However as the amount of Ad_3 increased the formation of agglomerates due to interaction of various Ad_3 particles occurred. The formed agglomerates was overlapping Ad_3 surface, making it unavailable for adsorption. Some adsorption sites also got obstructed with this over-crowding of Ad_3 particles resulting low adsorption capacity with increasing dose of Ad_3 [51,52], Similar result was also reported by Bansala, et al., 2009 [53] in an attempt to exploit pretreated rice husk for adsorbing Cr (VI) from its aqueous solutions.

3.3.5. Contact Time Optimization

Finally the contact time optimization was done by conducting adsorption experiment at all achieved optimized condition i.e. 2.25 g adsorbent dose, 10°C temperature, 200 rpm agitation speed and 6 pH. The effect of contact time under said optimized conditions is given in Figure 10. The adsorption was 86.0, 92.3, 95.7, 97.8, 99.2 for contact time 0.5, 1.0, 1.5, 2.0 and 2.5 hr. respectively. The adsorption found to be increasing with increasing contact time and complete adsorption (100%) was achieved for contact time 3.0 hr. and above. During initial period of adsorption the effective adsorption was observed as large surface area of Ad_3 with all vacant adsorption sites was available for adsorption. The less prominent adsorption was observed as the contact time increased. The reason behind the same was as the time lapse the available adsorption sites on the surface of Ad_3 got saturated due to bonding with Cu ion. At this stage the adsorption became independent of adsorption sites on surface and dependent on the migration of Cu ions for further adsorption on adsorption sites within pores of Ad_3 [54,55,56]. Research conducted by previous workers Arivoli, et al., 2009 [57], Dorriset al., 2000 [58] also confirms the same trend in adsorption with respect to contact time. The minimum time observed to achieve complete adsorption was 3 hr. and hence optimized as the optimum time for the considered adsorption.

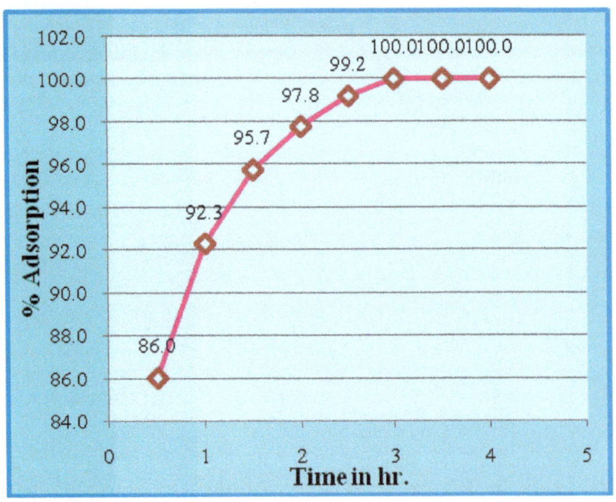

Figure 10. Effect of contact time on adsorption of Cu by Ad_3

3.3.6. Thermodynamic Study

Table 5 is showing the estimated values of standard Gibb's free energy at various temperature. The ΔG^0 at temperature 10, 20, 30, 40, 50 and 60°C was calculated to be -458.68, -316.04, -102.00, 42.14, 370.24 and 585.02 J/mol respectively. The ΔG^0 was found to be increasing with increasing temperature but remained negative upto 30°C and thereafter positive. The adsorption process found to be feasible and spontaneous (exergonic) [59] up to 30°C as ΔG^0 was negative and then became non-spontaneous (endergonic) as ΔG^0 was positive [60,61] for all the temperature above 30°C. Meena et al., 2005 [62] also found the negative and positive value of ΔG^0 while removing various heavy metals (Cd, Cu, Ni & Zn) using carbon aerogel as an adsorbent. ΔG^0 was found to increase gradually with increasing temperature, indicating spontaneity decreases with the rise in temperature.

Table 5. Standard Gibb's Free Energy of Adsorption at Different Temperature

Temperature(°C)	C_e(mg/L)	K_C	lnK_C	ΔG^0 =-RT lnK_C (J/mol)
10	27.1	1.2152	0.19	-458.68
20	26.3	1.1385	0.13	-316.04
30	25.2	1.0413	0.04	-102.00
40	24.5	0.9839	-0.02	42.14
50	23	0.8712	-0.14	370.24
60	22.1	0.8095	-0.21	585.02

The slope and Y-intercept of line obtained by ploting of lnK_C versus 1/T is presented in Figure 11. With increasing 1/T, lnK_C also found to increase linearly. The slope and Y-intercept of the obtained line was used in Van't Hoff equation to calculate the values of ΔH^0 and ΔS^0 Figure and is given in Table 6.

Table 6. Standard Enthalpy Change (ΔH^0) and Entropy Change (ΔS^0) of Adsorption

Equation of line with Corelation coefficient	ΔH^0 (J/mol)	ΔS^0 (J/mol K)
y = 775.5x - 2.525 R^2 = 0.984	-6.4475×10^3	-20.99

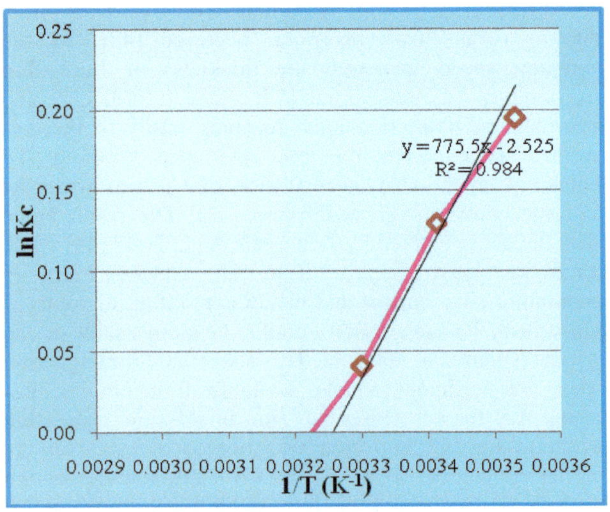

Figure 11. Plot of lnK_C versus 1/T

The value of ΔH^0 and ΔS^0 was found to be -6.4475x10^3 J/mol and -20.99 J/mol K respectively. Both ΔH^0 and ΔS^0 were negative in the given adsorption. The exothermic nature of the said adsorption can be again authenticated by negative value of [63]. The negative value of ΔH^0 and ΔS^0 reveals the adsorption reaction was favorable for enthalpy but unfavorable for entropy. The favoring of adsorption for enthalpy and not for entropy also confirms the reaction became less spontaneous as temperature increased. The negative value of ΔS^0 imply less randomness at the solid/solution interface during adsorption and the system became more ordered through adsorption process i.e. probability of favorable adsorption [64]. The result (both negative ΔH^0 and ΔS^0) agree with the findings of Liang et al., 2010 [65] while removing Cu ions using orange peels as adsorbents.

4. Conclusion

The activated carbon prepared from the waste of pulse processing industry (*Cajanus cajan* Husk) by the process of carbonization and activation with sulfuric acid had the potential to adsorb metal. The optimization of adsorption condition further enhanced efficiency of the prepared activated carbon. In an experimental batch adsorption the optimum condition found to remove complete Cu ions from its 50 ml, 50 ppm solution applying 2.25 g of developed activated carbon was 10°C temperature, 200 rpm agitation speed, 6 pH and 3 hr. contact time. The thermodynamic study of the said adsorption reveled the adsorption was exothermic, feasible and favorable, spontaneous at low temperature (10-30°C) and non-spontaneous at high temperature (40-60°C). The outcome of work may be utilized for designing adsorption process to control water pollution due to heavy metals particularly copper, as one of the possible value addition to Tur dal processing waste.

References

[1] Tumin, N.D., Chuah A.L., Zawani, Z. and Rashid, S.A., "Adsorption of copper from aqueous solution by elaisguineensis kernel activated carbon", *Journal of Engineering Science and Technology*, 3 (2). 180-189. Aug. 2008.

[2] Sud, D., Mahajan, G. and Kaur. M.P., "Agricultural waste material as potential adsorbent for sequestering heavy metal ions from aqueous solutions-A review", *Bioresource Technology*, 99 (14). 6017-6027. Sep. 2008.

[3] Wan Ngah, W.S. and Hanafiah M.A.K.M., "Removal of heavy metal ions from wastewater by chemically modified plant wastes as adsorbents: A review" *Bioresource Technology*, 99 (10). 3935-3948. Jul. 2008.

[4] Friedman, M., Waiss Jr., A C., "Mercury uptake by selected agricultural products and by-products". *Environmental. Science and Technology*, *6* (5). 457-458. May. 1972

[5] Kjellstrom, T., Shiroishi, K. and Erwin, P.E., "Urinary beta./sub 2/-microglobulin excretion among people exposed to cadmium in the general environment" *Environmental Research*, 13 (2), 318-344. Apri. 1977.

[6] Environment (Protection) Rules, *General standards for discharge of environmental pollutants*, Ministry of Environment and Forests, New Delhi, 1986, 5.

[7] Ideriah, T.J.K., David, O.D. and Ogbonna, D. N., "Removal of heavy metal ions in aqueous solutions using palm fruit fibre as adsorbent" *Journal of Environmental Chemistry and Ecotoxicology*, 4 (4). 82-90. 2012.

[8] Demirbas, E., Kobya, M., Senturk, E. and Ozkan, T. "Adsorption kinetics for the removal of chromium (VI) from aqueous solutions on the activated carbons prepared from agricultural wastes", *Water SA*, 30 (4). 533-539. Oct. 2004.

[9] Babel, S., and Kurniawan, T.A., "Low-cost adsorbents for heavy metals uptake from contaminated water: a review", *Journal of Hazardous Materials,* 97 (1-3). 219-243. Feb. 2003.

[10] Bhatnagar, A. and Sillanpaa, M., "Utilization of agro-industrial and municipal waste materials as potential adsorbents for water treatment—A review", *Chemical Engineering Journal*, 157 (2-3), 277-296. Mar. 2010.

[11] Ahalya, N., Kanamadi, R.D. and Ramachandra, T.V., "Cr (VI) and Fe (III) removal using Cajanuscajan husk", *Journal of Environmental Biology*, 28 (4). 765-769. Oct. 2007.

[12] Saeed, A. and Iqbal, M., "Bioremoval of Cd from aqueous solution by black gram husk (Cicerarientinum)". *Water Research*, 37 (14), 3472-3480. Aug. 2003.

[13] Ahaly, N. Kanamadi, R.D. and Ramachandra T.V., "Biosorption of chromium (VI) from aqueous solutions by the husk of Bengal gram (*Cicerarientinum*)", *Electronic Journal of Biotechnology*, 8 (3). 258-264. Dec. 2005. [Online]. Available: http://www.ejbiotechnology.info/content/vol8/issue3/full/10/. [Accessed Sep. 8, 2013].

[14] Saeed, A., Iqbal, M. and Akhtar, M.W., "Removal and recovery of lead(II) from single and multimetal (Cd, Cu, Ni, Zn) solutions by crop milling waste (black gram husk", *Journal of Hazardous Materials*, 117 (1). 65-73. Jan. 2005.

[15] Ahalya, N., Kanamadi, R.D. and Ramachandra, T.V., "Biosorption of iron (III) from aqueous solutions using the husk of *Cicerarientinum*", Indian Journal of Chemical Technology, 13 (2). 122-127. Mar. 2006.

[16] Ahmed, D., Hassan, M., Khalid, H.N., Khatoon,S., Jamil, N. and Fatima, H.B., "Biosorption of Chromium (VI) a low cost Milling Agro Waste Black Gram (Cicer Arientenum) Husk by Flash Column Process" *Journal of Scientific Research*, XXXIX (2), Dec. 2009.

[17] Kale, A.A., Burungle, A.S., Deshpande, N. R. and Kashalkar, R. V., "Kinetics and thermodynamics study of biosorption of Hg^{2+} by sulphonated biomass of *Cicerarientinum*-batch studies", *Der Pharma Chemica*, 2 (6).117-124. Dec. 2010.

[18] Panda, G.C., Das, S.K., Chatterjee, S., Maity, P.B., Bandopadhyay, T.S., and Guha, A.K., "Adsorption of cadmium on husk of *Lathyrussativus*: Physico-chemical study" *Colloids and Surfaces B: Biointerfaces*, 50 (1). 49-54. Jun. 2006.

[19] Saeed, A., Iqbal, M. and Holl, W.H., "Kinetics, equilibrium and mechanism of Cd^{2+} removal from aqueous solution by mung bean husk", *Journal of Hazardous Materials*, 168 (2-3). 1467-1475. Apri. 2009.

[20] Rao, M.M., Ramana, D.K., Seshaiah, K., Wang, M.C. and Chang Chi, S.W., "Removal of some metal ions by activated carbon prepared from *Phaseolus aureus* hulls", Journal of Hazardous Materials, 166 (2-3). 1006-1013. Jul. 2009.

[21] Ramana, D.K.V., Jamuna, K., Satyanarayana, B., Venkateswarlu, B., Rao, M. M. andSeshaiah, K., "Removal of heavy metals from aqueous solutions using activated carbon prepared from *Cicerarietinum", Toxicological & Environmental Chemistry*, 92 (8), 1447-1460. Jan. 2010.

[22] El-Ashtoukhya, E.-S.Z., Amina, N.K. and Abdelwahabb, O., "Removal of lead (II) and copper (II) from aqueous solution using pomegranate peels as a new absorbent", *Desalination*, 223 (1-3). 162-173. Mar. 2008.

[23] Sivaraj, R., Rajendran, V. and Gunalan, G.S., "Preparation and Characterization of Activated Carbons from *Parthenium* Biomass by Physical and Chemical Activation Techniques", *E-Journal of Chemistry*, 7 (4), 1314-1319. Oct. 2010. [Online]. Available: http://dx.doi.org/10.1155/2010/948015. [Accessed Aug. 22, 2013]

[24] El-Sayed, G.O., Aly. H.M. and Hussein, S.H.M., "Removal of Acrylic dye Blue-5G from Aqueous solution by Adsorption on Activated Carbon Prepared from Maize cops" *International Journal in Chemistry and Environment*, 1 (2). 132-140. Oct. 2011.

[25] Bureau of Indian Standards IS 877: 1989, *Activated carbons, powdered and granular-Methods of sampling and test* (Second Revision). BIS, New Delhi, P 1-7.

[26] Bureau of Indian Standards IS 14767: 2000, *Determination of specific electrical conductivity of soils-Method of test*, BIS, 2000, 1-4.

[27] CEFIC, *Test methods for activated carbon*. 1986, (European Council of Chemical Manufacturers Federation/ European Chemical Industry Council, 9-43.

[28] Parate, V.R. and Talib, M.I., "Study of Metal Adsorbent Prepared from Tur Dal (*Cajanuscajan*) Husk: A Value Addition to Agro-waste" IOSR Journal of Environmental Science, Toxicology and Food Technology, 8 (9). 43-54. Sep. 2014.

[29] Ibrahim, S., Wang, S. and Ang, H.M., "Removal of emulsified oil from oily wastewater using agricultural waste barley straw" *Biochemical Engineering Journal*, 49(1), 78-83. Mar. 2010.

[30] Hasan, S.H., Singh K.K., Prakash, O., Talat, M. and Ho, Y.S., "Removal of Cr (VI) from aqueous solutions using agricultural waste 'maize bran" *Journal of Hazardous Materials*, 152 (1). 356-365. Mar. 2008.

[31] Dash, S.N. and Murthy, R, C.V., "Preparation of carbonaceous heavy metal adsorbent from *shorearobusta* leaf litter using phosphoric acid impregnation". *International Journal of Environmental Science*, 1 (3). 296-313. 2010.

[32] Nuhoglu, Y. and Malkoc, E., "Thermodynamic and kinetic studies for environmentaly friendly Ni (II) biosorption using waste pomace of olive oil factory", *Bioresource Technology*, 100 (8). 2375-2380. Apr. 2009.

[33] Wahi, R., Ngaini, Z. and Jok, V U, "Removal of Mercury, Lead and Copper from Aqueous Solution by Activated Carbon of Palm Oil Empty Fruit Bunch" World Applied Sciences Journal 5 (Special Issue for Environment): 84-91. 2009.

[34] Wang, F.Y., Wang, H. and Ma, J.W., "Adsorption of cadmium (II) ions from aqueous solution by a new low-cost adsorbent—Bamboo charcoal", *Journal of Hazardous Materials*, 177 (1-3), 300-306. May. 2010.

[35] Garg, U.K., Kaur, M.P., Garg, V.K. and Sud, D., "Removal of Nickel (II) from aqueous solution by adsorption on agricultural waste biomass using a response surface methodological approach". *Bioresource Technology*, 99 (5). 1325-31. Mar. 2008.

[36] Bhattacharya, A.K., Naiya, T.K., Mandal, S.N. and Das, S.K., "Adsorption, kinetics and equilibrium studies on removal of Cr (VI) from aqueous solutions using different low-cost adsorbents", *Chemical Engineering Journal*, 137 (3). 529-541. Apri. 2008.

[37] Ye, H., Zhang, L., Zhang, B., Wu, G. and Du, D., "Adsorptive removal of Cu(II) from aqueous solution using modified rice husk", *International Journal of Engineering Research and Applications (IJERA)*, 2 (2). 855-863. Mar. 2012.

[38] Boonamnuayvitaya, V. Chaiya, C. Tanthapanichakoon, W. and Jarudilokkul, S., "Removal of heavy metals by adsorbent prepared from pyrolyzed coffee residues and clay", *Separation and Purification Technology*, 35 (1). 11-22. Feb. 2004.

[39] Sciban, M., Kalasnja, M. and Skrbic, B., "Modified softwood sawdust as adsorbent of heavy metal ions from water", *Journal of Hazardous Material*, 136 (2). 266-271. Aug. 2006.

[40] Amarasinghe, B.M.W.P.K. and Williams R.A., "Tea waste as a low cost adsorbent for the removal of Cu and Pb from wastewater", *Chemical Engineering Journal* 132 (1-3). 299-309. Aug. 2007.

[41] Jadhav., D.N. and A.K. Vanjara, "Adsorption of kinetics study: removal of dyestuff using sawdust, polymerized sawdust and sawdust carbon-II", *Indian Journal of Chemical Technology*, 11 (1). 42-50. Jan. 2004.

[42] Kavitha, D. and Namasivayam, C., "Experimental and kinetic studies on methyl blue adsorption bycoir pith carbon", Bioresource Technology,. 98 (1). 14-21. Jan. 2007.

[43] Sharma, Y.C. Prasad, G. and Rupainwar, D.C., "Adsorption of Cd onto tamarind seed", *International Journal of Environmental Studies*, 37, 183-191. 1991,

[44] Zouboulis. A.A.I., E.G. Rousou, K.A. Matis and Hancock, I.C., "Removal of toxic metals from aqueous mixtures. Part 1: Biosorption", *Journal of Chemical Technology and Biotechnology*, 74 (5). 429-436. May 1999.

[45] AjayKumar, A.V., Darwish, N.A. and Hilal, N., "Study of various parameters in the biosorption of heavy metals on activated sludge" *World Applied Sciences Journal*, 5. 32-40. 2009.

[46] Banerjee, K., Ramesh, S.T., Gandhimathi, R., Nidheesh, P.V., and Bharathi K.S., "A novel agricultural waste adsorbent, watermelon shell for the removal of copper from aqueous solutions", *Iranica Journal of Energy & Environment* 3 (2). 143-156. Jan. 2012.

[47] Moussavi, G., Barikbin, B., "Biosorption of chromium (VI) from industrial wastewater onto pistachio hull waste biomass", *Chemical Engineering Journal*, 162 (3). 893-900. Sep. 2010.

[48] Jain, A.K., Gupta, V.K. and Bhatnagar A., S., "Utilization of industrial waste products as adsorbents for the removal of dyes", *Journal of Hazardous Materials*, 101 (1), 31-42. Jul. 2003.

[49] Yu, JX., Li, BH., Sun, XM., Yuan, J., Chi, RA., "Polymer modified biomass of baker's yeast for enhancement adsorption of methylene blue, rhodamine B and basic magenta" *Journal of Hazardous Materials*, 168 (2-3). 1147-1154. Sep. 2009.

[50] Gupta, S. and Babu, B.V., "Removal of toxic metal Cr (VI) from aqueous solutions using sawdust as adsorbent: Equilibrium, kinetics and regeneration studies" *Chemical engineering Journal*, 150 (2-3). 352-365. Aug. 2009.

[51] Namasivayam, C., Prabha D. and Kumutha, M., "Removal of direct red and acid brilliant blue by adsorption on to banana pith", *Bioresource. Technology*, 64(1). 77-79. Apr.1998.

[52] Ibrahim Mohammed, M.N., Nagh Wan, W.S., Norliyana, M.S., Daud Wan, W.R., Rafatullah, M., Sulaiman, O. and Hashim, R., "A novel agricultural waste adsorbent for the removal of lead (II) ions from aqueous solutions", *Journal of Hazardous Materials*, 182 (1-3). 377-385. Oct. 2010.

[53] Bansala, M., Garg, U., Singh, D. and Garg, V.K., "Removal of Cr (VI) from aqueous solutions using pre-consumer processing agricultural waste: A case study of rice husk" *Journal of Hazardous Materials*, 162 (1). 312-320. Feb. 2009.

[54] Radhika, M., and Palanivelu, K, "Adsorptive removal of chlorophenols from aqueous solution by lowcost adsorbent-Kinetics and isotherm analysis", *Journal of Hazardous Materials*, 138(1). 116-124. Nov. 2006.

[55] Namasivayam, C., Muniasamy, N., Gayathri, K., Rani, M., and Renganathan, K., "Removal of dyes from aqueous solution by cellulosic waste orange peel", *Journal of Hazardous Materials*, 57 (1). 37-43. Jul. 1996.

[56] Ahmad, A., Rafatullah, M., Sulaimanb, O., Ibrahima, M.H., Chiia, Y.Y. and Siddique, B.M., "Removal of Cu (II) and Pb (II) ions from aqueous solutions by adsorption on sawdust of Meranti wood", Desalination, 247 (1-3). 636-646. Oct. 2009.

[57] Arivoli, S., Nandhakumar, V., Saravanan, S. and Nagarajan, S., "Adsorption dynamics of copper ion by low Cost activated carbon" *The Arabian Journal for Science and Engineering*, Vol. 34 (1A). 1-12. Jan. 2009.

[58] Dorris, K.L., Yu, B., Zhang, Y., Shukla, A. and Shukla, S.S. "The removal of heavy metal from aqueous solutions by sawdust adsorption-removal of copper", *Journal of Hazardous Materials*, 80 (1-3). 33-42. Dec. 2000.

[59] Abdelwahab, O., 2008. Evaluation of the use of loofa activated carbons as potential adsorbents for aqueous solutions containing dye. Desalination, 222, 357-367.

[60] Sathishkumar, M., Binupriya, A.R., Kavitha, D. and Yun, S.E., "Kinetic and isothermal studies on liquid-phase adsorption of 2,4-dichlorophenol by palm pith carbon", *Bioresource Technology*, 98 (4). 866-873. Mar. 2007.

[61] Tan, I.A.W., Ahmad, A.L. and Hameed, B.H., "Adsorption isotherms, kinetics, thermodynamics and desorption studies of 2,4,6-trichlorophenol on oil palm empty fruit bunch-based activated carbon", *Journal of Hazardous Materials*, 164 (2-3). 473-482. May. 2009.

[62] Meena A.K., Mishra, G.K., Pai, P.K., Rajgopal, C. and Nagar, P.N., "Removal of heavy metal ions from aqueous solutions using carbon aerogel as an adsorbent", *Journal of Hazardous Materials*, 122 (1-2): 161-170. Jun. 2005.

[63] Banat, F., Al-Asheh, S. and Al-Makhadmeh, L., "Evaluation of the use of raw and activated date pits as potential adsorbents for dye containing waters", *Process Biochemistry*, 39 (2). 193-202. Oct. 2003.

[64] Ye, H., Zhang, L., Zhang, B., Wu, G. and Du, D., "Adsorptive removal of Cu(II) from aqueous solution using modified rice husk", *International Journal of Engineering Research and Applications (IJERA)*, 2 (2). 855-863. Mar. 855-863.

[65] Liang, S., Guo, X., Feng, N. and Tian, Q., "Isotherms, kinetics and thermodynamic studies of adsorption of Cu^{2+} from aqueous solutions by Mg2+/K+ type orange peel adsorbents", *Journal of Hazardous Materials*, 174 (1-3). 756-762. Feb. 2010.

18

Extraction and Preservation of Cashew Juice Using Sorbic and Benzoic Acids

Ngozi O .Kabuo, Moses Ojukwu*, Gloria C Omeire, J.C Ibeabuchi

Department of Food Science and Technology, Federal University of Technology, Owerri, Imo State, Nigeria
*Corresponding author: macmosesforall@yahoo.com, moses.ojukwu@futo.edu.ng

Abstract The extraction and preservation of cashew juice using sorbic and benzoic acid was carried to determine a more method for preserving cashew juice under a period of one month. Cashew juice was produced and preserved by five different methods viz – untreated (control) as sample (A_1); pasteurized juice only (A_2); pasteurization and preserved with sorbic acid (B_1); pasteurization and preserved with benzoic acid (B_2) and a combination of sorbic and benzoic acids coupled with pasteurization at 95°C. All the samples were stored in sterilized transparent bottles at 4°C and 28°C for 30 days. Storage stability and the microbial count were determine to obtA1n a suitable method for preserving cashew fruit (Anacardium Occidentale) juice. Sensory evaluation was carried out using 9 point hedonic scale by 20 panelist and statistical analysis on the sensory data was carried out to determine significant differences of the samples at 95% ($P \leq 0.05$) confidence level. The results obtA1ned showed that a combination of sorbic and benzoic acids together with pasteurization was capable of preventing the growth of microorganisms, increase shelf-life, inactive enzymatic actions in the juice and reduce the harsh taste of cashew juice for 30 days of storage.

Keywords: *cashew (Anacardium Occidentale), sorbic acid, benzoic acid, preservation, microbial count*

1. Introduction

A fruit juice is the unfermented, undiluted extractible fluid content obtA1ned from the expression of fruit. Fruit juices are tasteful food beverages which provide energy and contA1n considerable amount of important vitamins (especially vitamin C) and mineral (Duckworth, 1979). Juices which include single strength and concentrates are important means of utilizing fresh fruits, which provides and important source of nutrients in a convenient and most refreshing form (Duck worth, 1979).

There has been a shift over the past three decades in the forms in which fruits are consumed. The fruits consumed. The fruits consumed in processed forms has increased significantly while the rate of consumptions of various fruits in their unprocessed forms have declined greatly due to the concentration of people at the urban regions than the rural areas. The problems associated with freshly harvested fruits such as deterioration have increased the technologies of making the fruits accessible to people living in the urban areas in their processed forms. The processed fruits juices share was about 43% in 1950 but today, it has increased to more than 70% (Ohler,1988). The increased convenience of processed fruits, their all year round avA1lability and improved uniform quality have been the major factors influencing this shift. In most developing countries of the world, fruits are normally wasted and only small quantities are utilized by industries or consumed due to poor storage facilities, poor transportation system, handling equipment (like in green apples) and poor preservation methods. For example cashew fruits has a high rate of deterioration due to prone attack by insects, rodents and micro organisms such as yeast and mould which makes the fruit unavA1lable to the people in other parts of the country.

Efforts have been made by technologists all over the world to providing methods of processing tropical fruits and vegetable through pre-processing storage techniques, preparative processes, extraction, pasteurization, packaging and preservations using chemical and natural preservatives (Ihekoronye, A.I. and Ngoddy, 1985). Akinwale (2000), analyzed some physico-chemical properties of some tropical fruits and found out that cashew apple juice contA1ned the highest amount of vitamin C with 203.5mg/100ml which is more than 300% higher pasteurization which will inactivate enzymes as well as mellow down the harsh taste of the juice. Also suitable condition for the juice will be determined in order to maintain the shelf-life of the juice.

It is hoped that the end of this study, a high quality cashew juice having good taste, appearance and shelf life stability will be produced. The result of this work will also encourage cashew farmers as this will boost cashew fruit production and motivate processors to go into cashew juice processing.

2. Material and Methods

2.1. Materials Procurement

The Cashew fruits used in this work were obtA1ned from a cashew plantation at Okigwe, Imo State, Nigeria. Other materials include Sorbic acid, benzoic acid and packaging materials. Sterilized bottles were used and sterilized equipment.

The chemicals (Sorbic acid and benzoic acid) were obtA1ned from food science and Technology in FUTO and other equipment like refractor meter, Hydrometer and pH meter.

2.2. Material Preparations

The cashew fruits (apples with nuts) on arrival within the same day there where harvested, in the laboratory were thoroughly washed with distilled water before the nuts were carefully detached from the apples. The apples were then rewashed to remove any form of dirt that were found between the nuts and the apples. The muslin filter wrapped in an aluminum foil was sterilized at a temperature of 121°C and pressure of 16psi for 20mins in an autoclave. The apples were sorted to remove the unwholesome ones and were rewashed using distilled water, allowed to drA1ned and then reweighed.

2.3. Production of Juice From Cashew Apple

A modified extraction method was used in the extraction fo the cashew juice.

Approximately, 2500g of thoroughly washed cashew apples were crushed using a grater and the juice was extracted using a mechanical screw press (Duchscher crub press model NG, 1 usine de Wedker).

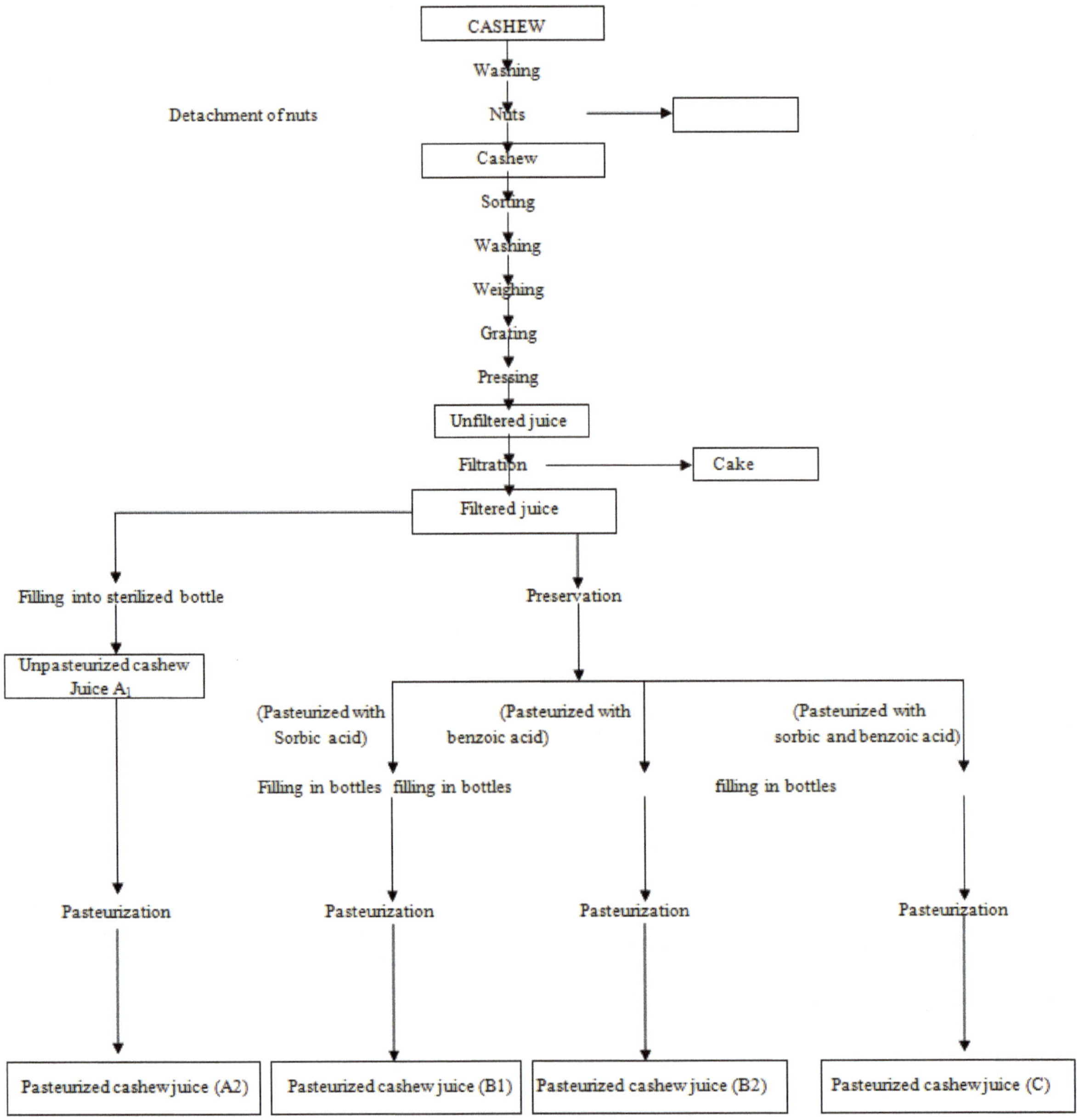

Figure 1. The production flow diagram of cashew juice processing, using different preservation methods

The juice obtA1ned was filtered using the sterilize muslin cloth both the juice and the cake were weighed and the weights were recorded. A pure juice sample was obtained which weighed approximately, 1900g. Different samples were made out of this juice which the first sample was aseptically obtA1ned and stored in a sterilized bottle, stored without preservatives and pasteurization. Another sample was made which only mild heat treatment was applied to it before storage while other samples were made with addition of sorbic acids and Benzoic acid respectively and also pasteurized. A combination of both sorbic and benzoic acid at 0.1% based on the total weight of the juice in equal ratio was also so use to obtA1n another sample.

The preservatives were first dissolved with a little portion of juice before adding to the remA1ning juice and thoroughly homogenized. The juices were immediately filled in sterilized bottles (autoclaved). Corked and pasteurized at a temperature of 95°C for 7mins, cooled and stored at different temperatures (room temperature and frozen temperature of 40°C and 28°C for the respective samples.

The production flow diagram of the cashew juice processing is shown in Figure 1.

3. Analysis on the Cashew Juice

3.1. Determination of pH of the Cashew Juice

The pH of the cashew juice was determined by using a standard pH meter (PHYWE, range 0-14). The pH of 7 after which the bulb was dipped in another buffer solution and the reading on the scale was adjusted. The process was repeated with another known pH (Tap water) to standardized the pH meter. Each time, the bulb was dipped into another solution it was thoroughly rinsed with distilled water after the calibration of the pH meter, the bulb was dipped into the juice samples and the pH reading of the juice was repeated 3 times and the average were recorded.

3.2. Determination of the Cashew Juice Colour

This was done using spectrophotometer method. The absorbance values of the juice was determined at 450nm and 600nm wave length using a spectrophotometer. Before the analysis, the juice was diluted with distilled water and was used as blank.

The measurement was made by inserting the coveter (glass tube) contA1ning the samples into the light path of the spectrophotometer which measured the intensity of light at various wave lengths transmitted by the solution.

The intensity of light were determined by the electric detector, which converted radiant energy to electric energy. The diluted juice samples which had the absorbance value at 450nm was used as a standard and the other filled with the juice were inserted into the instrument and knob was switched off before taking the reading of the absorbance on the electronic scale.

3.3. Determination of the Specific Gravity of the Juice

The specific gravity of the juice samples were measured by the use of hydrometer. The hydrometer was dropped into the juice sample when it was cooled at 20°C after extraction in a measuring cylinder and was allowed to float. The readings were taken directly from the hydrometer scale by reading the upper meniscus.

3.4. Determination of Total Soluble Solids (Brix Level) of the Cashew Juicie

Soluble solid content was measured as percentage sugar by the use of a refractometer. The sample was poured on the sample holder and covered. The readings was taken directly from the scale when viewed from the eye-piece after adjusting the knob until there was a clear demarcation between yellow and light by a red light which the red line was at the centre of the cross. The analysis was carried out for the respective samples and recorded.

3.5. Determination of Titratable Acidity of the Fruit Juice.

This was measured by using 0.1N NaOH solution and phenolphthalein indicator. The juice sample (25ml) was diluted to about 25ml with distilled water. One hundred milliliter solution of the sample was titrated with 0.1N NaOH per 100ml of original solution. This values were then converted to percentage titratable acidity using a standard formular.

3.6. Determination of the Microbial Counts of the Fruit Juice

The number of micro organisms present in the fruit juice were determined using the pour plate count method. The pipettes and plates were sterilized by autoclaving at 121°C before use and hygienically stored. The agar (potatoe dextrose) was prepared according to the instructions on the bottle. The sample were serially dilute the 5th dilution, then agar was allowed to cool to 15°C. 0.1ml of each of the dilution were poured in the plates and then nutrient agar was poured on the inoculated samples. The plates were covered immediately and gently shaken, then placed at 37°C for 24-48hrs. the microbial growth on each of various sample plates were countedand recorded. This was repeated at one week interval for (4) weeks.

Total count = Initial dilution x subsequent x Amount of plated = Dilution Factor.

3.7. Sensory Evaluation

Sensory evaluation is unique source of product information concerned with measuring the response of people to products in terms of appearance, aroma, taste, texture and after and after taste without benefit of label, pricing or other imagery (Iwe 2002).

The sensory evaluation was carried out by 20 panelists composed of 10 male and 10 females selected within the university, rating samples on 9 – point hedonic scale. This was a comparison test between the un-pasteurized and pasteurized cashew juice. Cashew juice preserved with sorbic acid compare toe cashew juice preserved with benzoic acid and finally comparing juice preserved with both sorbic acid and benzoic acid at the ration and other respective samples. The evaluation was based on quality parameters such as taste, flavor, mouth-feel, colour, and overall acceptance. The panelists were requested to compare the samples by observation and testing, indicating their feelings. The means of the scores by the judges were tested by analysis of variance for significant differences between their respective juice samples.

4. Results and Discussion

4.1. pH Value of Cashew Juice Stored at Different Temperature

the pH value of the various samples of the juice were shown in Table 1. The pH values of freshly juice treated with the preservatives (sorbic acid, benzoic acid and the combination) were the same and they have lowered pH than the pH (4.10) then samples contA1ning preservatives were slightly more acidic than the samples contA1ning no preservatives. Also, during storage (for 30 days) the pH of unpasteurized samples increased as result of microbial activities in the juice. The high pH of the control sample was an indication of fermentation which shows the effect of microbial activities in the juice. The high pH for the control sample was an indication of fermentation which shows the effect of microbial activities in the juice and sourness of the juice due to high pH. Sample A_2 (juice preserved by pasteurization only) had an increase in pH both at ambient temperature and refrigeration temperature (4°C). This increase was from 4.10 to 4.38, which shows the preservative effect of pasteurization and chilling effect on the juice. Pasteurization destroyed most of the enzymes and microorganism in the juice and the chilling effect made the environment to be unfavorable for the activities of these enzymes during the 30 days storage.

The pH of sample A_2 (pasteurized only) stored at 28°C increased to 5.20. this might be because Uof microbial actions in the juice during storage. The pH of the pasteurized sample were not as high as that of the untreated (Sample A1) stored under the same temperature over the same period. This shows the differences between the effect of pasteurized juice and unpasteurized juice. The samples which were preserved using sorbic and benzoic acid, B1 and B2 respectively did not show much increase in pH when stored for 30days at low temperature (4°C) B1 slightly increased from 4.00 to 4.10 at 4°C and to 4.13 at 28°C. Sample C is a combination of both preservatives (sorbic and benzoic acids) showed no change in pH during the period of storage at 4°C and 28°C. The pH of sample remA1ned unchanged for the period of storage both at 4°C and 28°C. But sample pH of B1 and B_2 were not affected during the cold storage but slightly increased at 28°C.

Table 1. MEAN pH OF CASHEW JUICE SAMPLES STORED AT DIFFERENT TEMPERATURE

Sample (s)	fresh juice (28°C)	Cashew juice stored for 30 days	
		(4°C)	(28°C)
Untreated juice-control (A_1)	4.10	4.50	6.40
Pasteurized juice only (A_2)	4.10	4.38	5.20
Juice preserved with Sorbic acid + pasteurization (B_2)	4.00	4.10	4.13
Juice preserved with Benzoic acid + pasteurization (B_2)	4.00	4.00	4.10
Juice preserved by Sorbic acid + benzoic acid + pasteurization C)	4.00	4.00	4.00

4.2. Total Soluble Solid of Cashew Juice (Brix Level)

Total soluble solids of cashew juice samples stored at different temperatures for 30 days were shown in Table 2. Sample A_1 and A_2 (untreated juice and pasteurization juice) greatly decreased in their brix level, from 13% to 2.1 for A_1 and 2.50 for A_2 respectively. This decrease could be attributed to the fermentation process due to actions of yeast and moulds in the juice. The reduction in brix level shows that micro organisms have used up the sugar in their activities contA1ning in the samples. Samples B1, B_2 and C had slight decrease in their total soluble solid content which was due to the inhibitory effect of sorbic and benzoic acids on the microorganisms that would have caused fermentation/spoilage of the juice. Pasteurization effect and low temperature storage of sample A_2 (pasteurized only) did not have a positive effect on the sample during storage. The brix level decrease from 13.10 to 5.8° brix and 2° 1° brix when stored at low temperature respectively.

Pasteurization of the juice with sorbic acid and benzoic acids mA1ntA1ned the percentage sugar of the juice when stored for 30 days.

Table 2. TOTAL SOLUBLE SOLIDS (°BRIX LEVEL) OF THE CASHEW OF STORAGE

Sample (s)	fresh juice (28°C)	Cashew juice stored for 30 days	
		(4°C)	(28°C)
Untreated juice-control (A_1)	13.1°	4.2°	2.1°
Pasteurized juice only (A_2)	13.1°	5.0°	2.5°
Juice preserved with Sorbic acid + pasteurization (B_2)	13.0°	11.4°	12.0°
Juice preserved with Benzoic acid +pasteurization (B_2)	13.0°	11.8°	12.2°
Juice preserved by Sorbic acid + benzoic acid + pasteurization C)	13.0°	4.00°	12.0°

There was some differences in Brix levels of samples stored in low temperature compared to room temperature. Sample B1 (treated with sorbic acid and pasteurization) and B_2 (treated with benzoic acid + pasteurization) showed an increase when stored in room temperature (B1increased from 11.4° to 12°, B_2 increased from 11.8 to12.2°) while low temperature storage reduced the percentage sugar in samples B1 and B_2 and C (11.4° , 11.8° and 11.9°) compared to ambient temperature. The samples preserved with combination of both preservatives had just a slight difference in brix level between low temperature and room temperature storage. This could be attributed to the combine effect of both preservative to inactivate the action of enzymes as well as mellow down the harshness of the juice.

4.3. Colour of the Cashew Samples

The colour of the juice on samples A_1 (pure juice, untreated) and A_2 (juice preserved by pasteurization only) became darker on daily bases during storage. Within the first two weeks of storage, the samples became very dark due to the action of non-enzymic browning reaction in the juice during storage.

For sample A_1 (pure juice, untreated) there were action of both enzymic browning and sample A_2 (pasteurized juice) darkening was suspected to be caused by Millard reaction. The reactions which takes place between amino acids and reducing sugars present in the juice decreasing the alpha-amino nitrogen content followed by undesirable colour, odour and flavor changes (Pribella and Betusowa, 1978). The amber-brown colour developed during processing of pear juice has been ascribed to poly phenol oxidize activity (Luh, 1980). Preservation of the juice by pasteurization only did not impact any preservative significance in colour of sample A_2 (pasteurized only) but sample A_1 which is the control, completely deteriorated due to Millard reactions and enzymatic activities.

The colour of sample B1 (preserved with sorbic acid + pasteurization), B_2 (preserved with benzoic acid + pasteurization only) and C (preserved with sorbic acid + benzoic acid + pasteurization) when tested by spectrophometric test for absorbance gave 0.18 for the 1st day. When stored for 30 days, sample B1, B2 and C recorded and absorbance value of 0.15, 0.15 and 0.16 respectively when stored under low temperature. Sample B_1, B_2 and C increased in the absorbance values when stored in room temperature compared to low temperature storage and recorded 0.40, 0.39 and 0.41 respectively. The appearance of samples B_1, B_2 and C did not record much changes but they mA1ntA1ned a shiny yellow colour during storage. This mA1ntenance of colour could be attributed to the pasteurization and the preservative effects on the juice samples which inactivated enzyme browning in the juice.

Table 3. THE COLOUR OF CASHEW JUICE SAMPLES UNDER 30 DAYS OF STORAGE

Sample (s)	fresh juice (28°C)	Cashew juice stored for 30 days	
		(4°C)	(28°C)
Untreated juice-control (A_1)	0.18	0.91	0.96
Pasteurized juice only (A_2)	0.18	0.72	0.78
Juice preserved with Sorbic acid + pasteurization (B_2)	0.18	0.15	0.40
Juice preserved with Benzoic acid +pasteurization (B_2)	0.18	0.15	0.39
Juice preserved by Sorbic acid + benzoic acid + pasteurization C)	0.18	0.16	0.41

4.4. Specific Gravity of Cashew Juice Samples Stored for 30 Days.

The specific gravity for fresh juice samples (A_1, A_2, B_1, B_2 and C) was 1.045. When the samples stored for 30 days, A_1 (pure) juice, (untreated) decreased from 1.045 to 0.076 at 4°C and 0.030 at 28°C. The specific gravity for A_2 was 0.80 at 4°C and 0.035 at 28°C, sample B1 (preserved with sorbic acid + pasteurization only) had a constant specific gravity for both 4°C and 28°C respectively. The samples B1 and B_2 had almost the same specific gravity through B_2 decreased from 1.04 to 1.30 when stored at low temperature. Sample C had a stable specific gravity when stored at low temperature for 30 days. The specific gravity remA1ned unchanged at 1.020 for both 4°C and 28°C. The changes that occurred in the specific gravities of the juice samples during storage could be attributed to the disintegration of fruit particles which settled at the bottom of the contA1ner, this was observed that the specific gravities of the stored juice sample under low temperature (28°C) in sample. This might be due to the fact that low temperatures improve the shelf life of food products.

Table 4. SPECIFIC GRAVITY OF CASHEW JUICE SAMPLE STORED AT DIFFFERENT TEMPERATURES

Sample (s)	fresh juice (28°C)	Cashew juice stored for 30 days	
		(4°C)	(28°C)
Untreated juice-control (A_1)	1.045	0.076	0.030
Pasteurized juice only (A_2)	1.045	0.080	0.035
Juice preserved with Sorbic acid + pasteurization (B_2)	1.040	1.020	1.020
Juice preserved with Benzoic acid +pasteurization (B_2)	1.040	1.030	1.040
Juice preserved by Sorbic acid + benzoic acid + pasteurization C)	1.40	1.020	1.020

4.5. Sensory Characteristics of Cashew Juices.

4.5.1. Colour

The result presented in Table 5, and appendices 2a and 2b showed that sample A_1 (Untreated juice, control), A_2 (pasteurized juice only), sample B1 (juice preserved with sorbic acid + pasteurization), sample B_2 (juice preserved with benzoic acid + pasteurization) and sample C (juice preserved with sorbic acid and benzoic acid + pasteurization) were significantly similar at P>0.05. Sample A_1, A_2, B1, B_2 and C had a mean score of 708^{9}, 7.4^{a}, $7..3^{a9}$, 7.2^{9} and 7065^{a}< 1.0242 (LSD) colour ranking at 5% confidence level.

4.5.2. Taste

Results showed in Table 5, Appendices 3a and 3b showed that sample A_1 (untreated juice control) was not significantly different from A_2 (juice preserved by pasteurization) and not significantly different from B_1 (juice preserved by sorbic acid and pasteurization only), sample B_2 (juice preserved by benzoic acid only) and C (juice preserved by sorbic acid + benzoic acid + pasteurization). The five sample were not significantly different from each other when evaluated. Ftab>Fcal(2.53>0.4080) at 5% confidence level.

Table 5. MEAN SCORES OF SENSORY ATTRIBUTES OF CASHEW JUICE SAMPLES PRESERVED FOR 30 DAYS

ATTRIBUTES	A_1	A_2	B1	B_2	C	LSD
COLOUR	7.8^a	7.4^a	7.3^a	7.2^a	7.6^a	1.0242
TASTE	7.35^a	7.0^a	6.9^a	6.7^a	7.45^a	
FLAVOUR	7.75^a	6.25^c	7.25^a	7.35^a	7.65^b	0.4933
MOUTHFEEL	6.75^a	6.1^a	7.5^a	6.0^c	7.0^a	0.56
GENERAL ACCEPTABILITY	$7.0b^c$	6.6^{bc}	7.2^b	6.4^c	7.8^a	0.6189

KEY
A, b, c, d, e; Means on the same row with similar superscripts are not significantly different at 5% confidence level.
A_1 Untreated juice, control
A_2 Pasteurized juice only
B_1 Juice preserved with sorbic acid + pasteurization
B_2 Juice preserved with benzoic acid + pasteurization
C Juice preserved with sorbic + benzoic + pasteurization.

4.5.3. Flavour

The scores given by the panel of judges on flavour shown in Table 5, appendices 4a and 4b showed that sample A_1 (pure juice, untreated) was significantly different from sample A_2 (pasteurized juice only) and also significantly different from B_1 (juice preserved with sorbic acid + pasteurization) but not significantly different from sample B2. Sample A_2 was significantly different from B_1 and significantly different from B_2 but not different from C. sample C (juice preserved with sorbic and benzoic acids + paste2 but significantly different from A2.

4.5.4. Mouth Feel

Table 5, appendices 1a and 1b showed results obtained from panelists the differences in the mouth feel of different samples. Sample A_1 (untreated juice, control) was similar to sample A_2, B_1 and sample C but different from sample B_2. Sample B2 (juice preserved with benzoic acid + preservation was significantly different from sample A_1, A_2, B_1 and C.

4.5.5. General Acceptability

On general acceptability, results presented in Table 6, Appendices 5a and 5b showed there was no significant difference between sample A_1 (pure juice untreated) and sample A_2 (juice preserved by pasteurization only). Sample A_1 and B_2 were also not significantly different fromsample C (juice preserved with sorbic acid + benzoic acid + pasteurization). Sample C was significantly different from sample A_1, A_2, B_1 and B_2 (juice preserved by benzoic acid + pasteurization) was significantly different from sample B1 (juice preserved with sorbic + pasteurization only).

Table 6. TOTAL MICROBIAL COUNTS IN CASHEW FRUIT JUICE SAMPLE WITH 30 DAYS OF STORAGE

Sample (s)	Week 1	Week 2	Week 3	Week 4
Untreated juice-control (A_1)	NG	0.23	0.57	0.71
Pasteurized juice only (A_2)	NG	0.1	0.30	0.41
Juice preserved with Sorbic acid + pasteurization (B_2)	NG	NG	0.11	0.11
Juice preserved with Benzoic acid +pasteurization (B_2)	NG	NG	0.10	0.11
Juice preserved by Sorbic acid + benzoic acid + pasteurization C)	NG	NG	0.10	0.10

KEY
NG NO GROWTH.

4.6. Total Microbial Counts in the Cashew Juice Sample

The total microbial counts in the cashew juice samples showed significant differences. Preservation by pasteurization, use of chemical preservatives and low temperature preservation has proven yet another point from the results obtained. Sample A_1 (control) showed a high level of deterioration because of the ability of yeast and moulds to grow at low temperature. Yeasts were major spoilage agents in chilled citrus juice and concentrates (Murdock and Hatchr, 1975).

During studies on thermal destruction of endogenous microbial populations in reconstituted orange juice, found that microbial population in reconstituted orange juice, found that microbial population in reconstituted orange juice sample incubated for 24 hours at 30°C were on predominantly yeast (Weihe et al; 1984) preservation of sample A_2 (juice preserved by pasteurization only) by pasteurization helped to control the growth of yeast and bacteria by inactivating them. From the result, the total count obtA1ned was 0.1 x 10^3 (cfu/ml) for sample A_2 compared to A_1 which had 0.23 x 10^3 (cfu/ml) during the second week of storage. There was no growth in sample B_1, B_2 and C in the second week. Within the 3rd week of storage of the sample, there was a sharp increase in the growth of microorganisms. Sample A_1 increased from 0.23 x 10^3 (cfu/ml) to 0.57 x 10^3 (cfu/ml) to the 37 x103 rd week while sample A_2 increased from 0.1 x10^3 (cfu/ml) to 0.30 x10^3 (cfu/ml). this differences between A_1 and A_2

were as a result of pasteurization effect on A_2 compared to A_1 (unpasteurized).

Sample B_1, B_2 and C showed a minimal growth in the 3rd week. Sample B1 had the total count of $0.1x10^3$ (cfu/ml). there was no significant differences in sample B_1, B_2 and C. the minimal growth development of microorganism was as a result of the inhibitory effects of the chemical preservatives in the juice samples. Within the 4th week of storage, the total count in sample A1 further increased from $0.57x10^3$ (cfu/ml) while sample A_2 increased from 0.30 to $0.41x10^3$ (cfu/ml). samples B_1 and C remA1ned unchanged while sample B_2 increased from 0.10 to $0.13x10^3$ (cfu/ml).

The situation in sample B_1 and C could be attributed to the selective inhibitory effect of sorbic acids on yeast and mould when found in food (Ihekoronye and Ngoddy, 1985).

5. Conclusion and Recommendation

5.1. Conclusion

Base on the results obtA1ned from the investigations on extraction and preservations of cashew juice using sorbic and benzoic acid, it may be concluded the sorbic acid and benzoic acid are good preservatives for cashew fruit juice. From the date obtA1ns on shelf-life stability, it showed that the colour, pH, specific gravity, total soluble solid (Brix level) and titratable acidity of the cashew juice was stable during storage when preserved with sorbic acid and benzoic acid. Pasteurization at 95°C for 7min of the juice inactivated the enzymes present in the juicy which causes browning. Sorbic and benzoic acids by its inhibitory ability on yeast and mould presented microbial growth in the juice during storage and also mellow the harshness yeastic taste of the cashew juice.

In the case of sensory evaluation, sample C (juice preserved with sorbic acid + benzoic acid + pasteurization) with ranking of 7.8^a on its general acceptability of the panelists was "very much like" due to good colour, taste, flavor and mouth feel, which made it to be significantly different from other sample. Sample A_1 (untreated juice, control) and sample A_2 (pasteurized juice only) deteriorated rapidly, showing a high microbial count within 30days of storage while sample B_1, (preserved with sorbic acid + pasteurization), B_2 (juice preserved with benzoic acid + pasteurization) and C (juice preserved with sorbic + benzoic acid + pasteurization) recorded a low microbial count within 30 days of storage.

5.2. Recommendations

The use of a combination of sorbic and benzoic acids with pasteurization in the preservation of cashew fruit (*Anacardium Occidentale*) juice is recommended.

For further research, recommendation of cashew processing with other fruits to make blends of juices due to its high ascorbic acid content.

References

[1] Akinwale (2000). Cashew apple juice: its use in fortifying the nutritional quality of some tropical fruits. Eur. Food Res. Technol., 211: 205-207.

[2] Duckworth, R.B. (1979) Fruit juice in "Fruit and vegetable" Duckworth, R.B (editor) pergamen Press New York.

[3] Ihekoronye, A.I and Ngoddy, P.O. (1985) Integrated Food Science and Technology for the Tropics. Macmillan Publishers Ltd; London PP293-300.

[4] Iwe M.O (2002). Handbook of sensory methods and analysis. PROJOINT Communications Services Ltd, Enugu. pp. 70-72.

[5] Luh, B. S. Nectar, pulpy juices and fruit juice blends in fruit and vegetable processing technology. In *Fruit and Vegetable JuiceProcessing Technology,* 3rd ed.; Nelson, P. E., Tressler, D. K., Eds.; AVI: Westport, CT, 1980; pp 471-476.

[6] Murdock, D. I. and W. S. Hatcher. (1975). Survey of Midwest grapeconcentrate plant for heat resistant mold. Unpublished lab. data. 8.

[7] Ohler, J.G. (1988) Cashew, communication 71, Department of Agricultural research, Koninkiljk Institute Voor de tropen, Amsteradam.

[8] Pribella, A., and Betusowa, M. (1978), *Frllchtsaft-lndllstrie,* 9(1), 15.

[9] WeiheE., Reinecke M. and Forssrnann W.G. (1984). Distribution of vasoactive intestinal polypeptide-like irnrnunoreactivity in the rnarnrnalian heart. lnterrelation with neurotensin- and substance P-like irnrnunoreactive nerves. Cell Tissue Res. 236, 527-540.

19

Some Physicochemical Characteristics and Storage Stability of Crude Palm Oils (*Elaeis guineensis* Jacq)

Goudoum Augustin[1,*], Makambeu Ndomche Anne[1], Abdou Bouba Armand[1], Mbofung Carl Moses[2]

[1]The University of Maroua, The Higher Institute of Sahel, Department of Agriculture, Livestock and Derived Products, Maroua, Cameroon
[2]The University of Bamenda, School of Technology, Department of Food and Bioresource Technology, Bambili, Cameroon
*Corresponding author: goudoumaugust@gmail.com

Abstract Oils constitute one of the essential components of balanced diet as good source of energy. The chemical and physical properties of oils are amongst the most important properties that determine the quality and help to describe the present condition of oils. The physicochemical characteristics and storage stability of crude palm oils producer traditionally and industrially in Cameroon were investigated and compared to reference oil manufactured by a leading manufacturing industry. The mean physicochemical parameters of crude palm oils (CPO) before storage were: moisture content (1.15 and 0.25%), free fatty acid (6.49% and 9.44%), peroxide (4.86 meq/kg and 5.70 meq/kg) respectively for traditional and industrial oils, and melting point (33°C). All the physicochemical parameters determined were significantly ($p<0.001$) higher than those determined for reference. Storage stability of both oils under 30°C showed faster deterioration than when stored at 20°C. The mean free fatty acid values (oleic acid) obtained for both oils when exposed at 30°C and 20°C respectively throughout the period of study (3 months) were 11.08% and 8.03% for industrial oil and 16.50% and 15.19% for traditional oil. Similarly, the mean peroxide values were 10.36 and 9.34 meq/Kg for industrial oil and 11.33 and 10.19 meq/Kg. The physicochemical properties of the CPO indicated that it is edible, drying and suggested its suitability for industrial purposes as well as the nutritional potentials, which could serve as an alternative food ingredient for unsaturated vegetable oil.

Keywords: *crude palm oils, storage, traditional, industrial, physicochemical parameters*

1. Introduction

Fats and oils whether the source is animal, vegetable or marine in origin represent the highest source of energy per unit weight that man can consume. Apart from being a source of reserved energy, fats deposit insulates the body against loss of heat and protects vital organs against mechanical injury [6]. They are important food source for man, and are also extensively used for nutritional, cosmetic, drug dispersant in therapeutics and industrial purposes and are used for supplying essential fatty acids such as linoleic and arachidonic acids [23].

In Cameroon, palm oil (*Elaeis guineensis*) meets 80% of total edible oil needs and it is estimated that 30% of crude palm oil (CPO) production is provided by none industrial oil mills. The quality is associated with the method of processing. In Cameroon, there are three types of oil palm processors, traditional (mostly manual), semi-mechanized and mechanized processors [18]. To a larger extent, smallholder/traditional processor dominate the sector accounting about 80% [20]. Smallholder processors maintain low level of hygiene in the processing mills [16,21]. These have resulted to low quality of CPO. Authors have reported the physicochemical and microbiological properties of CPO processed traditionally to be of poor quality. The semi-mechanized processor share about 50% of smallholder characteristics. The major processing activities irrespective of kind of processing that could affect the quality of CPO produced include bruises during transportation, fermentation prior to threshing, clarification and storage [14,19].

However, the consumption of CPO can also be detrimental to human beings, as CPO contains some components which are likely to enhance numerous reactions (hydrolysis, oxidation, etc.) involved in the degradation of this product. Moreover, these degradation reactions can also be initiated and/or accentuated by poor transportation and storage conditions [27] of the product as it is generally the case among small holders. The most effective degradation process of CPO is acidification which was already mentioned by Desassis in 1957 [18].

Previous studies tend to demonstrate that, there was a problem with the consumption of CPO with respect to food safety. The effect of processing methods and storage time on some physicochemical parameters of Cameroonian CPO was assayed. Based on the determination of these physicochemical parameters studies by Ngando et al. [17], CPO from small holders' extraction sites was of lesser quality as compared to that from industrial oil mills regarding food safety. Considering the fact that about 30%

of Cameroon's national production of CPO is provided by small holders, one can assume that the quality of this product which is freely available in local markets is subject to doubt.

The present study aimed to assess the storage quality and stability (regarding food safety) of CPO available producer by traditional and industrial method, as well as the physicochemical characteristics. For this purpose, the physicochemical parameters were chosen: free fatty acid, deterioration of bleachability index, moisture content, peroxide value and melting point. We decided to use them for our study as they were routinely used worldwide for the assessment of the quality of cooking oils.

2. Materials and Mehods

The monitoring of physico-chemical characteristics of CPO was made at two different temperatures during three months for three different samples.

2.1. Sample Collection

The oil samples were collected from the tanks after delivery by a partnership instead. Three types of oil were used for testing:

- Industrial oil representing the one that was extracted using an industrial process, delivered by a local company (Sample 1);
- Artisanal oil that is the one generally produced by small producers. The extraction is mostly done traditionally with rudimentary equipment (Sample 2).
- Palm oil stored in a tank that is an imported industrial oil in Cameroon, which is the reference (Sample 3);

Samples of these oils were collected, filtered to rid them of any impurities and stored at 20°C in small galvanized tanks (diameter: 20 cm, high: 40 cm, thickness: 0.6 cm and total volume 10 liters).

2.2. Physico-chemical Analysis of Palm Oil

The three samples of palm oil were analyzed before storage to determine their baseline characteristics. The various parameters tested and monitored throughout the storage are as follows: free fatty acid (FFA); deterioration of bleachability index (DOBI); Moisture content; the melting point and the peroxide value.

2.3. Determination of Free Fatty Acid (FFA)

Estimation of the percentage free fatty acids as oleic acid was done, following the method of Pearson [22]. A mass of 2.5 grams of the sample was weighed into a conical flask, 50 mL of neutralizing solvent (ethyl alcohol) was carefully added to the weighed sample in the conical flask then the flask was placed on a hot plate at 40°C, swirled gently and titrated with standard potassium hydroxide (0.5M). Values were calculated using the formula:

$$\text{Free Fatty Acid} = 25.6 \times \text{Molarity of KOH} \times \text{Volume of KOH} / \text{mass of sample.}$$

Where 25.6 in the formular for FFA determination is equivalence factors (e) for palmitic acid; the predominant fatty acid in palm oil.

2.4. Deterioration of Bleachability Index

Deterioration of bleachability index (DOBI) was measured using the PORIM Test Methods (1995) [1]. About 0.1 g of oil was weighed and dissolved in up to 25 ml n-hexane (95%). The oil solution was placed in a 1cm cuvette and the absorbance reading was taken at 446 nm and 269 nm by using the spectrophotometer of mark PharmaSpec MODEL UV-1700. DOBI value is defined as the ratio of the spectrophotometric absorbance at 446 nm to 269 nm.

2.5. Determination of Moisture

The moisture content was determined using AFNOR [2] method. A mass of 10grams of thoroughly mixed POME sample was weighed into a known mass of clean Petri dish which had been previously dried and cooled in a desiccator. It was placed in the oven for four (4) hours, allowed to cool to room temperature in a desiccator for 45mins and further weighed. This was repeated till a constant weight was obtained. The moisture and volatile matter content was expressed in percentage by mass using the formula:

$$\text{\% Moisture and Volatile content} = \left(\text{Mb} - \text{Md} / \text{Mb} - \text{M}\right) \times 100$$

Where M=Mass (g) of Petri dish
Mb = Mass (g) of Petri dish + sample
Md = Mass (g) of Petri dish + test sample after drying.

2.6. Determination of Melting Point

The melting point was determined using AOCSCc 3-25 [5] method. Capillary tubes were inserted into the blended oil samples to obtain a 10±2 mm long columns of oil sample, the capillary tubes were then sealed at one end using a Bunsen flame. The capillary tubes with the oil sample were kept in a refrigerator for three days during which the oil became solidified. With the help of a thread, the capillary tube with the sample was tie to a thermometer and inserted into a water bath at 30°C. The temperature at which the oil began to rise up in the tube was recorded as the melting point.

2.7. Peroxide Value Determination

The peroxide value was determined by titrating chloroform/glacial acetic acid/potassium iodide solution of the oil with an aqueous solution of sodium thiosulphate using starch as indicator [3]. About 10 g of oil was weighed into the 250 mL conical flask. A mixture of glacial acetic acid and trichloromethane chloroform (30 mL) was added in a ratio of 3: 2. 0.5 mL of saturated potassium iodide solution was also added. The mixture was properly shaken. 30 mL of water was added. The solution was filtrated with 0.01M sodium thiosulphate, while slowly adding the titrant with a continuous shaking until a yellow colour is shown. Approximately 5 mL of starch indicator was added to the titration process while shaking vigorously until a blue-black colour is discharged. A blank sample devoid of CPO was also analyzed using the same procedure. The peroxide value is expressed mathematically as follows:

Peroxide value (Meq Peroxide/kg) = (S – B) x M x 1000/Sample weight

Where S = Sample titrated

B = Blank

M = Molarity of sodium thiosulphate.

2.8. Statistical Analysis

The collected data were processed and analyzed with the XLSTAT software Version 2013. This software has permitted a variance analysis to compare the results obtained from the different treatments. Differences were considered significant from $p < 0.05$. The Beauvais-Pearson test was used to assess the degree of linear dependence amongst the various parameters assessed.

3. Result and Discussion

3.1. Physicochemical Characteristics of CPO Sample before Storage

The results of physicochemical properties before storage period of the three studied CPO oils obtained by different methods are presented in the Table 1.

Table 1. Physicochemical qualities of crude palm oil of three samples before storage.

	FFA (%)	DOBI	MIV(%)	MP (°C)	PV
Sample 1	6.49± 0.011^{a}	1.59± 0.005^{a}	1.15± 0,08^{a}	33± 0.5^{a}	4.86± 0.013^{a}
Sample 2	9.44± 0.023^{c}	2.47± 0.020^{b}	0.25± 0,023^{c}	33± 1^{a}	5.70± 0.015^{c}
Sample 3	4.46± 0.021^{b}	2.43± 0.010^{b}	0.12± 0,005^{b}	31± 0.5^{b}	2.52± 0.026^{b}

Averages followed by the same letter in the same column are not different significantly with P < 0.05 (Test of Duncan).
FFA : Free Fatty Acid ; DOBI : bleachability index ; MIV : moisture contents ; MP : melting point; PV : peroxide values

3.2. Evolution of the Free Fatty Acid during the Storage Period

Results of the shelf life of the CPO analyzed presented in Table 2 revealed variations in the different physicochemical properties study at two temperatures.

This table shows that the free fatty acid of CPO samples change significantly ($p < 0.001$) during each month for all stored samples. For sample 1, the FFA increased from 6.49 to 11.08% and from 6.49 to 10.12%, respectively for 30°C and 20°C temperatures after three months of storage. For sample 2, the FFA pass from 9.42 to 16.50% and from 9.42 to 15.19% respectively for 30 ° C and 20°C. With regard to sample 3, the FFA increased from 4.46 to 9.58% at 30°C and 4.46 from 8.03% at 20°C.

Table 2. Storage stability of free fatty acid of three crude palm oils stored under different temperature

	Storage at 30°C			Storage at 20°C		
Storage period (month)	**Sample 1**	**Sample 2**	**Sample 3**	**Sample 1**	**Sample 2**	**Sample 3**
0	6.49±0.011^{a}	9.42±0.023^{a}	4.46 ±0.021^{a}	6.49±0.011^{a}	9.42±0.023^{a}	4.46 ±0.021^{a}
1	8.93±0.013^{b}	12.05± 0.021^{b}	6.05±0.024^{b}	7.25±0.025^{b}	11.80±0.025^{b}	5.25±0.014^{b}
2	10.18±0.021^{c}	14.10±0.015^{c}	7.98±0.016^{c}	9.06±0.020^{c}	13.68±0.015^{c}	6.55±0.022^{c}
3	11.08 ±0.015^{d}	16.50±0.022^{d}	9.58 ±0.021^{d}	10.12±0.015^{d}	15.19±0.015^{d}	8.03±0.016^{d}

Averages followed by the same letter in the same column are not different significantly with P < 0.05 (Test of Duncan).

3.3. Evolution of the Deterioration of Bleach Ability Index Acid during the Storage Period

The Table 3, shows that the DOBI of CPO samples change significantly during each month for all stored samples ($p < 0.001$) at 30 and 20°C.

Table 3. Storage stability of the DOBI of three crude palm oils stored under different temperature

	Storage at 30°C			Storage at 20°C		
Storage period (month)	**Sample 1**	**Sample 2**	**Sample 3**	**Sample 1**	**Sample 2**	**Sample 3**
0	1.59±0.005^{a}	2.47±0.020^{a}	2.43±0.010^{a}	1.59±0.005^{a}	2.47±0.020^{a}	2.43±0.010^{a}
1	1.22±0.005^{b}	1.68 ±0.057^{b}	1.64±0.020^{b}	1.43±0.01^{b}	1.96±0.01^{b}	1.98±0.013^{b}
2	1.02±0.016^{c}	1.41±0.015^{c}	1.39±0.015^{c}	1.21±0.012^{c}	1.70±0.013^{c}	1.66±0.005^{c}
3	0.96±0.025^{c}	1.12±0.020^{d}	1.02±0.016^{d}	1.11±0.010^{d}	1.43±0.009^{d}	1.32±0.021^{d}

Averages followed by the same letter in the same column are not different significantly with P < 0.05 (Test of Duncan).

The CPO process industrially and traditionally exposed at 30°C for two month at, DOBI values decrease from 1.59 to 1.02 and 2.47 to 1.41 respectively and the values decrease to 0.96 and 1.12 respectively after three months. DOBI values for CPO process industrially stored at 20°C drop from 1.59 to 1.43 in the first month and to 1.11 in the third month. Similarly, the DOBI values for CPO process traditionally drop from 2.47 to 1.96 in the first month and recorded 1.43 in the third month. DOBI values between CPO stored at 30 and 20°C showed significant ($p<0.001$) difference.

3.4. Evolution of the Peroxide Index during the Storage Period

The Table 4, shows that the peroxide values of samples change significantly during each month for all stored samples ($p < 0.001$) at 30 and 20°C.

The initial PV was seen to be higher in the CPO process industrially and traditionally compared with the reference (sample 3) at two temperatures. This showed the relative oxygen uptake by the three oils under study. At the beginning of the experiment a peroxide value (PV) for sample 2 exposed to ambiance temperature (30°C) was 5.70 meq/kg and after two months the value rose to 9.23 meq/kg and further to 11.33 meq/kg after three months.

Table 4. Storage stability of peroxide values of three crude palm oils stored under different temperature (meq/kg).

	Storage at 30°C			Storage at 20°C		
Storage period (month)	**Sample 1**	**Sample 2**	**Sample 3**	**Sample 1**	**Sample 2**	**Sample 3**
0	4.86±0.013[a]	5.70±0.015[a]	2.52±0.026[a]	4.86±0.013[a]	5.70±0.015[a]	2.52±0.026[a]
1	6.32±0.021[b]	7.14±0.015[b]	3.70±0.024[b]	5.54±0.037[b]	6.75±0.016[b]	3.85 ±0.022[b]
2	8.24±0.033[c]	9.23±0.032[c]	5.69±0.011[c]	7.39±0.021[c]	8.12±0.023[c]	5.28±0.025[c]
3	10.36±0.03[d]	11.33±0.015[d]	8.20±0.030[d]	9.34±0.020[d]	10.19±0.26[d]	7.02 ±0.012[d]

Averages followed by the same letter in the same column are not different significantly with P < 0.05 (Test of Duncan).

The PV for CPO process industrially under the same condition as above rose from 4.86 meq/kg to 8.24 meq /kg and then recorded 10.36 meq/kg after three months. There is a significantly ($p<0.001$) higher difference between the three CPO.

PV for the CPO stored at 20°C did not show such a sharp increase after the first and second month of storage. During the first month, PV for sample 1 increased from 4.86 meq/kg to 5.54 meq/kg and increased further to 9.34 meq/kg two months later. That of sample 2 increased from 5.70 meq/kg to 6.75 meq/kg and in the next months recorded 10.19 meq/kg. Though there was a progressive increment in PV for the oils at the ambiance temperature up to three months of storage. The overall quality of the CPO samples assessed within this study was quite good regarding this parameter, as all the forty samples had PV below the 15 meq/kg maximal limit for cold pressed and virgin oils [8].

3.5. Evolution of the Melting Point during the Storage Period

The Table 5, shows that the melting point values of samples change significantly ($p<0.01$) during each month for all stored samples at 30 and 20°C.

Table 5. Storage evolution of meting point values of three crude palm oils stored under different temperature

	Storage at 30°C			Storage at 20°C		
Storage period (month)	**Sample 1**	**Sample 2**	**Sample 3**	**Sample 1**	**Sample 2**	**Sample 3**
0	33.83±0.28[a]	33.00±0.57[a]	32.16±0.28[a]	33.83±0.28[a]	33.00±0.57[a]	32.16±0.28[a]
1	32.00±0.00[b]	32.00±0.57[b]	31.33±0.57[b]	33.00±0.57[b]	33.00±0.50[a]	32.16±0.28[a]
2	32.16±0.28[b]	31.16±0.28[c]	30.00±0.00[c]	31.00±0.00[c]	31.00±0.057[b]	31.16±0.28[b]
3	31.33±0.57[c]	30.00±0.00[d]	29.16±0.28[d]	30.5±0.57[c]	28.00±0.57[c]	30.00± 0.00[c]

Averages followed by the same letter in the same column are not different significantly with P < 0.05 (Test of Duncan).

The CPO process industrially and traditionally exposed at 30°C for two month at 30°C, melting point values decrease from 33.83 to 32.16°C and 33.00 to 31.16°C respectively and the values decrease to 31.33 and 30.00°C respectively after three months.

Melting point values for CPO process industrially stored at 20°C drop from 33.83 to 33.00 in the first month and to 30.50°C in the third month. Similarly, the Melting point values for CPO process traditionally drop from 33.00 to 31.00°C in the first two months and recorded 28.00°C in the third month. Melting point values between CPO stored at 30 and 20°C showed significant ($p<0.01$) difference.

3.6. Correlation between Physicochemical Parameters of Different Crude Palm Oils during the Storage

In the optic to better elucidated the physicochemical phenomena proceeding within the palm oil samples during storage a correlation was established between the various physicochemical parameters (Table 6).

Table 6. Correlation between differents physicochemical parameters of crude palm oils during the storage

correlations	Sample 1	Sample 2	Sample 3
DOBI/ FFA	-0.901**	-0.912**	-0.907**
DOBI/ PV	-0.933**	-0.942**	-0.920**
FAA / PV	0.978**	0.960**	0.972**
Melting Point/PV	-0.854**	-0.844**	-0.823**
Melting point/DOBI	0.884**	0.901**	0.889**
Melting point /FFA	0.837**	0.840**	0.830**

** Signification at P< 0.01.

According to the results of correlation, there is a significant correlation ($P < 0.01$) between the physicochemical parameters. The correlation between the DOBI, FFA and the peroxide index is very significant ($P< 0.01$), but negative (- 0.933), so the DOBI decreases while the peroxide index and acidity increase. On the other hand the peroxide index and FFA increase simultaneously during storage (0.978). These correlations show that the physicochemical parameters decreased proportionally one to another.

4. Discussion

Sample 1 representing the industrial oil produced and delivered by a local company was a month old. This could explain the high FFA which is consecutive to the degradation. DOBI value of this oil also shows that the degradation is already started. The moisture of the sample 1 was 1.15%, which is higher than the norm of 0.3%. This may due to the fact that during storage, there is accumulation of water at the bottom of the storage tank due to the difference in density between water and oil. This accumulation of water could lead to an increase in the moisture content of the oil contained in the bottom of the storage tank. This increase in humidity can cause different damage occurring during storage of the oil. Sample 2, which is represented by oil traditionally produced by small holders, has a high levels of free fatty acid (9.44%) and peroxide indices (5.70). These values are higher than those of the reference (sample 1) that are respectively ≤ 5 and ≤ 3 for these two parameters.

This evolution of chemical parameters could be due to several factors such as the presence of water, the presence of oxygen, the presence of enzymes, the presence of heavy metals, microorganisms and high temperatures [6,13]. In the case of our analysis, a significant difference was observed between the values obtained at 30°C of the obtained at 20°C. The hydrolytic changes though not predominant, indicated that the formation of free fatty acids increased with increasing time of storage. The significantly ($p<0.001$) higher difference between the oils stored at 30°C and at 20°C may be due to the effect of sunlight and temperature differences as increase in temperature increases the rate of oxidation [6].

Many studies have pointed out the presence of a very active endogenous lipase in the mesocarp of the fruit of the oil palm [16,18]. This lipase is activated in the fruit at maturity upon wounding and/or bruising and is responsible for the hydrolysis of triglycerides and the liberation of FFA. In order to limit the action of the lipase, fresh fruit bunches (FFB) must be processed rapidly after harvest. This is the case in industrial oil mills where the harvested FFB are generally steam sterilized rapidly or at the very worst less than two days after harvest so as to inactivate the lipase, thus limiting subsequent FFA accumulation in the resultant CPO [13,18].

The decrease DOBI translate into lower carotenoid rate depends on the oxidation products. This decrease can be influenced by several factors. In fact, carotenoids are located within the internal membranes of chloroplasts which they are surrounded by a double membrane [11]. These carotenoids are fat-soluble, very sensitive to heat, but susceptible to enzymatic oxidation, chemical and photochemical [7,26]. During storage these carotenoids oxidize and gives rise to oxidation products [27]. These oxidation products are difficult to remove during refining. In addition, during refining, they polymerize under the action of high temperatures and make the dark refined oil [25,26]. A crude palm oil with low DOBI value (≤ 1.5) is difficult to refine because of the presence of oxidation products which are difficult to remove during refining [11]. This may cause instability of the color of palm olein obtained from palm oil degraded.

PV is used to assess the quality of cooking oils and fats through the measurement of the amount of lipid peroxides and hydroperoxides formed during the initial stages of oxidative degradation and thus, estimate to which extent spoilage of the oil has advanced. This parameter was chosen for our study as it is routinely used during food security controls to assess the quality of cooking oils. However, these peroxides are very unstable transitory products, and the determination of PV might not necessarily provide a correct estimate of the level of peroxidation, as it gives accurate information on the amount of peroxides and hydroperoxides but not on the secondary oxidation products. From another point of view, the overall low PV of CPO samples in our study is in the nature of things, as this oil is known to have a high oxidative stability. Thanks to its fatty acid composition with a balanced ratio of polyunsaturated/saturated fatty acids (51:49 w/w), CPO is less susceptible to oxidation and is widely used for frying of food [15]. Beside these visible harmful effects on the sensory quality of the oil, peroxidation also makes the oil dangerous for human health, as the free radicals generated by this process are proven to be carcinogenic [10,12,24].

Previous study has demonstrated that oil acidification and peroxidation processes are significantly enhanced by high moisture content of CPO samples at the outset [17]. Analyses showed there were no differences amongst samples for the three parameters assessed regarding the size of the market or information provided by the vendor on the origin, thus strongly suggesting that the CPO (randomly) sampled in our study were probably all processed by smallholders, as the control sample from an industrial oil mill had moisture content far below the 0.2% maximal limit.

Several factors could explain the decrease of the melting point. Indeed according Dandjouma et al. [9], changes at the melting temperature of the oils could be explained by the hydrolysis of triglycerides and the progressive oxidation observed during storage. The gradual hydrolysis of triglycerides materialized by an increase of the acid value of the oils, releases free fatty acids which are preferred substrates from oxidation. This oxidation leads firstly to saturated fatty acids a high melting point and on the other hand the fatty acid short chain low melting point fatty acids after cleavage at the unsaturations. Indeed, during storage, edible oils undergo physical and chemical changes under the influence of factors such as light, heat, trace metals and water. These changes give rise to secondary products called oxidation products.

5. Conclusion

The quality assessment of CPO produced by semi-mechanized and traditional processor in Cameroon showed that physicochemical parameters such as FFA, moisture content, DOBI and Peroxide value were high. The poor physicochemical properties are indications of poor handling and processing methods. This constraint can be addressed by further processing the oil. Removal of moisture through further processing could increase the quality of the oil.

Prior to the study, the physicochemical characteristics of the oil largely conformed to codex standards. Depending on the mode of storage, the physicochemical properties changed significantly with storage time. The highest changing property was that of peroxide in oils exposed at 30°C. Oils kept in tightly sealed containers and stored at 20°C exhibited little change.

References

[1] Abdul, H.N., M.H. Ng, Y.M. Choo, Ma, A.N, "Dry Heating of Palm Fruits: Effect on Selected Parameters", *American Journal of Engineering and Applied Sciences*, 5 (2). 128-131. 2012.

[2] AFNOR, "*Recueil de normes françaises. Corps gras, graines oléagineuses, produits dérivés*", AFNOR, Paris (France), 2ème édition. 1981.

[3] Aletor, V.A., G. A. Ikhena, and Egharevba, V. "The quality of some locally processed Nigerian palm oil: An estimation of some critical variables", *Food Chemistry*, 36. 311-317. 1990.

[4] AOAC, Association of Official Analytical Chemists, "*Official Methods of Analysis*", 14 ed,. Arlington, VA, 67. 503-515. 1984.

[5] AOCSCc 3-25, "*The determination of melting point in open capillary tubes (slip point) for palm oil products*", AOCS Press, Champaign, IL. 1989.

[6] Barku Atsu V.Y, H.D Nyarko and Dordunu P, "Studies on the physicochemical characteristics, microbial load and storage stability of oil from indian almond nut (Terminalia catappa l.)", *Food Science and Quality Management*, Vol. 8 (1). 9-17. 2012.

[7] Cheftel J.C., H. Cheftel et Besancon P, "*Introduction à la Biochimie et à la technologie des Aliments*", Volume 1 Edition technique et documentation 11 Rue Lavoisier, 75008 Paris, 420. 1983.

[8] Codex Alimentarius Commission/FAO/WHO food standards, "*Standard for named vegetable oils*", CODEX-STAN 210 Ed. FAO/WHO. 2011.

[9] Dandjouma A.K.A., C. Tchiegang, Parmentier M, "Evolution de quelques paramètres de qualité physico–chimique de l'huile de la pulpe des fruits de Canarium schweinfurthii Engl. au cours du stockage", *International journal of biological and chemical science*, 2(3). 249-257, 2008, 9. 2008.

[10] Eymard S, "*Mise en évidence et suivi de l'oxydation des lipides au cours de la conservation et de la transformation du chinchard (Trachurus trachurus) : choix des procédés*", thèse de doctorat, Université De Nantes. 28-38. 2003.

[11] Fatin Syakirah A., S. Rosnah, and Robiah Y, "The Effect of Storage Time of Chopped Oil Palm Fruit Bunches on the Palm Oil Quality. "ST26943"", *2nd International Conference on Agricultural and Food Engineering, CAFEi2014*". Agriculture and Agricultural Science Procedia 2.165-172. 2014.

[12] Frankel E.N, "*Lipid oxidation*", The Oily Press (vol. 10). Dundee, Scotland. 10. 1998.

[13] Gulla S. and Waghray K, "Effect of Storage on Physico-chemical Characteristics and Fatty Acid Composition of Selected Oil Blends", *J L S*, 3(1). 35-46. 2011.

[14] Izah, S.C., Ohimain, E.I, "Microbiological quality of crude palm oil produced by smallholder processors in the Niger Delta, Nigeria", *Journal of Microbiol. Biotechnol Res*, 3 (2). 30-36. 2013.

[15] Matthaüs B, "Use of palm oil for frying in comparison with other high-stability oils", *Eur. J. Lipid Sci. Technol.* 109. 400-409. 2007.

[16] Ngando Ebongue G.F., R. Dhouib, F. Carriere, P.H. Amvam Zollo, Arondel V, "Assaying lipase activity from oil palm fruit (Elaeis guineensis Jacq.) mesocarp", *Plant Physiol. Biochem*, 44. 611-617. 2006.

[17] Ngando Ebongue GF, E.A. Mpondo Mpondo, E.E.L. Dikotto, Koona P, "Assessment of the quality of crude palm oil from small holders in Cameroon", *JSPPR* 2. 52-58. 2011.

[18] Ngando-Ebongue G.F., E.A. Mpondo-Mpondo and Ewane M.A, "Some quality parameters of crude palm oil from major markets of Douala, Cameroon", *African Journal of Food Science*, Vol. 7(12). 473-478. 2013.

[19] Ohimain E.I., S.C. Izah, Abah S.O, "Air quality impacts of smallholder oil palm processing in Nigeria", *J. Environ. Protection*, 4. 83-98. 2013.

[20] Ohimain E.I., A.A. Oyedeji and Izah S.C, "Employment effects of smallholder oil palm processing plants in Elele, Rivers State, Nigeria", *Int. J. Appl. Res.* Technol,1(6). 83-93. 2012.

[21] Okechalu J.N., M.M. Dashen, P.M. Lar, B. Okechalu and Gushop T, "Microbiological quality and chemical characteristics of palm oil sold within Jos Metropolis, Plateau State, Nigeria", *Journal of Microbiology and Biotechnology Research*, 1(2). 107-112. 2011.

[22] Pearson D, "*The Chemical Analysis of Foods*", Churchill Living Stone, Edinburgh, London. pp 121-150. 1976.

[23] Rauken M.D and R.C. Kill, "*Fats and fatty foods*", In Food industries Manual, 23rd Edn.p.14. 1993.

[24] Rossel J.B, "*Measurement of rancidity*", In Rancidity in foods. Ed. By Allen JC and Hamilton RJ, UK. Aspen publishers. 22-51. 1990.

[25] Siew W.L., T.S Tang, Tan Y.A, "*Methods of Test for Palm Oil and Palm Oil Products*", PORIM Test Methods. 1995.

[26] Sivasothy K., R. Mohd Halim, Basiron Y, "A New System For Continuous Sterilization of Oil Palm Fresh Fruit Bunches", *Journal of Oil Palm Research*, 17. 145-151. 2005.

[27] Tagoe, S.M.A., M.J. Dickinson, Apetorgbor M.M, "Factors Influencing Quality of Palm Oil Produced at Cottage Industry Level in Ghana", *International Food Research Journal*, 19(1). 271-278. 2012.

Modeling and Evaluation of Thermal Diffusivity and Activation Energy of Potato slices in Forced Convection Multi Tray Solar Dryer

Srivastava A.K[1,*], Shukla S.K[2], Singh U.K[3]

[1]Faculty, Mechanical Engineering Department, SRMGPC, Lucknow U.P, India
[2]Centre For Energy Resource and Development (CERD), Mechanical Engineering Department, Indian Institute of Technology (Banaras Hindu University) Varanasi, India
[3]H.O.D Of Mechanical Engineering Department, K.N.I.T. Sultanpur
*Corresponding author: ak_srmcem@rediffmail.com

Abstract In the work being presented, the modeling and analysis of thermal diffusivity and activation energy of high moisture potato slices in multi tray solar dryer has been investigated. Thin layer drying experiments were conducted on potato slices kept in multi tray cabinet dryer with PCM (Lauric acid) as energy storage system on 14/3/2014 and 24/3/2014 under Indian climatic conditions. The experimental setup consists of a flat plate solar collector with a blower and a drying cabinet with PCM kept in a chamber. The study was performed in the drying temperature ranging from 300.2K to 332K and at 0.4 m/Sec air velocity. Seven common thin layer drying models were fitted to the experimental data and several statistical tools (R^2, RMSE and SSE) were used to adjudge the most appropriate model. Logarithmic model for effective moisture diffusivity and quadratic power model for activation energy were found as the most appropriate models for their computation. Effective diffusivity was found to vary from 1.17exp (-7) to 10.0889exp(-7)m^2/sec and activation energy was found to vary from 0.0525KJ/mole to 4.524742 KJ/mole. These findings are being presented in this paper.

Keywords: *moisture diffusivity. activation energy. moisture ratio, drying time*

1. Introduction

Solar drying is in practice since the time long time for the preservation of food and agriculture crops. This was done particularly by open sun drying under open sky. This process has several disadvantages like spoilage of product due to adverse climatic condition like rain, wind, moist and dust, loss of material due to birds and animals deterioration of the material by decomposition, insects and fungus growth. Also the process is highly labor intensive, time consuming and requires large area. With cultural and industrial development artificial mechanical drying came into practice. This process is highly energy intensive and expensive which ultimately increases product cost. Thus solar drying is the best alternative as a solution of all the drawbacks of natural drying and artificial mechanical drying. Solar dryers are used in agriculture for food and crop drying on the other hand for industrial drying processes dryers can be proved to be most useful device from energy conservation point of view. It not only saves energy but also saves lot of time, occupies less area, improves quality of the product, makes the process more efficient and protects environment also. Solar dryers overcome some of the major disadvantages of classical drying. Solar drying can be used for the entire drying process or for supplementing artificial drying systems, thus reducing the total amount of fuel energy required.

As a result of extensive literature survey, this has been observed that drying characteristics and associated parameters were always the keen interest of researchers and scientists specially in the area of food Science and Technology. Crank, 1975 [1] has derived the general solution of moisture movement of standard geometries such as slab, cylinder and sphere using appropriate boundary condition.

Most of researchers have estimated effective moisture diffusivity by using only first term of general solution (Zoyzas et al.1996 b) assuming a constant thermal diffusivity for the whole drying process. But in actual process, it is rarely constant. Thermal diffusivity of the food stuff varies considerably with moisture content, temperature and special coordinate. Such finding has been reported by Puyate and Lawrence, 2006 [3], and Karthanos et al.(1990) [4].

Alp Akin, Necdet Ozbalta and Ali Gunger, 2009 [5] has represented a report on equilibrium moisture content and equations for fitting sorption isotherms of capsicum annum. Water activity and moisture contents of capsicum annum in different salt solutions have been determined. The sorption isotherms of capsicum annum were determined within the range of 10% to 98 % relative humidity at three different temperature say 30°C, 45°C and 60°C using static gravimetric method. A new model between equilibrium moisture content and relative humidity was developed. Thin layer drying models for sweet potato in tray dryer has been investigated and presented by Olawale, A.S. and S.O.Omole, 2012 [6]. The drying rate for blanched and unblanched sweet potato was modeled by fitting eight models and Page model has been reported as the best for all samples.

Sarsavadia [7] developed a solar assisted forced convection dryer for dehydration of onion slices for the controlled conditions of drying air temperatures and airflow rates similar to those employed in commercial onion dehydration. The dryer was also facilated with recirculation of exhaust air. The total energy required for drying of onion slices increased with increase in airflow rate and decreased with increase in drying air temperature.

The effect of air temperature and airflow rate on the drying kinetics of gelidium Sesquipedable was investigated by Mohmed et al. [8] in convective solar drying.The drying process was conducted by using an indirect forced convection solar dryer. It consisted of a solar air collector, an auxiliary heater, a circulation fan and a drying cabinet. Drying was conducted at 40,50 and 60°C and thin layer convective solar drying of Gelidium Sesquipedable was investigated. From the obtained results it can be concluded that the main factor influencing drying kinetics was drying air temperature.

Ian C. Kemp et al. 2001 [9] has presented a review methods for processing experimental drying data obtained from drying kinetics rigs and pilot- plant trials. Different methods for fitting and smoothing drying curves are compared, aiming to generate curves that are usable in industrial design without losing vital information by over smoothing.

Abano E.E, Ma. H and W. Qu, 2011, [10] has investigated the influence of air temperature on the drying kinetics and quality of tomato slices. In particular, the effect of hot air temperature on the lycopene content, non-enzymatic browning, color and flavor changes during drying at various temperature in the range of 50-80°C was investigated. Drying time was considerably reduced from 1140 min to 540 min as the temperature increased from 50-80°C and lycopene levels of fresh tomato was observed fairly increasing.

Abdel basset Bessadok Jemai, 2013 [11] has investigated the drying kinetics of high water content materials such as agro food particles. Comparative study and analysis of agro food particles exhibiting non – Fickian mass transfer properties has been presented.

Therefore the aim of the work being presented may be given as-

i) To model moisture diffusivity and activation energy for high moisture potato crop.

ii) To find out best fit curve for moisture diffusivity and activation energy of potato crop.

iii)To calculate moisture diffusivity and activation energy for potato crop kept in three trays.

2. Materials and Methods

The potato purchased from a local market nearby I.I.T.B.H.U Varanasi, after cleaning were cut manually into slices of diameter 2.5cm and thickness 2.5mm. The samples were spread evenly in a single layer on three similar stainless steel wire mesh trays and were kept inside the drying chamber of forced convection solar dryer. Solar drying experiments were carried out on a clear sunny day.

2.1. Measurements

The tests on experimental setup of forced convection solar dryer were conducted for several days on the roof of Renewable Energy Laboratory, Mechanical Engineering Department, I.I.T BHU Varanasi during the month of March 2014 and observations recorded on 24/03/2014 have been taken for analysis purpose. During experiment, climatic parameters such as ambient air temperature, relative humidity of atmospheric air, wind speed and total solar radiation incident on collector surface were measured. Solar radiation was measured by the solarimeter. A solarimeter is a pyranometer type of measuring device which measures combined direct and diffused solar radiation.

Wind speed was measured using the anemometer with the range of 0.3–75 m/s. The accuracy of the anemometer is +2% + 0.2 m/s. The wind speed was always below 1 m/s and was recognized as small. Therefore the effect of wind speed and its direction was neglected.

The temperature of the air at the inlet and outlet of solar collector, at the exit of drying chamber and crop temperatures at each three respective trays were measured by calibrated copper–constantan thermocouples which have low-cost, acceptable accuracy and rapid response. The location of the thermocouples in the collector passage allowed the determination of temperature of entering and leaving air from collector. Thermocouple beads were put in the cylindrical passage 10 cm bellow. Atmospheric humidity and humidity inside the drying chamber were measured by humidity meter.

The sample weight loss was measured at regular intervals of time, using a precision Goldtech Brand electronic weighing machine maximum capacity 20kg and minimum 0.40gm. Moisture content (dry basis) was calculated from weight loss data and dry solid weight of the samples. Continuous drying was continued till the product was dried from initial moisture content of 85% to finally 12%. Moisture content of the samples based on wet basis have been measured by moisture measuring instrument. The dryer contributed to reduction in drying time in comparison with open sun drying.

The tests were conducted between 10:00 and 17:00 solar time. Hourly data recorded have been presented in Figure 1 to Figure 6.

3. Moisture Ratio

The moisture ratio M.R (kg of water/ kg of dry matter) was determined using the following equation:

$$M.R=(M-M_e)(M_i-M_e) \quad (1)$$

Where M.R = moisture ratio, M = moisture content at any specified time (% dry basis).

M_i = Initial moisture content (% dry basis), M_e = Equilibrium moisture content (% dry basis)

Literature reveals that Guggenheim-Anderson-de Boer (GAB) equation is considered as the most versatile model capable of application to situation over a wide range of water activity ($0.1 < aw < 0.9$) and to various materials. The GAB equation is probably the most suitable for process analysis and design of drying because of its reliability, it's simple mathematical form and its wider use. Parameter values of the GAB equation for potato crop may be given as-

Equilibrium moisture content

$$X = (b_0 b_1 b_2 a_w) / [(1-b_1 a_w).(1-b_1 a_w + b_1 b_2 a_w)] \quad (2)$$

where for potato crop, $b_0 = 8.7$, $b_1 = b_{10} \exp(b_{11} / RT)$

$$b_2 = b_{20} \exp(b_{21} / RT), b_{10} = 1.86 \exp(-5),$$
$$b_{11} = 34, b_{20} = 5.68, b_{21} = 6.75$$

and Water activity

$$a_w = p / p_w = R_{Heq} / 100 \quad (3)$$

During drying of potato slices, values of equilibrium moisture content as per GAB EQUATION were relatively small as compared to M and M_i. Therefore the equation No.(1) may be further modified as per (Doymaz. 2004) and may be given below-

$$M.R. = M / M_i \quad (4)$$

3.1.1 Effective Moisture Diffusivity and Activation Energy

The term effective moisture diffusivity may be defined as the rate of moisture movement without any consideration of type of mechanism involved. A complete drying profile consists of the first stage of drying, a constant rate period, and a falling rate period. In most applications the dominating stage is third period. It is frequently agreed that the mechanism of moisture within a hygroscopic solid during the falling rate period could be represented by one dimensional diffusion equation as a good approximation for most practical systems. Effective moisture diffusivity for potato slices kept in all three trays were computed using Fick's Equation of diffusion expressed as below-

$$\frac{\partial X}{\partial t} = \nabla.(Deff.\nabla X) \quad (5)$$

where x = moisture content (kg of water/kg of dry matter).

Following assumptions were made for getting the solution of above partial differential equation –

(i) the shape of material remains uniform during the drying.

(ii) thermal equilibrium exists at the interfacing of material surface and drying air.

(iii) the initial moisture content is uniform throughout the bulk of food stuff.

(iv) external mass transfer resistance has been assumed as negligible.

The solution of above differential equation as proposed by Crank(1975) may be given as-

$$M.R = \frac{M - Me}{Mi - Me} = \frac{8}{\pi 2} \sum_{n=1}^{\infty} \frac{1}{(2n-1)2} \exp\left\{-D_{eff} (2n-1)^2 \pi^2 t / 4 L^2\right\} \quad (6)$$

Drying time of potato slices being relatively very high, the first term of above equation as per Kingsly et al. (2007) may be given as-

$$\ln(MR) = \ln\left\{\frac{M - Me}{Mi - Me}\right\} = \ln(8/\pi^2) - \{-D_{eff} \pi^2 t / 4L^2\} \quad (7)$$

MR can be computed as given below-

$$MR = (8/2\pi) \exp\{-D_{eff} \pi^2 t / 4L^2\} \quad (8)$$

Effective moisture diffusivity may be computed from the expression given below-

$$\text{Slope of } \ln(MR) \text{ verses time curve} = \frac{\pi 2 Deff}{4L2} \quad (9)$$

According to (Babalis and Belessiotis.2004) as Arrhenius type of equation can be used for the computation of activation energy

$$D_{eff} = D_0 \exp(-E_a / RT) \quad (10)$$

$$\ln(D_{eff}) = \ln(D_0) - \left\{\frac{Ea}{RT}\right\}\left\{\frac{1}{T}\right\} \quad (11)$$

It is evident from equation No.11 that ln (D_{eff}) has been plotted against (1/ T) and slope of the curve i.e{ $\frac{Ea}{RT}$ } is the value of activation energy for the food stuff. Hence activation energy may be computed as-

$$\text{Slope of } \ln(D_{eff}) \text{ v/s } 1/T \text{ curve} = E_a / R \quad (12)$$

4. Experimental Observations

Experimental observations taken on 24/03/13 has been represented in graphical form from Figure 1 to Figure 6.

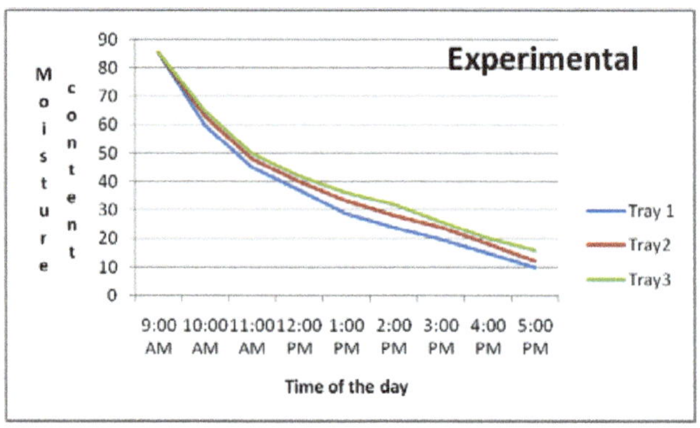

Figure 1. Moisture content of potato slices in tray1, tray2 and tray3 respectively

5. Results and Discussions

Moisture ratio on dry basis for each tray was computed by using experimental data presented in Figure 1 and formula (4). Equilibrium moisture content for potato slices of all three trays were computed by GAB Equation (2). The value of EMC being very small for each trays has been neglected in the formula (1) which was further modified and used as formula(4).

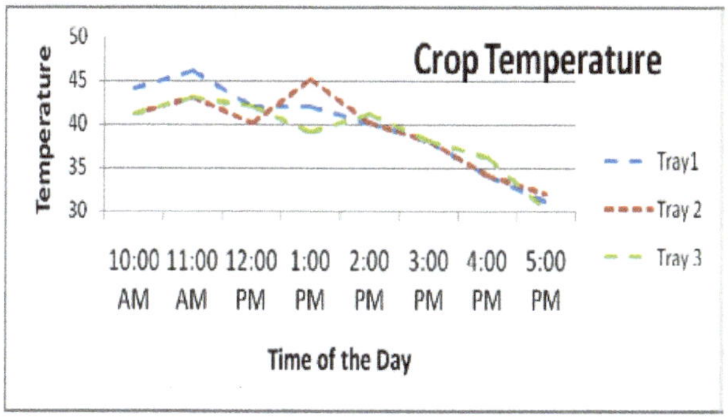

Figure 2 Crop Temperature of potato slices in tray1, tray2 and tray3 respectively

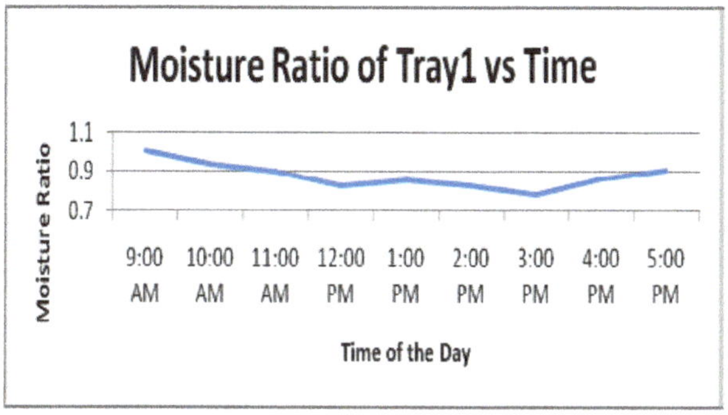

Figure 3. Moisture ratio of potato slices in tray1 (dry basis)

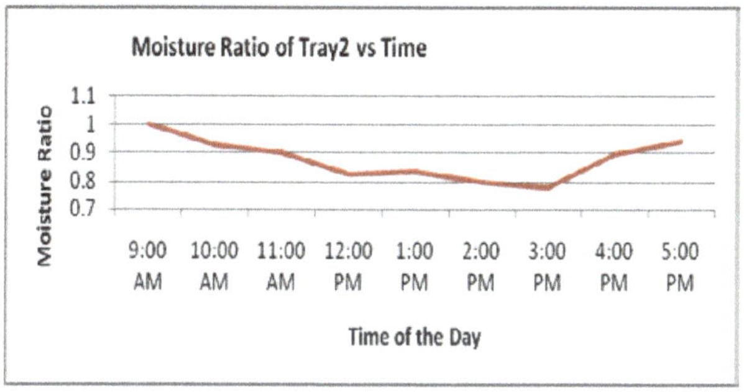

Figure 4 Moisture ratio (db) of potato slices in tray2

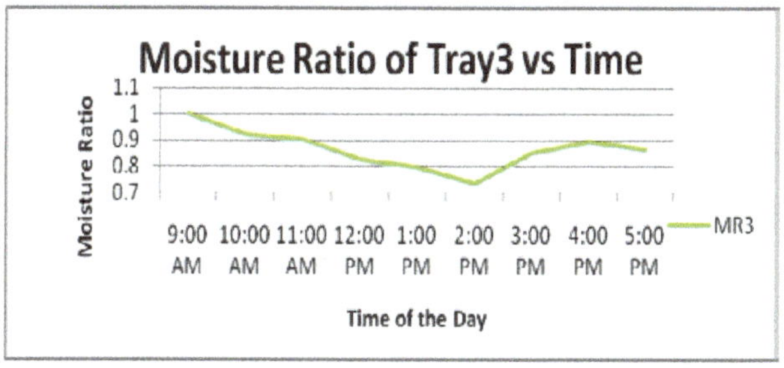

Figure 5 Moisture ratio (db) of potato slices in tray3

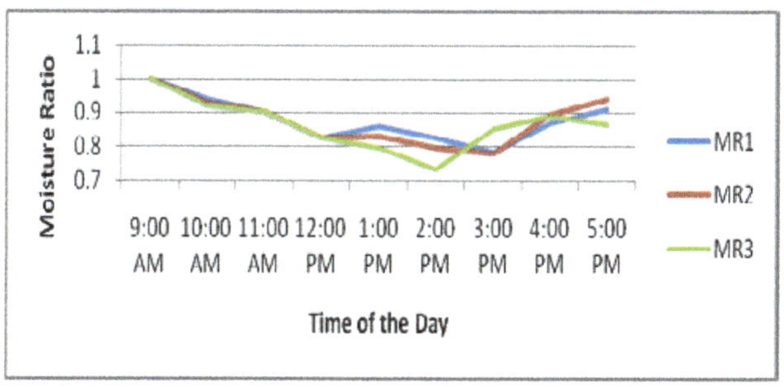

Figure 6. Moisture ratio (db) of potato slices in tray1, tray2 and tray3 respectively

Moisture ratio on dry basis for each tray has been computed and represented in Figure 3 to Figure 5. Further modeling. analysis and result discussion of moisture diffusivity and activation energy tray wise is being presented in 5.1 and 5.2 respectively.

5.1. Moisture Diffusivity Values and Models:

Best fit curves for ln(MR) verses time for each trays have been plotted and represented in Figure 7 to Figure 9 to predict the best possible thin layer drying behavior of potato slices. The logarithmic model was applied to best fit the ln(MR) verses time curve. The best fit logarithmic model with statistical fit parameters for each trays are given below-

Tray1: $\ln(MR) = -1.6101 \log(t) - 0.1107$
$R^2 = 0.9937$, RMSE = 0.09651

Tray1: $\ln(MR) = -1.497 \log(t) - 0.09579$
$R^2 = 0.9937$, RMSE = 0.09086

Tray1: $\ln(MR) = -1.432 \log(t) - 0.07983$
$R^2 = 0.9923$, RMSE = 0.09569

MATLAB 7.9 was used and slopes at each points of ln(MR) curve for each trays has been determined. Moisture diffusivity for potato slices of each trays were calculated by the formula (9) and results have been tabulated in Table 1.

Table 1. Result Table of Moisture Diffusivity (D_{eff}) and Activation Energy (E_a) for potato slices of Tray1, Tray2 and Tray3

Diffusivity for Tray1 (m2/s)	Diffusivity for Tray2 (m2/s)	Diffusivity for Tray3 (m2/s)	Ea (Tray1)	Ea (Tray2)	Ea (Tray3)
1.0088E-6	9.4340E-7	9.0286E-7	4.524734	4.524742	4.524734
5.0444E-7	4.7170E-7	4.5143E-7	3.885854	3.885863	3.885854
3.3629E-7	3.1446E-7	3.0095E-7	3.246975	3.246984	3.246975
2.5222E-7	2.3585E-7	2.2571E-7	2.608096	2.608096	2.608096
2.0177E-7	1.8868E-7	1.8057E-7	1.969217	1.969217	1.969217
1.6814E-7	1.5723E-7	1.5047E-7	1.330346	1.330338	1.330338
1.4412E-7	1.3477E-7	1.2898E-7	0.691464	0.691455	0.691461
1.2611E-7	1.1792E-7	1.1285E-7	0.052586	0.052573	0.052584

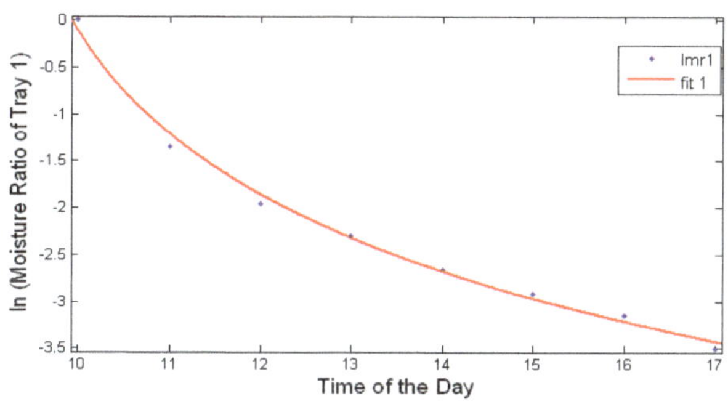

Figure 7. L_n(MR) verses Time curve For Potato Drying in Tray1

This study reveals that moisture diffusivity for tray1 varies from 1.2611exp (-7) m^2/ sec to 10.0889 exp(-7) m^2/ sec, for tray2. it varies from 1.1792 exp(- 7) m^2/ sec to 9.4306 exp(- 7) m^2/ sec and for tray3, D_{eff} varies from 1.12858exp(-7) m^2/ sec to 9.0286 exp (-7) m^2/sec. It can be observed from the results that the most effective factor which greatly affects the moisture diffusivity values in high moisture potato slices is the drying air temperature. Increase in air temperature caused increase in D_{eff} values which is evident from the result Table 1. Moisture diffusivity for tray2 and tray3 is observed in decreasing trend as compared to tray1 because air temperature gradually decreases for tray2 and tray3. Similar results regarding the effect of drying air temperature on moisture diffusivity during forced convection air drying have been reported for carrot slices (Aghbasshlo et al. 2009) and apricots (Doymaz et al. 2009).

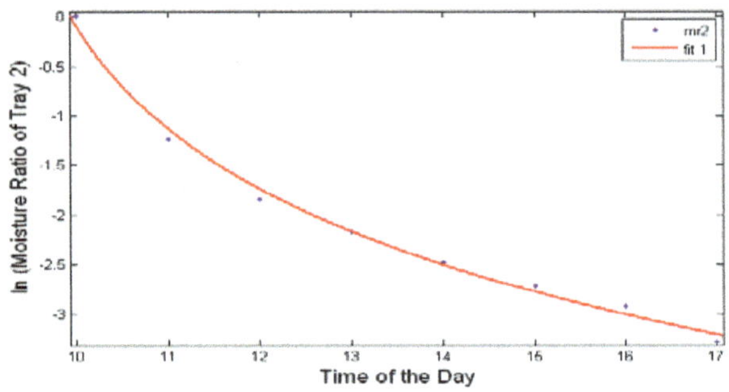

Figure 8. L_n(MR) verses Time curve For Potato Drying in Tray2

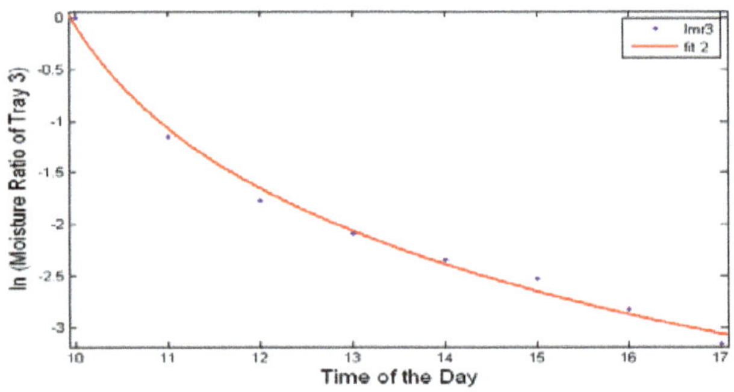

Figure 9. L_n(MR) verses Time curve For Potato Drying in Tray3

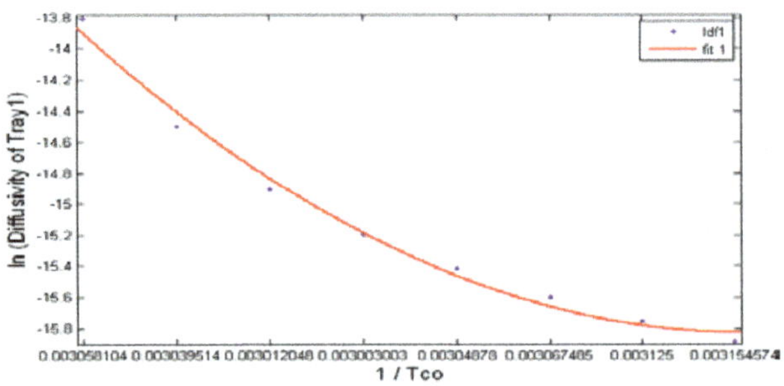

Figure 10. L_n(D_{eff}) against 1/T for thin- layer drying of high moisture potato slices Of Tray1

5.2. Activation Energy Values and Models

As a first step towards the computation of activation of activation Energy (E_a), ln(D_{eff}) vrses 1/T curves were plotted for each trays and reprsented in Figure 10, Figure 11 and Figure 12 respectively. A quadratic model was applied to best fit the curves of drying potato chips of each trays. The best fit quadratic power models for ln(Deff) v/s 1/T curve with statistical fit parameters as obtained for each trays may be given as-

Tray1. f(x) = $0.03842\ x^2 - 0.6211\ x - 13.32$
$R^2 = 0.9905$, RMSE = 0.0811

Tray2. f(x) = $0.03842\ x^2 - 0.6211\ x - 13.38$
$R^2 = 0.9905$, RMSE = 0.08114

Tray3. f(x) = $0.03842\ x^2 - 0.6211\ x - 13.43$
$R^2 = 0.9905$, RMSE = 0.08114

The most general best fit power model which describes the drying kinetics of potato may be expressed as f(x) = $0.03842\ x^2 - 0.6211x - 13.3$, $R^2 = 0.9905$.

As second step after getting best fit model, slopes at each points of ln(D_{eff}) curve for each trays have been determined and by using formula (12), activation energy (E_a) for potato slices of each trays has been computed and the result has been represented in Table 1.

Result Table 1 indicated the maximum and minimum value of activation energy as 4.524742 KJ/ mole and 0.052573 KJ/ mole.

Activation Energy of high moisture potato slices was relatively low as compared with other agricultural products. This is due to high moisture content, tissue of potato slices and starchy structure of potato slices. According to (Demirel and Turhan, 2003) less Activation Energy have been reported for banana slices as a result of high air temperature.

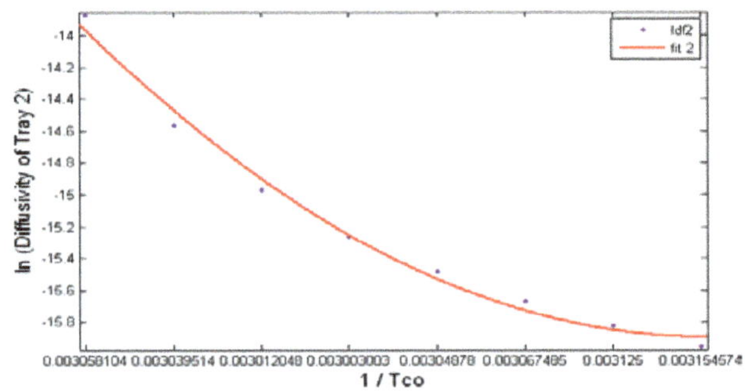

Figure 11. L_n(D_{eff}) against 1/T for thin- layer drying of high moisture potato slices of Tray2.

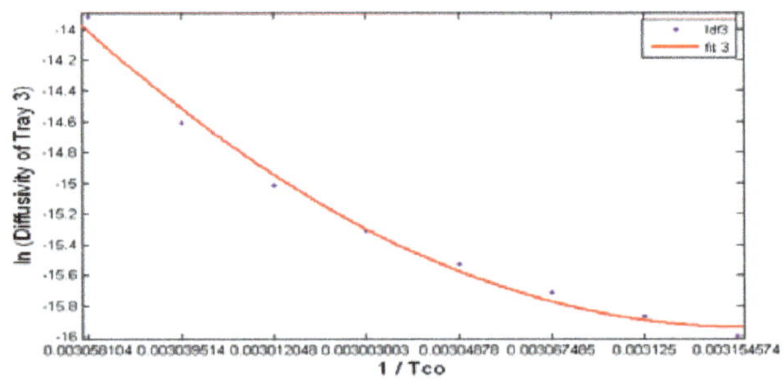

Figure 12. L_n(D_{eff}) against 1/T for thin- layer drying of high moisture potato slices of Tray3

6. Conclusion

Based on above studies and experimental results following conclusions can be made-

1) Logarithmic model has been found as the most appropriate mathematical model which may be used for the prediction of moisture diffusivity of potato slices of each tray in Indian climatic conditions.

2) Moisture diffusivity of tray1, tray2 and tray3 has been observed in decreasing trend. The result obtained is consistent and convincing as well with the decreasing trend of crop temperature of each trays and increasing

trend of moisture evaporation rate of each trays respectively.

3) Moisture diffusivity of each trays increases vertically in similar manner with respect to drying time.

4) Quadratic power model has been adjudged as the most appropriate mathematical model for predicting the activation energy of potato crops in Indian climatic conditions.

5) Experimental results indicate the maximum and minimum value of activation energy as 4.524742 KJ/ mole and 0.052573 KJ/ mole. Almost same average value of activation energy for potato slices of each tray has been found.

Nomenclature

D_{eff}	effective moisture diffusivity (m^2/ s)
t	drying time
$M.R$	moisture ratio (decimal)
M	moisture content at any specified time (dry basis)
M_e	equilibrium moisture content (dry basis)
M_i	initial moisture content (dry basis)
n	no. of terms taken
L	average half thickness of potato slice (m)
E_a	activation energy(KJ mole)
T	absolute temperature of air (K)
R	universal gas constant (8.314 Kj/mole K)
D_o	pre exponential factor.

Acknowledgement

The corresponding author hereby want to acknowledge the technical assistance provided by Renewable Energy Laboratory, Department Of Mechanical Engineering. IIT (BHU) Varanasi, U.P India for granting permission for the use of experimental set up along with all required tools, equipments and technical staff.

References

[1] Crank, J. 1975. *"The Mathematics Of Diffusion."* Vol. 2nd. London, U.K : Clarendon Press, Oxford.

[2] Zogzas, N.P., Maroulis, Z.B and Marinos- Kouris, D. 1996b. *"Effective Moisture Diffusivity Estimation From Drying Data. A comparison Between various Methods Of Analysis."* Drying Technology 14 (10): 2225-2253.

[3] Puyate, Y.T. and Lawrence, C.J.2006."For The Falling- Rate Period: *A Missing link At Moderate Drying intensity."* Chemical Engineering Science 61 (21) : 7177- 7183Sherwood's Models.

[4] Karthanos, V.T., Villalobos, G. and Saravacos, G.D. 1990. *"Comparison Of Two Methods Of Estimation Of The Effective Moisture Diffusivity From Drying Data."* Journal Of Food Sciences 55 (1): 218 – 231.

[5] Alp Akin, Necdet Ozbalta and Ali Gungor, 2009, *"Equilibrium Moisture Content And Equations For Fitting Sorption Isotherms Of Capsicum Annum."* GIDA (2009) 34 (4): 205- 211.

[6] Olawale, A.S and S.O. Omole, 2012, *"Thin layer drying models for sweet potato in Tray Dryer,"* Agric Eng Int: CIGR Journal, 14(2) : Manuscript No, 2060.

[7] Sarsavadia P.N, 2007, *"Development Of A Solar Assisted Dryer And Evaluation Of Energy Requirement For The Drying Of Onion."* Renewable Energy, 2007: 32:2529-47.

[8] Mohmed L.A, Kane CSE, Kouhila M, Jamali A, Mahrouz M and Kecharaou. N 2008," *Thin Layer Modeling of Gelidium Sesquipedale Solar Drying Process."* Energy Conversion Management, 2008: 49: 940-6.

[9] Ian C. Kemp, et al.2001, *" Methods For Processing Experimental Drying Data."* Drying Technology, 19(1), 15-34(2001).

[10] Abano E.E, Ma.H and W. Qu, 2011," *Influence Of Air Temperature On Drying Kinetics And quality Of Tomato Slices."* J.Food Process Technology 2: 123.

[11] Abdel basset Bessadok Jemai, 2013," *Characterizing The Drying Kinetics Of Water Content Agro Food Particles Exhibiting Non-Fickian Mass Transfer."* Chemical Engineering Transactions Vol 32, 2013.

[12] *Gupta* Akanksha, Shukla S.K. and Srivastava A.K. *Analysis of solar drying unit with phase change material storage systems* Int. J. Agile Systems and Management, 2013; 6(2): 164-174.

[13] S.V.Jangam, C.L.Law and A.S.Majumdar. 2010. *"Drying Of Foods, Vegetables and Fruits."* Volume 1.

[14] Shyam S. Sablani and Arun S. Majumdar," *Drying Of Potato, Sweet Potato and Other Roots."* C 2006 By Taylor & Francis.

[15] Dimitris Marinos- Kouris and Z.B.Maroulis. 2006. *"Transport Properties In The Drying Of Solids."* C 2006 By Taylor & Francis Group. LLC.

Some Quality and Micro-structural Characteristics of Soup Enriched with Debittered *Moringa Oleifera* Seeds Flour

Radha C.[1], Ogunsina B.S.[2,*], Hebina Babu K.T.[1]

[1]Department of Protein Chemistry & Technology, Central Food Technological Research Institute, Mysore, India
[2]Department of Agricultural & Environmental Engineering, Obafemi Awolowo University, Ile Ife, Nigeria
*Corresponding author: babatundeogunsina@gmail.com

Abstract Soup enriched with debittered moringa seeds flour (DBMS) was investigated for its quality and micro-structural characteristics following appropriate standard procedures reported in literatures. The results showed that soup sample enriched with 0.5% of DBMS had the highest overall sensory score for acceptability and its protein content was 4 g/100 g more than that of the control. Scanning electron micrographs showed soup sample enriched with DBMS as a dense, continuous agglomerated protein strands of irregular sizes and shapes. As the addition of DBMS increased, the viscosity and the body of the soup increased; the lightness "L" colour value of the soup reduced and the total colour difference (ΔE) increased. Enrichment of soup with DBMS promotes the use of moringa seeds as a protein ingredient; especially in the tropics where it is fast gaining acceptance as human food.

Keywords: *soup, debittered moringa seeds, microstructure, sensory properties*

1. Introduction

Awareness regarding the consumption of moringa products as food and nutriceutical keeps spreading in the tropics [1,2,3]. India leads the world's moringa seeds production with about 1.3 million metric tonnes produced over 380 km² land area. Major producing areas are in Andhra Pradesh, Karnataka and Tamil Nadu States [4]. Virtually all parts of moringa plant, from the leaves and flowers to the immature pods and kernels have been found useful for different industrial, medicinal and food purposes. In India, Thailand and the Philippines, moringa tree is commonly grown in home gardens and the leaves and immature pods are sold in the local markets as soup ingredient [5]. The oldest known use of moringa seeds flour is as substitute for aluminium sulphate (alum) in domestic water clarification [6]. The leaves have been explored as a feedstock for biogas production, organic manure, bio-pesticide and livestock feeds [5,6,7]. Moringa leaves and seeds are particularly rich in tocopherols, ß-carotene, proteins, vitamins, minerals and essential sulfur amino acids which are rarely found in daily diets [5,8,9]. On the other hand, Ferreira *et al.* [10] associated the long lasting bitter and sugary taste of moringa seeds to the presence of alkaloids, saponins, cyanogenic glucosides and glucosinolates. On this note, Ogunsina and Radha [11] successfully documented a non-chemical heat-assisted process for debittering and detoxifying moringa seeds and compared the functional properties of defatted and debittered moringa seeds flour with that of defatted soybean flour. The study revealed that debittered moringa seeds flour shows higher emulsification, foaming and gelation properties than soybean. Oliveira *et al.* [9] and Vasconcelos *et al.* [12] also found that the lipid content of moringa seed is higher than that of soybean and the oleic acid-rich moringa seed oil compares favourably with olive oil in terms of thermal, frying and oxidative stability [13]. In another baseline study, Ogunsina [14] found that debittered and roasted moringa kernels have a peanut-like aroma and taste. An investigation of the physicochemical and functional properties of full fat and defatted moringa kernels flour showed that it is a good source of vegetable protein for functional food products [8]. Consequently, Ogunsina *et al.* [15] studied the quality characteristics of bread and cookies enriched with debittered *Moringa oliefera* seeds flour (DBMS). It was found that bread with 10% DBMS flour and cookies with 20% DBMS grits had more protein, iron and calcium. It was submitted that incorporating moringa seeds in baked food products is a means of boosting nutrition in Africa and Asia where moringa is largely grown and malnutrition is prevalent [15]. Based on the foregoing, it is expected that future demand for moringa products will continue to increase in many regions of the world [7].

The preparation and consumption of soup as human food accompaniment must have begun around the time cooking of foods was invented [16]. A soup is typically prepared as a concoction of ingredients including pepper, tomatoes, onions, meat/fish, seasonings, spices and vegetables combined with other liquids or slurries; cooked together until an acceptable consistency, blend and desired taste are achieved. Sometimes, grain cereal or oilseed flours, potatoes and other additives are added as thickeners to improve the soup body; depending on the desired end product. Although, soup is a choice delicacy virtually everywhere; the recipe varies from place to place; depending on regional/traditional food habits, taste and local availability of specific ingredients [17]. Generally, the process of preparing a delicious soup is time consuming, laborious and painstaking; hence, semi-processed soup mix/powders have gained worldwide recognition in fast cookery. Being a liquid-food (usually served warm), a good soup powder should be re-hydratable and cookable within minimum time, retaining nutrients and palatability almost similar to the freshly cooked product [18].

New trends in food science now makes soup available in pre-processed form; as canned or dehydrated products permitting the inclusion of medicinal herbs, spices and adjustment of nutrients to meet specific dietary needs [17,19]. Soup mixes have been developed for a number of food products including: dika kernels; fish, moringa leaves; vegetables and mushroom [7,16,18,20,21,22,23]. In addition, groundnut, soybean, cowpea and crayfish soup powder are popularly eaten in Nigeria and other countries in the sub-Sahara region. In Asia and Middle East, cashew, almonds and pitachio are popularly used as food garnish, but hitherto there is no such information regarding moringa seeds. Given the growing awareness about the nutritional benefits of moringa seeds, this work investigates the microstructural and some quality characteristics of soup enriched with debittered *Moringa oleifera* seeds flour.

2. Materials and Methods

2.1. Sample Preparation

Dry *Moringa oleifera* seeds, Onion, carrot, ginger, garlic, green chilly, coriander leaves, pepper powder, seasoning and sodium chloride (Merck Co., Mumbai India) were purchased from Mysore market, Karnataka, India. The soup recipe was in accordance with a traditional method of curry preparation in Mysore, Karnataka, India. *Moringa oleifera* seeds were defatted, debittered, dried and powdered according to the method of Ogunsina and Radha [11] to obtain debittered moringa seed flour (DBMS). The ingredients: carrot, 14 g; onion, 24 g; green chilly, 2 g; coriander leaves, 6 g; powder pepper, 1 g; ginger, 1 g and garlic, 1.6 g were washed and chopped into small pieces and allowed to season for 2-3 mins. Debittered moringa seeds flour varied at 0, 0.5 and 1.0% w/w levels was added as pre-gelatinized solution and boiling continued for 5-7 mins until the desired consistency was reached. The soup samples were garnished with additional chopped coriander leaves and about 5 ml of vinegar was added to the preparation and it was stirred properly. The soup samples were dried at 50-55°C for 2 days in a cabinet drier; and afterwards, powdered by dry-milling in a kitchen blender to obtain instant soup mix [24]. Each soup mix sample was reconstituted in boiling water (1:100 w/v) and allowed to simmer for about 2 mins. The entire procedure may be summarized as shown in Figure 1.

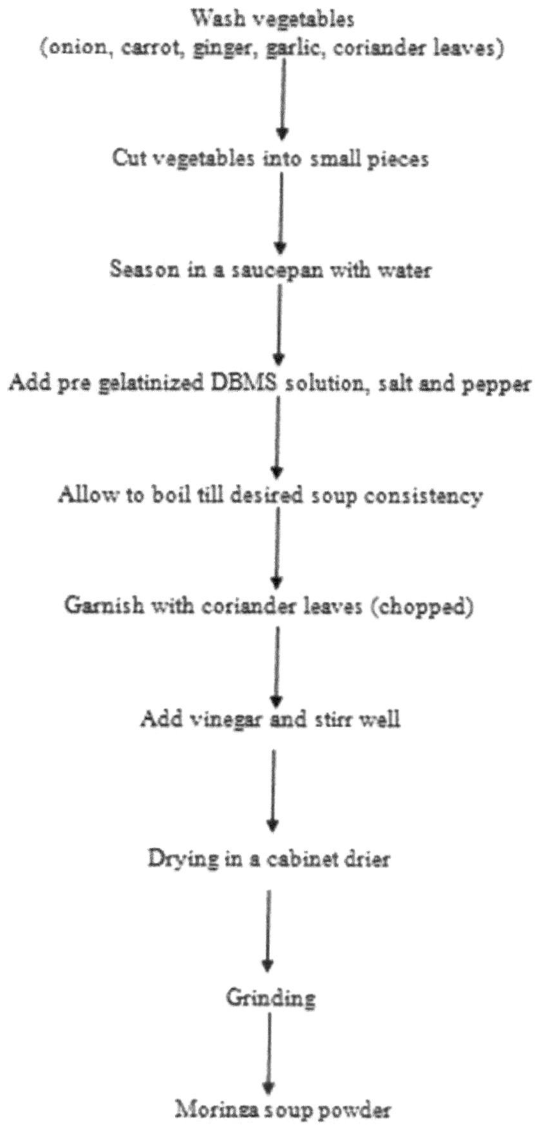

Figure 1. Procedure for preparing soup enriched with debittered moringa seeds flour

2.2. Protein Estimation

Percent nitrogen was estimated by micro-Kjeldhal method using an automated nitrogen distiller and protein content was calculated by multiplying the nitrogen value with 6.25. Determinations were carried out in Triplicates.

2.3. Sensory Evaluation

The soup samples were subjected to sensory evaluation by a taste panel of 20 people. The panelists have no prejudice against any specific food product. At the time of evaluation, soup samples were served in glazed china plates labeled with three-digit random codes and panelists were provided with distilled water to clean their mouth between samples. The order of serving of the soup samples was randomized so that bias due to the sequence of presentation was minimized. Twelve volunteers were screened and chosen as panelists based on their ability to identify the six basic tastes: sour, bitter, salty, pepper, onion, garlic and bitter. The panelists gave remarks about each sample; especially, considering colour, appearance, body and overall acceptability. A balanced 10-point hedonic rating was employed for all the attributes with 9–10 denoting "like very much", 7-8 "like", 5-6 "neutral", 3-4 "dislike", and 1-2 indicated "dislike very much." The booth area was under white fluorescent lighting at a temperature and relative humidity of 22±2°C and 50±5% respectively. The typical attributes for the soup samples were earlier identified in a preliminary session to profiling.

2.4. Scanning Electron Microscopy

The microstructural characteristics of the soup mix samples were studied by scanning electron microscopy. Scanning electron micrographs were taken using a scanning electron microscope (LEO 435 VP, Cambridge, UK). Dried and powdered soup samples were coated with gold using poloron coating system E-5000 within 2–3 min. The coating thickness was calculated as 200–300 nm, using the formula: $T = 7.5I\,t$, where I = current (mA), t = time (min), T = thickness (A°). The coated samples were transferred to the microscope where it was viewed at 15 kV and 9.75×10^{-5}torr using secondary electron image. A 35 mm Ricoh Camera (LEO 435 VP, Operator Manual Version V2.04) was used to capture the image.

2.5. Viscosity

Viscosity was determined using a viscometer (Model R1:3:M-3, Rheology International Ltd., Shannon, Ireland). About 80 mL of soup sample was filled into the 100 mL sample adaptor and the ASTM spindle size number 3 attached to the viscometer was immersed into the soup to the marked portion. Viscosity was measured at 100 rpm rotor speed under ambient temperature conditions (26±2 °C) and expressed as m.Pa.S.

2.6. Colour

Colour measurement of flour samples was carried out using a Labscan XE colour meter (Hunter Associates Lab. Inc., Reston, VA). Approximately 100 ml of each soup sample was served in a small glass bowl each with flat glass cover to provide uniform flat surface. The colour parameters measured were lightness (L), redness (+a), greenness (a), yellowness (+b) and blueness (-b). A white tile of known L, a, and b values were used as standard. Total colour difference (ΔE) was estimated using the following equation $\Delta E = (\Delta L^2 + \Delta a^2 + \Delta b^2)^{1/2}$ where ΔL, Δa and Δb were the respective differences between the standard plate.

3. Results and Discussion

The results show that the protein content of DBMS flour was about 45.22 g/100 g and that of the soup increased from 0.3 g/100 g, for the control; to 4 g/100 g, for soup sample S2 which was enriched with 0.5% DBMS. In India, the consumption of vegetable proteins in different food systems is very crucial because the population is largely vegetarian. Hitherto, soups and curries enriched with vegetable proteins are widely accepted as delicacies in different parts of the world and have been a proven way of enriching the diet of a teeming population of malnourished people. Instant soup mixes are popular among busy wives, restaurants and eateries around the world especially because it largely reduces problems associated with bulk handling during processing and the product has longer shelf life than liquid soup preparations. The utilization of a high protein seed such as moringa as a soup ingredient is a way of boosting the nutritional status of the low income group in Asia, Africa, Latin America where moringa has found favourable conditions for growth. The rapidly changing food habits in the tropics and emerging trends in the consumption of underutilized protein seeds complements the global drive towards ensuring that the world's poorest people are no longer malnourished [25].

The scanning electron micrographs of the flour samples are shown in Figure 2. Soup sample enriched with 0.5% DBMS shows a dense continuous agglomerated protein matrices of irregular sizes and shapes, whereas the control shows particles separated by large air vacuoles. Ogunsina and Radha [11] earlier submitted that the hydrothermal process involved in debittering may be associated with the increase in inter-protein bonding between the particles; because processing of proteins by moist heat can cause aggregation of particles. The relationship between structure and function in foods has long been studied by scanning electron microscopy. It is known that the microstructural characteristics of a particular food material are often linked with its texture, rheology, flavor, appearance and stability.

Figure 2. Scanning Electron Micrographs (SEM) of soup samples

LEO 435 VP, Cambridge model scanning electron microscope was used.
(a): SEM Soup (control) at cross section in1000X;
(b): SEM of Soup (control) at cross section in 500X
(c): SEM of Soup incorporated with 0.5% DBMS at cross section in 1000X and
(d): SEM of Soup incorporated with 0.5% DBMS at cross section in 500X

The results of sensory analysis of the soup samples are shown in Figure 3. The soup colours were generally buff, but the shade in S1 was lighter than that of S2 and S3. However, the buff colour increased as the addition of DBMS increased. The body of the soup samples also increased progressively from 6.7 to 8.6 as the incorporation of DBMS increased, making S3 thicker than the other two samples. This behaviour is further substantiated by the viscosity values of 273.5, 311.3 and 407 mPa.S which were obtained for S1, S2 and S3 respectively. It shows a progressive increase in viscosity as the incorporation of DBMS powder increased. Though the viscosity of a soup is often associated with the amount and quality of ingredients (with or without thickener), it measures the internal friction of the soup or its tendency to resist flow. The relationship between viscosity and shear rate is used for classifying foods as newtonian, non-newtonian, pseudoplastic or thixotropic. This finds relevance in food processing, quality control and sensory studies (26, 27); although organoleptic properties are largely a matter of personal perception and cannot be properly determined with analytical instruments.

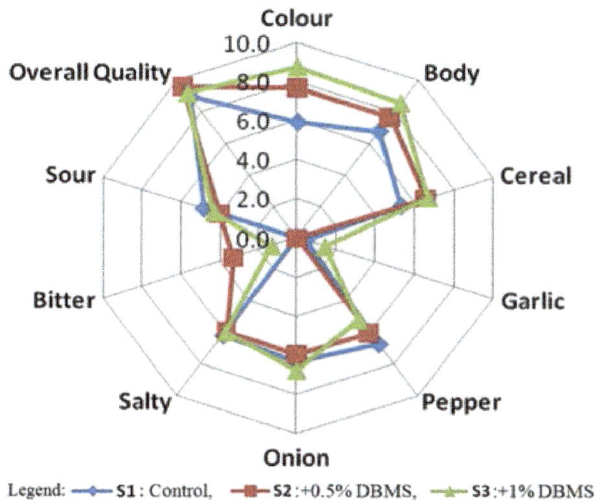

Figure 3. Sensory profile of soup samples

The three soup samples show an almost equal score for sourness and saltiness; although the score for S2 was lower and differs significantly (p≤0.05) from that of S1 and S3. It was further observed that the soup samples became less peppery as the incorporation of DBMS increased, showing soup sample S3 as the least peppery. Overall acceptability scores were 9, 9.2 and 9.6 for S1, S2 and S3 respectively; showing soup sample S2 as the most acceptable.

The hunter colour values of the soup samples were significantly different (p≤0.05) (Table 1). The total colour values (ΔE) of the samples were observed to increase as the incorporation of DBMS increased; showing significant differences (p<0.05). The lightness (L value) of soup S1 differs significantly (52.74) from that of S2 (45.73) and S3 (42.58). Higher "L" value observed for S1 indicates that it is lighter than the other two samples. This was also reflected in the sensory score which indicated a lighter shade of buff color for S1 than the other two samples. The value "a" which indicates redness was indicated for S2 and S3, whereas S1 showed a trace of greenness (-a). The value "b" which indicates yellowness was however found to be more for S2 and S3 compared to S1.

Table 1. Effect of incorporation of DBMS on the colour of moringa soup

Sample	L	a	b	ΔE
S1	52.74^{c}	-0.32^{a}	12.31^{a}	45.78^{a}
S2	45.73^{b}	3.69^{c}	18.12^{c}	53.06^{b}
S3	42.58^{a}	1.9^{b}	15.78^{b}	55.59^{c}

S1-Control; S2-0.5% of DBMS added; S3-1% of DBMS added
Values on the same column followed by different superscript letters are significantly different (p≤0.05). The colour parameters measured were lightness (L), redness (+a), greenness (a), yellowness (+b) and blueness (-b). A white tile of known L, a, and b values was used as standard. Total colour difference was estimated as: $\Delta E = (\Delta L^2 + \Delta a^2 + \Delta b^2)^{1/2}$ where ΔL, Δa and Δb were the respective differences between the standard plates.

4. Conclusions

Incorporating DBMS affects the quality and microstructural characteristics of moringa soup. The best overall acceptability was obtained at 0.5% optimal level of incorporation. Soup enriched with DBMS flour soup is richer in protein. Preparation of a DBMS enriched soup mix with acceptable sensory attributes is a means of promoting the consumption of moringa seeds as a valuable but underutilized protein resource.

Acknowledgements

Dr B.S. Ogunsina acknowledges the Fellowship of United Nations University from which interest to work on *Moringa oleifera* seeds emanated and the entire Management and Staff of Central Food Technological Research Institute (CFTRI), Mysore, India where the work was carried out.

References

[1] Fahey. J.W. (2005). *Moringa oleifera*: A review of the medical evidence for its nutritional, therapeutic and prophylactic properties. Part 1. Trees for Life Journal. http://www.tfljournal.org/article.php/20051201124931586. Accessed on January 4, 2014.

[2] Mehta, L.K., Balaraman, R., Amin, A.H., Bafna, P.A., Gulati, O.D. (2003). Effect of fruits of *Moringa oleifera* on the lipid profile of normal and hypercholesterolaemic rabbits. *Journal Ethnopharmacology,* 86(2-3), 191-195.

[3] Gilani, A.H., Aftab, K., Suria, A., Siddiqui, S., Salem, R., Siddiqui B.S. (1994). Pharmacological studies on hypotensive and spasmolytic activities of pure compounds from *Moringa oleifera. Phytotherapy Research,* 8(2), 87-91.

[4] Rajangam, J., Azahakia-Manavalan, R.S., Thangaraj, T., Vijayakumar, A. and Muthukrishan, N. (2001). Status of Production and Utilisation of Moringa in Southern India; International Conference on Development Potential for Moringa products; October 29th - November 2nd, Dar Es Salaam, Tanzania www.Moringanews.org/actes/rajangam_en.doc. Accessed January 4, 2014.

[5] Foidl, N., Makkar, H.P.S., and Becker, K. (2001). The potential of *Moringa oleifera* for agricultural and industrial uses. In: The Miracle Tree/The Multiple Attributes of Moringa (Ed. Lowell J Fuglie) CTA. USA. http://miracletrees.org/moringa.doc/the_potential_of_moringa_oleifera_for_agricultural_and_industrial_uses.pdf. Accessed on March 11, 2014.

[6] Siddhuraju, P. and Becker, K. (2003). Antioxidant properties of various solvent extracts of total phenolic constituents from three different agro-climatic origins of drumstick tree (*Moringa oleifera* Lam) leaves. *Journal of Agriculture and Food Chemistry,* 51(8), 2144-2155.

[7] Singh, H.P. (2011). Exploring R and D potential of moringa for nutrition and health care. *Indian Horticulture,* 56(1), 3-8.

[8] Ogunsina, B.S; Radha, C and Govardhan Singh, R.S. (2010). Physico-chemical and functional properties of full fat and defatted *Moringa oleifera* kernel flours. *International Journal of Food Science and Technology,* 45, 2433-2439.

[9] Oliveira, J.T.A., Silveira, S.B., Vasconcelos, K.M., Cavada, B.S., Moreira, R.A. (1999). Compositional and nutritional attributes of seeds from the multiple purpose tree - *Moringa oleifera* Lamarck. *Journal Science of Food and Agriculture,* 79(6), 815-20.

[10] Ferreira, P.M.P., Farias, D.F., Oliveira, J.T.A. and Carvalho, A.F.U. (2008). *Moringa oleifera*: Bioactive compounds and nutritional potential. *Revista de Nutrição,* 21(4), 431-437.

[11] Ogunsina, B.S. and Radha, C. (2010). A Comparative study of the functional and physico-chemical properties of debittered Moringa oleifera seeds and soybeans flours. *Ife Journal of Technology,* 19(1), 85-92.

[12] Vasconcelos, I.M., Maia, A.A.B., Siebra, E.A., Oliveira, J.T.A., Carvalho, A.F.F.U. and Melo, V.M.M. (2001). Nutritional study of two Brazilian soybean (*Glycine max*) cultivars differing in the contents of antinutritional and toxic proteins. *Journal of Nutrition and Biochemistry,* 12(1), 55-62.

[13] Ogunsina, B.S., Indira, T.N., Bhatnagar, A.S., Radha, C., Sukumar, D. and Gopalakrishna, A.G. (2014). Quality characteristics and stability of Moringa oleifera seed oil of Indian origin. *Journal of Food Science and Technology*, 51(3), 503-510.

[14] Ogunsina, B.S. (2009). Value addition to drumstick (*Moringa oleifera*) seeds using various adaptable technologies. A report of the United Nations University's Research fellowship Undertaken at Central Food Technological Research Institute, Mysore, India . pp 78.

[15] Ogunsina, B.S., Radha, C. and Indrani, D. (2011). Quality characteristics of bread and cookies enriched with debittered *Moringa oliefera* seed flour. *International Journal of Food Science and Nutrition*, 62(2), 185-194.

[16] Rahman, M. A. Saifullah, M. and Islam, M. N. (2012). Fish powder in instant fish soup mix. *Journal of Bangladesh Agricultural University*, 10(1), 145-148.

[17] Dalby, A. (2003). *Food in the Ancient World from A-Z.* Routledge, London pp307. www.foodtimeline.org/foodsoups.html. Accessed on July 18, 2015.

[18] Abeysinghe, C.P. and Illeperuma, C.K. (2006). Formulation of an MSG (Monosodiul Glutamate) free instant vegetable soup. *Journal of Natural Science Foundation,* 34(2), 91-95.

[19] Toussaint-Samat, M. (1992). *Food in History Barnes and Noble Books.* New York pp177. www.foodtimeline.org/foodsoups.html. Accessed on July 18, 2015

[20] Bamidele, O.P., Ojedokun, O.S. and Fasogbon, B.M. (2015). Physico-chemical properties of instant *ogbono* (*Irvingia gabonensis)* mix powder. *Food Science and Nutrition*, 3(4), 313-318,

[21] Wartha, G.C., Sharangdhar, S.T., Koli, J.M., Metar, S.Y. and Sharangdhar, M.T. (2013). Preparation of fish soup powder from tilapia fish. *Ecology, Environment and Conservation Paper,* 19(2), 425-428.

[22] Chandramouli, P., Divya, V.S., Bharathi; Sivakami, A.; Bharathiraja, B., Jayamuthunagai, J. (2012). Standardisation and nutritional analysis of soup powder prepared from *Moringa oleifera, Solanum trilobatum and Centella asiatica. International Journal of Future Biotechnology,* 1(1), 1-16.

[23] Abdel-Haleem, A.M.H. and Omran, A.A. (2014). Preparation of dried vegetarian soup supplemented with some legumes. *Food and Nutrition Sciences*, 5, 2274-2285.

[24] Warang, M. D., Mulye, V. B., Sapkale, P. H., Bondre, R. D. and Mohite, A. S., (2008). Fish soup powder production from *Protonibea diacanthus. The Asian Journal of Animal Science,* 3(1), 20.

[25] Charles, H., Godfray, J., Beddington, J.R., Crute, I.R., Haddad, L., Lawrence, D., Muir, J.F., Pretty, J., Robinson, S., Thomas, S.M. and Toulmin C. (2010). Food Security: The challenge of feeding 9 billion people. *Science,* 327, 812-818.

[26] Fellows, P. (2000). *Processing Technology Principles and Practice.* 2nd Edition, Woodhead Publishing Limited and CRC Press LLC, Washington DC. Pp 98.

[27] Ibanoglu, S. and Ibanoglu, E. (1998). Rheological characterization of some traditional turkish soups. *Journal of Food Engineering*, 35, 251-256.

22

Diffusion Coefficient Estimation in Shrinking Solids. A Case Study: Tomato

C. Martínez-Vera[1,*], I. Anaya-Sosa[2], M. G. Vizcarra-Mendoza[1]

[1]Departamento de Ingeniería de Procesos e Hidráulica, Universidad Autónoma Metropolitana-Iztapalapa.Apartado Postal 55-534, México D.F., 09340, México

[2]Escuela Nacional de Ciencias Biológicas, Instituto Politécnico Nacional. Unidad Profesional Lázaro Cárdenas, Prolongación de Carpio y Plan de Ayala s/n, Col. Santo Tomas C.P. 11340. Del. Miguel Hidalgo. México, D.F

*Corresponding author: cmv@xanum.uam.mx

Abstract In this work are presented experimental results of the drying kinetics for a solid that shrinks between 88% and 94% in volume during the drying process depending on the drying temperature. The drying process is modeled and simulated as an isothermal diffusional process taking in account the shrinkage of the dried material. Two falling-rate drying periods were considered in the drying process. Diffusion-moisture content relationships were determined for each drying period at each of the temperature levels at which the experiments were conducted. Arrhenius type expressions were obtained for the diffusion-temperature dependency from the diffusion-moisture content profiles averaged at each drying temperature considered in this study for each drying period.

Keywords: *diffusion coefficient estimation, shrinking solids, tomato drying, activation energy*

1. Introduction

Modeling the drying of shrinking solids as a fickian diffusional process presents some challenges. Taking in account the volume reduction increases the ill-conditioning of the non-linear equations that describe the process. The diffusion coefficient determination for cases as the one here considered are an open problem. Sometimes the diffusion coefficients are determined neglecting the shrinking effect (Demiray & Tulek, 2012 [1]). Other times the diffusion coefficients are determined departing from a mathematical solution of the problem not rigorously valid for the case considered: a truncated Fourier series solution in which the shrinking effect is included forcing the solution, and the diffusion coefficient to follow the experimental data (Hawlader, 2002 [2]; Giovanelli et al., 2002 [3]; Resende et al., 2007 [4]). Here we follow another approach: the drying-shrinking process is modeled rigorously taking in account the shrinkage and assuming a diffusion coefficient dependent on moisture content (Crank, 1975 [5]; Crank, 1984 [6]; Aguerre et al., 2008 [7]). The diffusion coefficient to be determined is an effective one that involves all moisture transport mechanisms that could be involved in the drying process. The moving boundary problem is transformed in a fixed boundary problem through a change of variable (Crank, 1975 [5]). The solids chosen as a case study are peeled tomato cubes. Tomato is a foodstuff with high moisture content that when is being dried presents a high volume reduction due to the water lost. A set of experiments were performed for drying of tomato mesocarp cut into cubes isothermically and the evolution of the moisture content was registered at several temperatures. The equilibrium moisture contents at the solid's surface were determined experimentally at each of the temperature levels analyzed in this study. The proper diffusion-moisture content relationship was determined fitting the parameters of the diffusion coefficient model assumed in order to give a good agreement with the experimental results by a trial and error procedure. The experimental and simulated drying kinetics curves and the solid's volume time evolution are presented for drying at 60°C, 70°C and 80°C in a moisture balance.

The main goal of the present study was to calculate the effective moisture diffusivity for the tomato mesocarp not departing from a solution to the diffusion equation no longer valid when the schrinking effect is taken in account but from a numerical solution to that equation when the volume reduction is considered. The activation energy was also evaluated.

2. Materials and Methods

2.1. Experimental

Fresh, mature and ripe red tomatoes (*Lycopersicum Esculentum*, cv. Roma) were purchased from a local market in México City. All samples were chosen as the same size and color. The tomatoes were washed and several longitudinal cuts were done along their surfaces.

The whole tomatoes were dipped in a 0.026% by weight $CaCl_2$ solution at 75°C for 60 s. After this treatment the tomatoes were peeled, the endocarp was eliminated by hand and the mesocarp (the fleshy part of the pericarp) was cut in cubes 1 x 1 x 0.7 cm^3. The initial moisture content of the tomato cubes was determined by the oven drying method, drying the solids in an oven at 95°C for 3 hours and, by gravimetry, the moisture content was found to be between 93 to 96% wb (Rivera, 2007 [8]). The dry mass was determined according to the methodology proposed by Lewicki (2002) [9] consisting in drying a sample in an oven at 95°C for 3 hours. The reported sphericity of cubes is 0.81 (Kunii & Levenspiel, 1991 [10]), which was confirmed by the oil displacement method). Drying experiments were performed in a moisture balance (Ohaus, MB200) with an accuracy of 0.01 g at three temperature levels: 60°C, 70°C and 80°C for 10800 s.

For modeling purposes the cubes were considered as spheres due to the fact that the cubes were rapidly deformed and converted in amorphous solids whose change of form could not be followed and whose superficial area is difficult to determine. Volume changes of individual samples were determined by the method of organic oil displacement in a graduated probet and from this volume the radius of the equivalent sphere was determined (initially it was 6.10 x 10^{-3} m). All the drying experiments were repeated twice. Each experimental run and it´s duplicate were not significantly different at 0.05 confidence level.

2.2. Mathematical Modeling

The solid particles are modeled as if they were spheres with moisture content gradients only in the radial direction. Assuming that the drying process is described by the diffusion equation (with an effective diffusivity that takes in account all possible moisture transport mechanisms) and that the solid volume reduction is proportional to the volume of water lost by the solid we describe the drying and shrinking of a solid particle under isothermal conditions taking in account the dependency of the volumetric moisture content on the time and volume variables and the dependency of the radius with time (Crank, 1975 [5]; Aguerre et al., 2008 [7]). The moving boundary problem is transformed into a fixed boundary problem through a variable change ($z=r/R$) in order to have variations of the dimensionless spatial variable between 0 and 1. The isothermicity condition leads to a Dirichlet condition at the border and the consideration of a spherical particle leads to a symmetry condition at the center of the particle. The resulting equations in terms of the volumetric moisture concentration, c, and dimensionless radial position are

$$\frac{dc}{dt}=\frac{1}{(z.R)^2}\frac{\partial}{\partial z}\left(z^2.D\frac{\partial c}{\partial z}\right)+\frac{1}{R}\frac{\partial c}{\partial z}z\frac{\partial R}{\partial t}, \quad (1a)$$

$$0\le z<1, t>0$$

$$\frac{\partial c}{\partial z}=0, z=0, \forall t \quad (1b)$$

$$c=c_{eq}, z=1, \forall t \quad (1c)$$

$$c=c_0, 0\le z<1, t=0 \quad (1d)$$

$$\frac{dR}{dt}=-\frac{D}{R\rho_w}\left[\frac{\partial c}{\partial z}\right]_{z=1} \quad (1e)$$

$$R=R_0, t=0 \quad (1f)$$

$$\bar{c}(t)=\frac{1}{V_p}\int_0^{V_p}c(r,t)dv \quad (1g)$$

where $\bar{c}(t)$ denotes the volumetric moisture content averaged over the volume at each time step and the diffusion coefficient-moisture content relationship is assumed to be of the next form

$$D(c(r,t))=d_1\exp\left[c(r,t)/d_2\right] \quad (2)$$

where the parameters d_1 and d_2 are fitting parameters to be determined for each experimental temperature.

The above given equations that describe the isothermal drying process (1a-1g) were solved discretizing the spatial variable by the finite differences technique (with a 160 points mesh) and the resulting system of non-linear ordinary differential equations together with the differential equation that describes the radius evolution (1e) were integrated by a proper method for stiff equations (Rosenbrock method (Press et al., 2007 [11])). The spatial moisture profiles inside the particle were numerically averaged at each time step (Equation (1g)) for obtaining the average moisture profile at each time step.

In order to fit the parameters of the diffusion-moisture model through a comparison of the average moisture content profile with the corresponding experimental profile at each level of temperature the experimental moisture content data (db) at each temperature were fitted by an exponential function

$$m=a_0+a_1\exp\left[-t/a_2\right] \quad (3)$$

where a_0, a_1 and a_2 are the parameters to be fitted by a.

3. Results and Discussion

Table 1. Initial and equilibrium moisture contents and dry masses

T [K]	m_0 [kg kg^{-1}]	m_{eq} [kg kg^{-1}]	m_{ds} [kg]
333	15.42	0.20	6.82 x10^{-4}
343	17.87	0.12	5.76 x10^{-4}
353	18.98	0.08	5.37 x10^{-4}

The experimental drying curves were obtained and the reduction in size was registered (the initial radius R_0 = 6.10 x 10^{-3} m and the final radiuses were: 2.96 x 10^{-3} m at 60°C, 2.65 x 10^{-3} m at 70°C and 2.41 x 10^{-3} m at 80°C). In Table 1 are reported the initial moisture content, the equilibrium moisture content and the solid's dry mass corresponding to each drying temperature. The experimental drying curves (db) in dimensionless form for drying at 60°C, 70°C and 80°C (the corresponding initial moisture contents were 93.90, 94.70 and 94.99 % wb at each temperature) are shown in Figure 1. In Figure 2 are shown the drying rate versus moisture content (db) curves. No constant rate period was observed in the drying of tomato mesocarp cubes. It can be observed that the drying rates are higher at higher temperatures and also that at 70°C and 80°C there are at least two drying periods and that at the lower temperature, 60°C, there are only one in the range of moisture content analyzed.

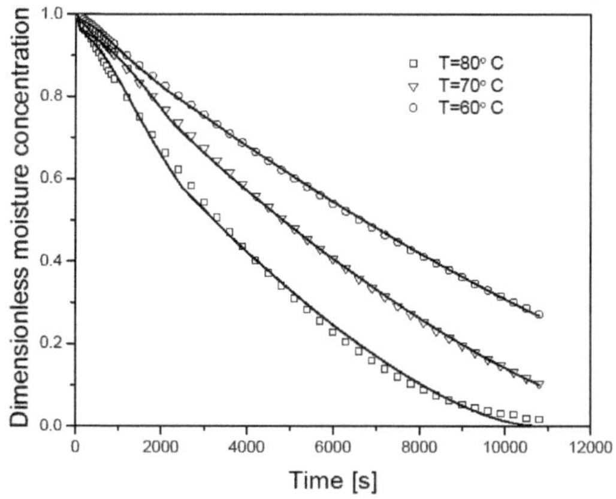

Figure 1. Experimental moisture content values (symbols) and simulated drying curves (continuous lines) obtained solving Equations (1a-1g) and Equation (2).

The parameters a_0, a_1 and a_2 that fit equation (3) to the experimental moisture content data (db) for drying at 60°C, 70°C and 80°C are given in Table 2. In all cases the regression coefficients were 0.999.

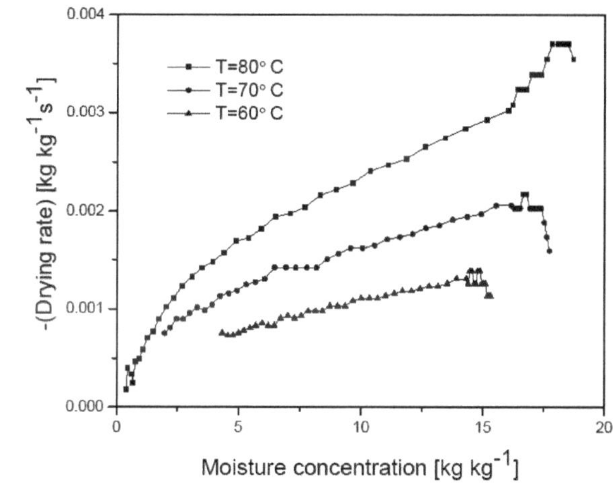

Figure 2. Drying rate versus moisture content

In order to obtain an acceptable agreement between the predicted average moisture content (simulation results) and the experimental data the constant d_l had to be adjusted as a step function for each temperature level changing abruptly at certain critical time (which corresponds to certain critical moisture content). It was not possible to obtain an acceptable prediction of the experimental results with a constant diffusion coefficient at each temperature.

Table 2. Constants for the expression fitting the experimental kinetic data

T [K]	a_0 [kg kg^{-1}]	a_1 [kg kg^{-1}]	a_2 [s]	R^2
333	-9.0464	24.5638	17699.8731	0.999
343	-9.7844	27.8552	12357.4720	0.999
353	-4.1033	23.3294	6082.9608	0.999

The parameter values for equation (2) at the three temperatures studied are reported in Table 3 where d_{1l} and d_{1h} denote the values of the parameter d_l fitted below and above the critical moisture content at which this parameter changes in a step form. These parameters were adjusted by a trial and error procedure on the program in which equation (1) was solved until a good agreement between the experimental and the simulated drying curves was obtained. The simulated profiles obtained with the parameters reported in Tables 1, 2 and 3 are shown in Figure 1 in which can be appreciated the good agreement among the simulation results (continuous line) and the experimental data (dots). The model predictions for the way in which the radius changes with time at the three temperature levels analyzed are shown in Figure 3.

Table 3. Parameters of the diffusion coefficient model (Equation 2)

T [K]	d_{1l} [m^{2}s^{-1}]	d_{1h} [m^{2}s^{-1}]	d_2 [kg m^{-3}]
333	1.05 x10^{-12}	1.30x10^{-12}	5 x10^4
343	1.45 x10^{-12}	1.78x10^{-12}	5 x10^4
353	1.80 x10^{-12}	2.60x10^{-12}	5 x10^4

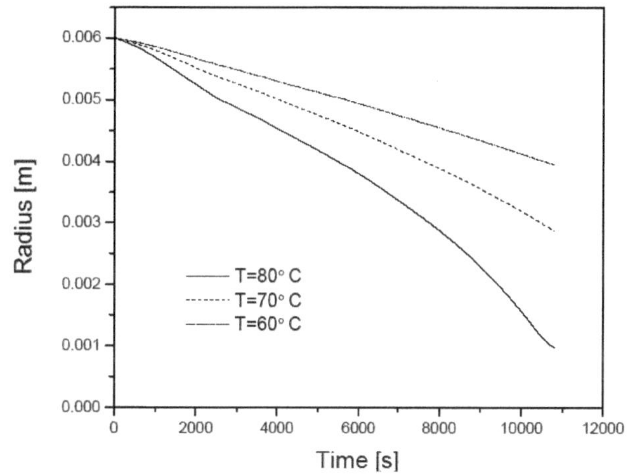

Figure 3. Model predictions of the radius reduction in the drying process

In Figure 4 are shown the profiles of the volumetric moisture content [kg/m^3] at the interface which increases continually due to the effect of the reduction in volume. Even though the dry base interfacial moisture content is constant the volumetric moisture content is changing due to the shrinking effect and, at the higher temperature, this effect is much more pronounced.

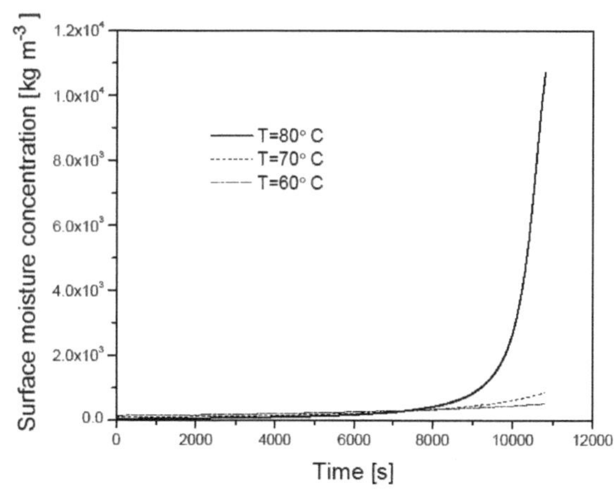

Figure 4. Evolution of the volumetric surface moisture content

The spatial averaged diffusion coefficients obtained for an instant of time above and below the moisture content value at which the diffusion changes drastically for each temperature are shown in Figure 5 where can be seen the moisture content corresponding to each temperature and the magnitude of the step change in the averaged diffusion coefficient. This information is summarized in Table 4 in which are reported the spatial average diffusion coefficient values, the moisture content at which they change at each drying temperature as well as the corresponding time.

Table 4. Average Diffusion coefficients and critical moisture contents

T [K]	D_l [m^2s^{-1}]	D_h [m^2s^{-1}]	m_{cr}[kg kg^{-1}]	t_{cr} [s]
333	1.32 x10^{-12}	1.63x10^{-12}	12.80	2200
343	1.82 x10^{-12}	2.23x10^{-12}	13.30	2250
353	2.25 x10^{-12}	3.25x10^{-12}	11.20	2500

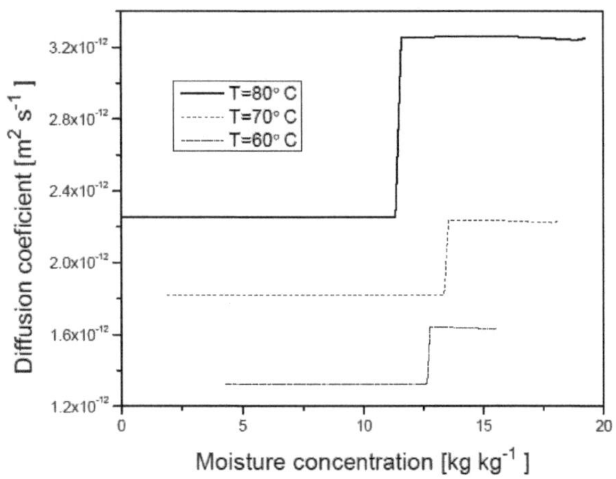

Figure 5. Spatial averaged diffusion coefficient versus moisture content.

Average diffusion coefficient values at each temperature level were obtained integrating Equation (2) with respect to the volumetric moisture content and dividing by the range of moisture contents considered (Equation (4)). The moisture content limits were chosen in such a way that the minimum and maximum volumetric moisture content values for each experimental run lay in between them.

$$\bar{D}(T)=\left(\frac{1}{c_2-c_1}\right)\int_{c_1}^{c_2} D(c)dc \tag{4}$$

The values of D averaged with respect to volumetric moisture content at each temperature level ($\bar{D}(T)$) obtained with Equation (4) are reported in Table 4. From the graphic of the natural logarithm of $\bar{D}(T)$ nversus $1/T$ (see Figure 6) for the high and low moisture content ranges (which are defined by the volumetric moisture contents below and above the critical moisture contents reported in Table 4) at each temperature level analyzed are obtained the activation energies and the pre exponential factors corresponding to each drying period postulated in this work. The fitted parameters for the Arrhenius model (Equation 5) leads to an activation energy for moisture contents above than about 12 kg kg^{-1} (db) E_{ah} = 33.710 kJ mol^{-1} and a pre exponential factor k_{0h} = 3.104 x 10^{-7} m^2s^{-1} (R^2 = 0.9951) and, for moisture contents below than 12 kg kg^{-1} (db), E_{al} = 26.127 kJ mol^{-1} and a pre exponential factor k_{0l} = 1.674 x 10^{-8} m^2s^{-1} (R^2 = 0.9898).

$$D(T)=k_0\exp\left[-E_a/R_gT\right] \tag{5}$$

The moisture diffusivity values reported in the literature for tomato (which could correspond to different geometries, temperatures, experimental conditions, etc.) range from 1.683 x 10^{-12} $m^2\,s^{-1}$ (Kingsly et al., 2007 [12]) to 2.3 x 10^{-9} $m^2\,s^{-1}$ (Giovanelli et al., 2002 [3]). The values obtained in this work are in the order of those reported by Kingsly et al. (2007) [12]. However it should be taken in account that, as pointed out by Liu et al. (2010) [13], food shrinkage produces a reduction in the distance required for the movement of water molecules, facilitating the moisture diffusion towards the outer surface and making the effective moisture diffusivity overestimated when the shrinking effect is not taken in account properly. Sometimes the effective diffusion coefficient is determined from the truncated Fourier series solution (for example Doymaz, 2006 [14]; Resende et al., 2007 [4]; Abano et al., 2011 [15]) or from a modification of this solution to the diffusion equation (Giovanelli et al., 2002 [3]) that is no longer valid when the diffusion coefficient is moisture content dependent or when the shrinking effect cannot be neglected (the Henderson and Pabis model is equivalent to taking in account just the first term of the Fourier series solution above mentioned).

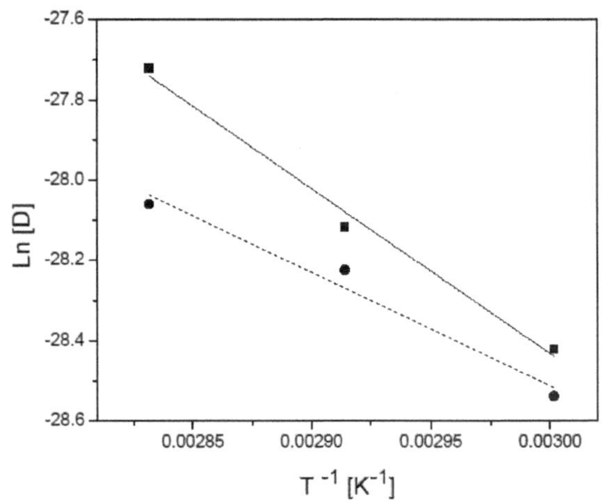

Figure 6. Fitting Arrhenius's models to the diffusion coefficients dependency on temperature for the high (spheres) and low (triangles) moisture content ranges at each temperature levelʼ analyzed (E_{ah} = 33.710kJ mol^{-1}; E_{al} = 26.127 kJ mol^{-1})

The presence of two different diffusion coefficient expressions for each drying air temperature implies two falling rate periods in the drying of pieces of tomato mesocarp. The existence of more than one falling rate drying period has been reported by other authors in drying some agricultural products. Chaijan et al. (2011) [16] found two drying falling-rate periods in the drying of high moisture corn. These authors reported expressions for the effective moisture diffusivity for each drying period at different air velocities. Taiwo et al. (2006) [17] found three falling rate periods in their study of drying of tomato slices. Also Hawlader et al. (1991) [2] reported two falling rate periods in the drying of tomato slices.

The results indicate that at the three temperature levels analyzed the diffusion coefficient is a weak function of moisture content (the diffusion-moisture content profiles appear as nearly constant at each temperature in the range above and below the critical moisture content) and it depends more strongly on temperature as can be appreciated in Figure 5. This fact could justify that the diffusion coefficient dependence on moisture content could be neglected when the drying process is not isothermic and the temperature range involved is wide enough. However the diffusion dependence on moisture content should be taken in account in the case of isothermic drying processes.

In this work activation energies were obtained for each drying falling rate period. Other authors, for example Chaijan et al. (2011) [16] have also determined activation energies for the two drying falling rate periods that they found for the drying of corn. The values of E_a obtained here (26.127 and 33.710 kJ mol^{-1}) are in the range of the values reported by Zogzas et al. (1996) [18] as typical for food stuffs and also are similar to the values reported for tomato by other authors. Demiray & Tulek (2012) [1] reported a value of 22.98 kJ mol^{-1} for the activation energy of slices of tomato (*Lycopersicum Esculentum*, cv. Rio Grande). Doymas (2007) reports activation energies of 17.40 and 32.94 kJ mol^{-1} for pre-treated and untreated samples of tomato halves (*Lycopersicum Esculentum*, cv. Mill) respectively. However, Purkayastha et al. (2011) [19] found a much higher activation energy (61.004 kJ mol^{-1}) for tomato slices (cv. Punjab Kesri). This authors attribute the discrepancy with the previously reported values by Doymas (2007) [14] to the effect of the pretreatment blanching method, which could lead to the gelatinization of starch granules or the concentration effect of sugars.

4. Conclusions

Based on the results above presented the following conclusions can be made: As expected the predicted shrinking rate is higher at higher temperatures. The model underestimates the shrinking rate at 60°C and 70°C and overestimates it at 80°C (the errors are: 33.2% at 60°C, 8.1% at 70°C and 56.3% at 80°C), which could indicate that the volume reduction is not equal to the volume of water lost as assumed in equation (1e). It should be mentioned that when the moisture diffusion coefficient is calculated departing from the series solution to the diffusion equation, valid when the volume reduction is negligible, there is no form of predicting the shrinking rate of the solid because the model does not contemplate the shrinking effect. In coincidence with other authors who worked in the drying of tomato, two falling-rate periods were postulated for the drying of peeled tomato cubes in this work.

Nomenclature

c	volumetric moisture content	kg m^{-3}
$\bar{c}$	average volumetric moisture content	kg m^{-3}
c_{eq}	equilibrium volumetric moisture content	kg m^{-3}
c_0	initial volumetric moisture content	kg m^{-3}
m	dry basis moisture content	kg kg^{-1}
m_{cr}	dry basis critical moisture content	kg kg^{-1}
m_{eq}	dry basis equilibrium moisture content	kg kg^{-1}
m_{ds}	dry solid mass	kg
m_0	dry basis initial moisture content	kg kg^{-1}
D	diffusion coefficient	m^2 s^{-1}
R	radius of the equivalent sphere	m
Rg	universal gas constant	Jmol^{-1} K^{-1}
R_0	initial radius of the equivalent sphere	m
t	time	s
V_p	volume of the equivalent sphere	m^3
z	dimensionless radial distance (=r R^{-1})	

References

[1] Demiray E., Tulek Y. Thin-layer drying of tomato slices in a convective hot air dryer. *Heat and Mass Transfer*, 48, 841-847. 2012.

[2] Hawlader M. N. A., Uddin M. S., Ho J. C., Teng A. B. W. Drying characteristics of tomatoes, *Journal of Food Engineering*, 14, 259-268. 1991.

[3] Giovanelli G., Zanoni B., Lavelli V., Nani R. Water Sorption, drying and antioxidant properties of dried tomato products, *Journal of Food Engineering*, 52, 135-141. 2002.

[4] Resende O., Correa P. C., Jarén C., Moure A. J. Bean moisture diffusivity and drying kinetics: a comparison of the liquid diffusion model when taking into account and neglecting grain shrinkage. *Spanish Journal of Agricultural Research*, 5 (1), 51-58. 2007.

[5] Crank J. *The Mathematics of Diffusion*, 2nd Edition, Clarendon Press, Oxford, 1975.

[6] Crank J. *Free and Moving Boundary Problems*, Clarendon Press, Oxford, 1984.

[7] Aguerre R. J., Tolaba M., Suarez C. Modeling volume changes in food drying and hydration, *Latin American Applied Research*, 38, 345-349. 2008.

[8] Rivera B. D. *Evaluación de la deshidratación de jitomate (Licopersicum Esculentum) con atemperado en lecho vibrofluidizado, considerando la retención de licopeno como parámetro de calidad*, Master´s thesis. Escuela Nacional de Ciencias Biológicas, Instituto Politécnico Nacional, México, 2007.

[9] Lewicki P. P., Le H. V., Pomaranska-Lazuka W. Effect of pretreatment on convective drying of tomatoes. *Journal of Food Engineering*, 54, 141-146. 2002.

[10] Kunii D., Levenspiel O. Fluidization Engineering, Second edition, Butterworth-Heinemann, USA, 1991.

[11] Press W. H., Teukolsky S. A., Vetterling W. T., Flannery B. P. *Numerical Recipes: The art of Scientific Computing*, Third edition, Cambridge University Press, New York, USA, 2007.

[12] Kingsly A. R. P., Singh R., Goyal R. K., Singh D. B. Thin-Layer Drying Behaviour of Organically Produced Tomato, *Am. Journal of Food Technology*, 2, 71-78. 2007.

[13] Liu X., Chen J., Liu M., Li Z., Tao Y., Zhu D. "The Effect of Biological Material's Tissue Shrinking on Moisture Diffusivity during Hot-air-drying". *Proceedings of the World Automation Congress*, Kobe, Japan, 2010, Pages 255-260.

[14] Doymaz I. Air-drying characteristics of tomatoes, *Journal of Food Engineering*, 78, 1291-1297. 2007.

[15] Abano E., Ma. H, Wu. Q. Influence of Air Temperature on the Drying Kinetics and Quality of Tomato Slices, *Journal of Food Processing and Technology*, 2. 2011.

[16] Chaijan, A. R., Parian A. J., Esna-Ashari M. Modeling of moisture diffusivity, activation energy and bspecific energy consumption of high moisture corn in a fixed and fluidized bed convective dryer. *Spanish Journal of Agricultural Research*, 9 (1), 28-40. 2011.

[17] Taiwo A., C., Sikiru A. R., Ojo A. Drying characteristics and sorption isotherm of tomato slices. *Journal of Food Engineering*, 73, 157-163. 2006.

[18] Zogzas N, P., Maroulis Z. D., Marinos-Koulis D. Moisture Diffusivity Data Compilation in Food Stuffs. *Drying Technology*, 14 (10), 2225-2253. 1996.

[19] Purkayastha M. D., Nath Amit, Deka B. Ch., Mahanta Ch. L. Thin layer drying of tomato slices. *J. Food Sci. Technol.*, 50 (4), 642-653. 2013.

Physicochemical, Structural and Rheological Properties of Chestnut (*Castanea sativa*) Starch

Marcio Schmiele[1,*], Georgia Ane Raquel Sehn[1], Valéria da Silva Santos[1], Thaís de Souza Rocha[2], Eveline Lopes Almeida[3], Elizabeth Harumi Nabeshima[4], Yoon Kil Chang[1], Caroline Joy Steel[1]

[1]Department of Food Technology, School of Food Engineering, University of Campinas, Campinas, Brazil
[2]Department of Food Science and Technology, State University of Londrina, Londrina, Brazil
[3]Department of Biochemical Engineering, School of Chemistry, Federal University of Rio de Janeiro, Rio de Janeiro, Brazil
[4]Institute of Food Technology, Campinas, Brazil
*Corresponding author: marcio.ufpel@gmail.com

Abstract Chestnuts have high starch content, which makes them an alternative source of starch for the food industry. Brazil is a country where the production of chestnuts has been increasing in recent years. The aim of this study was to extract starch from chestnuts (*Castanea sativa*), to characterize its physicochemical, structural and rheological properties, and to compare the results with corn starch. Chestnut starch presented light color, and granules smaller than corn starch, with various dimensions, suggesting a bimodal distribution. Chestnut starch showed 20.48% absolute amylose, higher amylopectin branched-chain length and B-type crystallinity. The infrared spectra of chestnut starch showed characteristic peaks at 1647, 1157, 1079, and 1018 cm^{-1}. Chestnut starch presented higher peak viscosity, breakdown and setback, and lower pasting and gelatinization temperatures than corn starch. The swelling power and the solubility of chestnut starch were significantly higher than those of corn starch. Chestnut starch showed characteristics of a gelling and thickening agent, with potential for use as an ingredient in the food industry, as an unconventional starch from an alternative source.

Keywords: *chain length, molecular structure, morphology, pasting properties*

1. Introduction

Species of chestnut tree were introduced in Brazil, and have adapted well to the climate in the higher altitudes of the south and southeast of the country, which resemble the temperate zone of the Northern Hemisphere. In São Paulo, there are commercial field of chestnut tree species including Chinese chestnuts (*Castanea mollissima*), Japanese chestnuts (*Castanea crenata*), American chestnuts (*Castanea dentata*) and, mainly, Portuguese chestnut (*Castanea sativa*) [1].

One of the main problems concerning trade of chestnuts in natura is related to its high moisture content (~50%) [2] and fungal development. To deal with the microorganism development, some countries use the frozen peeled or dehydrated to become an economic advantage which enables the diversification of products offered to consumers, including chestnut paste and flour, which are incorporated into a wide variety of food products [2,3].

According to reference [4], chestnut flour has about 6.0% moisture, 5.6% protein, 5.4% fat, 2.3 % fiber, 2.1% ash, and high starch content (78.6%). These authors reported chemical composition for chestnut starch obtained from *Castanea sativa*, Mill. for 0.83% protein, 1.51% fat, 1.09% fibers, 0.51% ash, and 96.06% starch, in dry basis (d.b.). Furthermore, the authors studied the viscoamylographic pattern of chestnut starch and determined paste clarity and elasticity, and gel strength, comparing the results with those of corn starch and cassava starch. They concluded that chestnut starch had properties that were intermediate between both conventional starch sources studied.

Considering the high starch content of chestnuts, and the need for a solution for the problem of preservation of chestnuts *in natura*, this study aimed to extract chestnut starch and characterize it for its physicochemical, structural and rheological properties, for its application as an ingredient in the food industry, using corn starch as the basis for comparison, due to the high use of this type of starch in the food industry. Furthermore, the use of analytical methods with high resolution could explaining better the chestnut starch performance for possible application.

2. Material and Methods

2.1. Raw Material

The chestnuts (*Castanea sativa*) were provided by CATI - Integral Technical Assistance Coordination (São Bento do Sapucaí, BRA) and regular corn starch (Amidex 3001) was donated by Corn Products (Mogi Guaçu, BRA).

2.2. Cleaning and Manual Dehusking

The healthy chestnuts were sanitized by immersion in chlorinated water (2 ppm) in a ratio of chestnut and water equal to 1:2 (w:v) for three minutes. Then, the chestnuts were dehusked manually and the inner pellicle was removed by abrasion under running water in a R6025 abrasive peeler (Bufalo, São Paulo, BRA) for four minutes at 1720 rpm.

2.3. Starch Extraction

The starch extraction from chestnuts followed the methodology proposed by [5]. Starch was extracted by grinding 500 g of chestnuts and 750 mL distilled water in a OBL10/2 blender (OXY, Santana de Parnaíba, BRA), at 35000 rpm for three minutes. The resulting material was filtered using an 88 μm sieve. The retentate was subjected to the extraction process twice more. An aliquot of 200 mL permeate was centrifuged in a FR22 refrigerated centrifuge (FANEM®, Piracicaba, BRA), for 10 minutes at 20°C and 1700 xg. The supernatant was discarded, the upper layer (containing fiber and protein) was removed manually with the aid of a spatula, and the pellet was re-suspended in 200 mL distilled water. This step was repeated four times. At the end, the starch was re-suspended in 100 mL of 99.5% ethanol, vacuum filtered on Whatman N°5 filter paper, and dried in a TE394/2 forced-air drying oven (TECNAL®, Piracicaba, BRA), at 40±0.2°C for 15 hours. The dried starch was ground in a blender until particle size was less than 250 μm.

2.4. Physicochemical Properties

2.4.1. Chemical Composition of Starch Samples

Chestnut and corn starch samples were characterized for moisture, protein, ash, and lipid contents according to AACC methods 44-15.02, 46 13.01, 08 01.01 and 30-25.01 [6], respectively. The total dietary fiber (TDF) was determined by AOAC method 991.43 [7]. Analyses were performed in triplicate and the results expressed as percentage. Digestible carbohydrates (starch and sugars) were calculated by difference.

2.4.2. Instrumental Color

Color measurements were performed in triplicate by the CIELab system, using a 45/0-L colorimeter (XE MiniScan, Reston, USA), through direct reading on the sample in the power form. The test conditions were: illuminant D65, observer angle of 10° and calibration mode Reflectance Specular Included.

2.5. Structural Properties

2.5.1. Scanning Electron Microscopy

The starch samples were fixed on stubs and coated with a 92Å thick gold layer in a SC7620 Polaron Sputter Coater (VG Microtech, Uckfield, GBR), and analyzed using a 440i scanning electron microscope (LEO Electron Microscopy, Oxford, GBR), with a 6070 dispersive energy X-ray detector (LEO Electron Microscopy, Cambridge, GBR). The accelerating voltage was 10kV and the beam current was 50pA.

2.5.2. High Performance Anion Exchange Chromatography with Pulse Amperometric Detection (HPAEC-PAD) for Analysis of Amylopectin Branched-Chain Length Distribution

The amylopectin of the chestnut and corn starch was debranched using isoamylase (Megazyme, 3U) enzyme according to [8] and the branched-chain length distribution was determined according to the method proposed by [9]. Briefly, it was used a ICS 3000 HPAEC-PAD system (Dionex Corporation, Sunnyvale, USA) equipped with an AS40 automatic sampler. Samples were filtered (0.22 μm membrane) and injected into the HPAEC-PAD system (20 μL sample loop). The flow rate was 0.8 mL/minute at 40°C. The standard quadruple potential (E) waveform was employed with the following periods and pulse potentials: E1=0.10V (t1=0.40 s); E2=2.00 V (t2=0.02 s); E3=0.60 V (t3=0.01 s); E4=0.10 V (t4=0.06 s). All eluents were prepared with ultrapure water (18 mΩ.cm) with N2 sparging. Eluent A was 150 mM NaOH and eluent B was 500 mM sodium acetate and 150 mM NaOH. The branched chains of amylopectin were separated using a Dionex CarboPacTM PA-100 guard column (4 mm x 50 mm) and a Dionex CarboPacTM PA-100 column (4 mm x 250 mm). The gradient of eluent B was 28% at zero minutes, 40% at 15 minutes, and 72% at 105 minutes. The data were analyzed using the Chromeleon software, version 6.8 (Dionex Corporation). The samples were analyzed in duplicate.

2.5.3. Absolute Amylose

The absolute amylose content was determined by using the Amylose/Amylopectin assay procedure of Megazyme (K-AMYL 07/11). The absorbance was measured by absorption using a DU-70 spectrophotometer (Beckman, Fullerton, USA), at 510 nm. Analyses were performed in triplicate, and the results were expressed in percentage.

2.5.4. X-ray Diffraction

The starch samples were conditioned at 100% relative humidity for seven days at room temperature in a desiccator containing 300 mL distilled water with 1% sodium azide to inhibit microorganism growth. The XRD patterns were obtained in a RU200B XRD equipment (Rigaku Rotaflex, Tokyo, JAP), with Cu rotating anode at 40 kV and 80 mA, and diffraction angles from 3° to 30° (2theta), with a step size of 0.02° at a scan speed of 2°/minutes. The relative crystallinity was calculated by the ratio of the total area and the area of the peaks, according to the method proposed by [10] using the software Origin Microcal Inc., version 6.0 (Northampton, USA), with a 5-point smoothing FFT Filter. The analysis was performed in triplicate, and the results expressed in percentage crystallinity.

2.5.5. Thermal Properties

The thermal properties of the starch samples were determined in a TA60 differential scanning calorimeter (Shimadzu, Kyoto, JAP), based on the methodology described by [11], with modifications. Aluminum pans containing 3 mg sample (d.b.) and 7 µL deionized water were hermetically sealed and equilibrated at room temperature for one hour before measurements. An empty pan was used as reference. The scanning temperature ranged from 30 to 95°C, at a heating rate of 5°C/minute. Based on the thermograms, the values of the To, peak temperature (Tp), temperature range (ΔT), and ΔH were obtained in triplicate, and the results were expressed in °C for To, Tp and ΔT, and in J.g-1 for ΔH.

2.5.6. Fourier Transform Infrared Spectroscopy

The methodology for obtaining the FTIR spectra was based on the study of [12]. The samples were dried in a TE-395 vacuum oven (Tecnal®, Piracicaba, BRA), at -600 mmHg and 50°C for 24 hours, ground in an agate mortar, and filtered through a sieve of 100 µm. The potassium bromide (KBr) pellets were prepared by mixing 2 mg sample and 200 mg KBr, and analyzed by a IRPrestige 21 Fourier Transform Infrared Spectroscope (Shimadzu, Kyoto, JAP). The absorbance ranged from 4000 to 400 cm^{-1} at a 4 cm^{-1} resolution and 40 scans. The proportion of crystalline area was also calculated according to the method described by [13].

2.6. Rheological Properties

2.6.1. Pasting Properties

The viscoamylographic profiles of the starch samples were determined according to the method 162 of ICC [14] using a RVA-4500 Rapid Visco Analyser (Warriewood, AUS), and the curves were analyzed by the software TCW3.15.1.255. The parameters evaluated were: pasting temperature, peak viscosity, breakdown, final viscosity and setback. The analysis was performed in triplicate, and the results expressed in °C for the pasting temperature and in cP for the other parameters.

2.6.2. Gel Strength

The gel strength of the starch samples was determined by the method proposed by [15], with modifications. The gels obtained from the pasting property analysis were poured into a cylindrical tube of polyvinyl chloride with 25 mm diameter and 20 mm depth, and stored at 7°C for 24 hours. The samples were kept at room temperature for 1 hour before analysis. The texture was determined in a TA-XT2i texture analyzer (Stable Micro Systems-Haslemere, GBR), with a load of 25 kg. The conditions were the following: pre-test, test and post-test speeds of 5.0, 1.0, and 1.0 mm.s-1, respectively; penetration distance 10.0 mm; detection limit 0.05N; and cylinder probe Derlin P/10. The analysis was performed with six replicates and the results expressed in N.

2.6.3. Swelling Power And Solubility

The swelling power and solubility were determined at 50, 60, 70, 80 and 90°C, according to the methodology described by [16], with modifications. Samples of 0.02 g (d.b.) were weighed in 50 mL conical centrifuge tubes. Then, 20 mL distilled water was added, and the tubes were placed in a water bath under mechanical stirring at 100 rpm. After 30 minutes of incubation, the tubes were centrifuged at 2200 xg for 15 minutes. The supernatant was collected and oven dried to constant weight to quantify the soluble fraction, and the results were expressed as percentages. The tubes containing the residual material were weighed to determine the swelling power. The analyses were performed in triplicate.

2.7. Statistical Analysis

The data were analyzed using the software Statistica 7.0 (Statsoft, Tulsa, USA) for analysis of variance and comparison between means by Student's t test, with a significance level of 5%.

3. Results and Discussion

3.1. Chemical Composition

The fat contents of chestnut and corn starch samples were 0.15±0.03 and 0.10±0.02%, respectively, and did not differ from each other. Both ash and total dietary fiber contents of chestnut starch were higher ($P \leq 0.05$) than those of corn starch, with ash contents of 0.05±<0.01 and 0.02±<0.01%, and total dietary fiber of 1.36±0.11 and 0.58±0.42 %, respectively. According to Demiate et al. [4], chestnut starch has characteristics similar to corn starch, especially with respect to lipids, ash and fiber. Reference [17] reported that purified starch showed protein content lower than 0.6%, which occurred with the starch samples of this study, with values of 0.14±0.02% for chestnut starch and 0.26±0.03% for corn starch. The total carbohydrate content of chestnut starch was 87.56%, similar to that obtained for corn starch (87.21%).

3.2. Instrumental Color

The color parameters of chestnut and corn starch samples were 99.11±0.30 and 99.58±0.43 for L*; 0.58±0.02 and 0.64±0.01 for a*; and 3.16±0.12 and 5.65±0.02 for b*, respectively. Although no difference was observed for the lightness of the starch samples, a* and b* parameters were statistically different ($P \leq 0.05$). Chestnut starch showed a greater tendency to red, probably due to the pigment found in the peel, while corn starch showed a greater tendency to yellow, probably because of corn pigments (zeaxanthin).

3.3. Morphology of the Starches Granule by Scanning Electron Microscopy

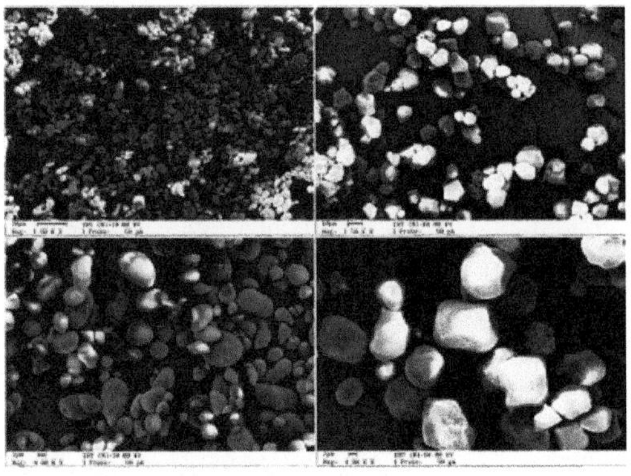

Figure 1. Micrographs of chestnut starch (left) and corn starch (right) at 1500 x (top) and 4000 x (bottom) magnifications

The micrographs of chestnut starch (Figure 1) showed oval and irregular shaped granules, presenting predominantly smooth surfaces, different to corn starch, which showed circular and polyhedral shaped granules, and porous surfaces. A few chestnut starch granules had holes or fissures on the surface, especially the larger ones. Chestnut starch exhibited a distinct population of smaller granules with a mean diameter from 2 to 5 μm, and a population of larger granules with mean diameters from 10 to 20 μm, suggesting a bimodal distribution. The corn starch sample showed granules of more uniform size, characteristic of a modal distribution.

3.4. Amylopectin Branched-Chain Length Distribution and Absolute Amylose Content

Chestnut and corn starch samples exhibited a profile of amylopectin branched-chain length distribution with two populations of different sizes (Figure 2). The first population showed a peak at DP 12 for both samples, and the second population showed peaks at DP 43 and DP 42 for chestnut and corn starches, respectively.

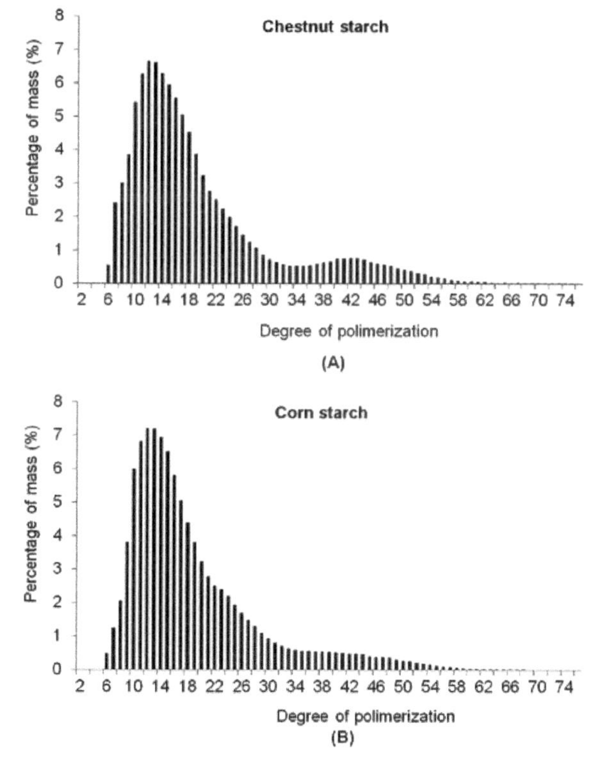

Figure 2. Amylopectin branched-chain length distribution of chestnut starch (A) and corn starch (B)

The amylopectin chains length was classified as proposed by [18], in which the amylopectin branched-chains A, B1, B2 and B3 occur at DP 6-12, 13-24, 25-36 and ≥37, respectively (Table 1). Chestnut starch showed a higher number of long (DP 25-36) and very long chains (DP ≥37), resulting in a higher DP (19.60) when compared to corn starch (18.75). The highest DP of chestnut and corn starches was 76 and 68, respectively. Chains at DP 18-21 represent the total length of the lamellar structure [19]. Thus, the amylopectin from chestnut starch has more branched-chains that participate in two or more clusters than corn starch.

The absolute amylose content was found to be 20.48±0.51% for chestnut starch and 19.94±0.56% for corn starch, and these values did not differ from each other.

Table 1. Amylopectin branched-chain length distribution of chestnut starch and corn starch in mass-basis

Starch	Degree of polymerization (DP)					
	6-12	13-24	25-36	≥37	Average	Highest detectable
Chestnut	27.96^{a}	50.42^{b}	10.50^{b}	11.12^{a}	19.60^{a}	76
Corn	27.61^{a}	52.78^{a}	12.16^{a}	7.45^{b}	18.75^{b}	68

Means with different letters in the same column differ statistically from each other ($P \leq 0.05$).

3.5. X-ray Diffraction

Figure 3 shows the XRD patterns for chestnut and corn starches samples. Chestnut starch showed single peaks at diffraction angles of 5, 8, 15 and 17° at 2theta, and a double peak between 22 and 24° at 2theta, which is characteristic of B-type polymorphism. Corn starch showed A-type polymorphism with single peaks at 15 and 23° at 2theta, and a double peak between 17 and 18° at 2theta, in agreement with the findings of [20]. The B type polymorphic profile is characteristic of starch containing a high population of very long amylopectin chains at DP>37 [21], which was observed for chestnut starch (Table 1). Starch presenting A-type crystallinity has smaller amylopectin branched-chain length than the B-type crystallinity [22], thus explaining the results of this study. The percentage of crystallinity of chestnut starch was 35.26±0.10%, which was similar to corn starch (36.44±0.51%).

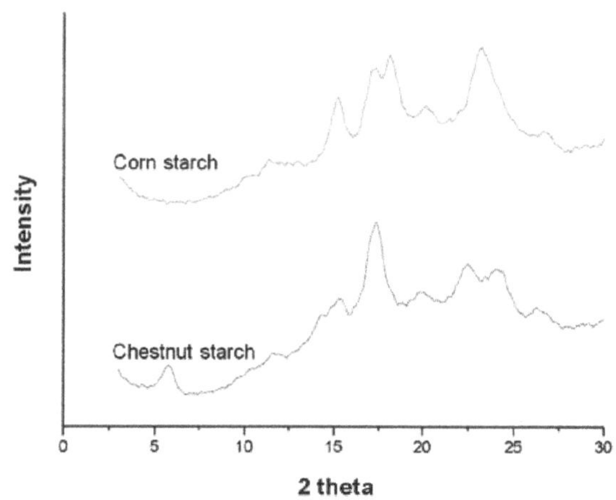

Figure 3. X-ray patterns of chestnut starch and corn starch

3.6. Fourier Transform Infrared Spectroscopy

The measurements occurred in the bands between 1800 and 450 cm^{-1}, since the organic compounds in foods have transmittance in this region [23]. The spectra of the chestnut and corn starches samples were very similar (Figure 4), differing only by the presence of a peak at 1190 cm^{-1} (indicated by the arrow) observed in the chestnut starch spectrum. Both spectra exhibited peaks characteristic of starch at wavelengths of 1647 and 1649 cm^{-1} for chestnut and corn starch, respectively, corresponding to the stretching (scissoring) of two O—H groups absorbing a water molecule. Similar bands presenting peaks at 1157, 1079 and 1018 cm^{-1} for chestnut starch, and at 1160, 1080 and 1010 cm^{-1} for corn starch were observed, and are due to the C—O bond stretching [24].

This technique is very sensitive to structural changes at the molecular level, such as the conformation of starch chains and crystallinity, unlike the XRD technique, which is related to the packing of double helices within an organized crystalline structure [25]. The absorbance band at 1047 cm^{-1} is sensitive to the amount of crystalline structure and the band at 1022 cm^{-1} is characteristic of the amorphous starch. Therefore, the ratio of the heights of bands at 1047 and 1022 cm^{-1} expresses the relationship between the amounts of crystalline to amorphous structures in starch [25,26]. It was observed that the ratio of the crystalline to amorphous structure was 0.83±<0.01 for chestnut starch and 0.89±<0.01 for corn starch, indicating that chestnut starch showed more amorphous structure than corn starch ($p<0.001$). Higher values for the ratio between the crystalline and amorphous structures are the result of a greater structural organization of the starch granule [13].

In XRD, there was no difference in the percentage of crystallinity between the samples, which is in disagreement with the result found in FTIR, possibly due to the greater sensitivity of infrared spectroscopy in relation to XRD.

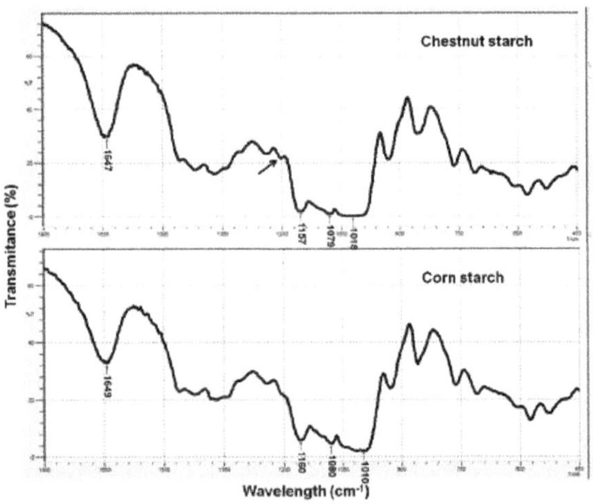

Figure 4. Fourier Transform Infrared Spectroscopy of chestnut starch and corn starch

3.7. Thermal Properties

Chestnut starch showed a lower T¬0 than corn starch. Both the T¬0 and the Tp of chestnut starch were statistically different ($P\leq 0.05$) from corn starch (Table 2). High T0 are related to a more densely packed crystalline structure and higher molecular order [27]. By FTIR, it was found that chestnut starch had a slightly lower molecular order than corn starch, which may have contributed to its lower gelatinization temperature. Moreover, when the absolute amylose content is similar, starch presenting B-type polymorphism such as chestnut starch, has a lower gelatinization temperature than starch presenting A-type polymorphism, which may be due to the greater amount of water molecules in the B-type polymorphism, making the double-helices unit cell more loosely packed than that in A-type polymorphism [19].

The extent of crystalline perfection is reflected in the T0 range and in the ΔH obtained by DSC. Greater ΔH expresses greater crystalline perfection and homogeneity [28]. These results suggest that despite a slight difference in the molecular order of the starch, this difference was not enough to influence the ΔH values, which were statistically equal.

Table 2. Thermal properties of chestnut starch and corn starch

Starch	T_{on} (°C)	T_{peak} (°C)	ΔT (°C)	ΔH ($J.g^{-1}$)
Chestnut	56.27±0.67^b	60.20±0.51^b	9.15±0.64^a	13.49±0.56^a
Corn	62.89±0.08^a	67.16±0.06^a	9.13±0.18^a	13.64±0.62^a

Mean of three replicates ± standard deviation; means with different letters in the same column differ statistically from each other ($P \leq 0.05$); Ton - onset gelatinization temperature; Tpeak - peak temperature; ΔT - temperature range; ΔH - enthalpy variation.

3.8. Pasting Properties and Gel Strength

The pasting temperature of chestnut starch was lower ($P\leq 0.05$) than that of corn starch (Table 3). Pasting temperature is influenced by several factors, including the size and shape of the starch granules, amylopectin branched-chain length distribution, degree of polymerization of amylopectin, presence of phosphate monoester derivatives and lipids [29].

Chestnut starch had higher viscosity values than corn starch, and these were statistically different ($P\leq 0.05$). The amylose content, lipid content, and amylopectin branched-chain length distribution influenced the starch pasting properties [19,30]. Chestnut starch had the highest average amylopectin branched-chain length, mainly due to the higher number of very long chains (DP≥37), which may have contributed to the higher viscosity values observed for this starch. The granule swelling and molecular rigidity are some key parameters to controlling starch pasting properties. The breakdown observed is the result of the disruption of the granular structure of the gelatinized starch due to mechanical stirring and high temperature. Reference [31] observed, for rice starches, that the higher the ratio of long chains of amylopectin, the lower the breakdown, since they help maintain the structure of the gelatinized starch, but this did not occur in the present study.

Chestnut starch showed higher setback, which may also be due to the higher number of long chains of amylopectin, since the absolute amylose content did not differ between both starch samples. Setback could be related to gel strength, however, the gel strength of both starch samples was similar, possibly due to temperature differences during measurements (50°C for RVA and room temperature for gel strength).

Table 3. Pasting properties and gel strength of chestnut starch and corn starch

Starch	Peak viscosity (cP)*	Breakdown (cP)*	Final viscosity (cP)*	Setback (cP)*	Pasting temperature (°C)*	Gel strength (N)**
Chestnut	2665±16^a	670±12^a	3121±20^a	1127±11^a	67.07±0.05^b	1.55±0.18^a
Corn	1690±10^b	469±4^b	1694±4^b	483±7^b	83.15±0.07^a	1.53±0.08^a

*Mean of three replicates ± standard deviation; **Mean of six replicates ± standard deviation, means with different letters in the same row differ statistically from each other ($P \leq 0.05$).

3.9. Swelling Power and Solubility

The swelling power of both starch samples increased with increasing temperature (Figure 5). Chestnut starch showed a greater increase in swelling power as a function of a higher temperature (2.23 g/g at 50°C, and 35.45 g/g at 90°C), as compared to corn starch (2.04 g/g at 50°C and 12.50 g/g at 90°C). The increase of the swelling power with temperature is a result of the increased mobility of starch molecules, which facilitates water diffusion, favoring swelling and solubility [32]. High amounts of long amylopectin branched-chains increase the gyration radius of the amylopectin molecules, helping to keep viscosity [31]. The gyration radius is related to the volume occupied by the amylopectin molecule in solution [33]. The higher ratio of long amylopectin branched-chains of chestnut starch may have resulted in a greater gyration radius of amylopectin molecules, which contributed to the higher swelling power as compared to corn starch.

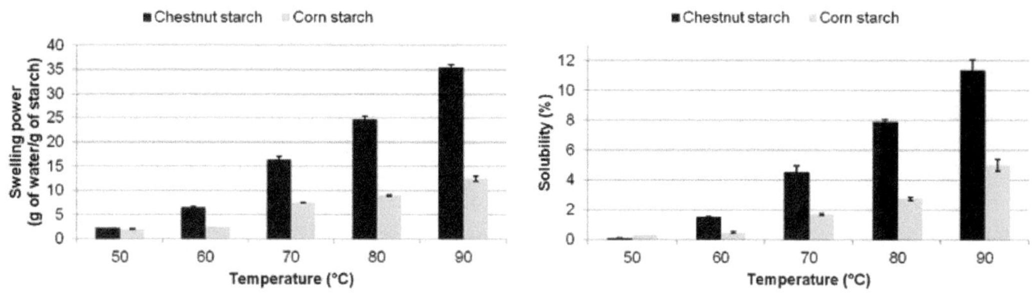

Figure 5. Swelling power and solubility of chestnut starch and corn starch

Gel formation requires hydration and swelling of starch granules, which occurs predominantly in the amorphous region of starch [32]. In FTIR analysis, chestnut starch presented higher percentage of amorphous area, which facilitated its hydration, resulting in higher swelling power.

Solubility is expressed as the percentage (in weight) of starch molecules leached into the aqueous medium after heating [16]. Chestnut starch presented higher solubility than corn starch from 60°C to 90°C. A significant increase in the solubility of chestnut starch was observed above 60°C, while corn starch showed an increase in solubility only after 70°C. This behavior is probably due to the pasting temperature and the gelatinization temperature of the starch when swelling begins, forming a disorganized structure of the starch granules, allowing linear molecules to leach from the granules. Thus, the chestnut starch granule seems to be more fragile than that of corn starch.

4. Conclusion

Chestnut starch has structural characteristics and physicochemical properties different from those presented by corn starch. Chestnut starch showed B-type polymorphism with granules of different sizes suggesting a bimodal distribution of the granules. The absolute amylose content of chestnut starch was similar to corn starch. Chestnut starch had higher pasting viscosity, swelling power and solubility than corn starch, as well as lower pasting temperature and gelatinization temperature, suggesting facility in cooking processes. Thus, chestnut starch presented thickening characteristics, with potential for application in food and non-food industries, and also for use as a raw material in studies of modified starches.

Acknowledgement

The authors thank the National Council of Technological and Scientific Development (CNPq) and the Coordination for Improvement of Higher Education Personnel (CAPES) for granting scholarships, and the Integrated Technical Assistance Coordination (CATI) and Corn Products for the donation of chestnuts and corn starch, respectively.

References

[1] Bueno, S.C.S. *Castanha: da Europa para o Brasil e do Brasil para o mundo 2009*. Available: <http://www.infobibos.com/Artigos/2009_4/CastanhaPortuguesa> [Accessed March. 15, 2015].

[2] Guiné, R.P.F. and Fernandes, R.M.C., "Analysis of the drying kinetics os chestnuts". *Journal of Food Engineering*, 76, 460-467, 2006.

[3] Koyuncu, T., Sedar, U. and Tosin, I. "Drying characteristics and energy requirement for dehydration of chestnuts (*Castanea sativa* Mill)". *Journal of Food Engineering*, 62, 165-168, 2004.

[4] Demiate, I.M., Oetterer, M. and Wosiacki, G. "Characterization of chestnut (*Castanea sativa*, Mill) starch for industrial utilization". *Brazilian Archives of Biology and Technology*, 44, 69-78, 2001.

[5] Bello-Pérez, L.A., García-Suárez, F.J., Méndez-Montealvo, G., Nascimento, J.R.O., Lajolo, F.M. and Cordenunsi, B.R. "Isolation and characterization of starch from seeds of Araucaria brasiliensis: a novel starch for application in food industry". *Starch/Stärke*, 58, 283-291, 2006.

[6] AACCI "American Association of Cereal Chemists International". *Approved methods*, St. Paul., 11th ed., 2010.

[7] AOAC "Association of Official Analytical Chemists". *Official methods of analysis of Association of Official Analytical Chemists.* Gaithersburg: AOAC International, 17th ed. 2000.

[8] Wong, K.S. and Jane, J. "Effects of pushing agents on the separation and detection of debranched amylopectin by high-performance anion-exchange chromatography with pulsed amperometric detection". *Journal of Liquid Chromatography*, 18, 63-80, 1995.

[9] Moraes, J., Alves, F.A. and Franco, C.M.L. "Effect of ball milling on structural and physicochemical characteristics of cassava and peruvian carrot starches". *Starch/Stärke*, 65, 200-209, 2013.

[10] Hayakawa, K., Tanaka, K., Nakamura, T., Endo, S. and Hoshino, T. "Quality characteristics of hexaploid wheat (*Triticum aestivum* L.): properties of starch gelatinization and retrogradation". *Cereal Chemistry*, 75, 576-580, 1997.

[11] Franco, C.M.L., Wong, K.S., Yoo, S.H. and Jane, J.L. "Structural and functional characteristics of selected soft wheat starches". *Cereal Chemistry*, 79, 243-248, 2002.

[12] Weber, F.H., Clerici, M.T.P.S., Collares-Queiroz, F.P and Chang, Y.K. "Interaction of guar and xanthan gums with starch in the gels

obtained from normal, waxy and high-amylose corn starches". *Starch/Stärke*, 61, 28-34, 2009.

[13] Ji, Y., Zhu, K., Zhou, H. and Qian, H. "Study of the retrogradation behaviour of rice cake using rapid visco analyser, Fourier transform infrared spectroscopy and X-ray analysis". *International Journal of Food Science and Technology*, 45, 871-876, 2010.

[14] ICC "International Association for Cereal Science and Technology". *Rapid pasting method using the Newport Rapid Visco Analyser*. Standard n° 162, 1996.

[15] Bourne, M. C. "Texture profile analysis". *Food Technology*, 32, 62-66, 1978.

[16] Leach, H.W., McCowen, L.D. and Schoch, T.J. "Structure of the starch granule-Swelling and solubility patterns of various starches". *Cereal Chemistry*, 36, 534-544, 1958.

[17] Appelqvist, I. A. M. and Debet, M. R. M. "Starch-biopolymer interactions: a review". *Food Reviews International*, 13, 163-224, 1997.

[18] Hanashiro, I., Abe, J. and Hizukuri, S. "A periodic distribution of chain length of amylopectin as revealed by high-performance anion-exchange chromatography". *Carbohydrate Research*, 283, 151-159, 1996.

[19] Jane, J.L., Chen, Y.Y., Lee, L.F., Mcpherson, A.E., Wong, K.S., Radosavljevic and M., Kasemsuwan, T. "Effects of amylopectin branch length and amylose content on the gelatinizationg and pasting properties of starch". *Cereal Chemistry*, 76, 629-637, 1999.

[20] Saibene, D. and Seetharaman, K. "Segmental mobility of polymers in starch granules at low moisture contents". *Carbohydrate polymers*, 64, 539-547, 2006.

[21] Hizukuri, S., Kaneko, T. and Takeda, Y. "Measurement of the chain length of amylopectin and its relevance to the origin of crystalline polymorphism of starch granules". *Biochimica et Biophysica Acta*, 760, 188-191, 1983.

[22] Tester, R. F., Karkalas, J. and Qi, X. "Starch-composition, fine structure and architecture". *Journal of Cereal Science*, 39, 151-169, 2004.

[23] Barbosa, L.C.A. *Espectroscopia no infravermelho na caracterização de compostos orgânicos*, Editora UFV, Viçosa, 2011.

[24] Han, F., Liu, M., Gong, H., Lü, S., Ni, B. and Zhang, B. "Synthesis, characterization and functional properties of low substituted acetylated corn starch". *International Journal of Biological Macromolecules*, 50, 1026-1034, 2012.

[25] Van-Soest, J.J.G., Tournois, H., Wit, D. and Vliegenthart, J.F.G. "Short-range structure in (partially) crystalline potato starch determined with attenuated total reflectance Fourier-transform IR spectroscopy". *Carbohydrate Research*, 279, 201-214, 1995.

[26] Chung, H.J., Hoover, R. and Liu, Q. "The impact of single and dual hydrothermal modifications on the molecular structure and physicochemical properties of normal corn starch". *International Journal of Biological Macromolecules*, 44, 203-210, 2009.

[27] Srichuwong, S., Sunarti, T.C., Mishima, T., Isono, N. and Hisamatsu, M. "Starches from different botanical sources I: Contribution of amylopectin fine structure to thermal properties and enzyme digestibility". *Carbohydrate Polymers*, 60, 529-538, 2005.

[28] Tester, R. F. "Properties of damaged starch granules: composition and swelling properties of maize, rice, pea and potato starch fractions in water at various temperatures". *Food Hydrocolloids*, 11, 293-301, 1997.

[29] Yoo, S.H., Lee, C.S., Kim, B.S. and Shin, M. "The properties and molecular structures of gusiljatbam starch compared to those of acorn and chestnut starches". *Starch/Stäke*, 64, 339-347, 2012.

[30] Jane, J.L. *Chemical and functional properties of food saccharides*. CRC Press, Boca Raton, 2003, pp. 81-101.

[31] Han, X.Z. and Hamaker, B.R. "Amylopectin fine structure and rice starch paste breakdown". *Journal of Cereal Science*, 39, 279-284, 2001.

[32] Lawal, O. S. "Studies on the hydrothermal modifications of new cocoyam (*Xanthosoma sagittifolium*) starch". *International Journal of Biological Macromolecules*, 37, 268-277, 2005.

[33] Yoo, S.H. and Jane, J.L. "Molecular weights and gyration radii of amylopectins determined by high-perfomance size-exclusion chromatography equipped with multi-angle laser-light scattering and refractive index detectors". *Carbohydrate Polymers*, 49, 307-314, 2002.

Optimisation of Kernels Preparation Conditions Involved in the Press Extraction of Shea (*Vitellaria paradoxa* Gaertner F.) Butter

Mohagir A. M.[1,*], Bup N. D[2], Abi C. F.[3], Kamga R.[4], Kapseu C.[4]

[1]Faculty of Pure and Applied Science, University of N'djamena, Chad
[2]Higher Institute of the Sahel, University of Maroua, Cameroon
[3]Department of Chemistry, Higher Teachers' Training College, University of Yaounde1, Cameroon
[4]National School of Agro-Industrial Science, University of Ngaoundere, Cameroon
*Corresponding author: amohagir2003@yahoo.fr

Abstract In this study, different conditions of press extraction of shea butter were optimised. Response Surface Methodology (RSM) using the Doehlert experimental design has been employed in the optimisation. The independent variables considered were roasting time (0-60 min), roasting temperature (50-300°C), cooking time (0-300 min) and cooling time (0 -60 min). Four-second order polynomial models were generated. The models were well described the responses: extraction yield, acid value, peroxide value and Lovibond red colour of the process with satisfactory fits in terms of absolute average deviation and coefficient of determination. The optimum conditions were found to be 10-15 min roasting time, 160-225°C roasting temperature, 140-275 min cooking time and 30-40 min of cooling time. The optimum responses were 45.7% extraction yield, 19.3 acid value, 5.4 peroxide value and 1.6 Lovibond red colour. The statistical relation between the four independent variables and the process responses was well described.

Keywords: *optimisation, RSM- Doehlert experimental design, shea, kernels, butter, pressing, physicochemical characteristics*

1. Introduction

Shea butter obtained from kernels of shea tree (*Vitellaria paradoxa* Gaertner F.) fruit, is highly appreciated for its multi uses. It is used as; cooking and frying fat, Cocoa Butter Substitute (CBS) in chocolate and confectionary, useful base for preparing ointments to treat inflammations, rashes on children, ulcers, for wrinkles and dry skin, as cosmetic moisturizing cream, an antiseptic for the treatment of wounds, as hair oil and rub for rheumatic pains [1]. Shea butter deteriorates when extracted inadequately, with the principal decomposition reaction being oxidation [2]. In the traditional aqueous extraction method of shea butter which is very predominant in the Central African region and is the main source of supply of local markets with shea butter, the seed is ground, boiled in water and the oil is scooped off the top of the mixture [3]. This procedure is tedious, labour-intensive, time consuming and gives relatively low extraction yields of about 20% [4]. Moreover, the produced butter is of an inferior quality; yellow, brown and dark coloured, in poor hygienic condition and with high acid and peroxide values [5]. Womeni *et al.* (2006) studied the effect of cooking (boiling water) and drying of shea nuts on the quality of solvent extracted shea butter. They reported that temperature and time of thermal treatment has a significant influence on the yield and the quality of shea butter, but miss control of optimal parameters could lead to dangerous chemical modifications (colour, odour, **nutritional values**, etc) of the shea butter [6]. Oil content of shea kernel is about 35 – 60 % [4].

Press extraction is influenced by a number of variables including the type of oil seeds, the roasting and cooking conditions (temperature, time), the cooling time, the pressing force and the volume of oilseed to be pressed. Operating at elevated temperature enhances the oil yield [7]. However, high temperature usually leads to the deterioration of oil, with the principal decomposition reaction being oxidation [8]. The optimum conditions of a manual press for extracting oil from oilseeds have been studied [9,10] as an alternative for the traditional aqueous methods. Such studies could help in introducing many changes in its design and thus could reduce the hard efforts exercised by women during the processing of shea butter.

Optimisation studies using response surface methodology (RSM) were deeply investigated [11,12]. Furthermore, the optimum conditions could assist designers to manufacture simple unit operations that could limit or eliminate the tedious practice of shea butter extraction (roasting, grinding and cooking pre-treatments). Then absolute average deviation (AAD) and coefficient of determination (R^2) could draw to investigate the adequacy of the proposed models.

The present study involves optimisation of some parameters that are likely to affect extraction yield. The general practice of determining these optima is by one-variable-at-time approach. One of the disadvantages of this approach is that it does not include interaction effects among the variables and is unable to determine the true optimum conditions. In order to overcome this problem, optimization studies were done using **response surface methodology** RSM [13,14]. RMS is a collection of mathematical and statistical technique that is useful for modeling and analyzing situations in which a response of interest is influenced by several variables, especially if there is a need to optimize the responses of a process [15]. Doehlert matrix as an experimental design represents a uniform distribution of experimental points in space of coded variables. It is used particularly when there is a need to cover an experimental domain of any form of uniformly distributed points in order to explore the total domain (margins and interiors). Moreover, it permits to follow in a sequential manner in studying a response surface of second degree [16]. Polynomial equations with and without interaction could be proposed as models for the mentioned processes. A few studies have been reported on the thermal pre-treatment of shea kernels prior press extraction and since, shea butter is a promising multifunctional edible fat. It seemed to be important to study roasting and steam cooking processes of shea kernel. The objective of this study was to evaluate the influence of roasting time, roasting temperature, cooking time and cooling time on the extraction yield of shea butter using manual vertical screw press described by Kapseu *et al.* (2002) [17]. The second preoccupation was to assess the quality of the shea butter in terms of acid value, peroxide value and Lovibond red colour.

2. Materials and Methods

2.1. Sun Dried Shea Kernels

Sun dried shea kernels were purchased from Penie village in Chad in the second week of September, 2005. The kernels were winnowed manually to remove husk, fine peel, infected kernels and foreign matter and then the sample was pooled and packed in polythene bag and transported to the laboratory of applied chemistry in faculty of pure and applied science, university of N'djamena. After one week the kernels were characterised for moisture and the total lipid content using the methods described by AFNOR (1981) [18] and IUPAC (1979) [19], respectively.

All the chemicals used were of reagent grade.

Table 1. Doehlert experimental design of four independent variables employed for press extraction of shea butter.

Run No	Roasting time X_1 (x_1)	Roasting temperature (°C) X_2 (x_2)	Cooking time (min) X_3 (x_3)	Cooling time (min) X_4 (x_4)
1	30.5 (0.000)	175 (0.000)	150 (0.000)	30 (0.000)
2	60 (1.000)	175 (0.000)	150 (0.000)	30 (0.000)
3	0 (-1.000)	175 (0.000)	150 (0.000)	30 (0.000)
4	45.3 (0.500)	300 (0.866)	150 (0.000)	30 (0.000)
5	15.8 (-0.500)	50 (-0.866)	150 (0.000)	30 (0.000)
6	45.3 (0.500)	50 (-0.866)	150 (0.000)	30 (0.000)
7	15.8 (-0.500)	300 (0.866)	150 (0.000)	30 (0.000)
8	45.3 (0.500)	216.7 (0.289)	300 (0.816)	30 (0.000)
9	15.8 (-0.500)	133.3 (-0.289)	0.00 (-0.816)	30 (0.000)
10	45.3 (0.500)	133.3 (-0.289)	0.00 (-0.816)	30 (0.000)
11	30.5 (0.000)	258.3 (0.577)	0.00 (-0.816)	30 (0.000)
12	15.8 (-0.500)	216.7 (0.289)	300 (0.816)	30 (0.000)
13	30.5 (0.000)	91.8 (-0.577)	300 (0.816)	30 (0.000)
14	45.3 (0.500)	216.7 (0.289)	187.5 (0.204)	60 (0.791)
15	15.8 (-0.500)	133.3 (-0.289)	112.5 (-0.204)	0 (-0.791)
16	45.3 (0.500)	133.3 (-0.289)	112.5 (-0.204)	0 (-0.791)
17	30.5 (0.000)	258.3 (0.577)	112.5 (-0.204)	0 (-0.791)
18	30.5 (0.000)	175 (0.000)	262.5 (0.612)	0 (-0.791)
19	15.8 (-0.500)	216.7 (0.289)	187.5 (0.204)	60 (0.791)
20	30.5 (0.000)	91.8 (-0.577)	187.5 (0.204)	60 (0.791)
21	30.5 (0.000)	175 (0.000)	37.5 (-0.612)	60 (0.791)

x the coded value of variables
X the real value of variables.

2.2. Optimisation of Extraction Conditions and Statistical Analysis

The response surface methodology using Doehlert experimental matrix was used to optimise the extraction process. SigmaPlot version 9.01 and Mathcad softwares were used for statistical analysis (ANOVA), regression models and graphical optimisation. Besides, the fit of models was verified by the coefficient of determination (R^2) and the absolute average deviation (AAD). Four independent variables namely roasting time (X_1: 0 - 60 min), roasting temperature (X_2: 50 - 300°C), cooking time (X_3: 0 - 300 min) and cooling time (X_4: 0 - 60 min) were chosen. The ranges of independent parameters were selected based on literature review and preliminary studies. Twenty-one different experiments were presented in Table 1 according to the experimental design for the four parameters. The experiments were figured in coded (x) and real (X) values. The response functions (Y_i) measured were extraction yield (E_{ield}), acid value (Y_{Av}), peroxide value (Y_{Pv}) and Lovibond red colour (Y_{Colour}) of shea butter. The responses were related to the coded values (x_i) by the second order polynomial that shown in equation 1.

$$Y_i = b_0 + \sum_{i=1}^{k} b_i X_i + \sum_{i=1}^{k} b_{ii} X_i^2 + \sum_{i=1}^{k-1} \sum_{j=i+1}^{k} b_{ij} X_i X_j \quad (1)$$

The coefficients of the polynomial were represented by b_0 (constant term), b_i (linear effects), b_{ii} (quadratic effects) and b_{ij} (interaction effects). X_i and X_j are the independent variables.

The analyses of variance (ANOVA) were generated and the effect and regression coefficients of individual, quadratic and interaction terms were determined. The significances of all terms in the polynomial were judged statistically at a probability (*P*) of 0.001, 0.01 and 0.05. The regression coefficients were then used to make statistical calculation to generate contour map and response surface graphs from the regression models.

2.3. Determination of Extraction Yield

The yield of the oil recovered from the ground shea kernels (Y_{ield}) was calculated as a ratio of mass of the extracted oil to the mass of dry sample (equation 2), while the extraction yield (E_{ield}) was calculated from the amount of extracted oil to the total amount of oil available (L_c) in the sample (equation 3).

$$Y_{ield} = \frac{M_0\ (g)}{M_d\ (g)} * 100 \quad (2)$$

$$E_{ield} = \frac{Y_{ield}}{L_c} * 100 \quad (3)$$

Where M_o is the mass of oil recovered in gram, M_d is the mass of dry sample; L_c is the total lipid content.

2.4. Characterization of the Extracted Butter

The colour of shea butter was measured using Lovibond method described by Sun *et al,* 2001[20] and ISO 15305, 1998 [21], while acid and peroxide values were evaluated using the method described in AFNOR (1981) [18]. Each experiment in the mentioned analysis was replicated twice and the average value was used.

2.5. Roasting and Cooking of the Kernels

Figure 1, shows the steps involved in the press extraction process. In roasting process, previously cleaned, washed and dried sand was put in a steel dish, and then the dish put in an electrical oven. The oven was adjusted to the desirable temperature for one hour. In each experiment of the roasting process, the sample of sun dried shea kernel was mixed with the heated sand and stirred every two minutes for a length of time called roasting time. After roasting the sample was pounded in a wooden mortar until it became flourlike, and then sieved to 2 – 3 mm squared mesh, using standard sieves [22]. In the cooking process, the sieved sample was put in a Pyrex flask (100 ml capacity), firmly covered with aluminum sheet and then placed in a thermostatic water bath containing boiling water. The beaker and the boiling water were separated by a 4 - 5 cm distance such that only steam from the bath was used in cooking. After cooking, the sample was put in a crucible to cool at ambient condition (28 ± 2°C) for a length of time called cooling time.

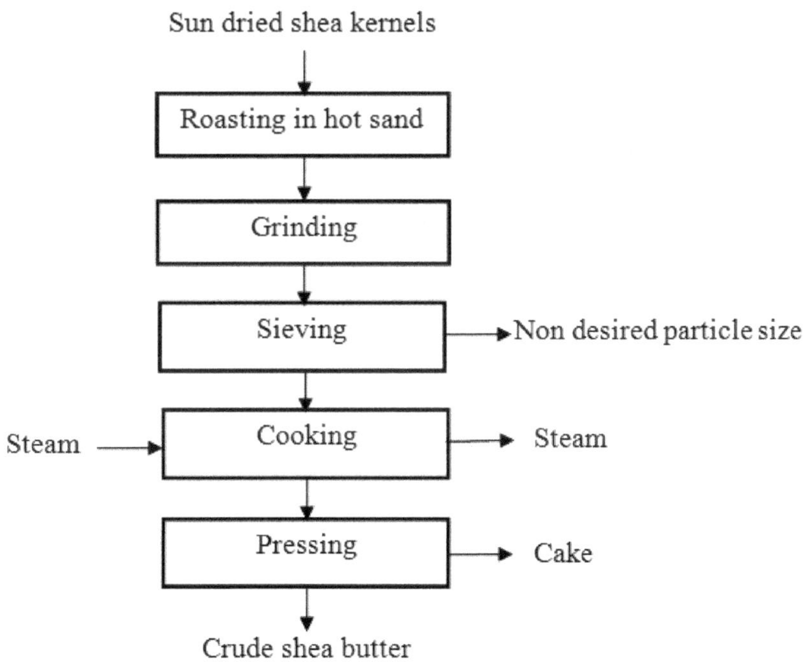

Figure 1. Flow sheet for the press extraction process of shea butter

2.6. Extraction Procedure

In each experiment of extraction process, 30 g of the cooked sample was introduced into small cotton bag of a known weight and then pressed. Pressing force was applied slowly up to the maximum degree according to the methods described by Kapseu *et al.*, 2002 [17]. After extraction, the oil was allowed to cool in a crucible for 30 minutes, then weighed and kept for further analysis.

Table 2. Influences of cooking time, roasting time, roasting temperature and cooling time on the responses of extraction process

Run No	Extraction yield (%)	Acid value	Peroxide value	Lovibond red colour
1	43.84 ± 0.47	19.76 ± 0.29	6.06 ± 0.21	2.17± 0.12
2	36.54 ± 0.29	19.66 ± 0.33	6.01 ± 0.37	9.87 ± 0.21
3	41.65 ± 0.47	19.67 ± 0.45	6.00 ± 0.17	0.33 ± 0.06
4	33.22 ± 0.37	14.32 ± 0.62	3.04 ± 0.56	70.00 ± 0.05
5	42.71 ± 0.24	19.64 ± 0.29	6.27 ± 0.03	0.38 ± 0.10
6	37.9 ± 0.34	19.81 ± 0.03	6.05 ± 0.27	0.23 ± 0.08
7	35.94 ± 0.15	19.82 ± 0.06	6.05 ± 0.07	11.45 ± 0.13
8	38.07 ± 0.08	19.76 ± 0.26	6.02 ± 0.30	4.23 ± 0.06
9	31.43 ± 0.05	19.90 ± 0.04	14.09 ± 0.35	3.22 ± 0.16
10	32.86 ± 0.53	17.12 ± 0.32	8.65 ± 0.78	1.38 ± 0.03
11	22.04 ± 0.20	14.45 ± 0.39	7.01 ± 0.47	70.00 ± 0.05
12	42.45 ± 0.42	19.74 ± 0.09	4.02 ± 0.14	2.17 ± 0.15
13	39.23 ± 0.07	19.73 ± 0.13	6.09 ± 0.52	0.22 ± 0.07
14	37.67 ± 0.48	19.62 ± 0.21	4.04 ± 0.28	4.45 ± 0.13
15	36.18 ± 0.03	19.72 ± 0.23	8.09 ± 0.33	0.28 ± 0.08
16	38.96 ± 0.35	19.88 ± 0.32	8.04 ± 0.31	0.30 ± 0.10
17	29.77 ± 0.09	14.45 ± 0.41	6.10 ± 0.07	70.00 ± 0.05
18	35.65 ± 0.31	19.74 ± 0.65	6.05 ± 0.06	2.62 ± 0.13
19	38.53 ± 0.28	19.83 ± 0.06	6.03 ± 0.20	3.25 ± 0.09
20	40.96 ± 0.16	19.79 ± 0.15	9.06 ± 0.26	0.23 ± 0.03
21	34.35 ± 0.18	19.79 ± 0.16	8.01 ± 0.55	2.20 ± 0.10

3. Results and Discussion

The experimental results for extraction yield, acid value, peroxide value and Lovibond red colour under different treatment conditions are shown in Table 2.

The results of the analysis of variance, goodness of fit and the adequacy of models are summarised in Table 3. The data showed a good fit with the equation 1, which were statistically acceptable at $P < 0.05$ level. The values of coefficient of determination for the extraction yield, acid value, peroxide value and the colour are 0.94, 0.90, 0.90 and 0.92 respectively. These values of R^2 showed that the proposed models are adequate. Mendenhall, 1975 [23] reported that the closer the value of R^2 to the unity, the better the empirical models the actual data. In fact, Joglekar and May 1987 [24] suggested that, for a good fit of a model, R^2 should be at least 80.0%. On the other hand, the low absolute average deviation (AAD) values for the extraction yield, acid value, peroxide value and Lovibond red colour (0.06, 0.15, 1.5 and 4.83 respectively) confirmed the adequacy of the models. Therefore, the models could be used to generate surface response curves to explain the influence of the independent factors on the responses studied.

Table 3. Regression coefficients, coefficient of determination (R^2) and absolute average deviation (AAD) for the four responses of the press extraction process

Coefficient	E_{ield} (%)	Acid value	Peroxide value	Colour
b_0	43.84***	19.800***	6.000***	2.2
b_1	-1.878	-0.800	-0.875	8.01
b_2	-4.262**	-1.986**	-1.862*	30.595**
b_3	6.122***	1.525*	-2.573**	-13.673*
b_4	1.732	0.838	-0.157	-10.123
b_{12}	1.207	-3.060	-1.588	33.776
b_{13}	3.987	2.738	5.158*	-9.634
b_{14}	-1.713	0.412	-2.014	-8.465
b_{23}	7.105*	2.171	1.976	-38.522
b_{24}	0.975	2.665	-2.303	-31.119
b_{34}	0.289	-1.719	1.324	14.911
b_{11}	-4.745	0.000	0.000	2.9
b_{22}	-4.949	-1.767	-0.917	23.468
b_{33}	-11.333**	-1.561	2.607	10.437
b_{44}	-7.121*	-0.395	1.062	5.609
R^2	0.94	0.90	0.90	0.92
AAD	0.06	0.15	1.50	4.83

*** $P < 0.001$, ** $P < 0.01$, * $P < 0.05$.

From Table 3 it is observed that roasting and cooling times had no significant influence ($P < 0.05$) on the responses. However, longer roasting and cooling times reduce the extraction yield and increase the colour of shea butter. There is neither an interaction nor quadratic significant ($P < 0.05$) influence of the independent parameters on the acid value and the colour of shea butter, but the interaction of roasting and cooking time increased the peroxide value of the shea butter at $P < 0.05$ level. The extraction yield was significantly reduced by the quadratic effect of cooking ($P < 0.01$) and cooling ($P < 0.05$) times, whereas the interaction of roasting temperature and cooking time significantly ($P < 0.05$) increased the extraction yield.

To aid visualisation, the effect of roasting time (X_1), roasting temperature (X_2), cooking time (X_3) and cooling time (X_4) on the extraction yield, acid value, peroxide value and the colour of shea butter are presented in Figure 2 - Figure 6 using response surface graphs.

3.1. Extraction Yield

The influence of roasting time, roasting temperature, cooking time and cooling time on the extraction yield of shea butter during the press extraction process is presented in Figure 2 (a, b, c). When cooking and cooling times were kept constant at central values; 150 and 30 min respectively, the response surface (Figure 2a) indicated that the extraction yield increased with an increase in roasting time up to 43% at 20 min and then dramatically decreased to 34.5% at 60 min. In the same vein, the yield was increased slightly from 40.5 to 41.5% at roasting temperature 148°C after which it sharply decreased to 31% at 300°C roasting temperature. The increase of extraction yield with an increase in roasting temperature was probably due to the denature of proteins and similar substances by heating. Heating decreases the affinity of oil for solid surfaces of seeds. Thereby increasing the rate at which oil leaves the surface into the extraction medium. On the other hand, the decrease in extraction yield after 20 min of roasting time could be attributed to thermal polymerisation and decarboxylation of the oil as a result of high temperature.

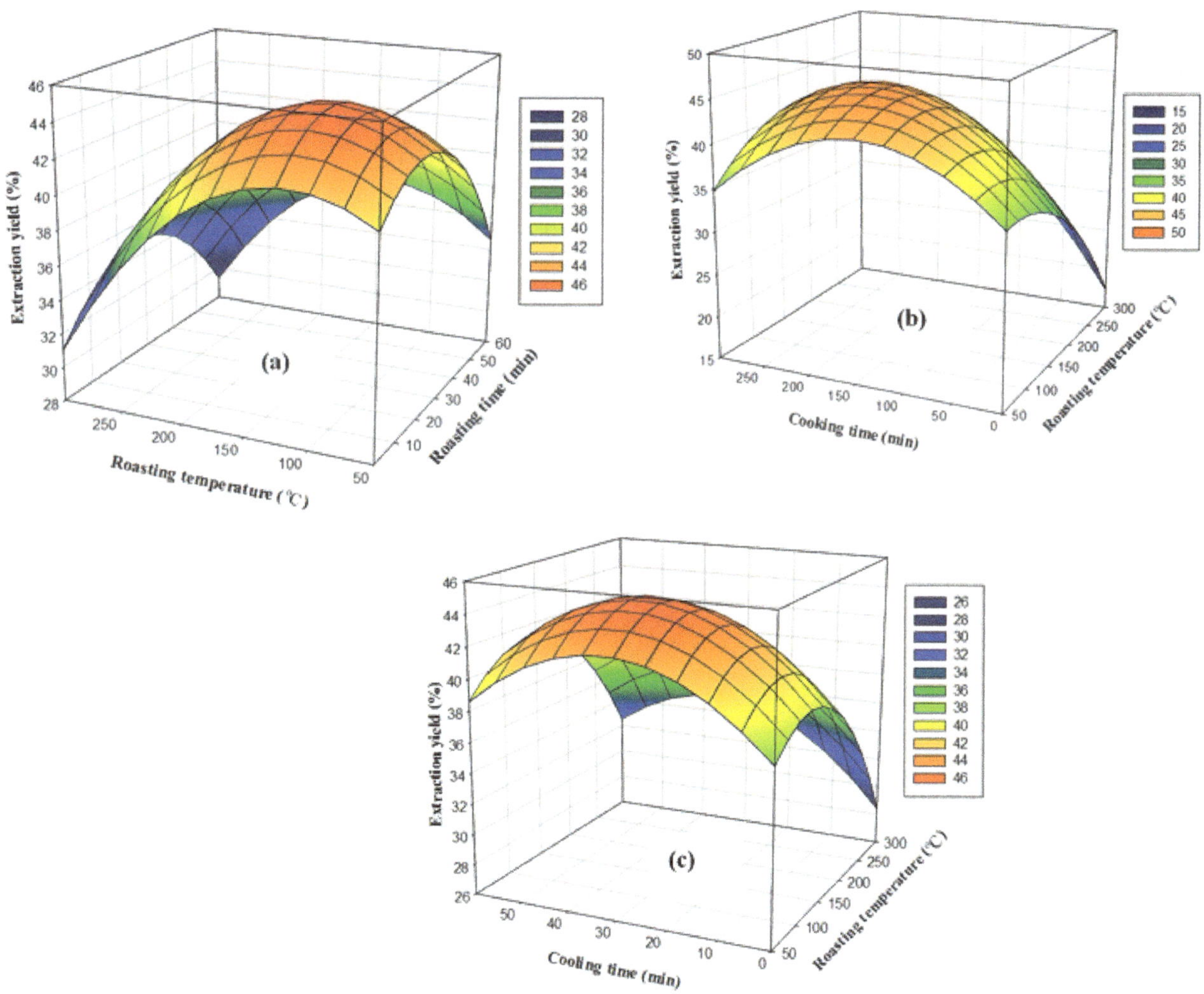

Figure 2. Response surface graph showing the variation of extraction yield of shea butter from sun dried shea kernels as a function of roasting temperature and roasting time (a), roasting temperature and cooking time (b), roasting temperature and cooling time (c)

3.2. Acid Value

Response surface plots and the model constants for acid value of press extracted shea butter are presented in Figure 3 and table III. The acid value was negatively related to the linear effects of roasting temperature and positively to the linear influence of cooking time at $P < 0.05$ confidence level. The interaction of roasting temperature and cooking time was positively related to the model constant (b_{23} = 2.2), but showed a negative effect on the acid value of shea butter. When the roasting and cooling times were maintained constant at the central points (Figure 3), it was observed that the acid value decreased rapidly as roasting temperature increased. The slight increase in acid value with the increase in cooking time could be attributed to chemical and enzymatic hydrolysis of oil in presence of heat and moisture, while the decrease in acid value with the increase of cooking time after 148 min could be related to deactivation of lipase by the vapour. This argument could be supported by the fact that the diffusion of heat from the heating medium (vapour) to the interior of the bulk of oilseed particle increases with time and so does the deactivation of the enzyme. The decrease in acid value with the increase in roasting temperature is probably due to polymerisation and decarboxylation as already mentioned before.

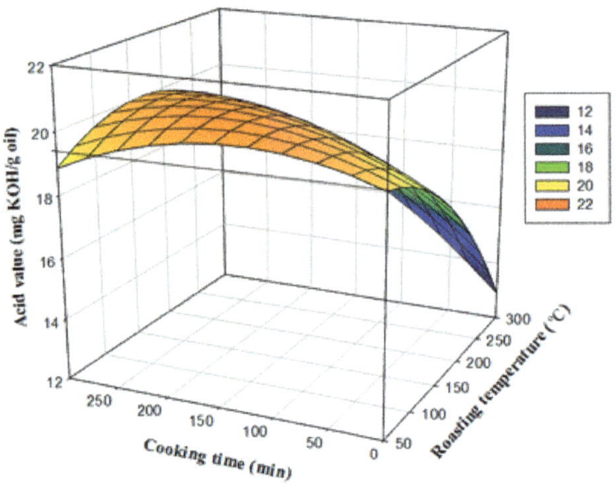

Figure 3. Variation of acid value of press extracted shea butter as a function of roasting temperature and cooking time

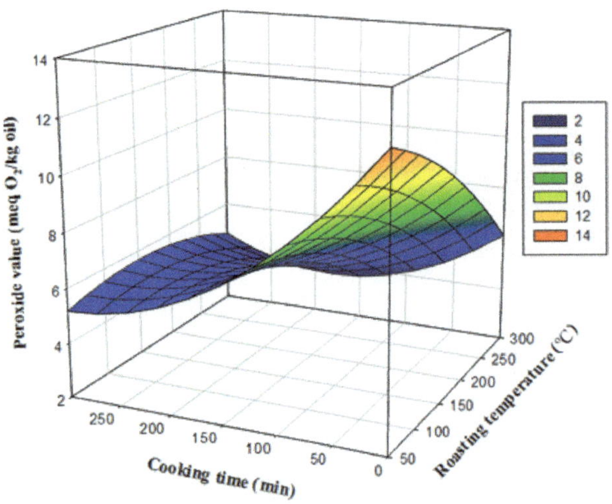

Figure 4. Variation of peroxide value of press extracted shea butter from sun dried shea kernels as a function of roasting temperature and cooking time

3.3. Peroxide Value

From Table 3, the model constants showed that, peroxide value was negatively related to the linear effects of roasting (b_2 = -1.9) and cooking (b_3 = -2.6) times at $P < 0.05$ confidence level, while the interactions of roasting temperature and cooking time (b_{23} = 2.0) and roasting time and cooking time (b_{13} = 5.2) were positively related to the peroxide value of shea butter. It should be noted that, negative value of model constant signifies a lowering effect of the peroxide value of shea butter. The decrease in peroxide value with an increase in roasting temperature and cooking time (Figure 4) could be due to decomposition of peroxides, which are unstable at high temperature.

3.4. Lovibond Red Colour

In table III, it is observed that, the linear terms of: roasting time, roasting temperature, the interactions of roasting time and roasting temperature; cooking time and cooling time and the quadratic terms of the four independent parameters studied had a negative influence on the colour of press extracted shea butter. On the other hand, the linear terms of cooking time, cooling time and the interaction effects of (roasting time and cooking time), (roasting time and cooling time), (roasting temperature and cooking time) and (roasting temperature and cooling time) had a positive effect on the colour of shea butter. It should be noted that the negative value of model constant indicates lowering of colour intensity of shea butter. The increase in colour index with an increase in roasting temperature (Figure 5) could be attributed to colour formation by both non-enzymatic browning reaction and phospholipids degradation during the roasting process or to the fact that colour reversion occurred. Sonntag, 1982 [25] reported that deep oxidation products (chromophore and 6-quinone), known as tocored were isolated from soybean oil. These products were responsible for colour reversion. It was seen with the necked eye that, when roasting temperature increased over 220°C without intervention of cooking process, a dark browning colour was formed.

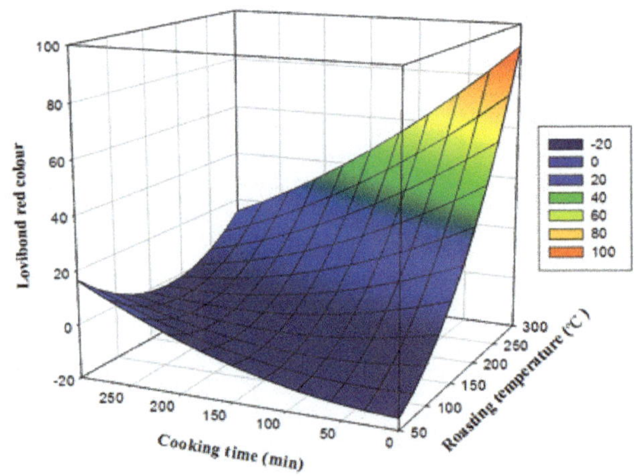

Figure 5. Variation of Lovibond red colour of press extracted shea butter as a function of roasting temperature and cooking time

3.5. The Optimum Condition of Extraction Yield

With the aim to point out the optimal conditions of press extraction process, a graphical optimisation was conducted using Sigmaplot software (SigmaPlot ®, 2004) [26]. Such a methodology consists of overlaying the curves of the four contour plots obtained from the Doehlert experimental design according to the specific criteria imposed. The optimum conditions were defined in order to get high extraction yield (> 45%) and minimum acid value, peroxide value (< 7%) and colour (0 - 10 Lovibond red colour) to meet nutritional and industrial economic requirements. Figure 6 shows the overlaying plot in which the purple shaded area represents the optimum domain satisfying the imposed criteria.

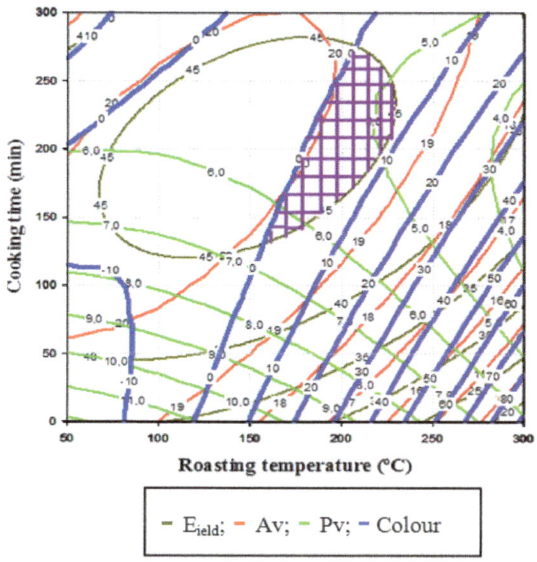

Figure 6. The optimum region identified by overlaying contour plots of the four responses (E_{ield}, Y_{Av}, Y_{Pv} and Y_{colour}) of press extraction process as a function of roasting temperature and cooking time

The optimum domain under these conditions were 10 - 15 min roasting time, 160 - 225°C roasting temperature, 140 - 275 min cooking time and 30 - 40 min cooling time. The corresponding responses were 45.7 % extraction yield, 19.3 (mg KOH/g oil) acid value, 5.4 (meq O_2/kg oil) peroxide value and 1.6 Lovibond red colour. These results confirmed that these optimum conditions had not affected the acid value of shea butter extracted from raw sun dried shea kernels (19.8). To confirm these results, additional laboratory studies permitted us to take the following as optimum points: roasting time (10 ± 3 min), cooling time (30 ± 5 min), roasting temperature (190 ± 2°C) and the steam-cooking time (200 ± 5 min).

The shea butter obtained in these conditions, could belong to category 1 reported by Mégnanoul *et al,* (2013) [27] which used in cosmetics and pharmaceutical industries, if the optimum cooking conditions stated by Bup *et al.,* (2009) [28] are used to cook the shea nut before processing, in order to improve the acid value.

4. Conclusion

Response surface methodology using Doehlert experimental design was successfully employed in the optimisation of press extraction process. Four second order polynomial models with satisfactory validation in terms of coefficient of determination (R^2) and absolute average deviation (AAD), were generated to describe the press extraction process. The optimum press extraction conditions were found to be 10 - 15 min roasting time, 160 - 225°C roasting temperature, 140 - 275 min cooking time and 30 - 40 minute cooling time. The corresponding responses of these optimum conditions were: 45.7% extraction yield, 19.3 (mg KOH/g oil) acid value, 5.4 (meq O_2/kg oil) peroxide value and 1.6 Lovibond red colour.

Nomenclature

AAD	Absolute average deviation
$\mathbf{E_{ield}}$	Extraction yield (%)
Av	Acid value (mg KOH/g)
$\mathbf{L_c}$	Total lipid content (%)
m	Mass (g)
$\mathbf{M_c}$	Moisture content (%)
$\mathbf{M_o}$	Mass of oil recovered (g)
$\mathbf{M_d}$	Mass of dry sample (g)
$\mathbf{b_0}$	Constant term
bi	Linear constant effect
bii	Quadratic constant effect
bij	Interaction constant effect
$\mathbf{X_i}$ **, Xj**	Constant independent parameters
$\mathbf{x_i}$	Coded value of the independent parameter
Pv	Peroxide value (meq O_2/kg)
meq	milliequivalent
$\mathbf{Y_{Av}}$	Measured acid value
$\mathbf{Y_i}$	Response function
$\mathbf{Y_{ield}}$	Yield of the oil recovered (g/g)
$\mathbf{Y_{Colour}}$	Measured Lovibond colour scale
$\mathbf{Y_{loss}}$	Measured oil loss (%)
$\mathbf{Y_{Pv}}$	Measured peroxide value
$\mathbf{R^2}$	Coefficient of determination
P	Probability

Acknowledgment

The corresponding author herby wants to acknowledge the financial assistance partially provided by the French Government through EGIDE program (Centre Français pour l'Accueil et les Echanges Internationaux) and University of N'Djamena, Chad.

References

[1] Kapseu, C., Dirand, M., Jiokap, Y., Parmentier, M., and Dellacherie, J, "Acides gras et triglycerides du beurre du karité du Cameroun", in *Séminaire international sur le séchage et sur la valorisation du karité et de l'aiélé, à l'Ecole Nationale Supérieure des Sciences Agro Industrielles*, 135-143. 1999.

[2] Adoum, D, "Technique artisanale d" extraction du beurre de karité à Moyen Chari sud du Tchad", in *séminaire atelier oléagineux et petites technologies II*, 92-96.1996.

[3] Womeni, H.M, "Identification et analyse des opérations critiques de préparation des fruits, graines et amandes de karité (*Butyrospermum parkii* (G. Don) Kotschy)", Etude de leur influence sur la qualité du beurre. Ph.D thèse, ENSAI, Université de Ngaoundéré, Cameroun, 248p, 2004.

[4] Tano-Debrah, K., and Ohta, Y, "Enzyme-assisted aqueous extraction of fat from kernels of shea butter tree (*Butyrospermum parkii*)", *J. Am. Oil. Chem. Soc.*, 71. 979-983. 1994.

[5] Mohagir, A.M., Kapseu, C., Kamga, R., Tchiegang, C., Kayem, J., and Fon Abi, C, "Effects of Some Parameters on the yield, Acid

and Peroxide Values of Aqueous Extracted Shea Butter", *International Journal of Applied Chemistry*, 1. 49-56. 2005.

[6] Womeni, H. M., Ndjouenkeu, R., Kapseu, C., Mbiapo, F. T., Parmentier, M., and Fanni, J, "Effet de la cuisson et du sechage de noix de karité (*Butyrospermum parkii* (G. Don) Kotschy) sur la qualité du beurre", *Tropicultura*, 24. 175-182. 2006.

[7] Mpagalile, J.J., and Clarke, B.C, "Effect of processing parameters on small scale coconut oil expression efficiency", *Int. J. Food Sci. Nutr.*, 56. 125-132. 2005.

[8] Mpagalile, J.J., Hanna, M.A., and Weber, A, "Seed oil extraction using a solar powered screw press", *Industrial crops and products*, 25. 101-107. 2007.

[9] Lou, Z., Wang, H., Zhang, M., and Wang, Z, "Improved extraction of oil from chickpea under ultrasound in a dynamic system", *Journal of Food Engineering*, 98 (1), 13-18. 2010.

[10] Lee, S.W., Jeung, M.K., Park, M.H., Lee, S.Y., and Lee, J.H, "Effects of Roasting Conditions of Sesame Seeds on the Oxidative Stability of Pressed Oil during Thermal Oxidation", *Food Chemistry*, 118. 681-685. 2010.

[11] Rai, P., Majumbar, G.C., Dasgupta, S., and De, S, "Optimizing pectinaseuasge in pre treatment of mosambi juice for clarification by response surface methodology", *Journal of Food Engineering*, 64. 397- 403. 2004.

[12] Sachindra, N.M., and Mahendrakar, N.S, "Process optimization for extraction of carotenoids from shrimp waste with vegetable oil", *Bioresource Technology*, 96. 195-1200. 2005.

[13] Rostami, M., Farzaneh, V., Boujmehrani, A., Mohammadi, M. and Bakhshabadi, H, "Optimizing the Extraction Process of Sesame Seed'S Oil Using Response Surface Method on the Industrial Scale", *Industrial Crops and Products*, 58. 160-165. 2014.

[14] Mani, S., Jaya, S. and Vandivambal, R, "Optimizationof Solvent Extraction of Moringa (MoringaOleifera) Seed Kernel Oil using Respons Surface Methodology", *Food and Bioproducts Processing*, 85. 328-335. 2007.

[15] Montgomery, D.C, *Design of analysis of experiments*, 5th ed., New York, John Wiley and Sons, 2001. 363-391.

[16] Mathieu, D., and Phan-Tan-Luu, R, *Approche méthodologique des surfaces de réponse*, in Plan d'expériences «Applications à l'entreprise », *Editions Technip*, 1995, 211-278.

[17] Kapseu, C., Avouampo, E., and Djeumako, B, "Oil extraction from Dacryodes Edulis (G.Don) H. J. Lam fruit", *Forest, Trees and Livelihoods*, 12. 97-104. 2002.

[18] AFNOR (Association Française pour la Normalisation), "Recueil des normes françaises, Corps gras, graines oléagineuses, produits dérivés", 2ième édition, Paris, 438p, 1981.

[19] IUPAC (International Union of Pure and Applied Chemistry), "Méthodes d'analyse des matières grasses et dérivées", 6 ième édition, 1ière partie 1.122, Institut des Corps Gras, Bordeaux, 1979.

[20] Sun, F., Zhau, D., and Zhau, S, "Determination of oil colour by image analysis", *J. Am. Oil Chem. Soc.*, 78. 749-752. 2001.

[21] ISO 15305, "Animal and vegetable fats and oils determination of Lovibond colour, Inspection of vegetable oils: methods for identification of transparency, colour, odour and taste", *Chinese National Standard GB 552585*. 1998.

[22] Mohagir, A.M., Kamga, R., Kapseu, C., and Abi, C.F, "Optimization of Some Pre-treatments involved in the Press Extraction of Shea (Vitellaria paradoxa Gaertner F.) Butter", *Asian Journal of Applied Sciences*, 2 (4). 372-384. 2009.

[23] Mendenhall, W, *Introduction to probability and statistics*, 4th ed., North Settuate, M. A., Duxbury pres, USA, 1975.

[24] Joglekar, A.M., and May, A.T, "Product excellence through design of experiments", *Cereal Food World*, 32. 857- 868. 1987.

[25] Sonntag, N, *Analytical methods*, in Bailey's Industrial Oil and Fat Products, (Edt. Swern, D), John Wiley and Sons, U.S.A., Vol.2, 4 th ed., 1982. 431-488.

[26] SigmaPlot ® for windows, Version 9.01, Systatsoftware, Inc. USA, 2004.

[27] Mégnanou1, R.M., AKPA1, E., Sevérin1, K.A., and NIAMKÉ, S.L, "Definition of Optimal Processing Conditions for Proposing Shea Butter Sensorial Standard via Ivorian Consumers'criteria", *IJPAES*, 3(1).7-14. 2013

[28] Bup, N.D., Abi, C.F., Tenin, D., Kapseu, C., and Tchiegang, C, "Optimisation of the Cooking Process of Sheanut Kernels (*Vitellaria Paradoxa* Gaertn.) Using the Doehlert Experimental Design", *Food and Bioprocess Technology*, 5(1). 108-117. 2012.

25

Effect of Wheat Bran Supplement with Sugarcane Bagasse on Growth, Yield and Proximate Composition of Pink Oyster Mushroom (*Pleurotus djamor*)

M. T. Hasan[1], M. H. A. Khatun[2,*], M. A. M. Sajib[3], M. M. Rahman[4], M. S. Rahman[1], M. Roy[4], M. N. Miah[1], K. U. Ahmed[1]

[1]Department of Biochemistry, Sher-e-Bangla Agricultural University, Dhaka, Bangladesh
[2]Department of Food Technology & Rural Industries, Bangladesh Agricultural University, Mymensingh, Bangladesh
[3]Department of Food Technology, Engineering and Nutrition, Lund University, SE Lund, Sweden
[4]Department of Food Engineering and Tea Technology, Shahjalal University of Science and Technology, Sylhet, Bangladesh
*Corresponding author: suavekobita@gmail.com

Abstract This study evaluated the effect of supplementing different levels of wheat bran with sugarcane bagasse on the production of pink oyster mushroom (*Pleurotus djamor*) and find out their yield and proximate composition. The sugarcane bagasse was mixed at spawning with 0%, 10%, 20%, 30%, 40%, and 50% of wheat bran supplement and arranged in a complete randomized design with three replications, three spawn packets in each replication and six treatments. Result showed significantly ($P<0.01$) different, the highest mycelium running rate was observed on 40% and lowest on 0% due to using different levels of wheat bran supplement. Number of days from stimulation to primordia initiation and harvest, sugarcane bagasse supplemented with 30% wheat bran took longer time (5.5 days) for primordia and (5 days) to harvest, while 20% took shorter time (6.8 days) to full colonization. 10% supplement provided highest result in term of average number of primordia/packet (176.3), fruiting body/packet (77.6) and weight/fruiting body (5.3 g). The growth of pileus and stipe were significantly ($P<0.05$ (stipe length, pileus thickness) and $P<0.01$ (diameter)) different, being highest diameter on 30%, stripe length on 10% and pileus thickness on 40%. The yields of pink oyster mushroom showed significantly ($P<0.01$) different result, biological yield, economic yield, dry yield, biological efficiency and cost benefit ratio were obtained higher with 10% wheat bran supplement. The maximum moisture content was determined on 50%, dry matter, ash on 20% and protein content, crude fiber on 30%. 50% supplement was accounted for the highest amount of nitrogen, iron and phosphorus, whereas the quantity of calcium, magnesium and zinc were higher on 20%, Potassium on 40% and sulfur on 30% supplement. Thus, 10% wheat bran supplementation with sugarcane bagasse proved to be a viable option to produce pink oyster mushroom due to economical effectiveness while any supplementation above this level might reduce the yield of mushroom significantly.

Keywords: *wheat bran, sugarcane bagasse, pink oyster mushroom, proximate composition, supplementation, pleurotus djamor, biological efficiency*

1. Introduction

Bangladesh is an agrarian based country with increasing number of population; available land for cultivation is decreasing day by day in order to accommodate vast population. Using conventional agricultural methods it is hard to fulfill the demand of food supply for this increased number of population. Now, we have to increase intensive use of land to increasing crop production. But it is very difficult in countries like Bangladesh due to weather condition, natural calamities and other barriers. At this time, we could meet our protein need from fish as well as energy from rice which is two more staple food of Bangladesh. In last decade, the fish production decreased considerably and we had to meet our protein need from vegetable source i.e. pulse. But now-a-days this is also much more expensive and it's a high time to find out an alternative source of protein as well as to introduce methods to increase production per unit area with the vertical use of land. Mushroom, which is a highly nutritious, delicious, medicinal and economically potential vegetable can be a solution to this.

Mushrooms are large reproductive structures of edible fungi belong to the class of Basidiomycetes or Ascomycetes. Approximately 0.3 million varieties of mushrooms are identified. Among them which are fully edible and have no toxic effect are to be considered as edible mushroom. Out of 2000 species of prime edible mushrooms about 80 have been grown experimentally, 20 cultivated commercially and 4-5 produced on industrial scale throughout the world [1]. The vegetative part of mushroom consists of thread like long thin mycelium which under suitable condition forms fruiting body or sporocarps. This fruiting body is used as edible mushroom.

Oyster mushrooms are the easiest and least expensive commercial mushrooms to grow because they are well known for conversion of crop residues to food protein [2]. Oyster mushroom (*Pleurotus ostreatus*) is an edible mushroom having excellent fragrant and taste and its cultivation on crop residues is considered as potential source of income, an alternative food production, provision of employment, and for recycling of agricultural wastes. The market for mushrooms has been reported to be on a continuous growth due to the interest in their culinary, nutritional, health benefits and their potential for use in waste management [3]. Pink oyster mushroom comes with the scientific name *Pleurotus djamor* var. *roseus* that has a light to dark pink colored cap depending upon the strain and growing conditions.

Figure 1. Pink oyster mushroom (*Pleurotus djamor*) [4]

Primordia and young mushrooms are bright pink but become less intensely colored as the mushroom matures. The color disappears entirely after cooking. The Pink oyster mushroom is one of the fastest growing *Pleurotus* species and can readily colonize on any kind of agricultural waste including wheat or paddy straw, sawdust, sugarcane bagasse. The fruit body formation also takes very less time as compared to all other *Pleurotus spp*.

Report showed that oyster mushrooms are one of the most delicious foods due to their high nutritional value, very good taste and medicinal value [5]. The low calorie and cholesterol free mushroom diets also display certain medicinal properties. Mushroom reduces the diabetic on regular feeding [6]. It also reduces the serum cholesterol in human bodies which reduces hypertension [7]. Mushroom inhibits the growth of tumor and cancer. Edible mushrooms have been treated as important tool in modern medicine for their medicinal values. Oyster mushroom contains 19-35% protein on dry weight basis as compared to 7.3% in rice, 13.2% in wheat and 25.2% in milk [1]. It is rich in essential minerals and trace elements [8]. Mushrooms are source of Niacin (0.3 g) and Riboflavin (0.4 mg). Mushroom is a good source of trypsin enzyme. It is also rich in iron, copper, calcium, potassium, vitamin D and folic acid. Mushrooms are valuable health food, which are low in calories, high in vegetable proteins, zinc, chitin, fiber, vitamins and minerals. [9]

Substrate plays an important role in the yield and nutrient content of oyster mushroom. The substrates on which mushroom spawn (Merely vegetative seed materials) is grown, affects the mushroom production. [10] A remarkable variation observed in nutritional content of oyster mushroom in different substrates [11]. Mushroom cultivation has been shown to exploit the natural ability of fungi to bio-divert solid waste generated by industry and agriculture into food and therefore many agricultural and industrial wastes can be utilized as substrates for production of *Pleurotus species* [12]. The mushroom mycelia requires specific nutrients for its growth and the addition of supplements increases mushroom yield by providing specific nutrients for the mycelium growth [13]. Cereal bran rich in protein is usually added to the substrate in *P. ostreatus* cultivation to stimulate mycelia growth and increase the yield of mushroom [14]. The objective of this work was to evaluate the effects of wheat bran supplementation on the growth and productivity of pink oyster mushroom, with a view to know the physio-chemical characteristics of pink oyster mushroom (*Pleurotus djamor*).

2. Material and Methods

The experiment was carried out at the laboratory of Department of Biochemistry and Mushroom Culture House (MCH), Sher-e-Bangla Agricultural University, Dhaka-1207. *Pleurotus djamor*, the most cultivated pink oyster mushroom in Bangladesh was used to perform this study.

2.1. Sample Collection

Mother culture of Pink Oyster Mushroom was collected from National Mushroom Development and Extension Center (NAMDEC), Savar, Dhaka.

2.2. Varietal Characteristics of Pink Oyster Mushroom

Pink oyster mushroom (*Pleurotus djamor*) is characterized by the rapidity of the mycelial growth and high saprophytic colonization activity on cellulosic substrates. Their fruiting bodies are shell or spatula shaped with pink color. If the temperature increases above 32°C, its production markedly decreases.

2.3. Design of Experiments

Six different experiments with six treatments with three replications were conducted to achieve the desired objectives. The treatments were as follows:

T_1: Sugarcane bagasse supplemented + wheat bran@ 0% (Control)
T_2: Sugarcane bagasse supplemented + wheat bran @ 10%
T_3: Sugarcane bagasse supplemented + wheat bran @ 20%
T_4: Sugarcane bagasse supplemented + wheat bran @ 30%
T_5: Sugarcane bagasse supplemented + wheat bran @ 40%
T_6: Sugarcane bagasse supplemented + wheat bran @ 50%

The experiment was laid out in single factor Completely Randomized Design (CRD). The experiment included six treatments with three replications and three spawn packets in each replication.

2.4. Substrate Preparation

Spawn packets using different levels of supplements were prepared separately. With spawn preparing substrate;

different supplements (at the different rate on dry weight basis) and $CaCO_3$ (1 g per packet) was added. The measured materials were taken in a plastic bowl and mixed thoroughly by hand and moisture was increased by adding water. Moisture was measured by using the moisture analyzer and adjusted the moisture content at 65%.

2.4.1. Preparation of Spawn Packets

The mixed substrates were filled into 9×12 inch polypropylene bag at rate 750 g. The filled polypropylene bags were prepared by using plastic neck and plugged the neck with cotton and covered with brown paper placing rubber band to hold it tightly in place.

2.4.2. Sterilization, Inoculation and Mycelium running in Spawn Packets

The packets were sterilized about 1 hrs and then these were kept for cooling. After cooling, 5g mother spawn were inoculated into the packets in the laminar airflow cabinet and were kept at 20-22°C temperature until the packets become white with the mushroom mycelium. After completion of the mycelium running the rubber band, brown paper, cotton plug and plastic neck of the mouth of spawn packet were removed and the mouth was wrapped tightly with rubber band. Then these spawn packets were transferred to the culture house.

2.4.3. Cultivation of Spawn Packet

Two ends, opposite to each other of the upper position of plastic bag were cut in "D" shape with a blade and opened by removing the plastic sheet after which the opened surface of substrate was scraped slightly with a tea spoon for removing the thin whitish mycelial layer. Then the spawn packets were soaked in water for 15 minutes and invested to remove excess water for another 15 minutes. The packets of each type were placed separately on the floor of culture room and covered with paper. The moisture of the culture room was maintained 80-85% relative humidity by spraying water 3 times a day. The light around 300-500 lux and ventilation of culture house was maintained uniformly. The temperature of culture house was maintained 22°C to 25°C. The first primordia appeared 2-4 days after scribing depending upon the type of substrate. The harvesting time also varied depending upon the type of substrate.

2.4.4. Collection of Produced Mushrooms

Pink oyster mushrooms matured within 2-3 days after primordia initiation. The matured fruiting body was identified by curial margin of the cap. Mushrooms were harvested by twisting to uproot from the base.

2.5. Growth and Yield Analysis of Oyster Mushroom

The parameters measured were number of contaminated bags, number of days for substrate colonization, mushroom pileus diameter (cm), stipe length (cm), thickness of pileus and stripe (cm), total mushroom yield (g), biological efficiency (BE) and proximate composition and their mineral content.

2.6. Proximate Composition Analysis of Oyster Mushroom

The moisture content, dry matter, total fat, ash and carbohydrate content were determined by the standard AOAC method [15] and the estimation of total protein was made by method described by Ronald and Ronald [16]. Crude fibre was estimated by the process of Ranganna [17].

2.7. Elementary Composition Analysis

In order to investigate the mineral content of oyster mushroom the main elements including Nitrogen, Sulphur, potassium, magnesium, calcium, iron, zinc and phosphorus were determined based on the methods of AOAC [15,18].

2.8. Statistical Analysis of Data

The data for the characters considered in the present experiments were statistically analyzed following the Complete Randomized Design (CRD) and Randomized Complete Block Design (RCBD) method. The analysis of variance was conducted and means were compared following least significant difference (LSD) test at 1% and 5% level of probability for interpretation of results.

3. Results and Discussion

3.1. Mycelium Running Rate (cm)

Mycelium running rate per day (MRR) for each type of substrates was measured after the mycelium colony crossed the shoulder of the packet. The linear length was measured at different places of packet. The mycelium running rate of oyster mushroom greatly influenced with the supplement of wheat brans in different levels. The highest running rate was observed in 50% (0.96 cm) and in 0% (0.72 cm). The other treatments varied significantly over control.

3.2. Time Required from Stimulation to Initiation of Priomordia

The time from stimulation to primordia initiation ranged from 3.33 days to 5.50 days. The highest time from stimulation to primordia initiation was observed in 30% (5.50 days). Duration of primordia initiation to first harvest of oyster mushroom was significantly lower as compared to control where no supplement was used and the duration required for total harvest of oyster mushroom increased with the level of supplement used. In the present study, the time required for total harvest also decreased with the levels of supplements increased compared to sugarcane bagasse alone.

3.3. Time Required from Priomordia Initiation to Harvest

The lowest time from primordia initiation to harvest was in 20% (3.39 days) and the highest time from primordia initiation to harvest was observed in 30% (5.00 days) followed by 50% (4.50 days). Spawn running pinhead formation took 7-8 days and fruiting body formed after 3-5 days, sporocarps harvested after 10-12 days. A comparison of the data among different treatments is shown on Table 1.

Table 1. Growth and Time of full Colonization of Pink Oyaster Mushroom on Wheat Bran

Treatments	Mycelium running rate in spawn packet (cm)	Time from Stimulation to Priomordia Initiation (days)	Time from Priomordia Initiation to Harvest (days)
T_1 (0%)	0.72^{e}	3.333^{b}	4.220^{b}
T_2 (10%)	0.78^{d}	3.333^{b}	4.167^{b}
T_3 (20%)	0.87^{c}	3.387^{b}	3.390^{c}
T_4 (30%)	0.91^{b}	5.500^{a}	5.000^{a}
T_5 (40%)	0.96^{a}	4.000^{b}	4.333^{b}
T_6 (50%)	0.79^{d}	4.167^{b}	4.500^{ab}
CV (%)	0.71	13.57	6.76
Level of Significance	**	**	**
LSD (0.05)	0.028	0.9763	0.5241

Means followed by same letter significantly different at 1% or 5% level of significance. ^{NS} Not significant * Significant at 5% level; ** Significant at 1% level.

3.4. Average Number of Primordia/Packet

The highest average number of primordia/packet was observed in 10% (176.3) followed by 20% (159.0), 30% (148.0) and 50% (146.0) and the lowest average number of primordia/packet were in 0% (75.33). The number of primordia and the average yield significantly varied with the substrates used in production of oyster mushroom.

3.5. Average Number of Fruiting Body/Packet

The highest average number of fruiting body/packet was observed in 10% (77.67) followed by 30% (69.67) and the lowest average number of fruiting body /packet were in 50% (49.00). The number of primordia increased with the levels of supplement and continued up to a certain range and decline thereafter. In the present study the average number of fruiting body in creased up to 10 % of cow dung used as supplement and decreased thereafter.

3.6. Average Weight of Individual Fruiting Body

The average weight of individual fruiting body in different treatment ranged from 2.167 g to 5.283 g. The highest average weight of individual fruiting body was observed in 10% (5.283 g) and the lowest average weight of individual fruiting body was in 30% (2.167 g). The effect of supplementation on the weight of fruiting body was comparatively higher ranged from (5.02g to 7.01g) which may be due to environmental conditions or growing season [19]. Table 2 shows the effect of different levels of wheat bran with sugarcane bagasse on the yield contributing characters of the studied mushroom.

Table 2. Effect of different levels of wheat bran with sugarcane bagasse on the yield contributing characters

Treatments	Avg. no of primordia/packet	Avg. no of fruiting body/packet	Avg. wt. of individual fruiting body (g)
T_1 (0%)	75.33 b	55.33a	5.127 a
T_2 (10%)	176.3 a	77.67a	5.283 a
T_3 (20%)	159.0 a	64.33a	4.017 ab
T_4 (30%)	148.0 a	69.67a	2.167 c
T_5 (40%)	116.0 ab	65.33a	3.860 abc
T_6 (50%)	146.0 a	49.00a	3.150 bc
CV (%)	24.80	24.80	24.76
Level of Significance	*	ns	*
LSD (0.05)	61.72	28.68	1.772

Means followed by same letter significantly different at 1% or 5% level of significance.
^{NS} Not significant * Significant at 5% level; ** Significant at 1% level.

3.7. Effect on Dimension of Fruiting Body

The highest average length of stipe was observed in the treatment 10% (1.940 cm) and the lowest length and diameter in 50% (0.996 cm and 0.60 cm) respectively. The highest average diameter of stipe was observed in 30% (1.15 cm). The length of stipe of oyster mushroom on different substrates varied from 1.93cm to 2.97cm and the diameter ranged from 0.74cm to 1.05cm. 30% wheat bran supplement showed the highest average diameter (6.53 cm) of pileus and the lowest in 50% (5.68 cm). The highest average thickness of pileus was observed in 50% (0.616 cm) and the lowest in 10% (0.380 cm). The diameter of pileus ranged from 4.85 cm to 8.95 cm and thickness of the pileus ranged from 0.45cm to 0.70cm due to different substrates. The results are shown on Table 3.

Table 3. Effect of different substrates on the dimension of fruiting body of pink oyster mushroom

Treatments	Length of stipe (cm)	Diameter of stipe (cm)	Diameter of pileus (cm)	Thickness of pileus (cm)
T_1 (0%)	1.6201^{ab}	0.72^{d}	5.82^{d}	0.38^{b}
T_2 (10%)	1.940^{a}	0.86^{c}	5.98^{c}	0.46^{b}
T_3 (20%)	1.917^{a}	0.97^{b}	6.27^{b}	0.46^{b}
T_4 (30%)	1.393^{ab}	1.15^{a}	6.53^{a}	0.43^{b}
T_5 (40%)	1.200^{b}	0.82^{d}	6.34^{b}	0.61^{a}
T_6 (50%)	0.9967^{b}	0.78^{c}	5.96^{c}	0.45^{b}
CV (%)	0.6559	0.086	0.128	0.1286
Level of Significance	*	**	**	*
LSD (0.05)	23.84	1.03	1.01	14.75

Means followed by same letter significantly different at 1% or 5% level of significance.
^{NS} Not significant * Significant at 5% level; ** Significant at 1% level.

3.8. Biological Yield

The supplementation of sugarcane bagasse with wheat bran had great effect on biological yield. The highest biological yield was acounted for 10% (379.5 g) and lowest in 30% (154.7 g). The highest yield of Oyster mushroom (*Pleurotus ostreatus*) with the substrate composed of 20% rice husk. The highest biological yield was 247.3g/packet and the trend of economic yield corresponded with different supplements at different level.

3.9. Economic Yield

The supplementation of sugarcane bagasse with wheat bran increases the economic yield over control. The highest economic yield was recorded under treatment 10% (386.6 g) and lowest in 30% (151.9 g). The yield of *Pleurotus ostreatus* responded with the levels of supplements used with sawdust and increased with the level of supplementation and declined thereafter.

3.10. Dry Yield

The dry yield of the oyster mushroom, grown on sugarcane bagasse responded significantly in terms of dry yield with the different levels of supplement (wheat bran). The dry yield of mushroom was higher in 10% (12.47.40 g) and minimum in 0% (8.637 g). The diameter of pileus increased the quality and yield of mushroom and highest dry yield from mango sawdust and the range of dry yield from 4.28 g to 29.98 g.

3.11. Biological Efficiency

The highest biological efficiency was calculated in treatment 10% (50.59%) and lowest in 30% (20.62 %). The biological efficiency for different substrates ranged from 35.2 to 60.9 % and its differs due to combined supplements with basal ingredient that results better mushroom quality as well as Biological efficiency.

3.12. Cost Benefit Ratio

The highest cost benefit ratio was calculated in treatment 10% (5.02) and lowest 0% (ratio 4.22) was. The other treatments differed significantly in terms of cost benefit ratio. The performances of substrates were significantly differed based on benefit cost ratio, the highest cost benefit ratio of 6.50 with wheat straw. Results related to cost benefit ratio is provide on Table 4.

Table 4. Effect of different levels of wheat bran with sugarcane bagassee on the yield, biological efficiency and cost benefit ratio

Treatments	Biological yield (g)	Economic yield (g)	Dry yield (g)	Biological efficiency (%)	Cost benefit ratio
T_1 (0%)	156.2^c	154.1^c	8.637^f	36.893^b	4.22^f
T_2 (10%)	379.5^a	386.6^a	12.47^a	50.593^a	5.02^a
T_3 (20%)	299.2^b	293.2^b	9.700^e	39.890^b	4.71^c
T_4 (30%)	154.7^c	151.9^c	11.34^c	20.62^c	4.37^e
T_5 (40%)	255.4^b	252.0^b	10.70^d	34.053^b	4.86^b
T_6 (50%)	276.7^b	272.1^b	11.82^b	20.827^c	4.48^d
CV (%)	16.32	16.16	0.66	16.32	0.153
Level of Significance	**	**	**	**	**
LSD (0.05)	75.29	73.99	0.1286	10.04	1.70

Means followed by same letter significantly different at 1% or 5% level of significance.
^{NS} Not significant * Significant at 5% level; ** Significant at 1% level.

3.13. Effect on Proximate Composition

A graphical presentation showing the effects of different treatment on proximate composition is provide on Figure 2 as well as the data table can be seen on Table 5.

Table 5. Effect of different levels of wheat bran with sugarcane bagassee on chemical composition of pink oyster mushroom

Treatments	Moisture (%)	Dry matter (%)	Protein (%)	Lipid (%)	Ash (%)	Carbohydrate (%)	Crude fiber (%)
0%	90.3^{abc}	9.630^{cd}	19.44^f	5.870^b	7.890^f	54.70^a	20.24^e
10%	89.85^c	9.530^d	23.5^e	6.343^a	8.757^d	49.01^b	22.65^c
20%	90.33^{ab}	10.02^a	25.6^c	3.580^f	9.510^a	38.54^d	21.37^d
30%	90.10^{bc}	9.190^e	30.4^a	3.893^e	8.950^c	36.97^e	24.88^a
40%	90.52^{ab}	9.870^{ab}	27.8^b	4.353^c	9.177^b	42.55^c	23.67^b
50%	90.66^a	9.720b^c	24.5^d	4.193^d	8.230^e	33.45^f	21.23^d
CV (%)	0.29	0.85	0.41	1.62	0.68	1.70	0.40
Level of Significance	*	**	**	**	**	**	**
LSD (0.05)	0.4744	0.1522	0.1908	0.1409	0.1151	1.317	0.1627

Means followed by same letter significantly different at 1% or 5% level of significance.
^{NS} Not significant * Significant at 5% level; ** Significant at 1% level.

3.13.1. Moisture Content

The moisture content of the fruiting body shows significant difference. The moisture percent ranged from 89.85% to 90.66%. The highest moisture percent was observed in treatment 50% (90.66%) and lowest in 10% (89.85%). The moisture content of oyster mushroom was grown from 87 to 87.5% on different substrates.

3.13.2. Dry Matter Content

The dry matter percentage of the fruiting body shows significant difference. The highest dry matter was observed in 20% (10.02 %) and lowest in 30% (9.190 %). There are no significant differences among the treatments when cow dung used as supplement.

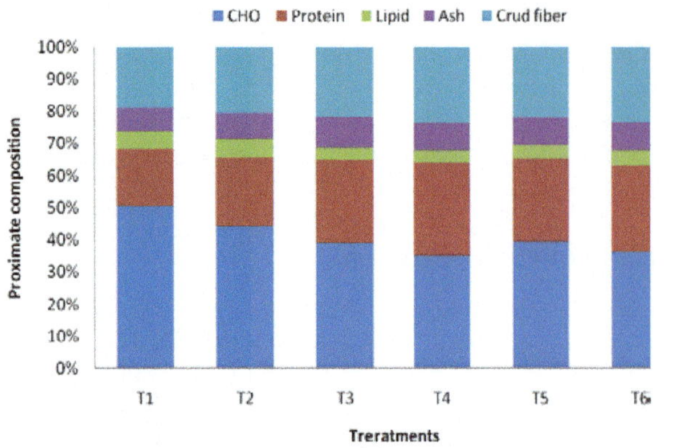

Figure 2. Effect of wheat bran supplements with sugarcane bagasse on the approximate composition of pink oyster mushroom (Dry weight basis)

3.13.3. Protein Content

The content of protein varied from 19.44 from 30.46% (w/w) in the mushroom grown on sugarcane bagasse with different levels of wheat bran. The highest content of protein was found in treatment 30% (30.46 %) and lowest in 0% (19.44 %). The protein content of oyster mushroom was 27.2% on an average.

3.13.4. Lipid Content

The lowest lipid percentage was accounted for treatment 20% (3.58 %) and highest in 10% (6.34 %). The result of the present study showed that the lipid content of the mushroom decreased as the supplements added with the substrates.

3.13.5. Ash Content

The highest percentage of ash was observed in the treatment 20% (9.51) and lowest in 0% (7.89). The range of ash content was 8.28 to 9.02% in *Pleurotus spp.* In the present study the ash content is as high as 12.80 might be due to the newly introduced varieties.

3.13.6. Carbohydrate Content

The lowest percentage of carbohydrate was counted under treatment 50% (33.45) and the highest in 0% (54.70). Related rresearch showed that carbohydrates content varies from 39.82 to 42.83% of in *Pleurotus spp [20].*

3.13.7. Crude Fiber Content

The highest percentage of crude fiber was found in 30% (24.88) and lowest in 0% (20.24). The findings of this present study is similar to other findings ranging from 22.87g/100g to 23.29g/100g of fiber in *Pleurotus spp [20].*

3.14. Effect on Elemental Content

3.14.1. Nitrogen Content

The highest percentage of nitrogen content (g/100gm) was counted under treatment 50% (4.99) and lowest nitrogen in 0% (3.59). The rest of the treatments were statistically similar in respect to percent nitrogen content. The findings of the present study match with other research that analyzed for various nutritional parameters and found that the range was 4.22 to 5.59 % of nitrogen on dry matter basis in fruiting bodies of oyster mushroom [21].

3.14.2. Phosphorus Content

The highest percentage of phosphorus content (g/mg/100gm) was counted under treatment 50% (1.077). The rest of the treatments were statistically similar but the lowest in 30% (0.913). But research found 0.97% phosphorus, in oyster mushroom grown on sugarcane bagasse based substrates [11].

3.14.3. Potassium Content

The highest proportion of potassium content (g/100gm) was counted under treatment 40% (1.567) and lowest 0% (1.190). On the other hand literature shows that 1.3% potassium found in oyster mushroom grown on sawdust based substrates [11].

3.14.4. Calcium Content

The maximum percentage of calcium content (mg/100g) was counted under treatment 20% (23.54) and in 40% (21.19). The findings of the present study match with the study that found 22.15 to 33.7 mg/100g of calcium in different oyster mushroom varieties [20].

3.14.5. Magnesium Content

The highest percentage of magnesium content (mg/100g) was counted under treatment 20% (21.05) and lowest in 0% (18.45). The rest of the treatments were statistically similar but differed significantly over control in respect to percent magnesium content. Literature shows that 13.4 to 20.22 mg/100g of magnesium in different oyster mushroom varieties [19]. (2).

3.14.6. Sulfur Content

There was no statistical difference among the treatments in terms of percent sulfur content. But the highest percentage of sulfur was counted under treatment 30% (0.04300) and the lowest in 0% (0.02300).

3.14.7. Iron Content

The maximum percentage of iron content (g/100g) was counted under treatment 50% (43.58) and lowest in 0% (40.52). 33.45 to 43.2 mg/100g of iron are found in different oyster mushroom varieties in other research [20].

3.14.8. Zinc Content

The highest percentage of zinc content (g/100g) was counted under treatment 20% (16.98) and lowest iron in 30% (14.45). This was more or less similar to other studies [20]. A comparison of the results can be retrieved from Table 6.

3.15. Correlation Study

A significant and positive correlation between average number of fruiting body and biological yield was observed when wheat bran was supplemented with sugarcane bagasse (Figure 3). The relationship showed a quadratic equation as $y = 3.223x+48.75$ ($R^2 = 0.143^{**}$). Where y = biological yield and x = average number of fruiting body. The majority of total variation in biological yield of the oyster mushroom can be explained by this equation. The R^2 value indicated that 14.30% of biological yield of Pink oyster mushroom (*Pleurotus djamor)* was attributed to the average number of fruiting body.

Table 6. Effect of different levels of wheat bran with sugarcane bagassee on elemental contents of pink oyster mushroom

Treatments	N (%)	P (%)	K (%)	Ca (%)	Mg (%)	S (%)	Fe (%)	Zn (%)
0%	3.59^{d}	0.95^{bc}	1.19^{d}	21.24^{e}	19.29^{d}	0.023^{a}	40.52^{d}	14.75^{d}
10%	4.23^{c}	1.07^{a}	1.55^{a}	23.08^{b}	18.45^{e}	0.032^{a}	42.30^{b}	16.55^{b}
20%	4.56^{b}	1.01^{ab}	1.39^{c}	23.54^{a}	21.05^{a}	0.030^{a}	43.57^{a}	16.98^{a}
30%	4.07^{c}	0.91^{c}	1.44^{bc}	22.15^{d}	20.24^{b}	0.043^{a}	41.54^{c}	14.45^{e}
40%	4.85^{a}	1.04^{a}	1.56^{a}	21.19^{e}	19.44^{c}	0.035^{a}	42.19^{b}	15.56^{c}
(50%)	4.99^{a}	1.07^{a}	1.51^{ab}	22.78^{c}	20.34^{b}	0.025^{a}	43.58^{a}	14.65^{de}
CV (%)	2.36	3.95	3.83	0.19	0.38	7.15	0.33	0.87
Level of Significance	**	**	**	**	**	**	**	**
LSD (0.05)	0.1908	0.08136	0.09965	0.08136	0.1409	0.05753	0.2508	0.2441

Means followed by same letter significantly different at 1% or 5% level of significance.
^{NS} Not significant * Significant at 5% level; ** Significant at 1% level.

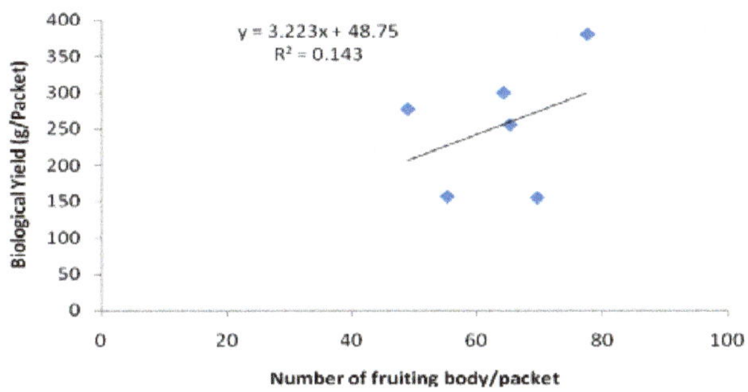

Figure 3. Relationship between average numbers of fruiting body with biological yield as influenced by different levels of wheat bran supplemented with sugarcane bagasse

A highly significant and correlation between average weight of individual fruiting body and economic yield was observed when wheat bran was supplemented with sugarcane bagasse (Figure 4). The relationship between average weight of individual fruiting body and economic yield could be expressed by the equation $y = 32.79x + 122.6$ ($R^2 = 0.188$*) where y = economic yield and x = average weight of individual fruiting body. The R^2 value indicated that 18.80% of economic yield of pink oyster mushroom (*Pleurotus djamor*) was attributed to the average weight of individual fruiting body.

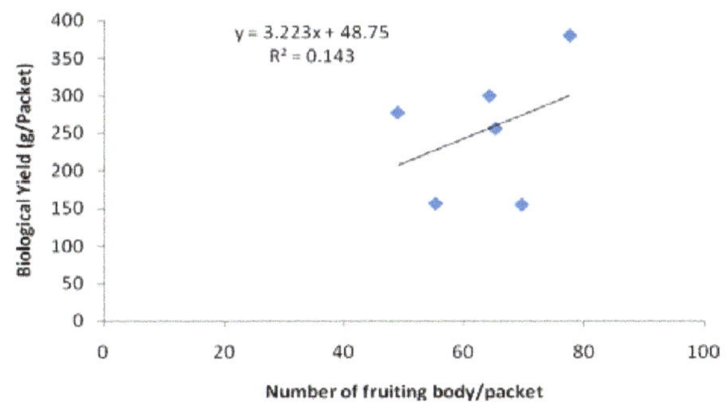

Figure 4. Relationship between average numbers of fruiting body with biological yield as influenced by different levels of wheat bran supplemented with sugarcane bagasse

4. Conclusion

The effect of 10% wheat bran as supplement with sugarcane bagasse were higher on growth, yield, biological efficiency and benefit cost ratio (BCR) of pink oyster mushroom (*Pleurotus djamor*) that required shorter days to full colonization as compared to the other treatments . It also proved to be better in terms mycelial running rate, length of stipe, average weight of individual fruiting body as well as average number of primordia and fruiting body. Sugarcane bagasse as supplement with wheat bran also had a great effect on proximate composition and mineral content of *Pleurotus djamor*. In this experiment more than one treatment performed better in case of benefit cost ratio. Therefore, 10% wheat bran supplemented with sugarcane bagasse can be recommended to pink oyster mushroom growers. In addition, it will be an economically effective due to highest yield and abundant availability throughout the year.

References

[1] Chang, S.T. and Miles, P.G. (1988). Edible Mushroom and their cultivation. CRC Press, Inc. Boca Raton, Florida U.S.A. pp. 27, 83, 88.

[2] Banik, S. and Nandi, R. (2004). Effect of supplementation of rice straw with biogas residual slurry manure on the yield, protein and mineral contents of oyster mushroom. *Industrial Crops and Products*. 20(3): 311-319.

[3] Amin, M.A. (2004). Studies on mycelium, spawn and production of certain edible mushrooms. M.S. Thesis, Department of Biotechnology, BAU, Mymensingh.

[4] Wikipedia (The free encyclopedia). (2015). *Pleurotus djamor*. Available from https://en.wikipedia.org/wiki/Pleurotus_djamor.

[5] Shah, Z. A., M. Ashraf and M. Ishtiaq. (2004). Comparative Study on Cultivation and Yield Performance of Oyster Mushroom (*Pleurotus ostreatus*) on Different Substrates (Wheat Straw, Leaves, Saw Dust). *Pakistan Journal of Nutrition* 3 (3): 158-160.

[6] Anderson, J. W. and Ward, K. (1979). High Carbohydrate high fiber diets for insulin- treated man with diabetes mellitus. *American Journal of Clinical Nutrition* 32:2313.

[7] Rahman, M.H., Ahmed,K.U., Roy,T.S., Shelly, N. J. and Rahman, M. S.(2012). Effect of wheat bran supplements with rice straw on the proximate composition of oyster mushroom (P*leurotus ostreatus*). *Bangladesh research publications journal* 7(4): 306-311.

[8] Chandha, K.L. and Sharma, S.R. (1995). Advances in Horticulture. Mushroom, Malhotra Publication house, New Delhi.13: 649.

[9] Alam, S. M. and R. Saboohi. (2001). Importance of mushroom. Available from http://www.mushroomworld.com.

[10] Klingman, A.M. (1950). Hand book of mushroom culture. CRC Publishing co. J. B. Kenneth Square, Pennsylvania, USA.

[11] Sarker, N.C., Hossain, M.M.,Sultana, N., Mian, I.H., Karim, A.J.M.S. and Amin, S.M.R. (2007). Impect of different Substrates on Nutrient Content of *Pleurotus ostreatus* (Jacquin ex Fr.) Kummer. *Bangladesh Journal of Mushroom.* 1(2): 35-38.

[12] Zadrazil, F., & Brunnert, F. (1981). Investigation of physical parameters important for the solid state fermentation of straw by white rot fungi. *European Journal of Applied. Microbiology and Biotechnol.*, 11, 183-188.

[13] Oei, P. (1996). *Mushroom cultivation*. Tool Publications, Leiden, The Netherlands.

[14] Siddiqui, M. A., & Khan, S. M. (1989). Some studies on the cultivation of oyster mushroom (Pleurotus spp.) on ligno-

cellulosic by-products of textile industry. *Proceedings of the 12th International Congress on the Science and Cultivation of Edible Fungi* (pp.121-128). Braunschweig, Germany.

[15] AOAC. (1990). Official Methods of Analysis. 15th ed., Association of Official Analytical Chemists, Artington, Virginia, USA.

[16] Salma, I. J., Sajib, M. A. M., Motalab, M., Mumtaz, B., Jahan, S., Hoque, M. M., Saha, B. K. (2015). Comparative Evaluation of Macro and Micro-Nutrient Element and Heavy Metal Contents of Commercial Fruit Juices Available in Bangladesh. *American Journal of Food and Nutrition 3(2):56-63.*

[17] Sajib, M. A. M., Jahan, S., Islam, M. Z., Khan, T. A. and Saha, B. K. (2014). Nutritional evaluation and heavy metals content of selected tropical fruits in Bangladesh. *International Food Research Journal* 21(2): 609-615.

[18] Sajib, M. A. M., Hoque, M. M., Yeasmin, S. and Khatun, M. H. A. (2014). Minerals and heavy metals concentration in selected tropical fruits of Bangladesh. *International Food Research Journal*, 21(5): 1731-1736.

[19] Bhuyan, M. H. M. B. U. (2008). Study on Preparation of Low Cost Spawn Packets for the Production of Oyster Mushroom (Pleurotus Ostreatus) and its Proximate Analysis. M.S. Thesis, Department of Biochemistry, SAU, Dhaka-1207.

[20] Alam, N., Khan, A., Hossain, M.S., Amin S.M.R. and Khan, L.A. (2007). Nutritional Analysis of dietary Mushroom *Pleurotus florida* Eger *and Pleurotus sajorcaj (*Fr.) Singer. *Bangladesh Journal of Mushroom.* 1(2): 1-7.

[21] Moni, K. H., Ramabardan, R. and Eswaran, A. (2004). Studies on some physiological, cultural and post harvest aspects of Oyster mushroom *Pleurotus ostreatus* (Berk). *Trop. Agril. Res.* 12: 360-374.

26

Effect of Preservation Methods of Oil Palm (*Elaeis guineensis*) Sap on Health Status of Male Wistar Rats

Theophilus Maduabuchukwu Ikegwu[1,*], Gabriel Ifeanyi Okafor[2], Izuchukwu Shedrach Ochiogu[3]

[1]Department of Food science and Technology, Faculty of Applied Science and Technology, Ibrahim BadamasiBabangida University, Lapai, Niger State, Nigeria
[2]Department of Food Science and Technology, Faculty of Agriculture, University of Nigeria, Nsukka, Enugu State, Nigeria
[3]Department of Veterinary Obstetrics and Reproductive Diseases, Faculty of Veterinary Medicine, University of Nigeria, Nsukka, Enugu State, Nigeria
*Corresponding author: tmikegwu@ibbu.edu.ng

Abstract Male Wistar rats (30) which were split into five groups of six rats each were administered four different forms of oil palm tree (*Elaeis guineensis*) sap samples by gavage based on 1.5% of their weekly body weights. The group 1 which served as control received only water, group 2 received pasteurized palm sap (PPS), group 3 received market palm wine (MPW), and group 4 received frozen palm sap (FPS), whereas group 5 received fresh palm sap (FrPS). Chemical compositions of the sap samples were determined. Normal feed and water were fed *ad libitum*. The mean body weight and feed intake of each rat group was measured every three (3) days. Blood samples were collected weekly for determination of random and fasten blood glucose. After 2 months of treatment, the bloods of each male rat group samples were collected for determination of the liver and kidney functions, while the palm sap products were subjected for microbial analyses. The sap samples content were analysed [1]. The palm sap was evaluated for its effect on the mean body weight and vital organ weight of albino rats. Results showed that palm sap had effect in reducing the mean body weight of the exposed rat groups compared to the control which could be attributed to the low alcohol content and probiotic effect of the fresh palm sap on the treated rat group. The spleen and kidney of the control rat group showed significant decrease compared to the treated rat groups, while the liver of the rat groups could be compared favourably. The percentage increases observed in spleen and kidney of the exposed rat groups was attributed to the presence of alcohol in the palm sap. However, since the liver of the palm sap treated rat groups were comparable to the control; it showed that the presence of alcohol in the palm sap administered to the Wistar rats were within normal limits. There were no significant differences on the random and fasting blood glucose of the rats which is an indication that palm sap drinking may not elevate the blood glucose levels in the blood stream. Results obtained also showed that palm sap contained mainly *Saccharomyces cerevisiea.* However, the presence of *Citrobacter freundi* in the MW could be associated with adulteration or exposure to unhygienic environment.

Keywords: *organ weight, body weight, palm sap, Liver, Kidney, Citrobacter freundi*

1. Introduction

The unfermented pale-yellow exudate from tapped unopened spathe of oil palm tree (*Elaeis guineensis*) is referred to as palm sap [2]. It is widely consumed as a refreshing and nourishing beverage, especially in West African countries. The fermented palm sap, called palm wine or toddy is the commonest form of the beverage. In some localities, the fresh sap may be bought undiluted and given to an elderly person as a mark of respect, or processed into different products such as caramel, sugar, spirit and recently, mixed fruit juice [3,4]. Palm sap is also usually pasteurized and bottled at industrial level in order to prolong its shelf life [4], although considerations for the total elimination of yeasts in fresh palm sap, believed to aid eyesight and reproductive system, had limited its industrial demand. Both males and females are predisposed to palm sap/wine drinking, although female folks drink less of the beverage than the male counterparts [5]. Consumption of the beverage (palm sap/wine) is a common feature in virtually all ceremonies such as traditional festivals, weddings and funerals where it is served as an indication of hospitality [5,6]. Traditionally, in south-eastern part of Nigeria, the dreg of palm wine is usually reserved for newly wedded men in order to boost their sexual performance and fertility. There is a dearth of scientific evidence to prove these claims. More so, the effects of different methods of preservation of this product on vital body organs and mean body weight need to be examined. This work was conceived to examine the effect of palm sap on vital body organs and mean body weight and also how different preservation methods some of the health indices of these rats.

2. Materials and Methods

2.1. Sample Source and Preparation

The sample was prepared using the method as in [1]. Fifteen (15) litres of freshly tapped oil palm sap (FrS) were collected from a palm wine tapper at Ajuona-Nsukka, Enugu State, in a cooler packed with ice blocks, and divided into three lots. Each lot (5 L) was subjected to either heat treatment (70°C for 40 min to obtain pasteurized palm sap [PS]), freezing (- 4°C till frozen [FS]), or left untreated to serve as fresh palm sap (FrS) along with five (5 L) litre of market palm wine (MW), and tap water that served as control. The commercial rat feed (Vital Pelletized Growers Feed) that was used, was manufactured by Grand Cereals and Oil Mills Ltd. (Jos, Plateau State, Nigeria) and contained cereals/grains, animal protein, vegetable protein, minerals, salts, essential amino acids, antibiotics, antioxidants, and vitamin premix. The nutrient composition as stated in the label included: crude protein, 14.50%; fat, 7.00%; crude fibre, 7.20%; calcium, 0.80%, and metabolizable energy, 2000 kcal/kg.

2.2. Experimental Animals

Thirty (30) male, healthy Wistar rats of 4 weeks old, weighing between 47.1–172.6 g, were obtained from the Department of Pharmacology and Toxicology, University of Nigeria, Nsukka. The rats were divided into five groups (1, 2, 3, 4, 5) of six rats each [1]. Each study group was split into two subgroups of three rats each, and housed in a stainless steel cage with plastic bottom grid and a wire screen top, in the animal house of the Department of Home Science, Nutrition and Dietetics, University of Nigeria, Nsukka. The room temperature and relative humidity for the rats were 23–24.5°C and 91–96%, respectively, with natural 12-h light–12-h dark cycle. The rats were served 80–100 g of the standard rat chow and at least 5 mL of water per day, ad libitum. The animals were acclimatized for 2 weeks before treatment. The group 1 rats were given normal feed and water only; group 2 rats were fed on PS in addition to the normal feed and water; group 3 rats were given normal feed, water, and MW; whereas groups 4 and five rats were given frozen palm (FP) sap and fresh (FrP), respectively, in addition to normal feed and water. The palm sap/wine was administered by gavage based on 1.5% of their weekly body weights. Each cage was supplied with 100 g of the specific diets. After three days, the remaining diets were weighed. The difference obtained between the two values was the amount of diet consumed by the rats for three days. Fresh supplies of diets were then added to the cages to maintain the total amount of 80 – 100 g. These processes were followed throughout the treatment period and the average taken after three days to determine the consumption rate per day. The male Wistar rats were weighed weekly with an electronic balance (Derive Instrument Company), and recorded. The male rats were humanely sacrificed after two months. The entire animal study was conducted in accordance with the Ethics and Regulations guiding the use of research animals as approved by the University of Nigeria, Nsukka.

2.3. Blood Glucose Test

The blood glucose of the rats was determined using Accu-check Active glucometer or blood monitoring system (Roche Diagnostics GmbH, Germany) [7]. Briefly, Scissors was used to cut a small portion of the rat's tail on weekly basis and blood droplets from the tail was dropped on the strip of the glucometer which displays the result within 5 seconds in mg/dl. The fasten blood glucose was determined using the same instrument but the male Wistar rats were fasted overnight before each determination.

2.4. Liver Function Tests

The parked cell volume (PCV) was evaluated by Microhaematocrit method as described [8]. The concentration of haemoglobin was determined by Cyanomethemoglobin method of Drabkin's haemoglobin reagent [9]. The red blood cell count, total white blood cell count and differential white blood cell count were determined using haemocytometer method [7]. The serum alanine amino transferase (ALT) and serum aspartate amino transferase (AST) were determined by colorimetric method for in-vitro determination of plasma ALT and AST in serum or plasma using QCA SGPT/ALT and QCA SGOT/AST test kits, respectively [10]. *Alkaline phosphate,* (ALP), was assayed using kits provided by Randox Laboratories Co; according to the method [11,12].

2.5. Kidney Function Tests

Serum creatinine was determined by the modified Jaffe method for the in-vitro determination of creatinine in serum, plasma or urine using the QCA creatinine test kit [13]. The bilirubin was determined by the in-vitro determination of total bilirubin in serum or plasma (QCA bilirubin test kit) [14].

2.6. Microbiological Analysis

The presence of microorganisms in the samples (both fresh and pasteurized) was determined by serial dilution of the samples into ten folds with sterile 0.1% media by the pour plate techniques [15] in triplicates. Bacterial counts were determined using nutrient agar (NA) and incubated for 24 h at 35°C while the yeasts and moulds counts were determined using Sabouraud Dextrose Agar (SDA) and 14% tartaric acid. The plates were incubated at 30°C for 3 days and then counted for microbial load.

2.7. Experimental Design and Statistical Analysis

The experiment was conducted in a completely randomized design [1]. Data generated were analysed using one way analysis of variance and mean separation was done by Duncan's New Multiple Range Test at 95% confidence interval using the SPSS version 17.0.

3. Results and Discussion

3.1. Results

The results obtained after going through the extensive chemical and biochemical analyses showed that the feed intake of rat group that had no treatment increased significantly ($p<0.05$) compared to the palm sap/wine treatment rat groups.

Table 1. Feed intake of Wistar rats treated with pasteurized and unpasteurized palm sap/wine

Treatment/ Feed intake	Group 1 (Water)	Group 2 (Pasteurized palm sap)	Group 3 (Palm wine)	Group 4 (Frozen palm sap)	Group 5 (Fresh palm sap)
Feed Intake (g)	78.00±6.71[a]	44.11±2.94[c]	43.78±2.40[c]	43.61±2.16[c]	53.44±3.27[b]
Body weight (g)	229.37±27.43[a]	154.12±28.55[b]	173.80±31.61[b]	215.87±6.31[a]	226.19±19.08[a]

Values are means± S.D (n = 6). Values with different superscripts within a row were significantly different ($p<0.05$) (Duncan separation). Mean value for two (2) months treatments.

The rat group treated with FrP sap showed high feed intake when compared with rat groups treated with PS, MW and FS. No significant ($p>0.05$) difference was observed in the rat groups treated with PS, MW and FrS. Though, the rat group treated with FrS showed high feed intake; this was not significant compared to rat groups administered with MW, PS and FS.

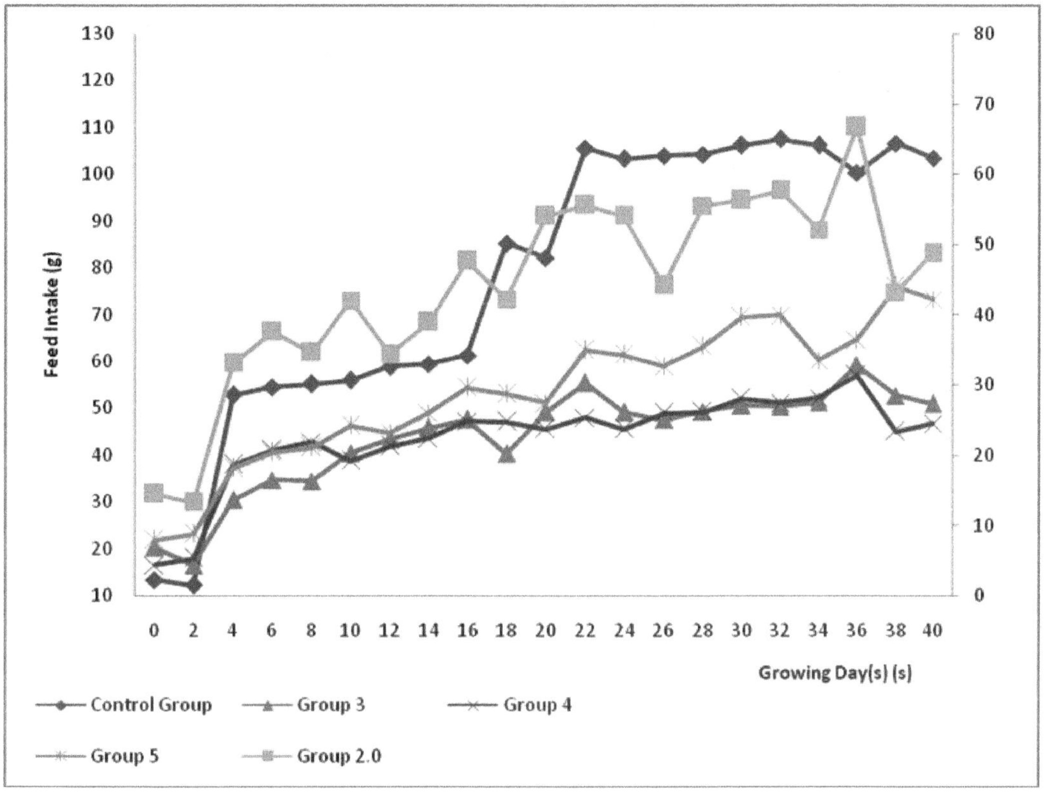

Figure 1. Weekly mean body weight of rats administered with treated palm sap/wine samples

Group 1 – Rats treated with normal feed + water
Group 2 – Rats treated with normal feed + water + pasteurized palm sap
Group 3 – Rats treated with normal feed + water + palm wine
Group 4 – Rats treated with normal feed + water + frozen palm sap
Group 5 – Rats treated with normal feed + water + freshly tapped palm sap.

The results in Table 1 showed the weekly mean body weight of the different groups of rats. The rat group that had no treatment (control) was found to have the highest body weight, which was 32.7% higher body weight than the least body weight (rats treated with PS). However, the decrease in weight was not significant ($p>0.05$) for group of rats treated with FS and FrP, but significant ($p<0.05$) for group of rats treated with MW and PS. The rat group administered with FrS showed uniform weekly mean body weight increase throughout the period of the study.

Table 2. Organ and body weight of rats treated with different and palm sap/wine samples

	Treatment Groups				
Some vital body organs and Weight	1 (Water)	2 (Pasteurized palm sap)	3 (Palm wine)	4 (Frozen palm sap)	5 (Fresh palm sap)
Spleen (%)	0.38±0.08[b]	0.50±0.11[a]	0.44±0.08[ab]	0.36±0.01[b]	0.44±0.08[ab]
Liver (%)	3.21±0.27[a]	3.39±0.34[a]	3.52±0.18[a]	3.13±0.20[a]	3.35±0.65[a]
Kidney (%)	0.66±0.05[b]	0.71±0.08[ab]	0.76±0.08[a]	0.69±0.0[ab]	0.70±0.06[ab]

Values are means± S.D (n = 6). Values with different superscripts within a row were significantly different ($p<0.05$).

Key:
Group 1 – Rats treated with normal feed + water
Group 2 – Rats treated with normal feed + water + pasteurized palm sap
Group 3 – Rats treated with normal feed + water + palm wine
Group 4 – Rats treated with normal feed + water + frozen palm sap
Group 5 – Rats treated with normal feed + water + freshly tapped palm sap.

The increase or decrease in organ weights of the male Wistar rats was studied. Results obtained showed that spleen and kidney of treated rat groups increased in weight while the liver decreased in weight compared to the rat group that had no treatment (Group 1). Increases were also observed in the rats administered with PS in their spleen and kidney. The increase in the organ weight in group 2 treated rats was highly significant ($p<0.05$) when compared with the group 1, and significant ($p<0.05$) when compared with the MW or FrS. The spleen of group 4 rats administered to FS was comparable to group 1 (control).

In the study of the random and fasting blood glucose, it was found that there were no significant differences ($p>0.05$) in the random and fasting blood glucose levels of the rats among the groups. Random blood glucose is indicative of a recent food intake and therefore should have higher reference values than the fasting glucose test. Fasting refers to refraining from eating or drinking any liquids other than water for eight to twelve hours and is used as a test for diabetes. During this period, glucagon (a hormone) is stimulated which increases the plasma glucose levels in the body. The stimulation of glucagon is counterbalanced by the production of insulin which helps to maintain the glucose levels, if the host is not diabetic. Therefore, people with diabetes either could not produce enough insulin to rebalance their blood sugar (type 1 diabetes) or that their body could not be able to use the produced insulin effectively enough (type 2 diabetes). The fasting blood sugar test is usually used to test the effectiveness of different medication or dietary changes on people already diagnosed as diabetic.

Table 3. Effect of palm sap/wine administration on the random and fasting blood glucose of Wistar rats

Blood glucose	Treatment Groups 1 (Water)	 2 (Pasteurized palm sap)	 3 (Palm wine)	 4 (Frozen palm sap)	 5 (Fresh palm sap)
Random blood glucose (iu/l)	109.80±5.22^a	113.60±7.70^a	113.20±10.76^a	116.80±5.12^a	115.6±17.01^a
Fasting blood glucose (iu/l)	93.80±17.85^a	92.20±22.79^a	85.00±13.66^a	95.80±13.88^a	84.80±11.61^a

Values are means± S.D (n = 6). Values with different superscripts in the same row were significantly different ($p<0.05$).
Key:
Group 1 – Rats treated with normal feed + water
Group 2 – Rats treated with normal feed + water + pasteurized palm sap
Group 3 – Rats treated with normal feed + water + palm wine
Group 4 – Rats treated with normal feed + water + frozen palm sap
Group 5 – Rats treated with normal feed + water + freshly tapped palm sap

The study of the haematological parameters in Table 4 showed that the rats treated with MW and FrS showed signs of adverse health effect when compared with the rat group that had no treatment. There were significant ($p<0.05$) differences in terms of parked cell volume (PCV), neutrophil, red blood cell and lymphocytes counts. There were also significant ($P<0.05$) differences in rat groups treated with FS and PS when compared to the control rat group in terms of their haemoglobin levels. Results obtained also showed that there were no significant ($p>0.05$) differences in the white blood cell and lymphocyte counts between group 4, group 5 rat groups and group 2 together with group 3 treated rat groups.

Table 4. Effect of palm sap/wine treatments on haematological parameters of rats

Parameters		Group of rats/feed 1 (Water)	 2 (Pasteurized palm sap)	 3 (Palm wine)	 4 (Frozen palm sap)	 5 (Fresh palm sap)
PCV (%)		36.33±1.86^a	31.50±1.38^b	32.50±2.74^b	30.33±1.37^b	32.83±2.14^b
HB (g/dl)		12.72±0.30^a	9.48±1.33^c	10.82±1.45^b	8.57±0.32^c	11.65±1.35^{ab}
RBC (x10^6 mm^3)		5.79+0.51^a	4.99+0.23^b	5.16+0.54^b	5.05+0.05^b	5.64+0.60^a
WBC (x 10^3 mm^3)		9.15±2.27^a	10.43±1.18^a	9.23±0.19^a	10.28±4.57^a	8.90±1.58^a
Differential leucocyte counts (%)	**N**	19.83±2.79^c	33.75±12.15^a	21.50±4.76^c	29.50±2.43^b	21.67±1.51^c
	E	0.67±0.51^a	0.33±0.82^b	0.17±0.41^b	-	-
	M	-	-	-	-	-
	B	-	-	-	-	-
	L	79.83±1.72^a	78.25±1.25^a	79.25±2.81^a	76.33±5.24^a	79.20±1.93^a

- Values are means± S.D (n = 6). Values with different superscripts within a row were significantly different ($p<0.05$). - Not detected.

Key:
Group 1 – Rats treated with normal feed + water
Group 2 – Rats treated with normal feed + water + pasteurized palm sap
Group 3 – Rats treated with normal feed + water + palm wine
Group 4 – Rats treated with normal feed + water + frozen palm sap
Group 5 – Rats treated with normal feed + water + freshly tapped palm sap
PCV – parked Cell Volume
RBC – Red Blood Cell count
HB – Haemoglobin
WBC – White Blood Cell count
E – Eosinophil
N – Neutrophil
M – Monocytes
B – Basophile
L – Lymphocytes

The clinical chemistry/biochemistry of the male rats was equally studied. It was therefore evident from the results that rat groups treated with PS, FrS and FS exhibited significant differences ($p<0.05$) in the levels of aspartate amino-transferase (AST), alanine amino-transferase (ALT), and alkaline phosphatase (ALP) activity, compared to the rat group that had no treatment (Table 4). The increases in serum aspartate amino-transferase (AST) and alanine amino-transferase (ALT) in the rat group treated with MW was significant ($p<0.05$) when compared to the rat group that had no treatment, but Alkaline phosphatase (ALP) activity was not significant ($p>0.05$) compared to the non-treatment rat group.

It has been reported that ethanol induces changes in the serum liver and kidneys [16]. In this study there were significant ($p<0.05$) differences in most of the liver marker and enzymes, such as aspartate amino-transferase (AST), alkaline amino-transferase (ALT), and alkaline phosphatase (ALP) activity, which are indicators of liver diseases. A rise in liver marker enzymes such as alkaline phosphatase (ALP), aspartate amino-transferase (AST) and alkaline amino-transferase (ALT) indicates lower levels of serum albumin showing reduced liver function [17]. There were no significant ($p>0.05$) differences in the levels of creatinine, except in the group treated with FrS, but significant differences existed in the levels of bilirubin between group 2 and other groups. The level of bilirubin was least in the rat group treated with PS, which could be as a result of breakdown of haemoglobin in the liver as shown in Table 5.

Table 5. Effect of palm sap/wine treatments on selected biochemical parameters of rats

Clinical chemistry	Groups				
	Group 1 (Water)	Group 2 (Pasteurized palm sap)	Group 3 (Palm wine)	Group 4 (Frozen palm sap)	Group 5 (Fresh palm sap)
AST (iu/l)	74.50 ± 2.45^{ab}	60.50 ± 4.12^{b}	49.00 ± 0.96^{c}	55.67 ± 3.91^{bc}	82.17 ± 3.87^{a}
ALT (iu/l)	18.17 ± 3.87^{b}	18.50 ± 2.74^{ab}	16.33 ± 3.14^{b}	19.50 ± 4.28^{a}	23.00 ± 3.79^{a}
ALP (iu/l)	73.83 ± 1.91^{a}	61.67 ± 11.41^{b}	70.33 ± 5.20^{a}	64.67 ± 6.71^{b}	74.00 ± 2.79^{a}
Creatinine (mg/dl)	0.88 ± 0.12^{a}	0.82 ± 0.16^{a}	0.83 ± 0.16^{a}	0.83 ± 0.14^{a}	0.63 ± 0.18^{b}
Bilirubin (g/dl)	0.46 ± 0.09^{bc}	0.37 ± 0.06^{c}	0.54 ± 0.12^{ab}	0.63 ± 0.09^{a}	0.62 ± 0.17^{a}

Values are means± S.D (n = 3). Values with different superscripts within a row were significantly different ($p<0.05$).
Key:
Group 1 – Rats treated with normal feed + water
Group 2 – Rats treated with normal feed + water + pasteurized palm sap
Group 3 – Rats treated with normal feed + water + palm wine
Group 4 – Rats treated with normal feed + water + frozen palm sap
Group 5 – Rats treated with normal feed + water + freshly tapped palm sap
AST – Aspartate Amino Transferase
ALT – Alanine Amino Transferase
ALP – Alkaline Phosphatase.

Table 6. Microbial Isolation and counts in the palm sap/wine samples

Treatments	Pasteurized palm sap	Market palm wine	Frozen palm sap	Fresh palm sap
Yeast/Mould count (Cfu/ml)	ND	$3.78 \pm 0.23 \times 10^{8a}$	$1.27 \pm 0.02 \times 10^{8b}$	$1.76 \pm 0.07 \times 10^{8b}$
Organism present	ND	*Saccharomyces cerevisiae* and *Citrobacter fruendii*	*Saccharomyces cerevisiae*	*Saccharomyces cerevisiae*

Values are means± S.D (n = 3). Values with different superscripts within a row were significantly different ($p<0.05$). ND – Not detected.

Traditionally, palm sap/wine is known to contain yeast. The high demand of the product results from the presence of yeast believed to improve the eyesight of those that indulges in it. It was observed during laboratory investigation that there were microbial growth in the MW, FS and FrS, while no microbial growth was found on the PS. Microorganisms found in the MW, with higher microbial counts, (3.78×10^{8}cfu/ml) were *Saccharomyces cerevisiae* and *Citrobacter freundii* while only *Saccharomyces cerevisie* was isolated in the culture of FS and FrS.

3.2. Discussion

The investigation of the food intake of the male Wistar rats subjected to different palm products showed that the groups of rats responded differently to feed intake. Palm sap/wine is basically sugar product with high presence of glucose and fructose. The part of the brain that is responsible for food and water intake is the hypothalamus. The high glucose (glucose is regulated by phosphogluctokinase) intake can led to elevated insulin in the central nervous system which inhibits food intake in animals, including non-human primate and functions as a negative feedback signal of recent energy intake and body adiposity [18]. The body does this when it absorbs high fructose (fructose can continually enter the glycolytic pathway) intake leading to reduced insulin and leptin production (leptin is poorly secreted in the adipose tissues due to insulin production in the liver). Decreases in circulating leptin concentrations correlate with increased sensations of hunger during prolonged energy restriction in women [19], and leptin administration can reduce appetite in humans [20]. Insulin production is therefore attributed to high carbohydrate and glucose intake in the body system.

Research on the weight gain of male Wistar rats was conducted. The result obtained was consistent with Screeranjitkumar [21] that administration of toddy decreased the weight gain of dams. Increases in the weekly mean body weights were uniform throughout the period of study for the non-treated rat group, but declined as the rat advanced in age. This could be as a result of the yeast present in the palm sap and the low content of ethanol in the sample. Swathy et al. [22] reported that body weight and testis weight of rat decreased when administered with ethanol. It is, therefore, not surprising the weight gain of rats administered with different palm products were reduced since the carbohydrate intake were subsequently metabolised to alcohol in the animals' body by the *Saccharomyces cereviseae* present in the palm sap/wine.

It is important to state that the Significant ($p<0.05$) differences in the organ weight of the male Wistar rats could be owing to the presence of high carbohydrate (glucose and fructose) intake in the treated rat groups. Lalitha et al. [23] asserted that an increase in organ weight was as a result of elevation of cholesterol content of the rats, which was in agreement with the findings of other researchers [21]. Although researches on animal study attributed the abnormal increase in organ weight to the presence of alcohol, it is necessary to state that palm sap/wine contains alcohol, and that palm sap metabolism by inherent microbial organisms results in the production of different categories of alcohol depending on the length of storage. According to Nayanatara et al. [24], alcohol fed rats showed significant increase in the weight of liver. The result of this study was in agreement with Klesges et al. [25] who showed that alcohol increases the metabolic rate of rats, thus causing more calories to be burned rather than stored in the body as fat, while Colditz et al. [26] reported that the consumption of sugar decreased as the consumption of alcohol increased, thereby leading to a loss in weight. There were no significant ($p>0.05$) differences in the organ weights of rats treated with frozen sap and freshly tapped palm sap when compared to the control group. The high content of *Saccharomyces cerevisiae* in the two groups could have contributed to the complete metabolism of food substrates in the organs. However, the spleen of rats administered with fresh palm sap were significantly ($p<0.05$) elevated from 0.38% to 0.44%, while that of the pasteurized palm sap treated rats increased from 0.38% to 0.50%.

The results of the fasting and random blood sugar showed that they were within the normal range (70 – 126 mg/dl). This implied that consumption of these products could not at 1.5% of the body weight of Wistar rats lead to high blood sugar. This finding was consistent with the report of Matawalli et al. [27] that alcohol fed rats showed no adverse health effect when administered at normal dosage. It was evident from the study that palm sap/wine treatment can not cause hypoglycaemia. The high activity of probiotic bacteria (*Zymomonas mobilis*) and yeasts have been associated with breaking down of reactive oxygen species that may be responsible for elevating blood sugar levels.

The haematology of the male Wistar rats was studied. The research study showed inconsistency with the report of Matawalli et al. [27] that alcohol fed rat administered with normal dosage shows no adverse health effect. The significant ($p<0.05$) difference in the haematological parameters of rats administered palm sap/wine when compared to the rat group that had no treatment may be as a result of chronic administration of palm sap/wine to the rat groups. The decrease in the level of haemoglobin could be a sign of anaemia in the treated male rats. WHO [28] reported that haemoglobin cut-off values that indicate anaemia vary with physiological status (e.g. age and sex), and is caused by iron deficiency. So far, no study examining the effect of palm sap/wine in vivo has been made on rats' serum parameters and haematology as the treatment model. Because of this reason, there is no scientific evidence to compare the results. The results of the haematological parameters were in agreement with the result of the blood glucose (Table 3) of the rats fed with FS and PS, though the increases observed in them were not significantly ($p>0.05$) different when compared to the control rat groups. The leucocytes counts, however, could not give specific information which necessitated the differential leucocytes counts. Neutrophil is majorly responsible for phagocytosis of pathogenic microorganisms during the first few hours after their entry into tissues [29]. There was no significant ($p>0.05$) difference in the neutrophil counts between group 5, 3 and the group 1 treated rats, but group 2 treated rats was significantly higher ($p<0.05$) compared with the group 1.

The group 4 and 5 rats treated with FS and FrS had the highest values of bilirubin. Bilirubin is formed by the breakdown of haemoglobin in the liver, bone marrow and spleen [17]. The bilirubin concentrations were within the normal limits of 0.4 – 1.7 d/dl [30,31]. The slight variations in bilirubin concentrations in the treated groups, except PS group were indicative of non–adverse effects on haemoglobin metabolism. An increase in plasma bilirubin results in jaundice. It has been shown that raised bilirubin indicates failure of the liver excretory function and increased breakdown of red blood cells [31]. This observation could not be confirmed from the results in Table 5, which showed that the treated rat groups manifested some signs of ill-health due to chronic administration of palm sap/wine based on 1.5% of the body weight per ml.

The research also examined the clinical chemistry of the male Wistar rats in order to find the effect of the treatment on the enzyme liver and kidney makers. A decreased serum alkaline phosphatase (ALP) may be due to zinc deficiency, hypothyroidism, vitamin C deficiency, folic acid deficiency, excess vitamin D intake, low phosphorus levels (hypophosphatasia), celiac disease, malnutrition with low protein assimilation, insufficient parathyroid gland function, and pernicious anaemia and vitamin B insufficiency [32]. There were no significant ($p>0.05$) differences between the serum creatinine levels in rat groups 1, 2, 3 and 4 but significant ($p<0.05$) differences exists between them and group 5 treated rats. High levels of serum creatinine are an indication of liver disease, since it shows low filtration rate of the kidney. Creatinine levels differ among individuals and groups according to age and sex. Normal levels of creatinine in the blood are approximately 0.6 to 1.2 mg/dl in adult males and 0.5 to 1.1 mg/dl in adult females [33]. The significant ($p<0.05$) low level of creatinine in the group 5 rats administered with fresh palm sap was an indication of reduced muscular mass in the treated rat group, since the level (0.63 mg/dl) falls within the normal range.

The significant increase in the neutrophil count in group 2 treated rats may be an indication of increased pathogenic organisms that necessitated increases in phagocytosis so as to eliminate the effect of pathogens in the body. Basophile and monocyte counts were not detected in the treated rat groups. Monocyte is responsible for the defence of tissue against microbial agents while basophil counts increase upon sensitization to an antigens (or allergen) [34] [16]. White blood cell is responsible for fighting germs and harmful microorganisms. Lymphocytes are responsible for humoral antibody formation and cellular immunity. The results therefore suggested that the entire rat group showed signs of immunostimulatory effect.

Probiotics can be described as live microorganisms that, when administered in adequate amounts confer a health benefit on the host. The classification of palm wine as a probiotic beverage was as a result of the presence of

Lactic acid bacterials, *Bifidobacteria* and *Saccharomyces boulardii*. However, *Saccharomyces cerevisiea* has not been characterized as a probiotic microorganism. The presence of alcohol in the palm products were elaborated by the presence of inherent *saccharomyces cerevisiae* while the presence of *Citrobacter freundii* in the MW confirms environmental contamination of the product and contributed in higher percentage weight of liver in the male animals compared to the other treated rat groups. The MW also contained the highest level of microbial count which is significant ($p<0.05$) compared to either the FS or the FrS and may be attributed to contamination by *Citrobacter freundii* while alcohol where not detected in the pasteurized palm sap. The rat treated palm sap products showed significant decrease ($p>0.05$) in the weight gain. The gain in weight for the FrS treated rat group was gradual and progressive while the control rat group progressed rapidly and fall. The progressive increase in the weight gain of rats by the FrS treated group may not be unconnected with the low alcohol content and the probiotic effect of the *Saccharomyces cerevisiea.* The weight of PS treated rat group was the lowest. The increase in the body weight of rats treated with FrS and FS could be attributed to the activities of probiotic yeasts in them compared to the body weight of rats treated with either MW or the PS. Research has shown that probiotic palm yeasts in MW by their activities increase the amino acids content of the feeding especially lysine, methionine and cysteine [35]. The significant increase in the number of microbial organisms in the MW was as a result of growth of *Citrobacter freundii. Citrobacter freundii* is an environmental microorganism which has been associated with the contamination of marine fish. The presence of this bacterium is suggestive of sewage contamination [36]. *Citrobacter freundii* is also associated with nitrogen fixation and is also associated with phage mediated interactions [37,38]. *Citrobacter freundi*i has not been isolated in palm wine by other researchers. Therefore, its presence in palm wine bought from the market is indicative of microbial contamination from dirty environment.

4. Conclusion

It was evident from the results obtained that palm sap drinkers may not be diabetic as a result of drinking palm sap samples. Efforts should be made to protect palm sap from contamination either by environmental microorganism or adulteration in order to exclude the presence of *Citrobacter freudi* or any other microorganism that may be injurious to health from contaminating palm sap products. The research was able to correlate the decrease in body weight to the effect of drinking palm sap products. It could therefore be asserted that the decrease in body weight resulting from drinking palm sap products could be attributed to the fact that drinkers takes less food after drinking. Since palm sap is made up of basically glucose and lactose, it is not surprising that lovers of such products rarely eats food since palm sap products supplies energy to their system and thus, they fills fulfilled. However, the drinkers of FrS sample proved otherwise, since it was comparable to the control. It may be concluded that prolonged drinking of palm sap could lead to exposure of the body to pathogenic organisms. The increase in the neutrophil count in Wistar rats treated with palm saps could be as a result of increased attack of pathogenic microorganism leading to increased response to the defence of the body against such attacks. Also, chronic administration of the palm sap samples manifested in the significant differences observed with the haematological parameters observed when compared with the control rat groups. Therefore, chronic administrations of palm sap products have effect in exposing the body to health challenges by decreasing the liver and kidney functions, and reduction in mean spleen, kidney and body weights.

Acknowledgements

We wish to thank the Department of Veterinary Medicine and Toxicology, University of Nigeria, Nsukka for availing us of the opportunity to work in the Laboratory for the determination of haematological and clinical chemistry parameters. Our special thanks go to the late palm sap/wine taper, Mr. Ugwuanyi Joseph of Ejuona village, Nsukka. May we also acknowledge the Technologists in the Department of Food Science and Technology, University of Nigeria, Nsukka for their contributions during the course of the research work.

Conflict of Interest Statement

The authors have no conflict of interest to declare.

References

[1] Ikegwu, T.M., Okafor, G.I. and Ochiogu, I.S. (2014). Effect of preservation methods of oil palm sap (*Elaeis guineensis*) on the reproductive indices of male wistar rats. *Journal of Medicinal Food,* 17(12): 1368-1374.

[2] Ikegwu, T. M, 2011. *Effects of preservation methods of oil palm (elaeis guineensis) sap on selected reproductive indices of male wistar rats.* Unpublished MSc. Thesis of the Department of Food Science and Technology, University of Nigeria, Nsukka.

[3] Dalibard, C, 1999. Overall view on the tradition of tapping palm tree and prospects for animal productions. Livestock Research for Animal Development. *International Relations* 11(1):1-2. http://www.lrrd.org/lrrd11/1/dali111.htm.

[4] Ikegwu, T. M, 2008. Production of Mixed-fruit Juices from Blend of Sour soup, Pineapple and Oil Palm Sap. An unpublished B.Sc. Project in the Department of Food Science and Technology, University of Nigeria, Nsukka.

[5] Enekwechi, E.E, 1996. Gender differences in motivation for alcohol use among Nigerian University students. *Journal of Alcohol and Drug Education*, 41(2):1-10.

[6] Bennett, L. A., Campillo, C., Chandrashekar, C. R. and Gureje, O, 1998. Alcoholic beverageconsumption in India, Mexico, and Nigeria-A cross-cultural comparison. *Alcohol Health and Research World,* 22(4), 243-252.

[7] Schalm, O.W., Jaivi, N.C. and Caroll, E.J., 1975. Veterinary haematology, 3rd ed. Lea and Febiger, London, 385-390.

[8] Coles, E.H, 1986. *Veterinary Clinical Pathology*, 4th Edition. W.B. Saunders Co. Philadelphia, pp. 498-517.

[9] Kachmar, J.F, 1970. *Determination of blood haemoglobin by the cyanomethaemoglobin procedure.* In: Tietz NW Ed. Fundamentals of Clinical Chemistry, W.B. Sanders Company, Philadephia, 268-269.

[10] Reitman, S. and Frankel, S, 1975. Colorimetric methods for determination of serum transaminase activity. *Am. J. Clin. Path.*, 10, 262.

[11] Klein, F., Mahlandt, B.G., Lincoln, R.E., Dearmon, I.A. and Fernelius, A.L, 1960. Immunization as a factor affecting the course of septicemic anthrax. *Science* 133:1021-1022.

[12] Babson, L.A., Greeley, S.J., Coleman, C.M. and Phillips, G.D, 1966. Phenolphthalein monophosphate as a substrate for serum alkaline phosphatase. *Clin. Chem.,* 12: 482-490.

[13] Blass, K. G., Thiebert, R. J. and Lam, L. K., 1974. Study of mechanism of JAFE reactions. *Clin. Chem.* 12(7): 336-343.

[14] Doumas, B.T., Perry, B.W., Sasse, E.A. and Straumfjord, J.V, 1973. Standardization in bilirubin assays: evaluation of selected methods and stability of bilirubin solutions. *Clin. Chem.* 19: 984-993.

[15] AOAC: *Official Methods of Analysis.* 15th ed. Association of Official Analytical Chemists, Washington DC, 2010.

[16] Raj, A., Praveen, K.V., Varghese, S., Mukkadan, J.K. and Joseph, P.K, 2009. Biochemical effects of feeding soft drink and ethanol. *Indian Journal of Experimental Biology*, 47:333-337.

[17] Edem, D.O, 2009. Haematological and histological alterations induced in rats by palm oil – containing diets. *European Journal of Scientific Research,* 32(3): 405-418. http://www.eurojournals.com/ejsr.htm.

[18] Elliott, S.S., Keim, N.L., Stern, J.S., Teff, K., and Havel, P.J, 2002. Fructose, weight gain, and the insulin resistance syndrome. *Am. J. Clin. Nutr.* 76:911-922.

[19] Keim, N.L., Stern, J.S. and Havel, P.J, 1998. Relation between circulating leptin concentrations and appetite during a prolonged, moderate energy deficit in women. *Am. J. Clin. Nutr.,* 68:794-801.

[20] Westerterp-Plantenga, M.S., Saris, W.H., Hukshorn, C.J., Campfield, L.A, 2001. Effects of weekly administration of pegylated recombinant human OB protein on appetite profile and energy metabolism in obese men. *Am. J .Clin. Nutr.* 74:426-434.

[21] Screeranjit-Kumar, C.V., Lal, J.J., Indira, M. and Vijayemal, P.L, 1998. Postnatal metabolic changes in the pups of rats exposed to toddy (palm wine) during pregnancy and lactation. *Food Chemistry*, 62(2):149-155.

[22] Swathy, S.S., Panicker, S. and Indira, M, 2006. Effect of exogenous solanium on the testicular toxicity induced by ethanol in rats. *Indian Journal of Physiology and Pharmacology*, 50(3):215-224.

[23] Lalitha, T., Kumar, K., Ramakrishnan, C.V. and Telang, S. D, 1988. Effect of maternal alcohol consumption on the lipid composition of central nervous system in the offspring. *Journal of Neurochemistry,* 50(5):1346-1351.

[24] Nayanatara, A.K., Nagaraja, H.S., Ramaswamy, C., Bhagyalakshmi, K., Bhat, M.R. and Harini, N, 2009. Estimation of tissue lipid peroxidation level and organ weight in litters of wistar rats exposed to prenatal alcohol ingestion. *Journa of Physiology and Biomedical Science*, 22:44-47. http://www.j-pbs.org/pdf/221/JPBS081009.pdf.

[25] Klesges, R.C., Mealer, C.Z. and Klesges, L. M, 1994. Effects of alcohol intake on resting energy expenditure in young women social drinkers. *American Journal of Clinical Nutrition*, 59: 805-809.

[26] Colditz ,G.A., Giovannucci, E., Rimm, E.B., Stampfer, M.J., Rosner, B., Speizer, F.E, 1991. Alcohol intake in relation to diet and obesity in women and men: A paradox. *American Journal of Clinical Nutrition,* 54:49-55.

[27] Matawalli, A., Geidam, I.P. and Hajjagana, L, 2004. Effects of aqueous stem bark extract of *Momordica balsamina* linn on serum electrolytes and some haematological parameters in normal and alcohol fed rats. Pak. *J. Biol. Sci.,* 7(8): 1430-1432.

[28] *WHO*: 2001. Iron deficiency anaemia. Assessment, prevention and control. A guide for programme managers. WHO/NHD/01.3. Geneva.

[29] Aboderin, F.I. and Oyetayo, V.O, 2006. Haematological studies of rats fed different doses of probiotic, Lactobacillus plantarum, isolated from formative corn slurry. *Pakistan Journal of Nutriiton,* 5(2): 102-105.

[30] Vasudevan, D.M. and Sreekumari, S, 2005. Textbook of Biochemistry (For Medical students) 4th ed. Jaypee Brothers Medical Publishers (P) Ltd., New Delhi, India, pp. 56 – 57, 277, 485, 486, 502-503.

[31] Harrison, D.J., Foulis, A.K. and Burt, A.D, 2011. Disorders of the liver, gallbladder and pancreas. www.aboutgastro.com/digestive-system/disorders-of-liver-gallbladder-pancreas.

[32] Coresh, J., Astor, B.C. and McQuillan, G, 2002. Calibration and random variation of the serum creatinine assay as critical elements of using equations to estimate glomerular filtration rate. *Am. J. Kidney Dis*., 39: 920-929.

[33] Cunningham, A.B., and Wehmeyer, A.S, 1988. Nutritional value of palm wine from *Hyphaenecoriacea* and *Phoenix reclinata* (Arecaceae). *Economic Botany*, 42(3):301-306.

[34] Baker, F.J. and Silverton, R.E, 1985. Haemostasis and blood coagulation. In: FJ Baker, RE Silverton and CJ Pallister (Eds). Introduction to Medical Laboratory Technology. Sixth edition. Butterworth & Co. Sydney: 312-330.

[35] Bohoua, G.L, 2008. Effect of palm wine yeasts and yoghurt probiotics on the growth performance of broilers. *Livestock Research for Rural Development* 20(3):7-8.

[36] Ward, D.A., Hackney, C, 1991. Microbiology of marine food products. Van Nostrand Reinhold. Chapman and Hall, New York, p. 50.

[37] Hopwood, D. A., Chater, K. F., 1989. Genetics of bacteria diversity. Academic Press Ltd. London, 1989; p. 109-195.

[38] Fry, J.C., Day, M.., 1990. Bacterial genetics in natural environments. Chapman and Hall, London 1990; p. 44.

Assessment of the Effects of Microbial Fermentation on Selected Anti-Nutrients in the Products of Four Local Cassava Varieties from Niger State, Nigeria

Etsuyankpa M. B.[1,*], Gimba C. E.[2], Agbaji E. B.[2], Omoniyi K. I.[2], M.M. Ndamitso[3], Mathew J. T.[3]

[1]Centre for Preliminary and Extra-mural Studies, Federal University of Technology, Minna, Niger State, Nigeria
[2]Chemistry Department, Ahmadu Bello University, Zaria, Kaduna State, Nigeria
[3]Chemistry Department, Federal University of Technology, Minna, Niger State, Nigeria
*Corresponding author: myankpa@yahoo.com

Abstract The effect of microbial fermentation on the anti-nutrient composition of some cassava products was evaluated by the inoculation of the pulps of four local varieties of cassava, using microorganisms (*Saccharomyces cerevisae and Lactobacillus bulgaricus*). The results of the study revealed that microbial fermentation caused significant reductions ($P < 0.05$) in the cyanide (86%), tannins (73%), oxalate (61%), phytate (72%) and saponins (92%) contents of the cassava products. The results of the study suggest that cassava products can be nutritionally improved with *Saccharomyces cerevisiae* and *Lactobacillus bulgaricus* fermentations.

Keywords: *Saccharomyces cerevisiae, Lactobacillus bulgaricus, fermentation, anti-nutrients, microorganisms, inoculation*

1. Introduction

It was estimated that the world production of cassava roots stood at 184 million tonnes in 2002. Reports indicated that the production is highest in Africa where 99.1 million tonnes were grown, 51.5 million tonnes were cultivated in Asia and 33.2 million tonnes in Latin America and the Caribbean [1]. Of the countries of the world, Nigeria is found to be the largest exporting country of dried cassava with a total of 77% of world export in 2005, seconded by Vietnam, with 13.6%.. Indonesia was next with 5.8% and Costa Rica 2.1%.. World-wide cassava production recorded an increase of 12.5% between 1900 and 1988 [1,2]. In Nigeria the crop contributes over 50% of food intake and provides substantial percentage of food energy in the daily diet. Also, it significantly serves as livestock feed and industrial raw material for production of alcohol and acetones, starches for textiles and pharmaceutical industries over the years [3].

Among the world's staple crops, cassava has the lowest protein/energy ratio (7.4 mg protein/cal compared to 26 mg protein/cal for maize) and is deficient in iron, zinc, pro-vitamin A or β-carotene and vitamin E. This is coupled with the fact that it has one major problem; the roots and leaves of poorly processed cassava plants contain a substance that, when eaten, can trigger the production of cyanide.This is a serious problem for the 500 million people who rely on cassava as their main source of calories, among them are subsistence farmers in Sub-Saharan Africa [4,5].

Proteineous foods are usually known to be expensive and beyond the reach of most of the populace. Its scarcity has greater impact on children, whose physical and mental developments require nutritionally balanced diets. Malnutrition leads to stunting and underweight [6]. The utilisation of cassava is limited by its extremely low protein content and so the consumption of its products has been implicated in malnutrition. Low protein intake in Africa has been attributed to the increasing high cost of traditional sources of animal protein [7].

In many regions of the world over 30% of the population have insufficient calories in their diet to meet their nutritional needs. By 2020, the Food and Agricultural Organisation (FAO) [8] has estimated that one billion persons will be undernourished or receive insufficient calories in their diet. Nowhere is this problem more severe than in sub-Saharan Africa. [4].

Cyanogens are found in 3 forms in cassava: cyanogenic glucoside (95% linamarin and 5% lotaustratin), cyanohydrins, and free cyanide. These glucosides, typified by linamarin [alpha-Hydroxyisobutyronitrile-beta-D-glucose] or [2-(beta-D-Glucopyranosyloxy)-2-methylpropanenitrile] (molecular formular: $C_{10}H_{17}NO_6$) and lotaustralin [2-(beta-d-glucopyranosyloxy)-2-methylbutyronitrile] or [2-(-D-Glucopyranosyloxy)-2-methylbutyronitrile] (moleular formular: $C_{11}H_{19}NO_6$) are hydrolysed to hydrocyanic acid (HCN) by endogenous linamarase (beta-glucosidase or β-glucosidase) when cassava tissues are disrupted by cutting, grating, bruising or other mechanical means.

Different processing techniques exist to remove cyanogens and their effectiveness depend on the processing steps and the sequence utilized, and it is often time-dependent [9,10,11]. Cyanide is found to be the most toxic factor restricting the consumption of cassava roots and leaves. Indeed, cassava, particularly its bitter varieties, has a cyanide level higher than the Food and Agricultural Organisation/World Health Organisation (FAO/WHO) (1991) recommendations, which is < 10 mg cyanide equivalents/kg DM, to prevent acute toxicity in humans. Cassava leaves have a cyanide content ranging from 53 to 1,300 mg cyanide equivalents/kg of dry matter (DM), and cassava root parenchyma has a range of 10 to 500 mg cyanide equivalents/kg DM; both of these are much higher than what is recommended.

Antinutrients are natural or synthetic compounds that interfere with the absorption of nutrients They reduce the body's ability to absorb or use essential nutrients like vitamins and minerals, and are found in all plant foods, although the types and amounts vary tremendously from food to food. Examples of antinutrients include phytic acid (or phytates), lignans, saponins, phytoestrogens, oxalates, phenolic compounds (tanins), protease inhibitors. Lipase inhibitors, amylase inhibitors and trypsin inhibitors

Phytate (inositol hexakisphosphate) is a compound found in high abundance in cassava, with approximately 624 mg/100 g in roots. Phytic acid is able to bind cations such as magnesium, calcium, iron, zinc, and molybdenum and can, therefore, interfere with mineral absorption and utilization which may affect requirements. It may also bind proteins preventing their complete enzymatic digestion [12,13].

The polyphenol content (tannins) in cassava leaves increases with the maturity of the plant. Polyphenols are able to form insoluble complexes with divalent ions such as iron, zinc, and copper. They can also inactivate thiamin, bind certain salivary and digestive enzymes, and enhance secretion of endogenous protein. Consequently, they inhibit nonheme-Fe absorption, reduce thiamin absorption and the digestibility of starch, protein, and lipids, and also interfere with protein digestibility [14].

Oxalic acid and oxalates are found in many plants, particularly in members of the spinach family Oxalate is the conjugate base of oxalic acid which is capable of binding to metal ions such as Ca^{2+} and Mg^{2+} to form precipitates in the body and prevent their absorption. Oxalates are among the ant-nutrients affect the utilization of protein by forming complexes with proteins, which inhibit peptic digestion. Oxalate content ranges from 1.35 to 2.88 g/100 g DM for cassava leaf meal. The negative effect of oxalates on humans depends on the level of both oxalate and calcium in the cassava leaves [15,16].

Saponins have been found to be the bioactive component of ginseng responsible for its metabolic and potential health effects. Ginseng has been known for treatment of a number of different ailments, such as providing energy, anti-inflammatory effects and preventing fatigue. It also helps to improve cognitive function and erectile dysfunction of men. Saponins can also act on the central nervous system of humans with potential therapeutic effects. Cassava leaf meal has a steroidal saponin content ranging from 1.74 to 4.73 g/100 g DM, which compares to those found for soybeans (0.07 to 5.1 g/100 g DM), but is lower than those observed in alfalfa (5.6 g/100 g DM) and beet leaves (5.8 g/100 g DM). Saponin content increases in cassava leaf meal with plant maturity [9].

Of the series of available cassava processing methods, fermentation is the first major step in the production of products such as 'gari', 'fufu' and 'lafun'. Some of the objectives behind the fermentation include biological enrichment of the substrate in terms of protein, vitamins, essential amino acids and essential fatty acids. Others purposes for which fermentation is carried out are detoxification of food anti-nutrients, impartation of good aroma, flavour and texture as well as preservation of the fermented products [17]. It is one of the less expensive means of increasing the protein quality of cassava [17,18].

The modern processing techniques involve enhanced fermentation of cassava pulp by the addition of non-pathogenic microorganisms. The use of microorganisms to convert carbohydrates, lignocelluloses and other industrial wastes into foodstuffs rich in protein is possible due to some inherent characteristics [17].

Firstly, microorganisms have a very fast growth rate with high protein content (35 to 60%) [19]. Microorganisms produce an amount of essential nutrient(s) sufficient to correct or prevent deficiency in the consumer. Also, the nutrient(s) produced by microorganisms do not impart undesirable characteristics to food (changes in colour, taste, smell, texture, or cooking properties) and do not unduly shorten the shelf life.

Cyanide can poison a person by either inhalation or ingestion [20]. The normal range of cyanogens content of cassava tubers falls between 15 and 400 mg HCN/kg fresh weight. It was reported by [6] that cyanide levels were found in the range 6 to 370 mg/kg depending on the particular cultivar, growing conditions, (i.e. soil type, humidity, temperature) and the age of the plant [8,21].

The World Health Organisation (WHO) has set the safe level of cyanogens in cassava flour at 10 ppm and the acceptable limit in Indonesia is 40 ppm [22]. Phytates should be lowered as much as possible, ideally to 25 mg or less per 100 g or to about 0.03 percent of the phytate-containing food eaten. At this level, micronutrient losses are minimized [23]. Because retention of phosphorus decreases when phytate in food is 30 – 40% or more of total phosphorus.

The ability of tanins to form complexes with numerous types of molecules such as carbohydrates, proteins, polysaccharides, bacterial cell membranes, enzymes involved in protein and carbohydrates digestion accounted for their adverse effects on animal nutrition. Not only do they form complexes with proteins but also precipitate them. Tannin-protein complexes involve both hydrogen-bonding and hydrophobic interactions. Some tannins were found to produce toxic and anti-nutritional effects in monogastric and ruminant animals and cause reduced intake, lower nutrient digestibility and protein availability. Tannins are commonly found in fruits (grapes, persimmon, blueberry, etc.), tea, chocolate, legume forages (trefoil, etc.), legume trees (Acacia spp., Sesbania spp., etc.), and grasses. They contribute to many aspects of our daily lives as they are responsible for the astringent taste being experienced when consuming wine or unripe fruits, and for the enchanting colours seen in flowers and in autumn leaves [24,25,26].

This study examines the effect of microbial fermentation (using *Saccharomyces cerevisae and Lactobacillus bulgaricus*) on reducing the anti-nutrient compositions of the cassava products obtained by the inoculation of the pulps of four local varieties of cassava. This is geared towards the provision of enriched cassava products to the populace.

2. Materials And Methods

2.1. Materials

All chemicals used were of analytical grade. The inoculants used for the study were *S cerevisiae* and *L. bulgaricus*. They were obtained from the culture bank of the Department of Microbiology, Federal University of Technology, Minna, Nigeria.

Tuberous roots of four local varieties ('Baba Iya', 'Wahabi', 'Dan Warri' and 'Kpace Bokun') of cassava were obtained from the sampling farm cultivated at Wushishi in Kontagora Senatorial Zone of Niger State, Nigeria. The tubers were each processed into three different products- 'fufu', 'gari' and 'lafun'. These varieties were identified by a Research Fellow and, an agronomist, at the Niger State Agricultural Development Projects, Minna, Niger State, Nigeria.

2.2. Method of Processing the 'Gari'

The fresh roots were peeled and grated, after which the grated pulp was put in sacks (polypropylene) and the sacks placed under heavy stones for 4 days to expel excess liquid from the pulp while it was fermenting. The dewatered and fermented lumps of pulp were then crumbled by hand, and most of the fibrous matter removed. The remaining mass was sieved with traditional sieves of iron mesh. After being sieved, the fine pulp was then roasted in an iron pan over fire.This resulted in a finished white lumpy product called 'gari'. For the production of microbial treated 'gari', the pulp was seperately inoculated with the microbes prior to fermentation.

2.3. Method of Procssing the 'Lafun'

'Lafun' in Nigeria, 'cossettes' in Zaire and Rwanda, 'kanyanga' and 'mapanga' in Malawi and 'makopa' in Tanzania was processed by peeling the fresh root of cassava, followed by grinding and then draining the extracts of the pulp under heavy weight. Sun-drying of the pulp for 5 days until dried yielded 'lafun' The dried crumbs was then milled into flour. Meanwhle, cassava mash to be made into 'lafun' products were neither fermented nor inoculated to agree with the method used locally for preparationas as the fermentation process is usually skipped. In essence, therefore, lafun products provide room for comparison between the fermented products in one hand and the microbially fermented products on the other.

2.4. Method of Processing the 'Fufu'

The freshly peeled roots were soaked in water for 3 days so as to absorb water and get fermented and soft. After fermentation, the water was drained off, most of the fibre were removed and the roots pounded in a wooden mortar, until a soft mash was formed. Excess water was extracted from the mash by placing heavy objects such as rocks on top of the sacks containing the 'fufu'. In order to obtain a microbial fermented 'fufu', 1500 cfu each of *S cerevisiae* and *L. bulgaricus* per 1kg of cassava pulp were seperately introduced prior to fermentation.

2.5. Chemical Analysis of the Cassava Products

Triplicate samples of each cassava product from each of the four varieties were collected (n = 28) and the amount of cyanide in the samples was determined using the Alkaline Picrate Method described by [27]. The saponin content was estimated gravimetrically using a soxhlet extractor involving two different solvents (acetone and methanol) [28]. Determination of oxalate in the samples was carried out in accordance with the methods described by [29]. Phytate was determined in the samples using the method of anion-exchange [30].

2.6. Statistical Analysis

Data was reported as averages of triplicate determinations and analysed using Analysis of Variance (ANOVA). Duncan multiple test at 5% level of significance was used to determine the significant differences among the samples. The statistical package used was IBM SPSS Statistics version 20 (IBM Corp. Amonk. NY).

3. Results and Discussion

3.1. Effect of Inoculation on the Cyanide Content of the Products

From the results in Table 1 and Table 2, the cyanide contents of non-fermented 'lafun' products were significantly higher compared to the 'gari' and 'fufu' products that were subjected to natural and microbial fermentations. However, there were no significant differences ($p < 0.05$) between the values obtained in the natural and microbial fermented products except in the products obtained for 'Baba Iya' variety as shown in Table 1, which has been acclaimed by the farmers in the community to be the most toxic. For example, 'gari' and 'fufu' produced from 'Baba Iya' through natural fermentation had the values reduced from 5.92 ± 0.00 and 4.20 ± 0.08 mg/100 to 2.22 ± 0.04 and 1.84 ± 0.21 mg/100g respectively when treated with *L. bulgaricus*, and were reduced to 1.75 ± 0.05 and 1.10 ± 0.21 mg/100g respectively for 'gari' and 'fufu' products subjected to *S. cerevisae* fermentation.

These values obtained from this research work can be compared with the ones reported by [37], which indicates that the unfermented low-cyanide cassava flour (lafun) recorded 1.09 mg/100 g cyanide content and the *Rhizopus oryzae* and *Saccharomyces crevisae* fermented of the low-cyanide cassava flour were found to be 0.34mg/100g and 0.37 mg/100 g cyanide respectively. Similarly, the unfermented medium-cyanide cassava flour recorded 2.45 mg/100 g cyanide composition while the *Rhizopus oryzae* and *Saccharomyces crevisae* fermented of the medium-cyanide cassava flour contained 0.85 mg/100 g and 0.84 mg/100 g cyanide respectively. In another report, reference [38] shows that cyanide contents of cassava flour (lafun), garri and fufu were 6 ± 0.12ppm, 10 ± 0.13ppm and 5 ± 0.10 ppm respectively.

Table 1. Selected Anti-nutrient compositions of natural, microbial and non fermented cassava products of KB and BI varieties

Variety	product	Parameters				
		Cyanide µg/100g	Oxalate mg/100g	Phytate (mg/100g)	Tannins g/100g	Saponnins (%)
KB	Gari	2.77±0.15^{c}	9.44±1.21^{e}	297.50±0.45^{b}	7.53±2.40^{f}	5.06±0.00^{e}
	Lafun	3.48±0.03^{c}	9.71±0.25^{e}	267.50±8.05^{b}	8.62±1.00^{g}	4.49±0.01^{d}
	Fufu	1.05±0.08^{a}	8.29±0.50^{d}	252.41±7.41^{b}	6.91±0.09^{e}	2.25±0.10^{b}
LIKB	Gari	1.7±0.05a^{b}	9.03±1.23^{e}	201.00±4.70^{b}	6.42±1.14^{e}	4.21±0.02^{d}
	Fufu	0.98±0.13^{a}	8.20±0.28^{d}	214.22±3.32^{b}	5.89±2.11^{d}	1.89±0.00^{a}
SIKB	Gari	2.70±0.23^{c}	9.23±0.25^{e}	242.00±6.70^{b}	4.98±106^{c}	3.91±0.02^{d}
	Fufu	0.72±0.21^{a}	8.23±0.11^{d}	237..50.±8.8^{b}	5.23±0.82^{d}	2.21±0.12^{b}
BI	Gari	5.92±0.00^{e}	10.11±0.00^{f}	248.04±6.90^{b}	6.19±0.03^{e}	3.89±0.00^{c}
	Fufu	4.20±0.08^{d}	11.27±1.01^{g}	204.50±0.45^{b}	5.86±0.55^{d}	2.19±0.00^{b}
	Lafun	6.25±0.08^{f}	12.93±0.08^{h}	267.54±3.94^{b}	6.00±0.92^{e}	4.27±0.00^{d}
LIBI	Gari	2.22±0.04^{b}	6.23±0.25^{b}	177.00±4.70^{a}	4.13±1.34^{c}	2.45±0.00^{b}
	Fufu	1.84±0.21^{b}	6.01±0.15^{b}	196.50±7.80^{a}	4.81±2.40^{c}	3.48±0.00^{c}
SIBI	Gari	1.75±0.05^{b}	5.83±0.02^{a}	201.50.±8.8^{b}	5.00±0.84^{d}	4.44±0.00^{d}
	Fufu	1.10±0.21^{a}	5.11±1.00^{a}	164.54±3.84^{a}	5.11±0.61^{d}	4.22±0.01^{d}

Means with the same superscript within a column are not significantly different ($p > 0.05$)
BI = Baba Iya, LIBI = L. *bulgaricus* fermented, SIBI = *S. cerevisea* fermented, KB = Kpace Bokun, LIKB = *L. bulgaricus* fermented, SIKB = *S. cerevisea* fermented

Table 2. Selected anti-nutrient compositions of natural, microbial and non fermented cassava products of DW and W varieties

Variety	Product	Parameters				
		Cyanide µg/100g	Oxalate mg/100g	Phytate (mg/100g)	Tannins g/100g	Saponins (%)
W	Fufu	0.89±0.15^{a}	6.23±0.45^{b}	312.09±6.41^{c}	2.99±0.07^{a}	3.54±0.01^{c}
	Gari	2.31±2.99^{b}	9.01±0.07^{d}	310.50±8.30^{c}	3.89±1.00^{b}	2.05±0.00^{b}
	Lafun	6.34±0.05^{f}	12.93±0.08^{h}	278.50±4.45^{b}	4.89±0.58^{c}	5.69±0.01^{e}
SIW	Fufu	0.82±0.10^{a}	5.74±0.20^{a}	271.27±5.37^{b}	2.97±1.26^{a}	2.57±0.00^{b}
	Gari	0.78±0.02^{a}	8.48±1.11^{c}	291.90±8.00^{b}	2.71±0.02^{a}	2.21±0.02^{b}
LIW	Gari	1.92±0.04^{b}	8.70±0.25^{c}	304.58±3.90^{c}	2.03±1.04^{a}	2.00±0.00^{b}
	Fufu	0.86±0.21^{a}	6.14±0.00^{b}	244.10±6.44^{b}	2.44±0.00^{a}	2.48±0.00^{b}
DW	Gari	1.69±0.39^{b}	9.23±0.81^{e}	287.50.±8.8^{b}	7.42±0.72^{e}	4.07±0.00^{d}
	Fufu	0.99±0.06^{a}	5.28±1.81^{a}	298.04±6.90^{b}	7.18±0.58^{c}	3.02±0.01^{c}
	Lafun	4.22±1.44^{d}	9.57±0.25^{e}	248.50±0.45^{b}	5.32±1.07^{c}	3.24±0.01^{c}
LIDW	Gari	1.51±0.26^{a}	9.03±0.04^{e}	177.54±3.94^{a}	6.03±0.57^{d}	2.35±0.00^{b}
	Fufu	0.87±0.06^{a}	5.18±0.29^{a}	217.00±4.70^{b}	3.89±0.62^{a}	3.01±0.01^{c}
SIDW	Gari	1.51±0.05^{a}	9.40±0.01^{e}	226.70±7.80^{b}	4.22±1.81^{b}	2.11±0.00^{b}
	Fufu	1.18±0.10^{a}	5.21±0.65^{a}	201.50±8.81^{b}	5.86±0.94^{e}	2.98±0.02^{a}

Means with the same superscript within a column are not significantly different ($p > 0.05$) DW = Dan Warri, LIDW = *L. bulgaricus* fermented, SIDW = *S. cerevisea* fermented W = Wahabi, SIW = *S. cerevisea* fermented, LIW = *L. bulgaricus* fermented.

It was evident from the results that the level of reduction in cyanide level by the action of *L. bulgaricus*-aided fermentation was not significantly different ($P < 0.05$) from the ones brought about by *S. cerevisae*-facilitated fermentation (Duncan multiple test). However, the ability of the two organisms to bring about reduction of cyanide in 'fufu' and 'gari' products was found to be similar as presented in Figure 1, which contained the results expressed as percentage decrease in the cyanide content. The percentage decrease in cyanide content was highest in Baba Iya species that was *S. cerevisae*-treated 'gari' (70.43%) and *L. bulgaricus*-fermented 'gari' from 'Kpace Bokun' had 38.6% decrease in cyanide level. On the other hand, gari from Wahabi species that was 'inoculated with *S. cerevisae* recorded 43.2% decrease while *L. bulgaricus*-treated 'fufu' from Baba Iya also recorded a 56.2% decrease.

The reduction in cyanide level in the products is attributed to several factors including:

(i) Changes in the texture in the plant tissues which make it possible for vacuole-bound cyanogenic glycosides to diffuse and come in contact with membrane-bound linamarase.

(ii) Increase in β-glycosidase activity in cassava tissue, and

(iii) Utilisation of cyanogenic glycosides and the breakdown of products by fermentation microorganisms. When cyanogenic glycoside is exposed to linamarase, it is cleaved to produce hydrogen cyanide which evaporates or pressed out during cassava processing [31] and [32].

Cassava processing which involves grating, fermentation and roasting can lead to a reduction in the total cyanide more in 'gari' [33]. The results indicated that there was a significant decrease ($P < 0.05$) in cyanide content of the fermented products as a result of inoculation. The results of this investigation corroborate with the findings of other researchers [32], [34], [35] and [36], though the authors used different microorganisms for the fermentation.

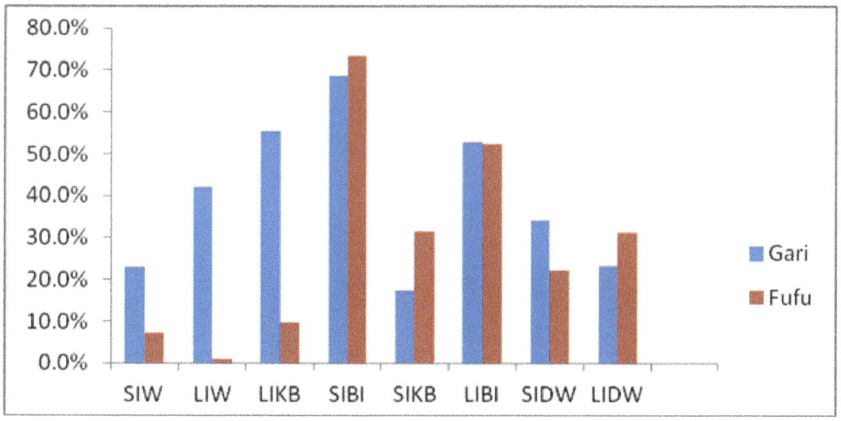

Figure 1. Percentage decrease of cyanide in 'garl' and 'fufu' samples due to inoculation

3.2. Effect of Inoculation on the Oxalate Levelof the Products

The decrease in oxalate content of the products was only significant amongst the natural and microbial fermented products produced from 'Baba Iya' variety (Table 1 and Table 2). Expectedly, the levels of oxalate in the uninoculated 'lafun' products were higher compared to the corresponding values of oxalate contents in both the natural and microbial fermented products. The results in the Table 1 and Table 2 indicated that oxalate content had values ranging from 11.27 ± 1.01 mg/100g for the naturally fermented 'fufu' from 'Baba Iya' to 5.11± 1.00 mg/100g dry sample for 'fufu' treated with *S. cerevisae* and from the same variety. Though, result data on the oxalate content of cassava products, is scanty, the available data showed that attempt made by [34] to investigate the effect of boiling and natural fermentation on the anti-nutritional factors of two cassava varieties (Qulle and Kello) collected from Southern Ethiopia.the results revealed that the level of oxalate was found to be 24.93±0.08 and 5.04±0.02, 86.18±0.10 mg/100g in raw Qulle and Kello samples respectively.

From the results of this research, reduction in oxalate was more pronounced in 'gari' products. This was attributed to processing method of 'gari' which involves grating, fermentation and roasting thereby leading to a reduction in oxalate.

In Figure 2, the oxalate level was presented as percentage derease in the products, the results indicated that reduction of oxalate was more in the 'gari' products. This was also linked to the processing method of 'gari' which involves grating, fermentation and roasting thereby leading to reductions in oxalate. The result further pointed out that the level of oxalate reduction by the action of *L. bulgaricus*-aided fermentation was not significantly different from the ones brought about by *S. cerevisae*-facilitated fermentation. There were also significant differences ($p < 0.05$) between the values of oxalate in the natural and microbial fermented products.

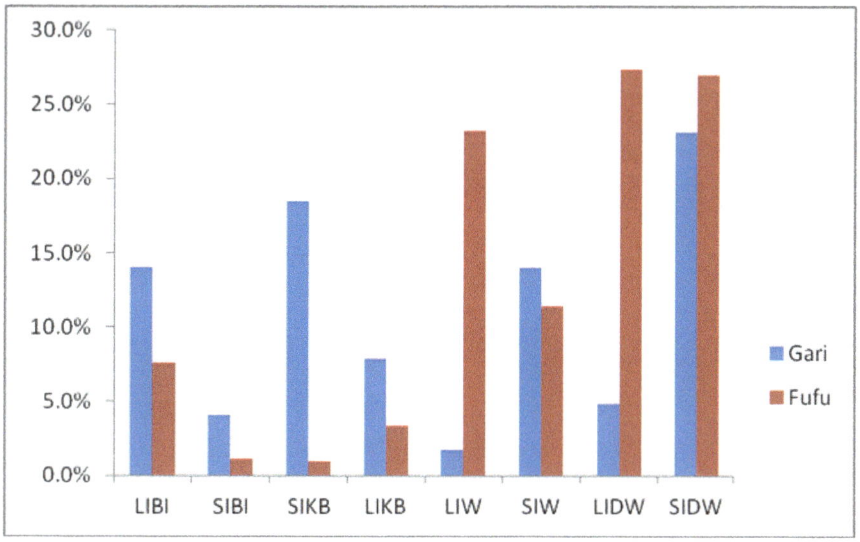

Figure 2. Percentage decrease of oxalate in 'garl' and 'fufu' samples due to inoculation

3.3 Effect of Inoculation on Thephytate Content of the Products

The phytate content in the products ranged from 177.00 ± 4.70 mg/100 g to 312.09 ± 6.41 mg/100g for 'gari' fermented with *L. bulgaricus* from 'Baba Iya' and 'fufu' naturally fermented and made from 'Wahabi' cultivars respectively (Table 1 and Table 2). Expectedly, the values recorded for 'lafun' products (the unfermented products) and the ones obtained for 'gari' and 'fufu' products being naturally inoculated were higher than the values obtained for 'gari' and 'fufu' products that were microbially treated. These values conformed to those obtained by [35] which ranged from 173.57 ± 0.56 mg/100g to 104.48 ± 0.68 mg/100g for Qulle flour sample and Kello varieties respectively. The results also agreed with the ones recorded by the work of [39] after fermenting cassava pulp with pure strains of some common saprophytes namely, *Aspergillus flavus, Aspergillus niger, Rhizopus oryzae* and *Saccharomyces spp.* The results of the studies showed that phytate content in microbial fermented cassava products decreased [flour (505.6 ± 4.2–748.5 ± 2.3 mg/100 g) and gari (373.5 ± 6.2–56.7 ± 5.2 mg/100 g)] when compared to naturally fermented cassava products [flour (705.1 ± 1.2 mg/100 g) and gari (633.9 ± 3.2 mg/100 g)] and unfermented cassava products [flour (874.4 ± 3.2 mg/100 g) and gari (662.8 ± 6.5 mg/100 g)].

Figure 3 presents the percentage decrease in the levels of phytate in the microbial fermented cassava products. The results showed that the highest percentage reduction was recorded in 'Kpace Bokun' cultivar used to produce 'fufu' and treated with *S. cerevicea* (38.24%) and the least reduction was in 'gari' from 'Wahabi' treated through *L. bulgaricus* fermentation (1.91%). Also, the ability of the organisms to detoxify 'fufu' products of phytate is higher than in the 'gari' in most cases. The decrease in the phytate content of the microbially fermented cassava products could possibly be attributed to the secretion of the enzyme phytase by microbes during fermentation. These enzymes are capable of hydrolysing phytate, thereby, decreasing the phytate content of the products [37].

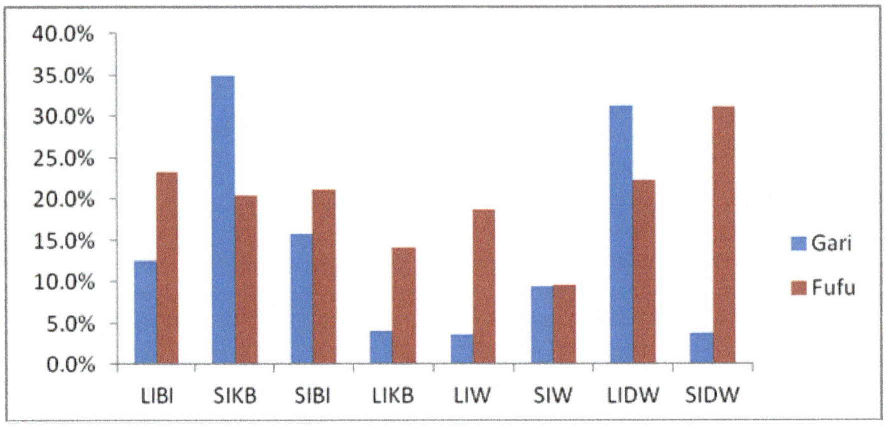

Figure 3. Percentage decrease of phytate in 'garl' and 'fufu' samples due to inoculation

3.4. Effect of Inoculation on the Tannin Composition of the Products

The amount of tannin presented in the Table 1 and Table 2 indicated that the levels of tannin ranged from 2.03 ± 1.04 mg/100g to 8.62±1.00mg/100g for 'gari' from 'Dan Warri' inoculated with *S. cerevisae* and 'lafun' from 'Kpace Bokun' not fermented respectively. There was a significant difference between the percentage obtained for the microbial treated products compared to the untreated ones. Levels of tannin in the 'lafun' products were the highest because the products were neither microbial fermented nor naturally fermented. Reference [37] carried out similar research work and the findings indicated thatt the unfermented low-cyanide cassava flour (lafun) had 0.4 ± 0.0% tanin content and the *Rhizopus oryzae* and *Saccharomyces crevisae* fermented of the low-cyanide cassava flour recorded 0.10 ± 0.0% and 0.10 ± 0.0 0.37% tanin respectively. As for the unfermented medium-cyanide cassava flour it was found to contain 0.4 ± 0.0% tanin content while the *Rhizopus oryzae* and *Saccharomyces crevisae* fermented of the medium-cyanide cassava flour recorded 0.2 ± 0.0% and 0.2 ± 0.0% tanin respectively.

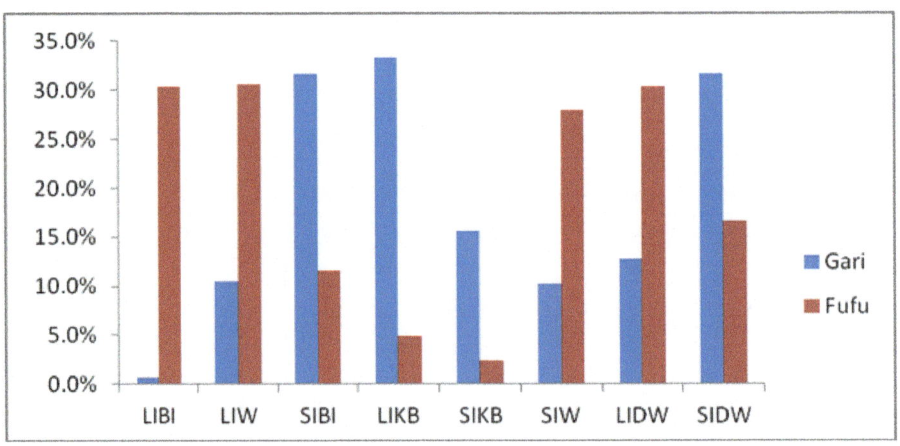

Figure 4. Percentage decrease of tannin in 'garl' and 'fufu' samples due to inoculation

The percentage loss in tannins occurred in all the cultivars fermented with *L. bulgaricus* and *S. cerevisae* as shown in Figure 4. For tannins, the reductions were markedly high in *S. cerevisae*-inoculated 'gari' from 'Kpace Bokun' (33.86%), 'Wahabi' (30.33%), *L. bulgaricus*-treated 'gari' from 'Wahabi' (47.81%) and 'Baba Iya' (33.27%) . Tannin contents of 'Dan Warri's 'gari' treated with *S. cerevisae* and that of 'fufu' of the same 'Dan Warri' decreased by 43.12% and 51.73 % respectively. Decrease in tannin content can be attributed to the degradation of polyphenols as a result of different microorganisms that are involved in fermentation and the corresponding microbial enzymes that are released during fermentation period [35]. From the results, it can be inferred that the two organisms were good at bringing about a reasonable reduction of tannin in the 'gari' products compared to 'fufu' products, while there was no appreciable difference between the abilities of the two organisms to reduce tannin contents in cassava products.

3.5. Effect of Inoculation on the Saponin Content of the Products

From Table 1 and Table 2, the result of saponin content showed that the levels obtained for the inoculated cassava products were higher than those values recorded for the uninoculated counterparts. This trend remained the same irrespective of the variety of cassava and the anti-nutritional component under consideration. There is no available data specifically on saponins in cassava products or even cassava tubers from literature.

The result of saponin content presented in Figure 5 indicated that the highest reduction in saponin occurred in *S. cerevisae*-inoculated 'fufu' from 'Baba Iya' (70.11%) and the least reduction was in 'gari' treated with *L. bulgaricus* produced from 'Wahabi' (2.43%). From these results it is obvious that *S. cerevisae* had the ability to detoxify casssava products more than the *L. bulgaricus,* since most of the treatments with L. *bulgaricus* recorded lower reductions in the anti-nutrient.

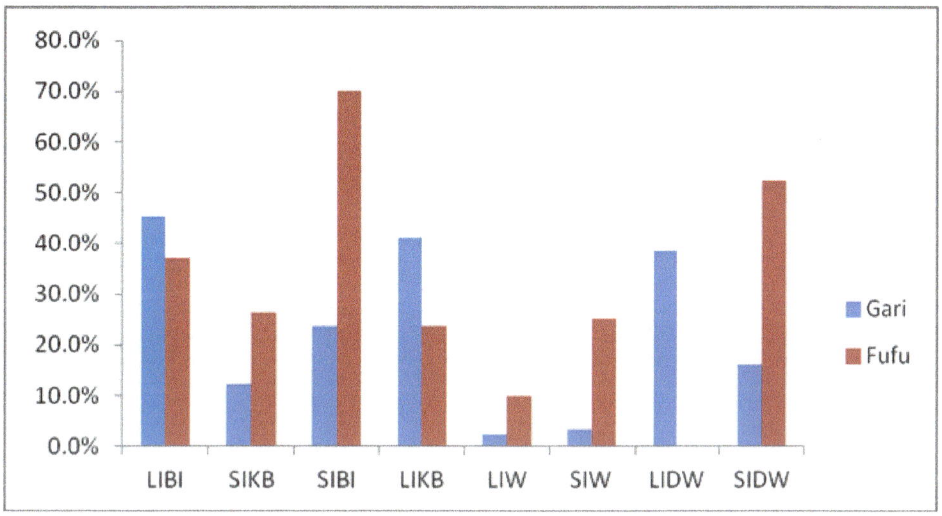

Figure 5. Percentage saponin decrease in 'garl' and 'fufu' samples due to inoculation

4. Conclusion

The inference drawn from the study are that inoculation of cassava pulp with *S. cerevisae* and *L. bulgaricus* during fermentation enhanced the nutritional status of the products obtained by drastically reducing the anti-nutrients components determined. The nature of organism used for fermentation significantly affected the quality of the products.

S. cerevisae had greater ability to a large extent (about 40%) to detoxify the cassava products than the *L. bulgaricus*. The two organisms (*S. cerevisae* and *L. bulgaricus*) showed greater potency to improve the quality of 'fufu' products than the quality of 'gari' products. There was no clear trend regarding varietal compositions between the improved and local varieties investgated. With the significant reductions in the anti-nutrients determined, health disorders associated with the consumption of excess anti-nutrients, particularly among people who are already malnourished could reduce drastically.

Aknowledgements

The success of this work owes much to the untiring efforts and constructive critcisms of Professor C. E. Gimba, Professor E. B. Agbaji and Dr K. I. Omoniyi. May they all be handsomely rewarded by the Almighty God.

I would like to acknowledge all the wonderful people who encouraged me to complete this work, especially Dr Muhammad Muhammad Ndamitso, Head of Chemistry Department, Federal University of Technology, Minna, Emimaru Ahmad of the Deparment of Petroleum Resources, Abuja office and a host of others too numerous to mention. Your contributions have been highly beneficial to the success of this work.

References

[1] Wanda, L.B.W., Diana, I.A., Jennifer, M.M. and Richard, T.S., "Cyanogenesis in cassava: the role of hydroxynitrile lyase in root cyanide production", *Plant Physiology,* 116 (4). 1219-1225. Apr. 1998.

[2] Odoemenem, I.U. and Otanwa, LB., "Economic analysis of cassava production in Benue State, Nigeria", *Current Research Journal of Social Sciences,* 3(5). 406-411. Dec. 2011.

[3] Ajakaye, O.A. and Ojo, S.O., "Productivity Analysis of Cassava Production in Ekiti State, Nigeria" Unpublished M-Tech Thesis. Federal University of Technology, Akure, Nigeria. 2009.

[4] Sayre, R1,, Beeching, J.R., Cahoon, E.B.,, Egesi ,C., Fauquet, C., Fellman, J., Fregene, M., Gruissem, W., Mallowa, S., Manary, M., Maziya-Dixon, B., Mbanaso, A., Schachtman, D.P., Siritunga, D., Taylor, N., Vanderschuren, H. and Zhang, P., |The BioCassava plus program: biofortification of cassava for sub-Saharan Africa", *Annual Review of Plant Biology* 2011.

[5] Sayre, R., Research News: Researchers get to the root of cassava's cyanide-producing abilities, Retrieved from http://www.acs.ohio-state.edu/units/research/achieve/search.htm, 2003.

[6] Hahn, S.K., Terry, E.R., Leuschner, K., Akobundu, I.O., Okali, C. and Lal, R., Field Crops Research; Cassava improvement in Africa. Science Direct, 2: 193-226, 1992.

[7] Akoja, S.S. and Mohammed, A.O., "Evaluation of the proximate, pasting and sensory characteristics of cassava flour (fufu) fortified with pigeon pea flour", *Pakistan* Journal *of Nutrition,*. 10 (2). 124-128. 2011.

[8] FAO Food and nutrition Papers – 24, Phytate. Inc: FAO Corporate Document Repository. *Roots, tubers, plantains and bananas in human nutrition,* 1990.

[9] Montagnac, J.A., Davis, C.R. and Tanumihardjo, S.A., "Nutritional value of cassava for use as a staple food and recent advances for improvement" *Comprehensive Reviews in Food Science and Food Safety*, 8(3): 181-194. Jun. 2009.

[10] Ayodeji, O.F., "Nutrient composition and processing effects on cassava leaf (*Manihot esculenta, Crantz*) Antinutrients", *Pakistan Journal of Nutrition,* 4 (1). 37-42. 2005.

[11] Giraud, E., Gosselin L. and Raimbault, M., "Degradation of cassava linamarin by lactic acid bacteria". *Biotechnology Letters*. 14 (7). 593-598. Jul. 1992.

[12] Norhaizan, M.E. and Nor Faizadatul, A.A.W., "Determination of phytate, iron, zinc, calcium contents and their molar ratios in commonly consumed raw and prepared food in Malaysia", *Malaysian Journal of Nutrition* 15(2). 213-222. Sept. 2009.

[13] Sarkiyayi, S. and Agar, T. M., "Comparative analysis on the nutritional and anti-nutritional contents of the sweet and bitter cassava varieties", *Advance Journal of Food Science and Technology* 2(6). 328-334. Jan. 2010.

[14] Adeniji, T.A., Sanni, L.O., Barimalaa, I.S. and Hart, A.D., "Anti-nutrients and heavy metals in some new plantain and banana cultivars", *Nigerian Food Journal*, 25(1). 171-177. 2007.

[15] Holmes, R.P. and Kennedy, M., Estimation of the oxalate content of foods and daily oxalate intake. *Kidney International,* (4). 1662-1667. Nov. 2000.

[16] Tracy, R,, Oxalates in spinach – Is oxalic acid a green smoothie health concern? Retrieved October 30, 2012 from http://www.incrediblesmoothies.com/green-smoothies/oxalates-spinach-oxalic-acid-health-con. (2007).

[17] Aro, S.O., "Improvement in the nutritive quality of cassava and its by-products through microbial fermentation", *African Journal of Biotechnology*, 7(25). 4789-4797. Dec. 2008.

[18] Andrew, R.O. and Peter, A.A., "Two-stage fermentation method for production of protein-enriched feed from cassava", *Enzyme and Microbial Technology*, 2(3) 241-243, Jul. 1980.

[19] Kobawila, S.C., Louembe, D., Keleke, S, Hounhouigan, J. and Gamba C., "Reduction of the cyanide content during fermentation of cassava roots and leaves to produce bikedi and ntoba mbodi, two food products from Congo", *African Journal of Biotechnology*, 4 (7). 689-696. 2005.

[20] Jack, B., France, B., Venkatesh, M. and Peter, R., Food fortification - the debate continues. Science in Africa. No. 75. Rretrieved, August 23, 2010, from http://scienceinafrica.com/old/2002/june/food.htm, 2002.

[21] Paula, A.C., Estevao, M., Mario, E., Fernando, M., Julie, C.M., Rezaul, H., and Howard, B. J., "Processing of cassava roots to remove cyanoge", *Journal of Food Composition and Analysis*, 18: 451-4 60. Apr. 2005.

[22] Emmanuel, N.M., "Cyanide content of gari", *Toxicology Letters,* 3(1). 21-24. Jan. 2002.

[23] Rami, N., "Whole grain toxicity - Phytic acid contained in popular foods", Retrieved July 14, 2011, from www. FoodMatters.tv/Documentary, 2010.

[24] Céline, P., Bernard, C., Ana-Belén, B., Stéphanie, C., Hélène, F., Javier, P., and Aude, V., "Tannin Oxidation: Intra- versus Intermolecular Reactions". *Biomacromolecules, 11* (9). 2376-2386. Jul. 2010.

[25] Kumar, R. /FAO, "Anti-nutritional factors, the potential risks of toxicity and methods to alleviate them" Retrieved July 20, 20134, from http://www.fao.org/docrep/003/t0632e/t0632e10.htm, Pp 145-160.

[26] Savolainen H., "Reactions of Tannins with Human Serum Proteoglycans". *Journal of Applied Toxicology* 17 (2): 133-135. Mar. 1998.

[27] Meredith, G.B., Sylvia, V.E. and Howard, J.B., "Picrate paper kits for determination of total cyanogens in cassava roots and all forms of cyanogens in cassava products", *Journal of the Science of Food and Agriculture,* 79. 593-601. Apr. 1999.

[28] Simone, C.B.G., Eloir, P.S. and Valquiria, L. B., "HPLC method to assay total saponins in Ilex paraguariensis aqueous extract", *Journal of the Brazilian Chemical Society*, 16 (4). Jul./Aug. 2005.

[29] Irawan, A.S., Suherman, D. and Tangendjaja B., "A Screening method for total oxalate determination in plant material by gas chromatography", *Journal of the Science of Food and Agriculture*; 39 (3). 233-238. Sept..2006.

[30] Harland, B.F., and Oberleas, D., "Anion-exchange method for determination of phytate in foods: collaborative study", *Journal of Association of Analytical Chemists*, 69(4). 667-700. Aug-Jul. 1986.

[31] Asegbeloyin, J.N. and Onyimonyi, A.E., "The effect of different processing methods on the residual cyanide of 'gari'", *Pakistan Journal of Nutrition*, 6 (2). 163-166. 2007.

[32] Eustace, A.I. and Dorothy, M.L., "Cyanide detoxification in cassava by-products by fungal solid fermentation", *The Journal of food Technology in Africa* 5 (2). 48-5. 2000.

[33] Akindahunsi, A.A., Oboh, G. and Oshodi, A. A., *Effect of fermenting cassava with rhizopus oryzae on the chemical composition of its flour and gari products*, International Atomic Energy Agency: The Abdus Salam International Centre for Theoretical Physics, 1999.

[34] Tilahun, A. T., Shimelis, A. E., & Gulelat, D. H., Effect of processing on physicochemical composition and anti-nutritional factors of cassava (*Manihot esculenta crantz*) grown in Ethiopia. *International Journal of Science Innovations and Discoveries*, 3 (2): 212-222, 2013.

[35] Ebuchi, O. A.T., Babalola, O. & Ahmed, Z., Phytochemical, nutritive and anti-nutritive composition of cassava (*Manihot Esculenta*) tubers and leaves. *Nigerian Food Journal*, 23: 40-46, 2005.

[36] Fagbemi, A.O. & Ijah, U. J. J., Microbial population and biochemical changes during production of protein-enriched 'fufu'. *World Journal of Microbiology and Biotechnology*, Published by Springer Netherlands. 22(6): 635-640, 2006.

[37] Oboh, G. & Elusiyan, C. A., Changes in the nutrient and anti-nutrient content of micro-fungi fermented cassava flour produced from low- and medium-cyanide variety of cassava tubers. *African Journal of Biotechnology,* 6 (18): 2150-2157. 2007.

[38] Orjiekwe, C.L., Solola, A., Iyen E. and Imade, S. "Determination of cyanogenic glucosides in cassava products sold in Okada, Edo State, Nigeria". *African Journal of Food Science*, 7(12). pp. 468-472, Dec. 2013.

[39] Oboh, G., Akindahunsi, A.A.and Oshodi A.A. "Dynamics of phytate-zinc balance of fungi fermented cassava products (gari and flour)", *Plant Foods for Human Nutrition*, 58(3), pp 1-7. Sept.. 2003.

Proximate, Anti-Nutritional and Phytochemical Composition of the Yellow Variety of the *Synsepalum Dulcificum* (Miracle Fruit) Berry

Njoku N. E.*, Ubbaonu C. N., Alagbaoso S.O., Agunwa I.M, Eluchie C.N.

Department of Food Science and Technology, Federal University of Technology, Owerri, ImoState, Nigeria
*Corresponding author: alagbaososerah@gmail.com

Abstract The pulp of freshly harvested mature *Synsepalumdulcificum* (miracle fruit) berries were analyzed for proximate, pH, titratable acidity, anti-nutritional and phytochemical content. The results indicated a moisture content of 45.12%, protein content of 2.48% and a carbohydrate content of 48.84%.The anti-nutrients content include Tannin 2.90 ± 0.64mg/100g, phytate 5.21 ± 0.92 mg/100g, glycosidic cyanide 0.03 ± 0.00mg/100g, steroid 1.56±0.03 mg/100g and oxalate 11.04 ± 0.29%. The anti-nutritional levels were all within the recommended safety limits. Some phytochemicals like flavonoids, resins and steroids were also identified. It is hoped that the knowledge of miracle fruit's nutritional content/importance will enhance its utilization in our diet.

Keywords: *berry pulp, phytochemicals, proximate, anti-nutrients, Synsepalumdulcificum*

1. Introduction

The miracle fruit, also known as 'miraculous berry' or 'miracle berry' is a fruit of a tropical plant, *Synsepalumdulcificum* and a member of the Sapotaceae family [1] which is native to West African (Plate 1). It is known as 'Agbayun' by the Yorubas', and called different names such as 'Uninaa', 'Uneonu', 'Unighini' and 'Uni' by the Ibos'jn Nigeria. The plant is commonly found growing in the wild, in fringes of virgin forests, though it also grows naturally on farms and around dwellings and in secondary bushes [2]. Its fruit has a relatively large seed which is encapsulated by a translucent pulp. This pulp is covered by a thin skin.

Plate 1. The *Synsepalumdulcificum* Tree

The plant has two varieties, the red and the yellow ripe berry varieites [1]. The red berry is not really sweet to taste by itself but miraculously gives a sweet taste when food, even a sour food is being eaten after taking or licking it, thus, it is aptly named 'miracle fruit' [1].

Although the miraculous unique sweet flavour-enhancing capacity of the fruit has been known since the 18th century [3], scientific investigations of the fruit were not initiated until 1964 [4], and even so, only some scanty work has been done on this berry. Noticeably, the few research reports available were focused on the red berry variety. There is paucity of information on the yellow berry variety which is more prevalent in Nigeria, especially the Eastern part of Nigeria. Although it is primarily used as a sweetener, for gari and palm-wine drinking, it is hoped that there are other uses it can be subjected to. However, the insufficient study and information on the fruit have aided the under- utilization of the fruit in modern diet. Consequently, the fruit is gradually but definitely going into extinction, making it quite rare to find in our environment. This work, therefore, aims to investigate the chemical (proximate) and anti-nutritional composition of this unique yellow berry as well as some of its phytochemicals as a means of observing its nutritional potential of meeting the macronutrients needs of its consumers.

2. Materials and methods

2.1. Materials Collection

Freshly harvested mature berries of *Synsepalumdulcificum* (miracle fruit) were obtained from Umuagwo in OhajiEgbema Local Government Area of Imo State, Nigeria. The samples were immediately transported to the laboratory. The pulp were extracted by scraping the fruits with clean stainless spatula.

2.2. Proximate Analysis

The analysis of proximate composition was done in triplicate trials using the procedure described by [5]. The moisture, ash, protein, fibre, fat and carbohydrate contents were all determined in triplicate trials.

2.3. Determination of Titratable Acidity

One gram of the sample was weighed out and macerated with 50ml of distilled water. The mixture was filtered with Whatman's filter paper, No 1. The filtrate was collected into a 50ml conical flask. One millilitre of phenolphthalein indicator was added to it. The solution was titrated with 0.1N Sodium Hydroxide until the colour turns pink. The reading on the burette was noted.

2.4. Determination of pH

Ten grams of the sample was mixed with 25ml of distilled water. It was stirred continuously to dissolve. The mixture was left to stand for 5 min. Zero point three millilitres was pipette out. The pH was read off a digital Hanner (pHS – 25) pH meter calibrated with pH 7 buffer.

2.5. Determination of the Anti-nutritional Content

The determination of tannin, glycosidic cyanide, steroid, saponin and alkaloid contents of the berry was done using the procedure as described by (6) and modified by [7] while the phytate and oxalate contents were determined following the procedure of (8) using a spectrum 21D spectrophotometer. Trypsin inhibitor content was determined using the spectrophotometric method described by [9] while the procedure of [10] was implemented for hemagglutinin.

2.5.1. Determination of Tannin Content

One gram of the sample was macerated with 50ml methanol. It was filtered with Whatman's No 1 filter paper. Five milliliters (5ml) of the filtrate was pipette out and 0.3ml of 0.1N ferric chloride in 0.1N hydrochloric acid (HCl) was added to it. Zero point three millilitre (0.3ml) of 0.0008M potassium ferricyanide was also added. The absorbance of the resulting solution was read at 720nm. Concentration of tannin was thus calculated:

$$Concentration = \frac{Absorbance\ of\ sample}{Gradient\ Factor} xDF \quad (2.5.1)$$

Where DF= Dilution Factor
Gradient Factor = Slope of the standard curve (Figure 1).

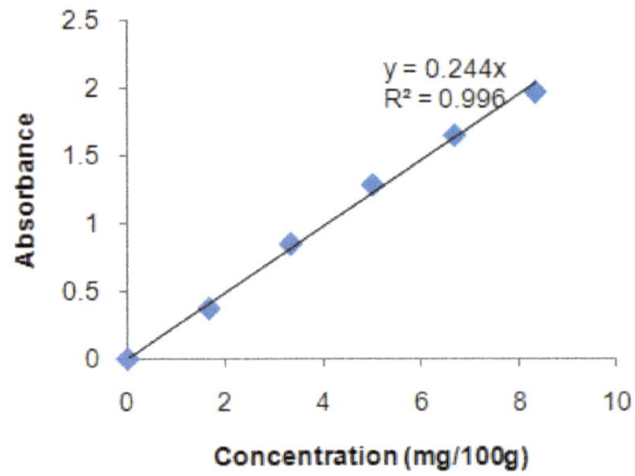

Figure 1. Standard Curve of Tannin Concentrations

2.5.2. Determination of Glycosidic Cyanide Content

One gram of the sample was macerated with 50ml of water. It was filtered with Whatman's No 1 filter paper. One milliliter (1ml) of the filtrate was pipette into a test tube and 4ml of alkaline picrate solution added to it. The solution was boiled for 5 min. It was then cooled in a water bath. The absorbance of the solution was read at 490nm.

$$Cyanide\ (mg/100g) = \frac{Absorbance\ of\ samplex\ Gradient\ Factor\ x\ DF}{Sample\ weight} \quad (2.5.2)$$

Where DF= Dilution Factor
Gradient Factor = Slope of the standardcurve (Figure 2).

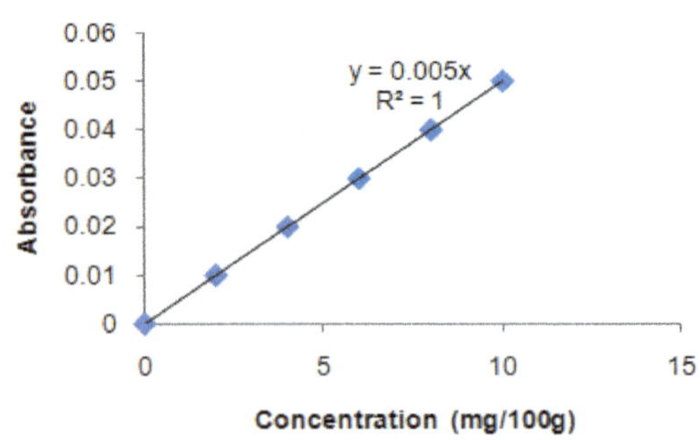

Figure 2. Standard Curve of Glycosidic Cyanide Concentration

2.5.3. Determination of Steroid Content

One gram of the sample was macerated with 20ml of ethanol. The solution was filtered with Whatman's No 1 filter paper. Two milliliters (2ml) of the filtrate was pipetted out and 2ml of cholesterol color reagent was added. The solution was left to stand for 35 min after which the absorbance of the solution was taken at 550nm.

$$\text{Concentration} = \frac{Absorbance\ of\ sample}{GradientFactor} xDF \quad (2.5.3)$$

Where DF= Dilution Factor
Gradient Factor = Slope of the standard curve (Figure 3).

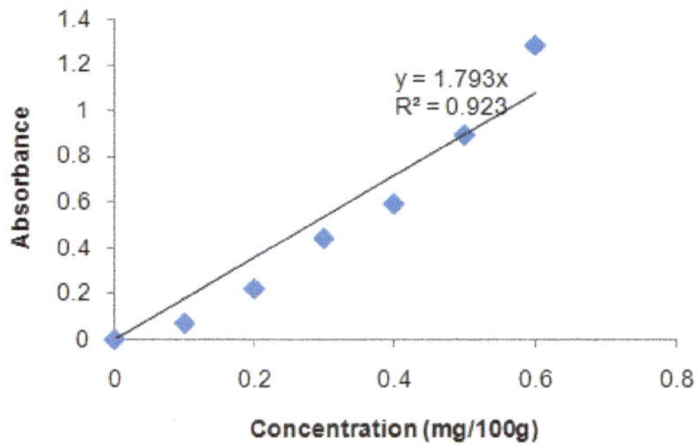

Figure 3. Standard Curve of Steroid Concentrations

2.5.4. Determination of Saponin Content

One gram of the sample was macerated with 10ml of petroleum ether. It was decanted into a beaker. Another 10ml of petroleum ether was added to the same sample and again, it was decanted into a beaker. The filtrates from the two decantation were combined and evaporated to dryness. The residue was dissolved with 6ml of ethanol. Two milliliters (2ml) of the resulting solution (mixture) was transferred (or pipette) into a test tube. A color reagent (chromogen) was added. The solution was allowed to stand for 30 min. The absorbance of the solution was read at 550nm.

$$\text{Concentration} = \frac{Absorbance\ of\ sample}{GradientFactor} xDF \quad (2.5.4)$$

Where DF= Dilution Factor
Gradient factor = Slope of the standard curve (Figure 4).

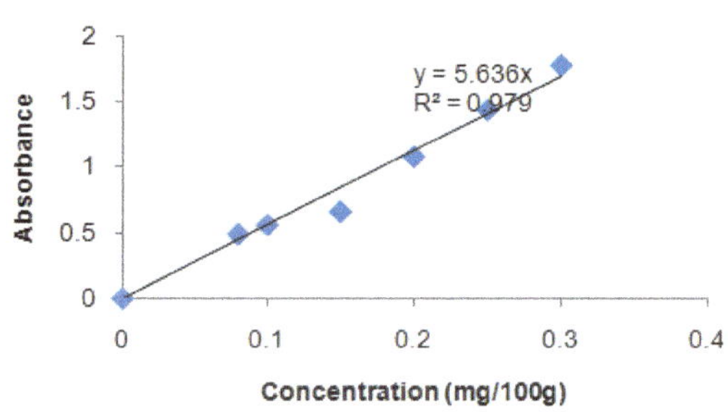

Figure 4. Standard Curve of Saponin Concentrations

2.5.5. Determination of Alkaloid Content

One gram of the sample was macerated with 20ml of ethanol. A twenty percent (20%) solution of sulphuric acid was added to the mixture in the ratio of 1:1. It was allowed to stay for 5 min. The mixture was filtered with a Whatman's No 1 filter paper. One milliliter (1ml) of the filtrate was pipette out and 5ml of 60% H_2SO_4 added. They were mixed together and allowed to stand for 3 hours. The absorbance of the resulting solution was taken at 565nm.

$$\text{Concentration} = \frac{Absorbance\ of\ sample}{GradientFactor} xDF \quad (2.5.5)$$

Where DF= Dilution Factor
Gradient Factor = Slope of the standard curve (Figure 5)

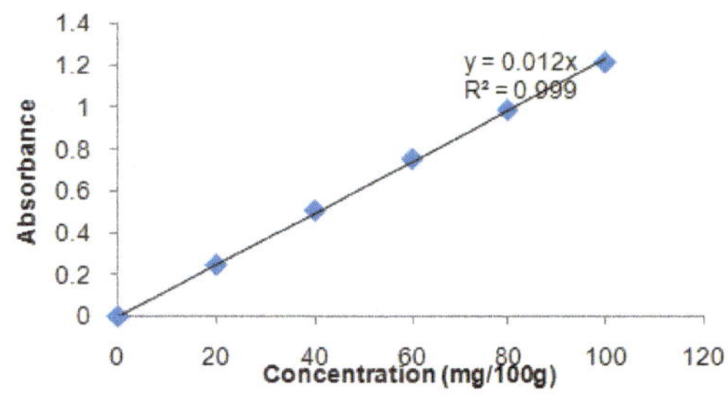

Figure 5. Standard Curve of Alkaloid Concentrations

2.5.6. Determination of Hemagglutinin Content

One gram of the sample was dispersed in a 10ml normal saline solution buffered at pH 6.4 with a 0.01M phosphate buffer solution. The solution was allowed to stand at room temperature for 30 min. It was then centrifuged and filtered to obtain an extract. Zero point one milliliter (0.1ml) of the extract diluent was pipette into a test tube and 1ml of trypsinzed albino rat blood was added. A blank containing only the blood cells was prepared in a test tube. This was used as the control. Both test tubes were allowed to stand at room temperature for 4 hours. One milliliter (1ml) of normal saline was added to both test tubes and allowed to stand for 10min. The absorbance of the two, sample and control were read at 620nm. The result was expressed as hemagglutinin units per milligram of the sample.

$$Hemagglutinin\ unit / mg = (b - a) xf \quad (2.5.6)$$

Where b= absorbance of test sample solution
a = absorbance of the blank control

$$f = experimental\ factor\ given\ by\ 1/w\ x\ V_f / V_a \quad (2.5.6.1)$$

Where w = weight of sample
V_f = total volume of extract
V_a = volume of extract used in the assay.

2.5.7. Phytate Content Determination

One gram (1g) of the test sample was weighed out into a 500ml flat bottom flask. The flask with the sample was placed in a shaker and extraction was done with 100ml of 24% Hydrochloric acid (HCl) for one hour at room temperature. The solution was decanted and filtered. Five milliliters (5ml) of the filtrate was pipetted out and diluted to 25ml with distilled water. Fifteen milliliters (15ml) of sodium chloride was added to 10ml of the diluted sample. The absorbance reading of the solution was taken at 520nm.

$$Concentration = \frac{Absorbance\ of\ sample}{Gradient\ Factor} xDF \quad (2.5.7)$$

Where DF= Dilution Factor
Gradient Factor = Slope of the standard curve (Figure 6)

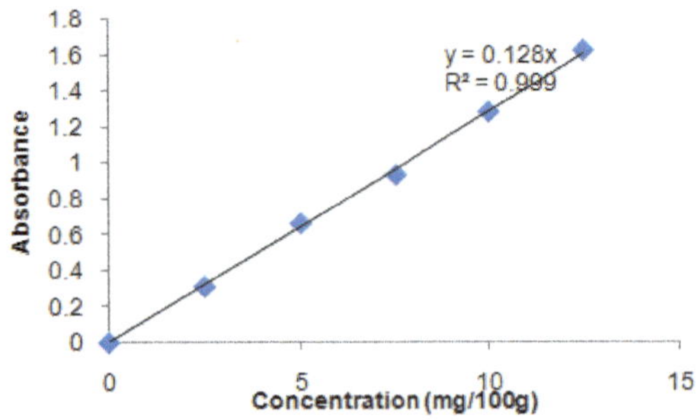

Figure 6. Standard Curve of Phytate Concentrations

2.5.8. Oxalate Content Determination

Two grams (2g) of the sample was weighed into a 300ml flask. Twenty milliliters (20ml) of 30% hydrochloric acid (HCl) was added to the sample and allowed to stand for twenty min. Forty grams of ammonium sulphate was also added and allowed to stand for another thirty minutes. The solution was filtered (with Whatman's No 1 filter paper) into a 250ml volumetric flask and made up to the mark with 30% HCl. Ten milliliters (10ml) of the filtrate was transferred into 100ml centrifuge tube. Thirty milliliters (30ml) of diethyl ether was added to it. The pH was adjusted to pH 7 with ammonium hydroxide. The solution was centrifuged at 10,000g for fifteen minutes. It was decanted into a 250ml conical flask and then titrated with 0.1M potassium tetraoxomanganate (IV) ($KMnO_4$). The volume used (titre) was noted.

The percentage oxalate was calculated thus:

$$\% \, oxalate = \frac{titrexmol \; KMnO_4 \; x \; DF(12.5) \, x \, 100}{Weight \; of \; sample} \quad (2.5.8)$$

Where DF is the dilution factor.

2.5.9. Trypsin Inhibitor Content Determination

One gram (1g) of the sample was weighed out and dispersed in 50ml of 0.5M sodium chloride (NaCl) solution. The mixture was stirred for 30 min at room temperature and centrifuged. The supernatants were filtered through Whatman No 41 filter paper. The filtrate was used for the assay. Standard trypsin was prepared using N-α- Bensoyl-DL-arginine-P-nitroanilide (BAPA) and used to treat the substrate solution. Into a test tube containing 10ml of the substrate and 2ml of the filtrate (extract), 2ml of the standard trypsin solution was added. A blank containing 2ml of the standard trypsin solution and 10ml of the substrate without the pulp sample was prepared and put in a test tube. The contents of the two test tubes were allowed to stand for 30min. Their absorbance were then measured spectrophotometrically at 410nm. The extent of inhibition of trypsin hydrolysis of the filtrate is used as a standard to measure the trypsin inhibition activites of the test sample. One trypsin unit inhibited (TUI) is equal to an increase of 0.01 in absorbance unit at 410nm. The trypsin inhibitor activity was expressed as the number of trypsin units inhibited (TUI) per unit weight (g) of the sample analyzed.

$$TUI/g = \frac{Absorbance \; of \; sample}{Absorbance \; of \; standard} x0.01F \quad (2.5.9)$$

Where b= absorbance of test sample solution

a = absorbance of the standard (uninhibited) sample or blank (control)

F = experimental factor, given by

$$F = 1/W \; x \; V_f / V_a \quad (2.5.9.1)$$

W = weight of the sample

V_f = total volume of the extract (filtrate)

V_a = volume of filtrate used in the analysis (assay).

2.6. Phytochemical Analysis

The tests were carried out based on procedures outlined by [6] and modified by [7].

2.6.1. Qualitative Phytochemical Analysis: They Include

Tests for carbohydrate, proteins, alkaloids, saponins and tannins were all carried out. Others include:

2.6.1.1. Test for Reducing Sugar

Five millilitres of a mixture of equal parts of Fehling's solution A and B were added to 5ml of aqueous of sample, and then heated in a water bath for 5min. Brick red precipitate showed the presence of reducing sugar.

2.6.1.2. Test for Glycosides

Five millilitres of dilute sulphuric acid was added to 1g of the sample in a test tube and boiled for 15min in a water bath. It was cooled and neutralised with 20% potassium hydroxide solution. Ten millilitres (10ml) of a mixture of equal parts of Fehling's solution A and B were added and boiled for 5min. A more dense brick red precipitate indicated the presence of glycoside.

2.6.1.3. Test for Flavonoid

Fifty millilitres of ethyl acetate was added to 1g of the extract and heated on a water bath for 3min. The mixture was cooled and filtered. Four millilitres of the filtrate was shaken with 1ml of dilute ammonium solution. The layers were allowed to separate. A yellow colour in the ammonium layer indicated the presence of flavonoids.

2.6.1.4. Test for Fats and Oil

One hundred milligrams of the extract was passed in between filter papers and the paper was observed. A control was prepared by placing 2 drops of olive oil on filter paper. Translucency of the filter paper indicated the presence of fats and oil.

2.6.1.5. Test for Steroids and Terpenoids

Nine millilitres of ethanol was added to 1g of the extract and refluxed for a few min. it was then filtered. The filtrate was concentrated to 2.5ml in a boiling water bath. Five millilitre of hot distilled water was added to the concentrated solution. The mixture was allowed to stand for one hour and the waxy matter was filtered off. The filtrate was extracted with 2.5ml of chloroform using separating funnel. One millilitre of concentrated sulphuric acid was carefully added to 0.5ml of the chloroform extracted in a test tube to form a lower layer. A reddish brown interface showed the presence of steroids. Another 0.5ml of the chloroform extract was evaporated to dryness in a waterbath. A grey colour indicated the presence of terpenoids.

2.6.1.6. Test for Acidic Compounds

One gram of the extract was placed in a clear dry test tube and sufficient water was added. The solution was warmed in a hot water bath and then cooled. A piece of water-wetted litmus paper was dipped in the solution and observed. Acidic compounds turned blue litmus paper red.

2.6.1.7. Test for Resins

One gram of the sample extract with chloroform and the extract solution was concentrated to dryness. The residue was re-dissolved in 3ml of acetone and 3ml of concentrated hydrochloric acid added. This mixture was heated in a hot water bath for 30min. a pink colour which changes to magenta red indicated the presence of resins.

2.6.2. Quatitative Phytochemical Analysis

For the quantitative analysis, the parameters measured include *Cyanogenic glycosides*, tannins, soluble carbohydrates and steroids. Others are:

2.6.2.1. Flavonoid Determination

One gram of the sample was weighed out and macerated with 20ml of ethyl acetate for 5min. it was filtered with Whatman No 1 filter paper. Five millilitre of the filtrate was pipette into a test tube and 5ml of dilute ammonia was added. The solution was shaken vigorously for 5min and then it was allowed to settle. The supernatant was collated and the absorbance measured using a spectrophotometer at 490nm.

2.6.2.2. Reducing Sugar Determination

One gram of the sample was macerated with 20ml of distilled water. It was then filtered with a filter paper. One millilitre of the filtrate was pipettedinto a testtube and 1ml of alkaline copper reagent was added. The solution was boiled in a fume cupboard for 5 min, and then cooled in a water bath. One millilitre of phosphomolybdic acid reagent and 7ml of water were added to the cooled solution. The absorbance was read at 420nm.

2.6.2.3. Glycosides

One gram of the sample was macerated with 20ml of water. Some quantity (2.5ml) of 15% lead acetate was added to the filtration and the solution was shaken vigorously. It was allowed to settle and the lower layer (infranatant) was collected and evaporated to dryness. The residue was dissolved with 3ml of glacid acetic acid. About 0.1ml of 5% ferric chloride was added. Some 0.25ml of conc H_2SO_4 was also added. The solution was vigorously shaken and put in a dark container for 2h. The absorbance was read at 530nm.

3. Results and Discussion

From Table 1, the moisture content of the berry sample (45.12%) was observed to be less than other berries such as blackberry (88.15%), cranberry (86.5%) and strawberry (89.9%) [11,12]. This suggests that its dehydration will require less energy and time with reduced nutrient loss. The moisture content of any food indicates its level of water activity and may thus, be used to measure its stability and susceptibility to microbial contamination [13]. The ash content (0.87%) was higher than the contents of blackberry (0.5%), blueberry (0.2%), strawberry (0.5%), raspberry (0.5%) and cranberry (0.2%) [14].

Table 1. The proximate composition, pH and titratable acidity of the pulp of *Synsepalumdulcificum* BERRY

COMPONENT	CONTENT
MOISTURE (%)	45.12
ASH (%)	0.87
FIBRE (%)	0.57
FATS (%)	2.1
PROTEIN (%)	2.48
CARBOHYDRATE (%)	48.85
pH	3.2
TITRATABLE ACIDITY (g/100g)	0.02

This gave an idea of the inorganic content of the sample, since the inherent minerals are contained in the ash. Miracle berry's relative higher protein value (2.48%) compared to most berries (blueberry-0.74%, cranberry-0.4%, strawberry-0.8%) and tropical fruits (banana-1% - 1.3%, orange-0.9%, apple-0.3%-0.4%, pawpaw-0.47-1.17%, guava-1% and avocardo pear-2%) [15,16] seem to suggest that a glycoprotein may be involved in its action of imparting sweetness in non-sweet food. The crude fiber content (0.57%) was lower than the content instrawberries (1.6%), blackberries (3.2%) and raspberries (4.7%)(14). As such, the sample may be desirable for incorporation in weaning diets. Emphasis has been placed on the importance of keeping fibre intakes low in the nutrition of infants and pre-school children (17). The fat content of the sample (2.10%) was higher than the values reported for most fruits(pineapple-0.1%, guava-0.4%, sweet orange-0.1% and banana-0.4%)excluding avocardo pear (17-20%) [15,18]. The pH of the pulp was found to be 3.2 which compares well with other berries such as blackberry (3.4), strawberries (3.3-3.4) and raspberry (3.4) [14]. However, the sample was more acidic than acai berry (5.21). Its titratable acidity was 0.0196g/100g (as citric acid). This value was lower than that of some fruits such as orange (0.139g/100g), banana (0.453g/100g), pawpaw (0.072g/100g) and mango (0.369g/100g) reported by [19].

Table 2. The phytochemical test result

PHYTOCHEMICAL	QUALITAITVE	QUANTITATIVE
Protein	+	
Carbohydrate	++	3.279mg/100g
Saponin	+	
Reducing sugar	++++	397.101mg/100g
Flavonoid	++	6.5842mg/100g
Steroid	+	
Tannin	++	
Alkaloid	++	
Glycoside	+	
Fats/oil	-ve	
Acidic compound	-ve	
Resin	-ve	

Qualitative analysis were carried out to ascertain the presence of the different phytochemicals in the berry pulp before quantitative analysis were carried out. Table 2 presents the quality and quantity of phytochemicals observed in the berry pulp.It has been observed that phytochemicals have positive health effects in human and animal bodies, especially in areas of coronary heart disease, ulcers, diabetes, high blood pressure, mascular degeneration, inflammation, infection and psychotic diseases [20]. The various phytochemicals observed are known to have beneficial importance in medical sciences.Reducing sugars were more abundant than other phytochemicals whereas fats and oil, acidic compounds and resins were not observed at all. Flavonoids, tannin and alkaloid were equally present. Flavonoid and tannins are classified as phenolic metabolites [21] which are known to give defense against infection. This is in agreement with the observation of [22] that berries may provide additional health benefits because they also contain high levels of a diverse range of phytochemicals/phytonutrients consisting predominantly of phenolic type molecules. Berry phenolics also have a wide range of biological properties which includes cell regulatory effects [22,23]. It is interesting to note that glycosides were present in the pulp, though in less quality than flavonoids because [23] observed that human bioavailability of flavonoids differs significantly depending on the type of glycosides that these molecules possess. Flavonoids not only have anticancer properties but have also been shown to have anti-allergic, cardiovascular protection properties, anti-inflammatory properties and antiviral properties [24,25]. Tannins and alkaloids are known to be antimicrobial, anthelmintic and antidiarrhoeal [26]. Tannins also possess astringent properties, hastening healing of wounds [28,29]. Pure isolated plant alkaloids and their synthetic derivatives, on the other hand, are used as basic medicinal agents for analgesic antispasmodic and bactericidal effects [27]. These properties indicate that the berry pulp may be an antioxidant source.

In Table 3, more quantitative phytochemical and anti-nutritional content of the sample was shown. The value of tannin in the test sample (2.90mg/100g) was higher than the tannin content of 0.029mg/100g in Africa star apple [13] but lower than the values of 7.5mg/g and 25.30mg/100g observed in the fruits analyzed by [30] and [31] respectively. The value is also lower than the lethal dose level of 30mg/kg (3mg/100g) observed by [32]. From medicinal point of view, polyphenols (to which tannin belongs) have been reported to act as antioxidants by preventing oxidative stress that causes diseases such as coronary heart disease, some types of cancer and inflammation [33]. The phytate value (5.21mg/100g) is lower than that in guava (0.8 mg/g), mango (0.86 mg/g) and pineapple (0.90 mg/g) [34]. Thus, the value of phytate in *Synsepalumdulcificum* berry might not pose any health hazard when related to a phytate diet of 10 – 60 mg/g consumed over a long period of time, which has been reported to decrease bioavailability of minerals in monogastric animals [35]. The berry's saponin content was low when compared to 17.80mg/100g observed by [36] in the African locust bean fruit pulp. Though high level of saponin has been associated with gastroenteritis manifested by diarrhoea and dysentery [37], acute poisoning is relatively rare [38]. It was however, reported that saponin reduces body cholesterol by preventing its re-absorption and suppressing rumen protozoan by reacting with cholesterol in the protozoan cell membrane (40). The value of trypsin inhibitor in the studied *sample* was 1.75 TUI/mg. However this is low when compared to a range of 6700 to 23300 TIU/g for cowpea [41] indicating that the test sample might not pose risk to health, especially considering the lethal dose of 2.50g/kg reported by [32]. The test sample glycosidic cyanide value was also lower than the 17.30mg/100g hydocyanic acid content of African locust bean fruit pulp and cyanogenic glycoside content of 0.730mg/100g for African star apple (*Chrysophyllumafricanum*) [13]. These values are below the accepted level of human consumption of 50-60mg/kg body weight/day. The Hemagglutinin value (5.10 HU/mg) was lower than the 200HU/mg and 100HU/mg recorded for groundnut oil and palm oil respectively by [32]. The value in the test sample was a lot lower than the lethal dose of 50mg/kg they observed. With an alkaloid content of 14.17mg/100g $\pm$ 0.68, this is higher than the content of alkaloid in groundnut oil (0.2%) and palm oil (0.16%) [32]. Consumption of some alkaloid-containing foods in excess of 20mg/100g sample may trigger off toxicological manifestations [32,42]. The pulp of *Synsepalumdulcificum* had a steroid value of 1.56mg/g $\pm$ 0.03. This was high compared to the 0.27g/100g and 0.07g/100g contents in Bitter leaf (Onugbo) and scent leaf (nchanwu) respectively [43]. Eating plant foods high in sterol is known to help lower the cholesterol level in the blood [44]. Hence, consumption of miracle berry could help to check the level of cholesterol in the blood.

Table 3. Other qunatitative phytochemicals and antinutrients in *Synsepalumdulcificum* berry pulp

ANTINUTRIENT	VALUE
TANNIN (mg/100g)	2.90
PHYTATE (mg/100g)	5.21
OXALATE (%)	11.04
SAPONIN (mg/100g)	0.65
TRYPSIN INHIBITOR (TUI/mg)	1.75
GLYCOSIDIC CYANIDE (mg/100g)	0.03
HEMAGLUTININ (Hemaglutinin Unit/mg)	5.10
ALKALOID (mg/100g)	14.17
STEROID (mg/100g)	1.56
SOLUBLE CARBOHYDRATE (mg/100g)	3.279
FLAVONOID (mg/100g)	6.5842
REDUCING SUGAR (mg/100g)	397.101

Values are means of triplicate determinations.

4. Conclusion

Proximate evaluation of the *Synsepalumdulcificum* berry pulp showed a substantial amount of carbohydrate, lipid and protein. The moisture content was also lower than the content in other berries. Interesting phytochemicals were observed with flavonoid, tannin and alkaloid having comparative quality presence. The anti-nutrients observed were all below their respective lethal doses. Thus, the berry pulp can be safely utilized as food and ingredient in food processing.

References

[1] Njoku, N.E.,Ubbaonu, C.N.Alagbaoso, S.O., Eluchie, C.N. and Umelo, M.C. (2015). Amino acid profile and oxidizable vitamin content of Synsepalumdulcificum berry (miracle fruit) pulp.*Food Science &Nutrition*. 3(3): 252-256.

[2] Opeke, I. K. (1984). Growth and morphogenesis of mature embryo of Capsella in culture. *Plant Physiology* 39: 691-699.

[3] Njoku, N.E. (2014). Some phytochemical components and anti-nutritional composition of the yellow variety of Synsepalumdulcificum berry (miracle fruit). *Proceedings of the 38th Annual Conference and General Meeting of NIFST*, Lagos. pp 147-148.

[4] Wilson, I. and Wilson D., (2006). Its a Miracle. *Rare Fruit Review* [online]. Available at: http://www.capetrib.com.au/miracle.htm [Accessed January 20, 2012].

[5] AOAC. (1994). Association of Official Analytical Chemists. Official Methods of Analysis. Washington, D.C.

[6] Harbourne, J. B. (1973). *Phytochemical Methods. A guide to Modern Technology ofPlant Analysis,* 2nd ed. Chapman and Hall, New York. Pp: 88-185.

[7] Trease, G. E. and Evans, W. C. (1996). Phenols and Phenolic Glycosides in Trease and Evans Pharmacology and Bikere. Tindall, London. Pp. 832-836.

[8] Pearson, D. A. (1976). Chemical Analysis of Foods. 7th Edition. Church Hill Livingstone, New York.1:422-511.

[9] Amtfield, S.D., Ismond, M.A.H. and Murray, E.D. (1985). The fate of anti-nutritional factors during the preparation of faba bean protein isolates using a micellization technique. *Canadian Institute of Fd. Sci. Tech. J.* 18 (2):137-143.

[10] Onwuka, G. I. (2005). Food Analysis and Instrumentation: Theory and Practice. Naphtali prints, Lagos. Pp 156-161.

[11] Chen, S. C. (1992).Physicochemical principles for the concentration and freezing of fruit juices. In: Fruit Processing Technology. AgScience, Inc: Florida Pp. 23-25.

[12] Konja, G. and Lovric, T. (1993). *Berry Fruit Juice Processing Technology*. Agscience, Inc: Florida

[13] Edem, C. A. and Dosunmu, M. I. (2011). Chemical evaluation of proximate composition, ascorbic acid and anti-nutrients content of African star apple (*Chrysophyllumafrcanum*) fruit. *IJRRAS* 9(1):146-149.

[14] Belitz, H.D., Grosch, W. and Schieberle, P. (2009). Springer fruits. *Food Chemistry*. Available at: http://www.123foodscience.com/food_chemistry/fruits.pdf. [Accessed January 20, 2012].

[15] Ihekoronye, A. I. and Ngoddy, P. O. (1985). *Integrated Food Science and Technologyfor the Tropics*. Macmillan Publishers Ltd, London. Pp. 65-193, 283-294.

[16] Nwofia, G. E., Ojimelukwe, P. and Eji, C. (2012). Chemical composition of leaves, fruit pulp and seeds in some *Caricapapya* (L) morphotypes. *International Journal ofMedicinal and Aromatic Plants*, 2 (1): 200-206.

[17] Eromosele, I.C. and Eromosele, C.O. (1993). Studies on the chemical composition and physio-chemical properties of seeds of some wild plants. *Plant Food Hum. Nutr.* 43: 251-258.

[18] ANON, (1960). Food Composition Tables. Rome: FAO. 1:16

[19] Gonzalez, M., Hernandez, Y. and Lobo, M. G. (2012). Determination of Vitamin C in tropical fruit: a comparative evaluation of methods. Plant Physiology Laboratory, Department of Tropical Fruits, Spain: Instituto Canario de Investigaciones Agrarias, Apdo. Available at http://www.icia.es/icia/download/postcosecha/vitamina%20c.pdf. [Accessed January 20, 2012].

[20] Premier, R. (2002). Phytochemical composition: A paradigm shift for food-health considerations. *Asia Pacific J ClinNutr.* 11(6): 197-201.

[21] Dillard, C.J. and German, J.B. (2000).Phytochemicals: Nutraceuticals and human health. *J Sci Food Agric.*, 80: 1744-1756.

[22] Seeram, N.P. (2006). Berries. In: D. Heber *et al.* (eds). Nutritional Oncology, 2nd ed. Academic Press, London, UK. p. 615-625

[23] Seeram, N.P. and Heber, D. (2006). Impact of berry phytochemicals on human health: Effects beyond antioxidation. In: C.T. Ho *et al.* (eds). Lipid Oxidation and Antioxidants: Chemistry, Methodologies and Health Effects. ACS Symposium Series. Oxford University Press.

[24] Manach, C. and J.L. Donovan. 2004. Pharmacokinetics and metabolism of dietary flavanoids in humans. Free Radical Research, 38:771-785.

[25] Cody, V., Middleton, V.E. and Harbourne, J.B. (1986). Plant flavonoids in biology and medicine: Biochemical, pharmacological and structure–activity relationships. New York: Alan Liss,

[26] Gabor, M. (1986). Anti-inflammatory and anti-allergic properties of flavonoids. In: Cody, V., Middleton, V.E., Harbourne, J.B., eds. Plant flavonoids in biology and medicine: Biochemical, pharmacological and structure–activity relationships. New York: Alan Liss.

[27] Tiwar, P., Kumar, B., Kaur, M., Kaur,G. and Kaur, H. (2011).Phytochemical screening and Extraction: A Review*Internationale Pharmaceutica Sciencia,* Vol 1, issue 1

[28] Stray, F. (1998). The national guide to medicinal herbs and plants. Tiger books International: London. Pp 12-16.

[29] Morton, J. (2001). Purple mombin fruits of warm climaes. Miami Publishers: New York, p.245

[30] Kozioc, M.J. and Marcia, M.J. (2004). Chemical composition nutritional evaluation and economic prospects of Spondias purpurea (Anaracardiaceae). *Econ Bot.*, 52: 373-380

[31] Bello, M. O., Falade, O. S., Adewusi, S. R. andOlawore, N. O.(2008). Studies on the chemical compositions and anti-nutrients of some lesser known Nigeria fruits. *African J. Biotechn.*7 (21): 3972-3979

[32] Rathod, V. S. and Valvi, S. R. (2011). Antinutritional factors of some wild edible fruits from Kolhapur District. *Recent Research in Science and Technology* 3(5): 68-72.

[33] Inuwa, H. M., Aina, V. O., Gabi, B., Aimola, I. and Amao, T. (2011). Comparative determination of antinutritional factors in groundnut oil and palm oil. *Advance Journal of Food Science and Technology*, 3(4): 275-279.

[34] Tapiero, H., Tew, K. D., Nguyen, B. G. and Mathe, G. (2002). Polyphenols: do they play a role in the prevention of human pathologies? *Biomed. Pharmacother* 56: 200-207.

[35] Suree, N., Surat, K. and Akekachai, N. (2004). Phytate and Fiber Content in Thai Fruits commonly consumed by diabetic patients. *J. Med. Assoc.* Thai 87(12): 1444-1446.

[36] Thompson, L. U. (1993). Potential health benefits and problems associated with anti-nutrients in foods. *Food Res. International.* 26:131-149.

[37] Gernah, D. I., Atolagbe, M. O. and Echegwo, C. C. (2007). Nutritional composition of the African locust bean *(Parkiabiglobosa)* fruit pulp. Nutritional Composition of African locust Bean. *Nigerian Food Journal*, 25 (1):190-196. Available at: www.ajol.info/journals/nifoj. [Accessed January 20, 2012].

[38] Awe, I. S. and Sodipo, O. A. (2001). *Nigerian Journal of Food Science and Nutrition. 3: 103-107.*

[39] Tannenbaum, S. R. (1979) *Nutritional and Safety Aspects of Food Processing*. New York: Marcel Dekker Inc. Pp. 265-295.

[40] Umaru, H. A., Adamu, R., Dahiru, D. and Nadro, M. S. (2007). Levels of antinutritional factors in some wild edible fruits of Northern Nigeria. *African J.Biotech.*6(16):1935-1938.

[41] Adewusi, S. R. A. and Osuntogun, B. A. (1991). Effects of cooking on tannin content, trypsin inhibitor activity and in-vitro digestibility of some legume seeds in Nigeria. *Nigeria Food Journal* 9: 139-145.

[42] Aletor, V. A. and Fetuga, B. L. (1987). Pancreatic and intestinal amylase (EC 3.2.1.1) in the rat fed haemagglutininextract.II Evidence of impaired dietary starch utilization. *J. Anim. Physiol. Anim. Nutr.* 57(3):113-117.

[43] Onyeka, E. U and Nwambekwe, I. O. (2007). Phytochemical profile of some greenleafy vegetables in South-East Nigeria. *Nigerian Food Journal* 25 (1): 67-72. Available at: www.ajol.info/journals/nifoj [Accessed May 21, 2012].

[44] Piironen, V., Toivo, J., and Lampi, A. M. (2002). New data for cholesterol contents in meat, fish, milk, eggs and their products consumed in Finland. *Journal of FoodComposition and Analysis* 15: 705-713.

Adoptive Trials on the Yam Pounding Machine for Cowpea Dehulling

Sanni L. A.[1], Ogunsina B. S.[1,*], Adegbenjo A. O.[2], Ajayi M. O.[1]

[1]Department of Agricultural and Environmental Engineering, Obafemi Awolowo University, Nigeria
[2]Department of Bioresource Engineering, McGill University Montréal, Quebec, Canada
*Corresponding author: babatundeogunsina@gmail.com

Abstract The yam pounding machine developed by Makanjuola (1975) was adopted for dehulling steeped cowpea grains. The machine simulates traditional method of dehulling steeped cowpea grains which involves rubbing the steeped grains between two abrasive surfaces until the seed coat becomes loose and gets detached from the endosperm. The main components include the dehulling chamber, the lid, beaters (inclined or vertical) and spline coupling. Performance evaluation of the machine was carried out considering three local varieties of cowpea (red drum, white drum and Ife brown), three steeping time (3, 5 and 10 mins) and two beater orientation (inclined and vertical). The results show that 'Ife brown' for which coefficients of dehulling and yield were 0.9902 and 0.9727, respectively indicated the highest dehulling efficiency (96.3%). With the inclined beaters, the coefficient of dehulling increased as steeping time increased; whereas, a decreasing trend was observed for other performance indices. With the vertical beaters, coefficient of dehulling and dehulling efficiency increased as steeping time increased; whereas, coefficient of wholeness decreased as steeping time increased. For the three varieties of cowpea, coefficients of dehulling and yield increased with steeping time; while a decrease was observed for coefficient of wholeness as steeping time increased. A machine of this nature is a significant development over existing methods of dehulling cowpea and will minimize the drudgery associated with small scale dehulling of cowpea. In addition, it presents opportunities for the manufacture of a dual purpose kitchen appliance with accessories suitable for yam pounding and cowpea dehulling.

Keywords: *adoption, yam pounding machine, cowpea, dehulling, performance evaluation*

1. Introduction

The need to augment the demand for animal proteins with alternative sources has gained serious global concern in the last few decades because animal proteins are far from the reach of low income earners. Given the rich amino acids profile of most tropical legumes and oilseeds, their consumption and utilization as functional food ingredients have been widely reported [1,2,3]. Among several other legumes, cowpea (*Vigna unguiculata*), popularly called beans is the most important in Central and South America, Southern Europe, Asia and Africa [4,5]. In the above continents, it is a major source of protein and an important source of revenue for farmers [6,7].

Adejumo *et al.* [2] and Taiwo [8] documented the different forms in which cowpea is largely consumed in Nigeria, namely: cooked or mashed beans/porridge, baked beans paste (*ekuru*), fried beans paste (*akara*), curdled beans slurry (*moinmoin*) and gelatinized beans stew (*gbegiri*). Processing of cowpea into any of these food products often require wet or dry dehulling, during which the tannin-rich seed coat is removed. Although, the use of attrition mill to reduce dry cowpea grains into coarse particles, followed by winnowing and dry milling into flour has been an age-long practice [9], it leaves the product partially dehulled, yielding a flour speckled with particles of the seed coat. The product appears unpleasant, unhygienic and often unsuitable for certain food preparations especially because its functional properties get impaired. Ehiwe and Reichert [10] documented the use of certain roller mills and abrasive dehullers. In the study, it was found that the attrition mill was found to be more suitable for dehulling grains with loose seed coats, while abrasive dehuller was suitable for grains with more tightly adhering seed coats. Few dehullers have been reported for cowpea by Adejumo et al. [2], Olotu et al. [11], Audu [12], Prabhakar and Bhole [13], Babatunde [14] and Reichert et al. [15], but they are rarely suitable for kitchen-scale applications. Total removal of seed coat from the endosperm as achievable by wet dehulling is quite rigorous but preferred because it reduces the tannin and phytate contents of cowpea to a very large extent Amonsou *et al.* [16]. Jackson and Varriano-Mattson [17] documented that wet dehulling of cowpea removes up to 98% of the seed coat, reduces cooking time by about 60%, enhances cooking quality and improves protein quality, palatability and digestibility; hence, it is a very vital step in the processing of cowpea for different food applications [18]. The process involves soaking the grains in water for a period during which the seed coat softens, loosens and becomes easy to remove by gentle abrasion on a grinding stone or by tender stir-pounding in a pestle and mortar [10]. Afterwards, complete separation of the seed coat from the endosperm is achieved by agitating and floating the marshy mixture in water. After repeated agitation (about 3-4 times) and decantation, the product is obtained as clean whole or split cotyledons. Despite the existence of various reports on the dehulling of cowpea, doing it on a kitchen-scale is still a manual process in many African countries till today. In this work, the yam pounding machine was modified to dehull cowpea on a kitchen-scale. The mechanism simulates the traditional method in which steeped cowpea grains are rubbed gently between two abrasive surfaces and agitated until the seed coat becomes loose and detached from the endosperm.

2. Methodology

2.1. Description of the Yam Pounding Machine

In 1975, Emeritus Prof. G. A. Makanjuola [19] developed a machine for preparing pounded yam (Figure 1a). The compact design comprises a cylindrical cast aluminum cup (diameter, 200 mm and height, 110 mm) fitted with four alternate metallic beaters mounted on a shaft (Figure 1b); extending vertically downwards into the cup through the centre of the cover lid (Figure 1c). A vertically installed single phase electric motor is fitted to drive the beaters. Cooked (hot) diced pieces of yam are fed into the cup which is held fixed to the machine base. As the electric motor drives the metallic beaters, the slices of cooked yam get comminuted vigorously until the starch granules gelatinize to form a thick glutinous dough (otherwise known as pounded yam) the colour of which may be white, cream or yellow (depending on the colour/variety of yam used).

The machine having gone through some modifications has become a commercial household appliance in Nigeria and some other places in sub-Saharan Africa where pounded yam is an important local delicacy.

2.2. Modifications Made on the Yam Pounding Machine for Cowpea Dehulling

The commercial version of the yam pounding machine manufactured by Addis Engineering Ltd. was used for this work. The metallic beaters (Figure 1c) were replaced with rubber-lined beaters and a variable speed electric motor was incorporated to regulate speed. The rubber-lined surface of the beaters, the cylinder wall and the steeped cowpea grains rub against each other until seed coat removal is achieved. During the process, some of the cowpea cotyledons become broken into pieces or splits. The clearance between the beaters and the cylinder wall was based on the axial dimensions of cowpea earlier reported by Yalcin [20]. The main components of the modified machine include:

i) *The dehulling chamber:* This is a cast aluminum cup which holds the steeped grains for dehulling. It has a top diameter and height of 200 mm and 110 mm respectively. The inner surface of the cup is lined with an abrasive rubber overlay to increase friction between the cowpea grains and cylinder wall.

ii) *The lid:* The aluminum disc-cover of the yam pounder was modified into a 200 mm circular disc made from 3 mm thick aluminium to serve as lid to cover the cup. The lid has grooves which gives a firm grip on the cup.

Beaters: The beaters are made of brass rods with surfaces lined with abrasive rubber material. There are two beaters which were oriented vertically and alternately at 45 and 60°. As the beaters rotate, the friction developed between the rubber-lined surfaces, the cylinder wall and cowpea grains lead to loosening and removal of the seed coat from the endosperm. The length, the internal and external diameters of each beater are: 50, 10 and 15 mm respectively.

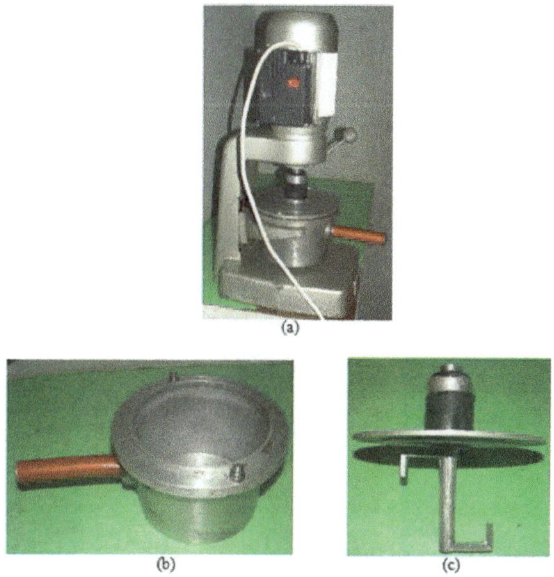

(a) The yam pounding machine (Makanjuola, 1975)
(b) The yam chamber/cup with handle
(c) The lid and beaters assembly

Figure 1. The yam pounding machine by Makanjuola (1975) and its main components [19]

iii) *Spline Coupling:* A 16 mm socket was used as a spline for coupling the central shaft which transmits torque from the electric motor to the beaters.

iv) *The electric motor*: A single phase electric motor (0.8 hp, 1410 rpm) drives the shaft upon which the beaters are mounted. Though 1410 rpm was found suitable for yam pounding, it was higher for dehulling of cowpea as it resulted in grains breakage. In a baseline study, a variable speed transformer was used to vary the speed and a speed limit of 200 rpm was found suitable for cowpea dehulling.

v) *Body/Frame:* The machine body is a welded and machined mild steel structure that holds all the components together as a compact unit.

vi) *Handle:* A 50 x 50 mm^2 wood was machined to a 35 mm diameter rod and fitted as a handle to hold the dehulling chamber.

The orthographic and sectional views of the kitchen appliance for dehulling steeped cowpea grains are presented in Figure 2a, Figure 2b, and Figure 3a, Figure 3b respectively showing the inclined and vertical beaters orientation.

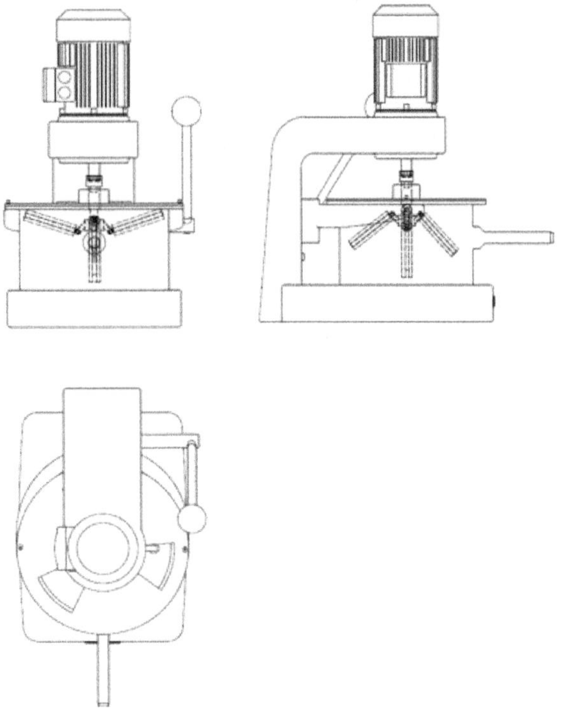

Figure 2a. Orthographic view of the cowpea dehuller with inclined beaters

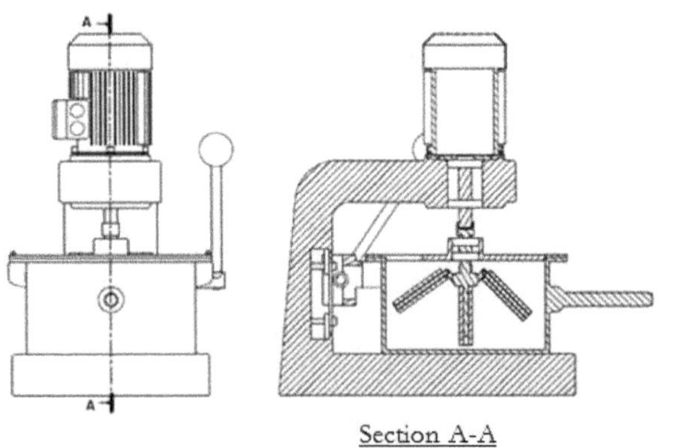

Figure 2b. Sectional view of the cowpea dehuller showing the details of inclined beaters

2.2. Performance Evaluation

About 25 kg each of three varieties of cowpea grains were obtained from the Obafemi Awolowo University teaching and research farm and Sabo market in Ile Ife. The endosperm and seed coat contents of three cowpea varieties (red drum, white drum and Ife brown) were determined following a modified method of Reichert *et al.* [15]. One hundred grams (100 g) of grains from each variety was soaked in distilled water for 10 mins, and afterwards, it was dehulled manually. The endosperm and the hull fractions were air-dried separately until a constant weight was obtained between two weighing successions. The weight of each fraction was determined per 100 g of sample.

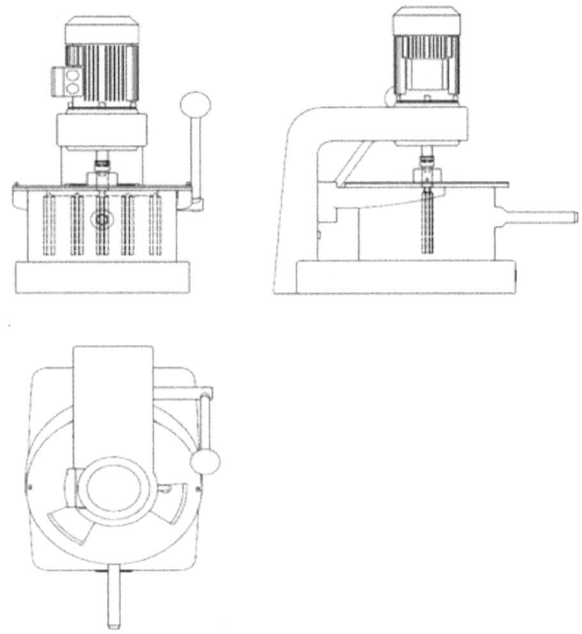

Figure 3a. Orthographic view of the cowpea dehuller with vertical beaters

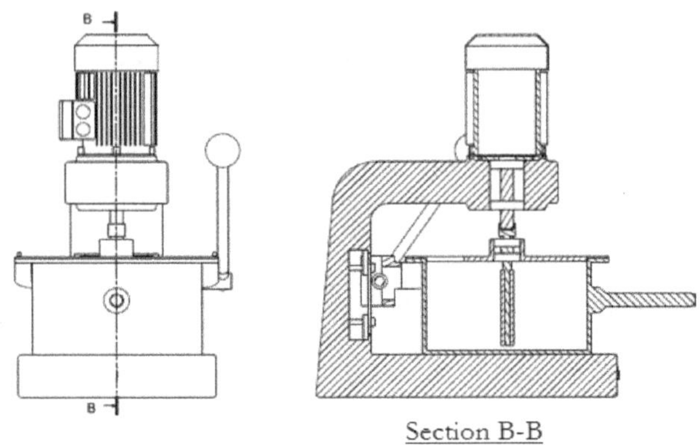

Figure 3b. Sectional view of the cowpea dehuller showing the details of vertical beaters

The machine was subjected to preliminary tests with each of the three varieties of cowpea (red drum, white drum and Ife brown) steeped for 5 mins using the inclined beaters and considering two resident times (10 and 15 s). The chamber was filled with water to ¾ of its volume and about 1 kg of steeped cowpea grains was fed into the cup and the machine was allowed to run for 10 s. The dehulling chamber was opened after each experimental run and the content was evacuated for analysis. The dehulled mass of grains was sorted into fractions as dehulled whole grains, dehulled split grains, partially dehulled grains, dehulled but broken pieces and un-dehulled grains. Each fraction was weighed in a precision electronic balance having an accuracy of 0.01 (GT 2100, Made in Germany). Machine performance indices: coefficient of wholeness, C_w; coefficient of dehulling, C_n; coefficient of yield, C_y; dehulling efficiency, E_d and the overall performance efficiency of the machine, E_o were calculated using equations (1) – (7) as documented by Lazaro [21].

$$C_w = \frac{M_{wk}}{M_{wk} + M_b} \tag{1}$$

$$C_n = \frac{M_p}{M_w} \quad (2)$$

$$C_y = \frac{Y_a}{Y_e} \quad (3)$$

$$Y_a = \frac{M_d}{M} \quad (4)$$

$$Y_e = (100 - M_w) \quad (5)$$

$$E_d = C_n \times C_y \quad (6)$$

$$E_o = C_n \times C_y \times C_w \quad (7)$$

where, M_{wk} = mass of whole kernels in the final product; M_b = mass of broken kernels in the dehulled grain sample; M_p = seed coat removed by the dehulling process; M_w = seed coat content of undehulled grain; M = mass of grain sample before dehulling. M_d = mass of the dehulled grain recovered; Y_a = actual yield obtained; Y_e = maximum expected yield of dehulled grains.

Following the procedure above, the machine was subjected to performance evaluation considering the three cowpea varieties listed above; three steeping times (3, 5 and 10 mins) and two orientation of beaters (inclined and vertical). The effect of beaters orientation and steeping time on the performance indices was investigated using 'Ife brown' (on the basis that it had the highest coefficient of dehulling). Furthermore, the effect of steeping time on the coefficients of dehulling, wholeness and yield for the three cowpea varieties was investigated.

3. Results and Discussion

Figure 4 shows the isometric drawing of the experimental machine and machine parts description; while pictorial views of the inclined and vertically arranged beaters are shown in Figure 5a and Figure 5b, respectively.

Item	Quantity	Part name	Material
1	1	Electric motor	Composite
2	1	Cover lid	Cast aluminum
3	1	The dehulling chamber	Cast aluminum
4	1	Spline coupling/Beater assembly	Mild steel/Rubber lined hollow cylindrical pipe
5	1	Lever to grip the cup	Stainless steel
6	1	Wooden handle	Wood
7	1	Machine base	Mild steel

Figure 4. Isometric drawing of the experimental machine showing machine parts

The seed coat and endosperm fraction of red drum, white drum and Ife brown cowpea varieties were found to be: 89.5, 11.5; 87.2, 12.3; and 84.9, 15.1 g/100 g, respectively. This shows that for the same weight of grains, Ife brown had the highest seed coat fraction while red drum had the least. Ehiwe and Reichert [10] affirmed that the amount of seed coat removed from the seed and yield are desirable indices in explaining the quality of dehulled grains.

Figure 5a. The rubber-lined inclined beaters

Figure 5b. The rubber-lined vertical beaters

From the results of preliminary tests (Table 1), it was observed that Ife brown for which coefficients of dehulling and yield were 0.9902 and 0.9727 respectively indicated the highest dehulling efficiency (96.3%). More whole grains were obtained for red drum and the machine was most efficient with white drum (E_o = 87.66%). Table 2 shows the effect of beaters orientation and steeping time on the performance indices. It was observed that for the inclined beater orientation, the coefficient of dehulling (C_n) increased as steeping time increased; whereas, a decreasing trend was observed for other indices. From statistical analysis, no significant difference was indicated by all the values. With the vertical beaters orientation however, the coefficient of dehulling and dehulling efficiency (E_d) showed an increasing trend as steeping time increased; whereas, coefficient of wholeness (C_w) decreased as steeping time increased. No significant changes were observed in the yield and machine performance efficiency. With higher values of C_n indicated for inclined beaters orientation, it may be inferred that this orientation is more suitable for dehulling steeped cowpea. Coefficient of dehulling is an index which defines the degree of hull removal achieved during the dehulling process. Although higher values of C_w were observed with vertical beaters orientation, wholeness of the grains may not be considered as an important quality index in cowpea dehulling. This is because milling or cooking is usually the next food process to which the dehulled grains will be subjected, depending on the desired food product. The effect of steeping time on C_n, C_w and C_y for the three cowpea varieties using the inclined beaters at 10 s resident time is shown in Figure 6. The results showed that for the three varieties of cowpea, coefficients of dehulling and yield increased with steeping time; whereas, a decrease was observed for coefficient of wholeness as steeping time increased. However, no difference was observed in the coefficients of dehulling and yield for the three varieties; hence, to preserve the wholeness of grains, 5 mins of steeping time was considered the most suitable.

Table 1. Results of preliminary tests using three varieties of cowpea, 5 mins steeping time and 10 s resident time with inclined beaters

Samples	C_n	C_w	C_y	E_d (%)	E_o (%)
Red drum	0.9728	0.9118	0.9672	94.09	85.79
White drum	0.9755	0.9260	0.9705	94.67	87.66
Ife brown	0.9902	0.8824	0.9727	96.32	84.99

C_n, C_w,and C_y represents the coefficients of dehulling, wholeness and yield; while E_d and E_o are dehulling efficiency and machine efficiency respectively.

Table 2. Effect of beaters orientation and steeping time on performance indices using Ife brown variety of cowpea

Steeping time (min.)	C_n		C_w		C_y		E_d (%)		E_o (%)	
	X	Y	X	Y	X	Y	X	Y	X	Y
3	0.9734	0.9696	0.9457	0.9586	0.9686	0.9647	94.28	93.54	89.16	89.67
5	0.9738	0.9720	0.9232	0.9256	0.9683	0.9662	94.29	93.92	87.05	86.93
10	0.9740	0.9721	0.9055	0.9139	0.9661	0.9657	94.10	96.57	85.21	88.26

X = inclined beaters orientation; Y = vertical beaters orientation; C_n, C_w,and C_y represent the coefficients of dehulling, wholeness and yield; while E_d and E_o are dehulling efficiency and machine efficiency respectively.

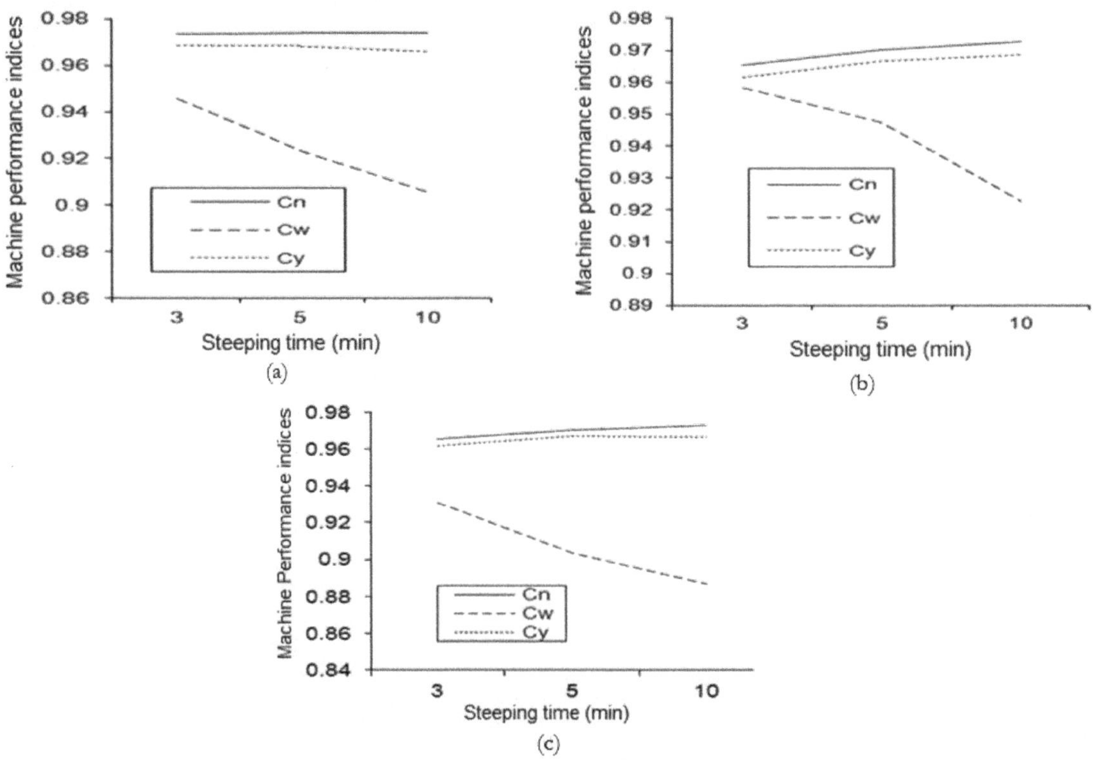

Figure 6. Effect of steeping time on the coefficients of dehulling, coefficient of wholeness and yield for three varieties of cowpea using the inclined beaters at 10 s resident time ((a) = Red drum, (b) = White drum, (c) = Ife brown, C_n = coefficients of dehulling, C_w = coefficients of wholeness and C_y = coefficients of yield))

4. Conclusions

An existing yam pounding machine has been modified for kitchen-scale dehulling of steeped cowpea grains. The modifications made include replacement of metallic beaters with rubber-lined beaters inclined at different orientations to create the required soft abrasion for the removal of seed coat of steeped cowpea grains. The machine indicated its highest dehulling efficiency as 96.3% for 'Ife brown' variety. This work, apart from minimizing the drudgery associated with kitchen-scale dehulling of steeped cowpea grains presents the possibility of developing the yam pounding machine into a dual purpose kitchen appliance with accessories suitable for the preparation of pounded yam and dehulling of steeped cowpea.

Acknowledgements

- Emeritus Prof. G.A. Makanjuola for granting permission to carry out this adoptive trial on the yam pounding machine.

- Prof M.O. Faborode for his motivation at different stages of the work.
- Addis Engineering Ltd., the commercial manufacturer of the prototype on which this modification was based.

References

[1] Odedeji, J.O. and Oyeleke, W.A. 2011. Proximate, physicochemical and organoleptic properties of whole and dehulled cowpea seed flour (*Vigna unguiculata*). *Pakistan Journal of Nutrition,* 10(12): 1175-1178.

[2] Adejumo, A.O.D., Oradugba, O.B., Ilori, T.A. and Adebisi, J.A. 2009. The performance evaluation of a kitchen model cowpea dehuller. In: Proceedings of 3rd International Conference of West African Society of Agricultural Engineers and 9th International Conference of Nigerian Institution of Agricultural Engineers. Ile-Ife, Nigeria. pp. 120-127.

[3] Gopalan, C., Rama Sari, B.V. and Balasubramanian, S.C. 2007. *Nutritive Value of Indian Foods*, National Institute of Nutrition, Indian Council of Medical Research, Hyderabad. pp. 20-30.

[4] Timko, M.P., Ehlers, J.D. and Roberts, P.A. 2007. *Cowpea*. In: Kole, C. (Ed.) Genome Mapping and Molecular Breeding in Plants, Vol. 3, Pulses, Sugar and Tuber Crops, Springer Verlag, Berlin .Heidelberg. pp. 49-67.

[5] Singh, B.B., Ehlers, J.D., Sharma, B., Freire Filho, F.R. 2002. Recent progress in cowpea breeding. In: Fatokun, C.A., Tarawali, S.A., Singh, B.B., Kormawa, P.M. and Tamo, M. Challenges and Opportunities for Enhancing Sustainable Cowpea Production. International Institute of Tropical Agriculture, Ibadan, Nigeria, pp. 22-40.

[6] Adipala, E., Ocaya, C.P. and Osiru, D.S.O. 2002. Effect of time of planting cowpea (*Vigna unguiculata* (L.) Walp.) relative to maize (*Zea mays* L.) on growth and yield of cowpea. *Tropicultura,* 20: 49-57.

[7] Langyintuo, A.S., Lowenberg-DeBoer, J., Faye, M., Lamber, D. and Ibro, G. 2003. Cowpea supply and demand in West Africa. *Field Crops Research*, 82: 215-231.

[8] Taiwo, K.A. 1998. The potential of cowpea as human food in Nigeria. *Technovation,* 18(6/7): 469-481.

[9] Akinjayeju, O. and Bisiriyu, K.T. 2004. Comparative studies of some properties of undehulled, mechanically dehulled and manually dehulled cowpea (*Vigna unguiculata* Walp. L.) flour. *International Journal of Food Science and Technology,* 39: 355-360.

[10] Ehiwe, A.O.F. and Reichert, R.D. 1987. Variability in dehulling quality of cowpea, pigeon pea and mung bean cultivars determined with the tangential abrasive dehulling device. *Cereal Chemistry,* 64(2): 86-90.

[11] Olotu, F.B., Aduba, J.J. and Olaoye, J.O. 2013. Effects of operative processing conditions on the performance of cowpea dehuller using selected cowpea varieties. *International Journal of Basic and Applied Sciences,* 2(3): 71-78.

[12] Audu, I., Oloso, A. and Umar B. 2004. Development of a concentric cylinder locust bean dehuller. *Agricultural Engineering International - the CIGR Journal.* Manuscript PM 04 003. Vol. VI.

[13] Prabhakar, S.P. and Bhole, G.N. 2000. Pre-treatment of pigeon pea grain for improvement of dehulling characteristics. *International Journal of Food Science and Technology,* 35: 305-313.

[14] Babatunde, O.O. 1995. Development and performance evaluation of a manually operated dehulling machine for soaked cowpea. *Journal of Agricultural Engineering and Technology.* 3: 1-7.

[15] Reichert, R.D., Lorer, E.F. and Youngs, C.G. 1979. Village–scale mechanical dehulling of cowpea. *Cereals Chemistry,* 56(3): 181-184.

[16] Amonsou, E.O., Houssou, P.A., Sakyi-Dawson, E. and Saalia, F.K. 2009. Dehulling characteristics, sensory and functional properties of flours from selected cowpea varieties. *Journal of the Science of Food and Agriculture,* 89(9): 1587-1592.

[17] Jackson, G.M. and Varriano-Mattson, E. 1981. Hard-to-cook phenomenon in beans: Structural changes during storage and inhibition. *Journal of Food Science,* 46(5): 1379-1385.

[18] Akinjayeju, O. and Ajayi, O.F. 2011. Effects of dehulling on functional and sensory properties of flours from black beans *(Phaseolus Vulgaris). Food and Nutrition Sciences*, 2: 344-349.

[19] Makanjuola G.A. 1975. The design, development and testing of a machine for preparing pounded yam and similar foods. *Journal of Appropriate Technology*, Intermediate Technology Publications 1(4): 9-10.

[20] Yalçın, I. 2007. Physical properties of cowpea (*Vigna sinensis* L.) seed. *Journal of Food Engineering.* 79:57-62.

[21] Lazaro, E.L. 1999. Improvement of the dehulling efficiency of sorghum and millet. Ph.D. Thesis. 6(2):106-14.

Evaluation of Some Chemical Composition, Minerals Fatty Acid Profiles, antioxidant and Antimicrobial Activities of Tulsi (*Ocimum sanctum*) from India

Sandip I. Vidhani[1], Vijay G. Vyas[1], Heena J. Parmar[1], Viren M. Bhalani[1], Mohammad M. Hassan[2,*], Ahmed Gaber[2], Baljibhai A. Golakiya[1]

[1]Elemental Cell, Food Testing Laboratory, Department of Biotechnology, Junagadh Agricultural University, Junagadh. India
[2]Scientific Research Center, Biotechnology and Genetic engineering Unit, Taif University, KSA
*Corresponding author: khyate_99@yahoo.com

Abstract The holy Tulsi, also known as the "queen of herbs" is the most sacred of all the herbs found in India. This sacred plant was found in almost every Indian household. The proximate, minerals and preliminary phytochemical analysis of *Ocimum sanctum* leaves were studied. The nutritional analysis of *Ocimum sanctum* shown high level of ascorbic acid and total carbohydrate i.e., 65.41 mg/100g and 39.58% in their leaves, Whereas the total phenol was found to be maximum (1.88 mg/g) in leaves. Leaves in present investigation, contains major nutrient like N (3.30%), P (1.10%), K (6.62 %), S (1.55 %) and Na (0.74%). Dry weight basis contains 20.64% total protein and 3.60% total fat. The oil of leaves also contains comparable amount of antioxidant as ascorbic acid, flavonoid and total phenol as well as linolenic acid, polyunsaturated fatty acid which was very good for health. According to our results, most of the identified compounds were biologically important. Further the *Oscimum sanctum* leaf possesses certain characteristics that can be ascribed to cultivation on a domestic plantation. Antimicrobial activity of Tulsi leaves extract was evaluated and the results shown that *E. coli*, *E. faecalis* were mostly susceptible to methanol extract than *S. aureus* and *A. hydrophila*. It can be suggested that *S. aureus* was the most resistant organisms to the concentrations of 20 and 40 mg/ml of the methanol extract of *Ocimum.* The results of this study indicated the possibility of using the leave extract of Tulsi (*Ocimum sanctum*) as a source of antibacterial compounds for treatment of infections caused by multi-drug resistant bacterial pathogens.

Keywords: *Ocimum sanctum, heavy metal, phytochemicals, antioxidant, antimicrobial, GC-MS*

1. Introduction

Ocimum sanctum in English Holy Basil, Tulsi (in Urdu) belongs to plant family *Lamiaceae.* It has made important contribution to the field of science from ancient times as also to modern research due to its large number of medicinal properties [1]. *Ocimum sanctum* has been described as of two types i.e., vanya (wild) and gramya (grown in homes). *Ocimum sanctum* has been used in India for around 5000 years and is acclaimed for its healing properties of the mind, body and spirit [2,3]. The use of plants as sources of medicines are human substance has been in vogue since antiquity [4]. Large numbers of plants are utilized in various systems of medicine practiced in India and local health traditions for the treatment of human diseases since time immemorial [5,6]. Tulsi is the legendary 'Incomparable one' of India, is one of the holiest and most cherished of the many healing and healthy giving herbs of the orient. Traditionally, Tulsi is taken in many forms: as an herbal tea, dried powder, fresh leaf, or mixed with honey or ghee. For centuries, the dried leaves of tulsi have been mixed with stored grains to repel insects [7]. Many research and studies suggest that tulsi may be a COX_{-2} inhibitor, like many modern painkillers, due to its significant amount of eugenol [4,5,8]. The chemical composition of Tulsi is highly complex, containing many nutrients and other biological active compounds [9,10]. These constituents significantly vary with time, cultivation process and storage. The nutritional and pharmacological properties of the whole herb in natural form, result from synergistic interaction of many different active phytochemicals, consequently, the overall effects of tulsi cannot be fully duplicated with isolated compound or extracts. Recent studies have shown new promising pharmacologically active chemical constituents from this ancient phytomedicine [4,5,11]. The present review summarizes the comprehensive information concerning the traditional use, ayurvedic properties and phytochemistry of tulsi.

It is to be found that the various *Ocimum* species are very much distinguished from each other. All the species are possessing different pharmacological activities science the huge variation in the chemical composition is there. The literature will serve as a guideline for the researchers in future work related to the complex phytochemistry of the genus *Ocimum*.

Heavy metals have bio-importance as trace elements but the biotoxic effects of many of them in human biochemistry are of great concern. They enter our bodies via food, drinking water and air [12]. Iron, zinc and copper are essential metals whereas cadmium, lead and mercury have no bio-importance [13].

The present study was achieved to evaluate the heavy metal deposition and phytochemical characterization, fatty acid profiles and antimicrobial activities of the leaves extract of *Ocimum sanctum* plant.

2. Materials and Methods

2.1. phytochemical Analysis of *Ocimum sanctum:*

The leaves was collected from three different places from local market of Junagadh and treated as replication. The amount of reducing sugar, total carbohydrates, true protein, Anthrone reagent and Folin-Phenol reagent was estimated. [14]. The phenol content was determined using methanolic extract [15]. Standard graph was prepared for quantification using gallic acid as a standard. Results of total phenol were expressed as mg of gallic acid equivalents per gm of fresh weight of sample. Total Ascorbic acid was quantified [16]. Total ascorbic acid was ex-pressed in mg per 100 gm leaves sample. Total flavonoid was estimated using 1 ml of methanolic extract in which 0.5 ml of 2% w/v $AlCl_3$ in methanol and 0.5 ml potassium acetate (120 mM) were added and incubated at room temperature for 30 minutes. Absorbance was read at 415 nm. Quercetin was used as a standard and the results were expressed as mg of quercetin equivalents per gm of fresh weight sample [17].

The dry leaves of *Ocimum sanctum* was used for oil estimated [18]. Oil was extracted from dry leaves by soxhlet extraction in hexane as solvent. The extracted oils were dried under reduce pressure in rotary evaporator to make free from solvent. Oils were stored at -20°C until prior to use for fatty acid profile. The nutritional data were expressed on dry weight basis.

2.2. Determination of Major Elements in *Ocimum sanctum:*

Plant samples were analysed for N by micro-Kjeldahj using Automatic Digestion, Distillation System (Vap 50s Gerhardt), Total P by vanado molybdo phosphoric yellow color method using spectrometer method and K by mix acid (perchloric acid and nitric acid) digestion and flame photometry [19].

2.3. Fatty Acid Profile of *Ocimum sanctum* Leaves Oil

The fatty acids profiles were determined by GC-MS. Fatty acid methyl esters were prepared using BF3 methanolic solution and extracted with hexane [20]. GC-MS analyses were performed with some modification [20]. GC-MS analyses were performed by Food Testing Laboratories using a Shimdzu model QP2010 quadruple mass spectrometer detector. The GC column was a DB-5, 30 m, 0.25 µm capillary. The initial column temperature was 60°C. The temperature program was 12°C per minute with one minute hold time when rich at 150°C. A final temperature was 240°C per minutes with hold time at five minutes and the mass spectrometer detector analyses. The ion source temp was 230°C. Interface temp was 240°C and the solvent cut time was 2 minutes. For the identification of the compounds the mass spectra of the samples were compared with those Mass spectral library as well as the fatty acids composition was quantified using appropriate standards.

2.4. Determination of Antimicrobial Activity of *Ocimum sanctum* Leaves Extract

The agar-well diffusion method was employed for determination of antibacterial activities [21]. The dry leaves of *Ocimum sanctum* was used for leaves extract [18]. Leaves extract was extracted from dry leaves by methanol 80 % as solvent. The extracts were dried to make free from solvent. Extracts were stored at 4°C until prior to use for antimicrobial activety (NCCLS, 2002) All bacteria were suspended in sterile water and diluted to~10^6 CFU/ml. 200 µl of the suspension was spread onto the surface of NA medium. Wells (4.6 mm in diameter) were cut from the agar with a sterile borer. The concentrations (20 and 40 mg /ml) of the *O. sanctun* leaves extract were applied. Negative controls were prepared using water. The artificial Streptomycin with a concentration of (0.8 µg/ml) was used as positive reference standards to determine the sensitivity of each microbial species tested and to compare the relative percent of antibacterial activity. The inoculated plates were incubated at 37°C for 24 h. Antibacterial activity was evaluated by measuring the diameter of inhibition zone (DIZ) of the tested bacteria. DIZ was expressed in millimeters [22].

2.5. Statistical Analysis

All experiments were repeated three times. The growth turbidity of the three tested strains treated with different concentrations of *O. sanctum* extract was calculated by one way ANOVA for the relationship between the growth turbidity and the concentrations of crude extract Both Microsoft Excel 2007 and SPSS (version 16) were used in such analysis.

3. Results and Discussion

3.1. Phytochemical Analysis of *Ocimum sanctum:*

The chemical composition of the *Ocimum sanctum* leaves and stem are given in (Table 1 and Figure 1). As can be seen, stem of leaves shown high level of nutrition value as compared to the leaves. The average percentage w/w of the ash content and the extractive values were determined. Total acidity was determined by simple titration method which shows the negligible quantity in leaves, while the total carbohydrate in leaves is quite high 39.58%. The nitrogen and protein were analyzed by Auto Kjedahl method and nitrogen and total protein value was

3.30 and 20.64% for leaves. These values were taken in triplicate but there was no change among these values having a very small number of standard deviation. Fat were extracted with 95% n-hexane by Soxhlet apparatus and found 3.60% crude fat in leaves. The reducing sugar, flavonoids, ascorbic acid and total phenol were also quantified and tabulated in Table 2, which shown maximum concentration of ascorbic acid that was 65.41 mg/g for leaves, while minimum concentration of total phenol was found 1.88 mg/g. The presence of phenol or alkaloid indicates that the use of the plants have harmless effect. The flavonoids in leaves 28.38 mg/g. The presence of flavonoids confirms that the plants has high antioxidant value, as well as justify its antimicrobial, anti-inflammatory, anti-mutagenic, antiviral and anti allergic actions these results was agree with those obtained by El-Awady [23].

Table 1. Preliminary phytochemical analysis of *Ocimum sanctum* leaves

Parameter	Concentration*
Total Protein %	20.64±1.47
Total Fat/oil %	3.60±0.08
Total Carbohydrate %	39.58±2.09
Reducing Sugar %	3.58±0.14
Total phenol mg/g	1.88±0.09
Ascorbic acid mg/100g	65.41 ±0.76
Total Flavanoid mg/g	28.38± 0.58

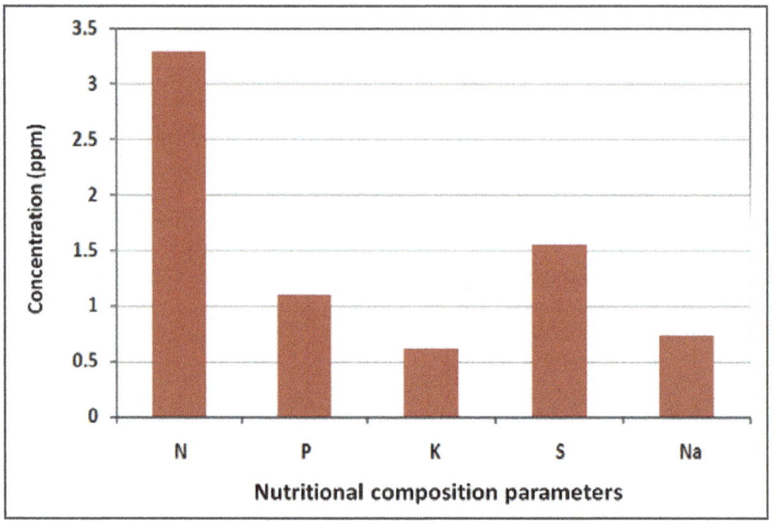

Figure 1. Chemical Nutritional compositions of *Ocimum sanctum* leaves

Table 2. Metal contents and heavy metal contents in medicinal plant *Ocimum sanctum*

Micro-nutrient	Concentration (ppm.)
Mn	61.75±2.06
Zn	32.38±1.42
Cu	14.48±0.72
Mo	0.58±0.21
Ni	5.64±0.37
Li	0.59±0.07
Al	Trace
Mg	Trace
Cd	ND
Cr,	3.67±0.020
Pb	1.17±0.045
As,	3.54±0.304
Hg	0.37±0.051

*data were expressed on dry weight basis.

3.2. Determination of Major and Minor Elements in *Ocimum sanctum:*

The micro (Co, Cu, Fe, Ni, Mn and Zn) and macro (K, Na Ca and Mg) minerals in *Ocimum sanctum* stem and leaves were determined (Table 2). The concentration of these elements reported as ppm on dry weight basis. Among micro minerals Mn and Zn have greater value as compare with other medicinal plants that was 61.75 and 32.38 ppm. while in macro minerals including K, Na, P, Mg of the leaves is same, which is 0.62, 0.74, 1.10 ppm and traces of Mg for leaves, respectively. The amount of heavy metals in the plants was analyzed to show the potential threat of their effects to the animals and human beings who consume them as such or their derived products. This study reports the investigation on the presence of five elements Cd, Cr, Pb, As and Hg (Table 2) in the plant leaves. The leaves were found to contain very high amount of detectable levels of Cr and As. *O. sanctum* leaf digest shown 3.67 and 3.54 ppm of Cr and As.

3.3. Fatty Acid Profile of *Ocimum sanctum* Leaves Oil

The major chemical compounds that were found in *O. Sanctum* extract using GC-MS analysis (shown in Figure 2 and Table 3) are Lauric acid, methyl ester (14.60 ppm), Myristic acid, methyl ester (54.64ppm), Palmitic acid, methyl ester (619.21 ppm), Palmitoleic acid, methyl ester (14.59 ppm), Stearic acid, methyl ester (279.03 ppm), Oleic acid, methyl ester (325.88 ppm), Linoleic acid, methyl ester (585.64 ppm), Linolenic acid, methyl ester (29.57), Arachidic acid methyl ester (102.84 ppm), and Behenic acid, methyl ester (77.84 ppm). The suitable extracts for respective compounds can be chosen on the basis of GC-MS analysis [21]. Eugenol is reported to possess Antimycotic, Antiviral, Desinsection, Antiparasitic, Antioxidant, Anticancer, and Antiinsect activities [24,25].

The current work provides preliminary information and methodologies for rapid quantifiable screening of chemical components and antioxidant potentials of *Ocimum sanctum[1]*. It is, therefore, important to study the total antioxidant properties and chemical profiling of *Ocimum sanctum*. Overall growth of the *Ocimum* species was found to be retarded by longer UV-B exposures. Slight enhancements in the contents of enzymatic and non-enzymatic antioxidants due to the low duration exposures may be the sign of recovery from oxidative stress condition. Antioxidants as Flavonoids, proline and ascorbate contents shown usually enhancements proving their potential for scavenging reactive oxygen species [7,9,10]. Methyl-Isoeugenol has the property of Antifungal activity, Nematicidal activity and Anti feedant activity [26,27]. Caryophyllene is well-known for its anti-inflammatory, cytotoxicity and antifungal activity [3,28,29]. *In vitro* screening of antioxidant activities in organ systems shown varied responses that might be explained on the basis of variable enzymatic contents in different organs. *O. sanctum* oil was proved to be better antioxidant for liver than muscle systems in oxidative stresses [25]. It was observed that there was an intimate relationship among the contents of natural antioxidant and recovery potential of plants from oxidative stress conditions in terms of antioxidant index [4,5,11].

Table 3. Chemical components of the essential oils distilled from *Ocimum sanctum* leaves.

Peak	R. Time	Tulsi (ppm)	Name
1	10.90	14.60	Lauric acid, methyl ester
2	14.30	54.64	Myristic acid, methyl ester
3	17.96	619.21	Palmitic acid, methyl ester
4	18.31	14.59	Palmitoleic acid, methyl ester
5	21.56	279.03	Stearic acid, methyl ester
6	21.91	325.88	Oleic acid, methyl ester
7	22.72	585.64	Linoleic acid, methyl ester
8	23.73	29.57	Linolenic acid, methyl ester
9	25.00	102.84	Arachidic acid methyl ester
10	28.45	77.84	Behenic acid, methyl ester

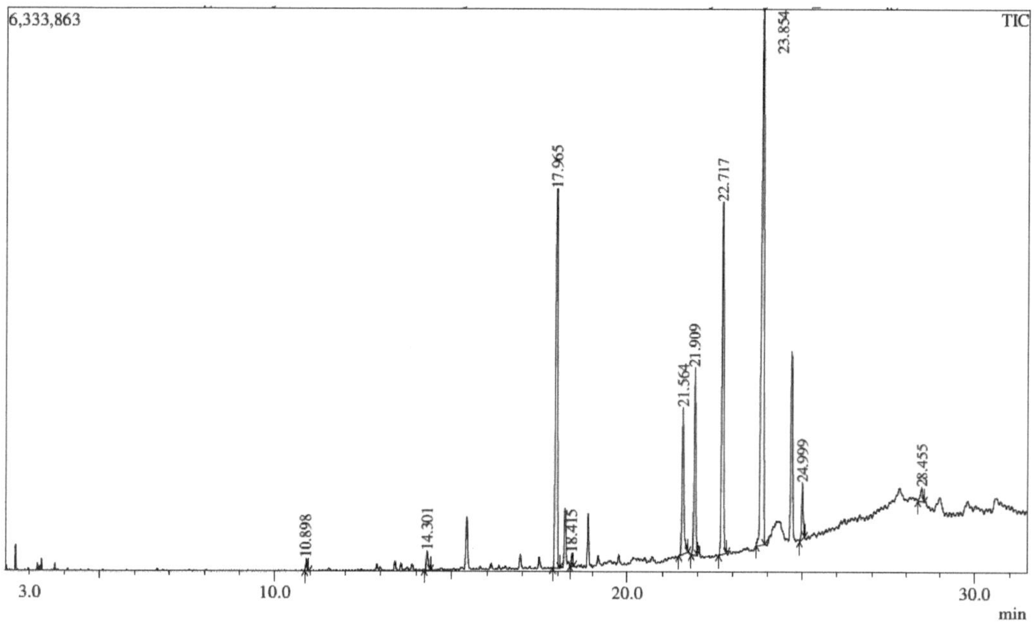

Figure 2. A typical gas chromatogram of the constituents of *Ocimum sanctum* oil

3.4. Antimicrobial activity of *Ocimum sanctun* extracts

Based on the growth inhibition zone diameter obtained by 20 and 40 mg/ml *Ocimum* methanol extract concentration, bacterial strains were tested. This method allows better diffusion of the extracts into the medium thus enhancing contact with the organisms. The antimicrobial activity of extracts of *O. sanctum* was used against four pathogenic organisms, *Escherichia coli*, *Staphylococcus aureus*, *Aeromonas hydrophila,* and *Enterococcus faecalis,* the data are presented in Table 4.

The results shown that the *Ocimum* extracts at final concentrations of 40 mg/ml was active against the four types of micro-organism, while, at the concentration 20 mg/ml for the methanol extracts was not efficient to kill some types of treated microorganism. On the other hand, the final concentration of 40 mg/ml methanol extract of *Ocimum* was effective to kill the treated microorganism. The results also indicated that the methanol was the best solvent for extracting antimicrobial substances from the leaves of this plant [23,30]. Interestingly, *E. coli*, *E. faecalis* were mostly susceptible to extracts than *S. aureus*, *A. hydrophila*. It can be suggested that *S. aureus* was the most resistant organisms to the concentrations of 20 and 40 mg/ml of the methanol extract of *Ocimum* (Table 4). Maximum activity of methanol extract was seen against *E. coli* (25 mm) at concentration 40 mg/ml. The plant extracts compared favorably with the standard antibiotic Streptomycin at a final concentration of 0.8 µg/ml. Similar reports of synergistic role of microbial inhibition on combined usage of plant extracts were also reported by many earlier workers [31,32,33]. Essential oil present in most of the *Ocimum* species is responsible for its antifungal, antibacterial and antiviral properties. Microorganisms develop resistance against various antibiotics and due to this an immense clinical problem develops in treatment of infectious diseases. Medicinal plants can be used to overcome this problem. Tulsi leaves have been reported to show strong antimicrobial activities against the *Klebisella*, *E. coli*, *Proteus*, *S. aureus*, *Shigella*, *P. aeruginosa* and *Vibrio cholera* [21,33].

Table 4. Diameter of inhibition zone (DIZ) in mm of four bacterial strains which caused by leaves extract of *Ocimum sanctum*

Isolates	DIZ of Streptomycin	DIZ of Methanol extract	
		20 (mg/ml)	40 (mg/ml)
E. coli	19.0+0.51	20.0±0.52	25.0±0.61
S. aureus	24.0+0.49	15.5±0.49	18.0±0.51
A. hydrophila	21.5+0.53	16.0±0.46	18.5±0.48
E. faecalis	16.5+0.42	19.0±0.51	21.0±0.54

The preliminary phytochemical parameters were studied not only in search of bioactive agents but also for starting products which uses in the synthesis of useful drugs [33]. The *Ocimum sanctum* plant was broadly used for the treatment of different diseases in third world countries, in the latest research on *Ocimum sanctum* found that it may have natural bioactive compounds which provide protection to animal against different diseases [14].

4. Conclusion

In present study we were achieved to evaluate the heavy metal deposition and phytochemical characterization, fatty acid profiles and antimicrobial activities of the leaves extract of *Ocimum sanctum* plant. We found that the nutritional analysis of *Ocimum sanctum* shown high level of ascorbic acid and total carbohydrate i.e., in their leaves, Whereas the total phenol was found to be low in leaves. Leaves also, contains major nutrient like N , P , K , S , and Na. The oil of leaves also contains comparable amount of antioxidant as ascorbic acid, flavonoid and total phenol as well as linolenic acid, polyunsaturated fatty acid which was very good for health. Antimicrobial activity of Tulsi leaves extract against *E. coli* and *E. faecalis* were mostly susceptible to methanol extract than *S. aureus* and *A. hydrophila*.

Acknowledgement

This research was supported by Food testing laboratory, Junagadh agriculture university, Gujarat, INDIA. We thank our Head of Department. Dr. B.A. Golakiya who provided insight and expertise that greatly assisted the research.

References

[1] Shafqatullah, M., Khurram, A., Khaliqurrehman Khan, F, A., Comparative Analyses of *Ocimum sanctum* Stem and Leaves for Phytochemicals and Inorganic Constituents, Middle-East Journal of Scientific Research, 13 (2): 236-240,2013.

[2] Kumar, A,. Shukla, R., Singh, P., Dubey, N,K., Chemical composition, antifungal and antiaflatoxigenic activities of *Ocimum sanctum* L. essential oil and its safety assessment as plant based antimicrobial, *Food Chem Toxicol*, 48(2):539-43, 2010.

[3] Fernandes, E,S., Passos, G,F., Medeiros, R., Anti-inflammatory effects of compounds alphahumulene and (−)-trans-caryophyllene isolated from the essential oil of cordia verbenacea, *European Journal of Pharmacology*, 569(3), 228-236, 2007.

[4] Rao, S,A,. Vijay, Y., Deepthi, T., Lakshmi, C,S., Rani, V., Rani, S., Antidiabetic effect of ethanolic extract of leaves of *Ocimum sanctum* in alloxan induced diabetes in rats, *Int J Basic Clin Pharmacol*, 2:613-6, 2013.

[5] Pattanayak, P., Behera, P., Das, D., Panda, S,K., *Ocimum sanctum* Linn. A reservoir plant for therapeutic applications: An overview. *Pharmacogn Rev*, 4(7):95-105, 2010.

[6] Saharkhiz, M,J., Alam Kamyab, A., Kazerani, N, K., Zomorodian, K., Pakshir, K., Rahimi,M,J., Chemical Compositions and Antimicrobial Activities of *Ocimum sanctum* L. Essential Oils at Different Harvest Stages, *Jundishapur J Microbiol.* Jan; 8(1): e13720, 2015.

[7] Khan, A., Ahmad, A., Akhtar, F., Yousuf, S., Xess, I., Khan, L,A., *Ocimum sanctum* essential oil and its active principles exert their antifungal activity by disrupting ergosterol biosynthesis and membrane integrity, *Res Microbiol*, 2010;161(10):816-23.

[8] Gaber, A., Hassan, M.M., El-Desoky S. E., Attia O. A., *In vitro* Antimicrobial Comparison of Taif andEgyptian Pomegranate Peels and Seeds Extracts. *J App Biol Biotech*, 2015; 3 (02): 012-017.

[9] Devendran, G., Balasubramanian, U., Qualitative phytochemical screening and GC-MS analysis of *Ocimum sanctum* L. leaves, *Asian J Plant Sci.*,1:44-8, 2011;

[10] Triveni, Kumar, K, Singh, A,K., Kumar, R., Gupta, V., Tripathi, K., *Ocimum sanctum* Linn: A review on phytopharmacology and therapeutic potential of Tulsi, *Int J Pharm Phytopharmacol Res*, 3:148-51, 2013.

[11] Gupta, S,K., Prakash, J., Srivastava, S., Validation of traditional claim of Tulsi, *Ocimum sanctum* Linn. as a medicinal plant, *Indian J Exp Biol*, 40:765-73, 2002.

[12] Lenntech., Water treatment and air purification. Water treatment, published by Lenntech, Rotterdamseweg, Netherlands. www.excelwater.com/thp/filters/Water-Purification.htm., 2008.

[13] Divrikli, U., Horzum, N., Soylak, M., Elci, L., Trace heavy metal contents of some spices and herbal plants from Western Anatolia, Turkey, *Int. J. Food Sci. Technol*, 41:712-716, 2006.

[14] Vivek, K.G., Surendra, K.S., Plants are natural antioxidants. *Nat. Prod. Rad.*, 5(4): 326-334, 2006.

[15] Malik, C,P., Singh, M,B., Plant Enzymology and Histo-Enzymology. Kalyani Publications, New Delhi, 1980.

[16] Ashwini, M., Janakiraman, B., Srinivasan, B., Shanmugaraj, B. M., Ramalingam S., Antioxidant activity in *in vivo* and *in vitro* cultures of onion varieties (Bellary and CO_3).,*Food and Nutrition Sciences*, 4, 918-923, 2013.

[17] Chanda, S., Dave, R., *In vitro* models for antioxidant activity evaluation and some medicinal plants possessing antioxidant properties, An overview, *Afr. J. Microbiol. Res*, 3:981-996, 2009.

[18] Association of Official Analytical Chemists (AOAC) Official Methods of Analysis of AOAC International.18th Edition. Maryland, USA: AOAC International. 2005.

[19] Fahmi, A. I., Nagaty, H.H., Eissa R.A. , Hassan, M.M., Effects of salt stress on some nitrogen fixation parameters in faba bean, *Pakistan Journal of Biological Sciences, 14: 385-391,* 2011.

[20] Viorica, M, P., Alexandra, G,, Diana, N,R., Delia, D, Camelia, M., Despina, B., Constantin, M., *Journal of Agroalimentary Processes and Technologies.*, 18 (2), 136-140, 2012.

[21] Gaber, A., Hassan, M,M., Dessoky, E,S., Attia, O, A., *In vitro* antimicrobial comparison of Taif and Egyptian pomegranate peels and seeds extracts, *Journal of Applied Biology & Biotechnology,* Vol. 3 (02), pp. 012-017, 2015.

[22] El-Tarras A. E., Hassan M. M. and El-Awady M. A. Evaluation of the genetic effects of the in vitro antimicrobial activities of *Rhazya stricta* leaf extract using molecular techniques and scanning electron microscope. *Afr J Biotech*, 12: 3171-3180, 2013.

[23] El-Awady M,A., Hassan ,M,M,, Abdel-Hameed, E,S., Gaber, A., Comparison of the antimicrobial activities of the leaves-crude extracts of *Moringa peregrina* and *Moringa oleifera* in Saudi Arabia, *Int.J.Curr.Microbiol.App.Sci*, 4(12): 1-9, 2015.

[24] Ou, H,C., Chou, F,I., Lin, T,M., Protective effects of eugenol against oxidized LDL-induced cytotoxicity and adhesion molecule expression in endothelial cell, *Food Chem Toxicol* , 44(9), 1485-595, 2006.

[25] Hussain, A., Brahmbhatt, K., Priyani, A., Eugenol enhances the chemotherapeutic potential of gemcitabine and induce anticarcinogenic and anti-inflammatory activity in human cervical cancer cells, *Cancer Biother Radiopharm*, 26(5), 519-27, 2011.

[26] Park, B,S., Lee, K,G., Shibamoto, T, Lee, S,E., Takeoka, G,R., Antioxidant activity and characterization of volatile constituents of taheebo (Tabebuia impetiginosa Martius ex DC), *Journal of Agricultural and Food Chemistry*, 51, 295-300, 2003.

[27] Dev, N.,. Das, A. K., Hossain, M. A., Rahman, S. M., Chemical compositions of different extracts of *Ocimum basilicum* Leaves, *J. Sci. Res*, 3 (1), 197-206, 2011.

[28] Sabulal, B., Dan, M., Caryophyllene-rich rhizome oil of Zingiber nimmonii from South India: chemical characterization and antimicrobial activity, Phytochemistry, 67 (22), 2469-2473, 2006.

[29] Ashour, M,L., El-Readi M, Youns M.,Chemical composition and biological activity of the essential oil obtained from Bupleurum marginatum (Apiaceae), *Journal of Pharmacy and Pharmacology*, 61(8), 1079-1087, 2009.

[30] Prakash, P., Gupta, N., Therapeutic uses of *Ocimum sanctum* Linn (Tulsi) with a note on eugenol and its pharmacological actions: A short review, *Indian J Physiol Pharmacol*, 49:125-131, 2005.

[31] Abeysinghe, P.D., Wanigatunge R,P., Evaluation of antibacterial activity of different mangrove plant extracts. *Ruh J of Sci*, 1:104-112, 2006.

[32] Kamegam, N., Karuppusamy, S., Prakash, M., Jayakuma,r M., Rajasekar, K., Antibacterial potency and synergistic effects of certain plant extracts against food borne diarrheagenic bacteria, *Int J of biomed and pharma sci*, 2(2): 88-93, 2008.

[33] Rivera, SEV., Escobar-Saucedo, M,A., Morales, D., Aguilar, C,N., Herrera, R,R., Synergistic effects of ethanolic plant extract mixtures against food-borne pathogen bacteria, *Afr J Biotech*, 13(5):699-704, 2014.

Proximate Composition, Fatty Acid and Mineral Contents of Four Freshwater Fish from Maga Lake (Far North Region of Cameroon)

Noël TENYANG[1,4,*], Roger PONKA[2], Bernard TIENCHEU[3], Fabrice TONFACK DJIKENG[4], Hilaire Macaire WOMENI[4]

[1]University of Maroua, Faculty of Science, Department of Biological Sciences, Maroua, Cameroon
[2]University of Maroua, The Higher Institute of The Sahel, Department of Agriculture, Livestock and By-Products, Maroua, Cameroon
[3]University of Buea, Faculty of Science, Department of Biochemistry, Buea, Cameroon
[4]University of Dschang, Faculty of Science, Department of Biochemistry, Dschang, Cameroon
*Corresponding author: noeltenyang@yahoo.fr

Abstract A study was conducted to explore the proximate composition, the fatty acids profile and the mineral contents of four freshwater fish namely mullet (*Liza falcipins*), carp (*oreochromis niloticus*), catfish (*Chrysichthys nigrodigitatus*) and kanga (*Heterotis niloticus*) collected from Maga Lake in Far North Region of Cameroon. It was been observed that moisture content (~78 g/100g fresh weight (FW) not varied between species ($P>0.05$). Nevertheless, ash content varied significantly ($P<0.05$) between species and carp presented a high ash content (9.04 g/100g dry weight (DW). The total lipid content was generally high, ranging from 18.88 to 55.19 g/100g DW. With regard to protein content, these fresh fish contained more than 39 g/100g DW. Significant differences ($P<0.05$) were observed in some fatty acids within the fish species analyzed. All these fish except carp were high in saturated, monounsaturated and polyunsaturated fatty acid, respectively. C16:0 and C18:1(n-9) were the predominant fatty acid in all fish analyzed (20-24 and 21-33 % repectively). The others major fatty acids detected were C18:2ω-6, C18:0 and C16:1ω-7. Although these four freshwater fish contained reasonable content of important polyunsaturated fatty acid such as eicosapentaenoic acid (EPA), docosahexaenoic acid (DHA) and arachidonic acids, catfish presented the highest content of polyunsaturated fatty acids (PUFA). These fish contained appreciable ratio ω-3:ω-6 suggesting that these fish especially carp could be used as a source of healthy diet for humans. The fish analyzed have a good supply of minerals and can be used for enhancing mineral intake and protecting population from mineral deficiency disease.

Keywords: *proximate composition, fatty acid, mineral content, Maga Lake fish, Cameroon*

1. Introduction

In Cameroun, like in most sub-saharan Africa countries, traditional fishing is practiced in almost all rivers, lakes, ponds and represents an important part of total fish captures. It is an important sector in the national strategies of fight against poverty and food safety. One of the most important producing sites is the lake of Maga. The nutritional and health benefit from the consumption of seafood is one of the reasons for the increasing consumer demand for fish and shellfish. The fats are needed in diets to absorb fat-soluble vitamins A, D, E and K from food; and for regulating body cholesterol metabolism [1,2]. The fish fats contain important PUFA like EPA, C20:5n-3, DHA, C22:6n-3, and arachidonic (C20:4n-6) acids which are not synthesised in the human body but their inclusion in human diets is essential [3]. PUFA contained in fish is known to regulate prostaglandin synthesis and hence induce wound healing [4]. ω-3 and ω-6 polyunsaturated fatty acids (PUFA) have been shown to have positive effects on cardiovascular diseases and cancers [5]. PUFA composition may vary among species of fish, even among fresh water and marine fish [6], it is important for human health, to increase the consumption of fish and its products [7,8]. Fish also contains significant amounts of essential amino acids and certain amino acids like aspartic acid, glycine and glutamic acid are also known to play a key role in the process of wound healing [9]. Fish contains micronutriments such as minerals. Micronutrients have been reported to play important role in mounting immune response and prevention or treatment of disease [10]. Deficiency in one of the micronutrients has important implications for the pathogenesis of diseases, their morbidity and severity [11].

In Cameroon, little studies have been carried out on the chemical composition of fish commercialized in the wouri river coast [12], but the composition of freshwater fish of

Maga Lake has received little attention. Thus, this study was carried out to determine the proximate composition, the fatty acid profile and the mineral content of four fresh fish species from Maga lake (Far North Region of Cameroon).

2. Material and Methods

2.1. Sample Collection

The fish samples chosen for this study are the commonly consumed in Far North, Cameroon. About 500 g of each of four fish species: mullet (*Liza falcipins*), carp (*oreochromis niloticus*), catfish (*Chrysichthys nigrodigitatus*) and kanga (*Heterotis niloticus)* were bought from Maga fish market around of the Maga Lake in Far North Region of Cameroon. Once collected, they were immediately transported to the laboratory in ice boxes for identification and analyses. Fresh fish were washed with tap water several times to remove adhering blood and slime. In the laboratory, the fish samples were separated and reduced to filet.

2.2. Proximate Composition

The moisture, lipid, protein and ash in samples were determined by the recommended AOAC methods [13]. The moisture content was determined by drying fresh sample in an oven at 105°C until constant weight. The lipid content of the powdered fish was determined with the help of soxhlet apparatus using the non- polar organic solvent hexane. Crude protein was determined by nitrogen determination using the Kjeldahl micro method and conversion of nitrogen to protein by the factor 6.25. Ash content determined by burning the organic components from the known weight of the homogenised dried fish muscle by using a furnace at 550°C for 48h.

2.3. Fatty Acid Composition

Lipids were extracted from the powdered fish and fatty acid composition of oil were investigated after conversion of their Fatty Acid Methyl esters (FAME) using boron trifluoride-methanol method. The lipids were saponified and esterified for fatty acid analysis by the method of Metcalfe, Schmitz, and Pelka [14]. The fatty acid methyl esters (FAMEs) were analyzed on a Hewlett-Packard (HP) 5880 gas chromatograph (GC) with a flame ionisation detector (FID). The esters were separated on a 50 m × 0.20 mm i.d. Wall-coated open tubular fused silica capillary column coated with Carbowax 20 M. Column injector and detector temperatures were 200 and 300°C, respectively. Carrier gas was helium; split ratio was 100:1. Identification was achieved by comparison to retention times of authentic standards, argentation TLC followed by GC of the bands separated by a degree of unsaturation, and mass spectrometry.

2.4. Mineral Determination

The method described by AOAC [15] was used for mineral analysis. The minerals (Ca, Mg, Na, K, Fe, Zn, Cu and Mn) were determined by atomic absorption spectrometer (Varian 220FS Spectr AA, Les Ulis, France). The sample was ashed at 550°C and the ash boiled with 10 ml of 20% HCl in a beaker and then filtered into a 100 ml standard flask. P was determined colorimetrically using the vanado molybdate method.

2.5. Statistical Analyses

All analyses were performed in three replications. Data on the composition of fish samples were analysed by one way analysis of variance (ANOVA) using Statistical Package for Social Science (SPSS 16.0 version). Differences between samples were tested according to Duncan test and considered to be significant at $P<0.05$.

3. Results and Discussion

3.1. Proximate Composition

The proximate analysis of the four freshwater fish is shown in Table 1. It appears that these fish are composed mostly of moisture (~80 g/100 g FW). These results are similar to those obtained by Zuraini *et al.* [16] and Tenyang *et al.* [12] in other fresh fish (~80.5 g/100 g FW) and were within the range cited by Yeannes and Almandos [17], for some freshwater fish (i.e. 70-85 g/100 g FW). This table revealed that ash content ranged from a low value 5.28 (kanga) to a high value 9.04 g/100g DW (carp). The high ash content obtained indicates that all these fish are rich in minerals. The ash content obtained in this investigation wa smilar to that found by Zuraina *et al.* [16] in three local Malaysian *Channa* spp. fish (1-1.8 g/100g DW).

The lipid content varied in all fish samples from about 18.88 to 55.19 g/100g DW (Table 1). Higher lipid content was characteristic of kanga, and lower contents were found for mullet fish. All these fish can be classified as a fat fish because their lipid content were above 10 g/100g DW [18]. The lipid content found in this study for carp, catfish and mullet are above the percentage reported by Tenyang *et al.* [12] (15.32 and 23.02 g/100g DW, respectively), in a work developed in Littoral Region of Cameroon. In another study, the total lipid content found in tilapias was 1.1 g/100g DW [19]. Still in Brazil, Souza *et al.* [20] reported values of 8.06% for total lipids in gutted tilapias. Lipid content is influenced by species, geographical regions, age, and diet [21,22]. Much of the crude lipid content is constituted by unsaturated fatty acids, which have important physiological functions according to previous studies [3].

Table 1. Proximate composition of four raw fish species collected in Maga Lake (Far North-Cameroon)

Fish Species	Moisture (g/100g FW)	Ash (g/100g DW)	Lipid (g/100g DW)	Protein (g/100g DW)
Mullet	81.64±1.79^{a}	8.72±0.06^{ab}	18.88±0.89^{d}	62.80±0.03^{a}
Carp	81.20±0.87^{a}	9.04±0.24^{a}	21.76±0.70^{c}	52.47±0.32^{b}
Catfish	78.46±5.96^{a}	7.32±1.27^{ab}	30.34±2.16^{b}	39.73±0.20^{c}
Kanga	78.59±1.26^{a}	5.28±0.28^{c}	55.19±1.58^{a}	52.30±2.00^{b}

Mean values in the same column with different superscript letters are significantly different ($P< 0.05$). FW: Fresh weight ; DW: Dry weight.

The freshwater fish uses in this study were found to be rich source of protein. The crude protein contents of these fish differed significantly among fish species (P< 0.05) (Table 1). Catfish was lowest in crude protein content (39.73 g/100g DW), when mullet was highest in crude protein (62.80 g/100g DW). The crude protein content of carp, mullet and catfish obtained in this study were comparable to those obtained by Tenyang *et al.* [12] in the same species (69.72 and 64 g/100g DW, respectively). Protein is necessary for building the structural components of human body, such as muscles and organs. The consumption of kanga can be encouraged because it has a higher protein content, which could help to prevent some problems due to protein deficiencies.

3.2. Fatty Acid Composition

A fish lipid contains long chains polyunsaturated fatty acids, which have a number of nutraceutical and pharmacological application and have all the essential fatty acids required by humans for normal growth and healthy life [3].

The compositions of fatty acids of the four raw fish species of Maga Lake are shown in Table 2. The results show that each fish species contained at least twenty-four fatty acids. Remarkable changes in the individual fatty acids within the species are noted. The most abundant fatty acid containing in all the fish for this study are palmitic acid (C16:0) ranging from 20 % to 24 % and oleic acid (C18:1 (n-9) ranging from 21% to 33% The high percentage of palmitic acid was found to carp fish, while the low percentage was obtained in kanga. Catfish contain the high level of oleic acid, while the low content was obtained in carp and kanga. The other major fatty acid was C18:2 (n-6). In particular, catfish has more C18:1 (n-9) than C16:0. These results are in agreement with the literature where C16:0 was reported as the most abundant fatty acid in six fish species commercialized in the wouri river coast in Cameroon and three freshwater fish species in Malaysian [12,16]. Meanwhile, Osman *et al.* [23] reported that C16:0 was not the most abundant fatty acid present in some Malaysian waters fish.

Table 2. Fatty acid composition (g/100g) of four raw fish species collected in Maga Lake (Far North-Cameroon)

Fatty acid as % of total fatty acids	Mullet	Carp	Catfish	Kanga
C12 :0, Lauric	ND	ND	1.03±0.08[a]	ND
C14.0, Myristic	3.37±0.02[a]	2.47±0.01[c]	2.84±0.04[b]	2.86±0.01[b]
C15 :0, Pentadecanoic	1.67±0.01[a]	0.96±0.04[b]	1.08±0.05[b]	1.92±0.02[a]
C16 :0, Palmitic	22.49±0.05[c]	23.99±0.02[a]	22.75±0.03[b]	20.07±0.03[d]
C17 :0, Heptadecanoic	3.04±0.02[c]	1.60±0.01[d]	3.12±0.02[b]	3.31±0.03[a]
C18 :0, Stearic	7.27±0.10[c]	6.37±0.05[d]	9.97±0.07[a]	8.92±0.03[b]
C20 :0, Arachidique	0.57±0.01[c]	0.65±0.05[c]	1.96±0.03[b]	2.63±0.10[a]
C22 :0, Behenic	0.25±0.05[c]	0.29±0.02[c]	0.38±0.07[b]	0.52±0.03[a]
C23 :0, Tricosanoic	0.96±0.06[b]	0.27±0.02[d]	0.38±0.05[c]	1.15±0.03[a]
C24 :0, Lignoceric	0.14±0.02[b]	0.10±0.01[c]	0.04±0.01[d]	0.28±0.03[a]
ΣSFA	**39.76±0.34[c]**	**36.70±0.23[d]**	**43.52±0.45[a]**	**41.66±0.31[b]**
C16 :1, Palmitoleic	7.41±0.10[b]	6.97±0.08[c]	7.02±0.05[c]	10.11±0.06[a]
C17 :1n Cis-10 heptadecenoic	1.15±0.04[a]	1.03±0.03[b]	1.15±0.01[a]	0.93±0.02[c]
C18 :1 (n-9), Oleic	24.60±0.19[b]	21.10±0.12[c]	32.96±0.23[a]	21.42±0.14[c]
C20 :1, cis-11 Eicosenoic	0.54±0.02[c]	0.89±0.09[b]	1.22±0.04[a]	ND
C22 :1, Erucic	0.06±0.01[c]	0.07±0.02[c]	0.18±0.01[a]	0.13±0.03[b]
ΣMUFA	**32.61±0.28[b]**	**30.06±0.34[c]**	**42.53±0.34[a]**	**32.59±0.25[b]**
C8 :2, Linoleic (ω-6)	12.36±0.12[a]	12.49±0.09[a]	3.77±0.07[c]	8.86±0.10[b]
C18 :3, ϒ-linoleic (ω-6)	0.55±0.01[c]	0.72±0.03[a]	0.18±0.01[d]	0.67±0.02[b]
C18 :3, α-linoleic (ω-3)	4.16±0.20[b]	11.15±0.15[a]	1.60±0.60[d]	3.26±0.26[c]
C20 :2, cis 11, 14 Eicosadienoic	0.43±0.03[c]	0.52±0.02[b]	0.77±0.07[a]	0.80±0.05[a]
C20 :3, cis-8, 11, 14 Eicosatrienoic (ω-6)	0.97±0.06[c]	2.37±0.11[a]	1.26±0.06[b]	0.70±0.04[d]
C 20 :4, Arachidonic (ω-6)	2.41±0.10[b]	1.21±0.05[d]	1.91±0.15[c]	4.35±0.13[a]
C20 :5, Eicosapentaenoic (EPA) (ω-3)	1.42±0.03[b]	0.66±0.06[d]	1.07±0.02[c]	1.66±0.10[a]
C22 :6, Docosahexaenoic (DHA) (ω-3)	2.13±0.09[c]	3.29±0.14[a]	1.06±0.05[d]	2.51±0.11[b]
ΣPUFA	**24.43±0.64[b]**	**32.41±0.65[a]**	**11.62±1.03[d]**	**22.81±0.81[c]**
Σω-3PUFA	**7.71±0.32**	**15.10±0.35**	**3.73±0.35**	**7.43±0.47**
Σω-6PUFA	**16.29±0.29**	**16.79±0.28**	**7.12±0.29**	**14.58±0.29**
ω-3 :ω-6	**0.47**	**0.90**	**0.52**	**0.51**
Other fatty acids	3.2	0.83	2.30	2.94

Mean values in the same row with different superscript letters are significantly different (P < 0.05).
ND: not detected; SFA : Saturated fatty acid ; MUFA : mono-unsaturated fatty acid PUFA : Polyunsaturated fatty acid.

A total saturated and unsaturated fatty acid composition of the four selected fish species is presented in Table 2. Catfish is recognized as a species, which has the higher percentage of saturated fatty acid (43.51 %). However, carp is a fish, which has the lowest saturated fatty acid content (36.70%). The high content of saturated fatty acid obtained in this work for these fish samples are in the line with those presented by Tenyang *et al.* [12] in Cameroon, and Gutierrez and Silvia [24] in Brazil on some marine fish. In generally, the MUFA contents for all fish species except the carp were much higher than the PUFA. Others researchers have also shown that freshwater fish have lower contents of PUFA compared to MUFA [25]. The trend is different when compared to an earlier study of freshwater fish, where the concentration of PUFA was higher than MUFA [23]. The difference can be attributed to the environment, age, genetic or nutrition of fish. Catfish presented the higher MUFA contents and lowest PUFA compared to others fish species in this study. The values for MUFA (30.06-42.53%) in the fish analysed, were higher than menhaden oil which is a marine fish oil (19,4%) reported by Osman *et al.* [23]. These fish species from Maga Lake in Far North region of Cameroon are sources of polyunsaturated fatty acid of the ω-6 family especially linoleic acid (C18:2) which is essential in human nutrition as these fatty acid are not synthesised in the body but are required in the tissue development (Table 2). These results were different when compared to an earlier study on some marine fish, where the concentration of PUFA ω-3 were higher than PUFA ω-6 [12], but were in agreement with the results presented by Fatima *et al.* [27] on some freshwater fish in Brazil. The difference can be attributed to the fact that marine fish staple diets are many zooplancton, which are rich in PUFA ω-3. The concentration of ω-3 PUFA of carp in this study (15.1%) was found to be much higher than those for others fish analysed. Catfish presented the low concentration of ω-3 PUFA. The value of ω-3 PUFA of catfish (3.73 %) in this work was lower than those reported by others researchers (14.73 %; Tenyang *et al.*, [12]). The predominant ω-3 PUFA present in these fish analysed were α-linolenic acid (C18:3ω-3), eicopentaenoic acid (C20:5ω-3; EPA) and docosahexaenoic acid (DHA). Carp fish presented the higher content of C18:3ω-3 and C22:6ω-3, while mullet and kanga presented the higher concentration of C20:5ω-3. The lowest value of DHA and EPA were obtained respectively in catfish and carp. The presence of EPA and DHA in all the fish analysed suggest that these fish can have a healing effect to alleviate muscle pain inflammation. DHA and EPA have been reported to have preventive effects on human coronary artery diseases [27]. DHA is essential for the development of the foetal brain and the eye retina [28]. Significant levels of EPA and DHA in fish species of this work indicated that those fish could be used to supplement essential fatty acids in the human diet.

The contents of arachidonic acid (C20:4ω-6) of fish analysed ranged from 1.21 to 4.35%. Carp fish presented the lower content of arachidonic acid (1.21 %), while kanga presented the higher content (4.35 %). Arachidonic acid presented in all the fish species in this work is a precursor for prostaglandin and thromboxane biosynthesis [29]. Arachidonic acid can facilite the blood clotting process and attach to endothelial cells during wound healing. The inclusion of the fish analysed in human diets might help in the wound healing process of the consumer.

The ratio of ω-3:ω-6 has suggested to be a useful indicator for comparing relative nutritional value of fish oils. It was suggested that a ratio of 1:1-1:5 would constitute a healthy human diet [21]. All four fresh fish species had the ratio of ω-3:ω-6 within the recommended ratio. A higher ratio was observed in carp, this means that carp has a higher nutritional value compared t other fish. Although these ratios were lower than those obtained by Osman *et al.* [21] for some marine fish (2.16-4.14) in Malaysia, there were higher than those presented by Farhat and Abdul [30] on three freshwater fish species in Pakistan (0.23-0.27).

3.3. Minerals Composition

3.3.1. Macroelements Contents

In Table 3, the macromineral contents of four raw fish species collected in Maga Lake in Cameroon are presented. There were significant differences (P<0.05) in the macromineral composition between the four freshwater fish. The macroelements analysis (Table 3) indicates high concentration of calcium (Ca) in all these four fish species (3733.49- 9777.48 mg/100 g DW). Carp is very rich in Ca while kanga has the lowest calcium value. The calcium obtained in this work is higher than those obtain by other authors [12,31]. Consumption of carp fish could be recommended to a subject who has problems of hypocalcaemia.

Table 3. Macro-nutrients contents (mg/100g DW) of four raw fish species collected in Maga Lake (Far North-Cameroon)

Fish species	Ca	Mg	K	Na	P
Mullet	6349.19±20.00^{c}	205.95±5.00^{a}	1044.31±14.18^{a}	110.67±2.67^{c}	3545.88±15.44^{b}
Carp	9777.48±17.48^{a}	211.44±2.22^{a}	781.67±11.00^{c}	141.33±2.03^{a}	3880.46±10.20^{a}
Catfish	8293.89±22.00^{b}	179.78±2.10^{b}	611.75±11.00^{d}	107.23±3.00^{c}	3539.48±20.00^{b}
Kanga	3733.49±25.00^{d}	149.80±4.10^{c}	1009.57±10.00^{b}	121.80±1.80^{b}	2065.70±20.00^{c}

Mean values in the same column with different superscript letters are significantly different (P< 0.05).

The magnesium (Mg) content oscillated between 149.80 (kanga) and 211.44 mg/100 g DW (carp). The Mg value obtained in catfish in this study is higher than those obtained by Beyza and Akif [31] in African catfish.

Calcium and magnesium plays a significant role in photosynthesis, carbohydrate metabolism, nucleic acids and binding agents of cell walls [32]. Calcium assists in teeth development [33]. Magnesium is essential mineral for enzyme activity, like calcium and chloride; magnesium also plays a role in regulating the acid-alkaline balance in the body [34].

Potassium (K) content ranged from 611.75 (catfish) to 1044.31 mg/100 g DW (mullet). These values are higher than K contents obtained from trout described by Wheaton and Lawson [35] (280-350 mg/100g). The K content for catfish in this work is lower than those obtained by Tenyang *et al.* [12] (683.3 mg/100 g DW) for the same fish. Mean while the K content of mullet reported by Tenyang *et al.* [12] (764.4 mg/100 g DW) is lower than those obtained on the same species of fish in this study. Sodium (Na) was found to be high ranged 107.23 (catfish) to 141.33 mg/100 g DW (carp) in our samples than those reported on others fish samples (30.8-91.7 mg/100 g DW) [12,31].

The Na/K ratio is very low in mullet fish (0.11), which is interesting from the point of view of nutrition, since the intake of sodium chloride and diets with a high Na/K ratio have been related to the incidence of hypertension [36].

Phosphorus (P) is a major constituent of bones together with calcium and magnesium. This mineral showed significant concentration variability between fish species ranging from 2065.70 to 3880.46 mg/100 g DW. Carp presented the highest value of P while kanga has the lowest content. The P concentration range obtained in this work is higher than the FAO [37] range of 68-550 mg/100 g DW, and other freshwater fish obtained by Alas *et al.* [38], (232-426 mg/100 g DW). The recommended dietary allowance for adults is 700 mg of P. About 20 g of catfish and mullet, 18 g of carp and 34 g of kanga fish in this study can contribute at 100 % of the daily requirement.

3.3.2. Microelements Contents

Micronutrient deficiencies are a major public health problem in many developing countries, with infants and pregnant women especially at risk [36,39]. Fish accumulate mineral in the head and viscera, so consumption of small fish which is eaten whole, can contribute significantly towards micronutrients intakes. Table 4 presented the concentration of zinc, copper, manganese and iron in four fresh fish species of Maga Lake obtained in this study.

Zinc (Zn) content ranged from 11.34 (carp) to 14.37 mg/100 g DW (mullet). These values were higher than those reported by Mogobe *et al.* [40] in Botswana for some common fish species (1.63-8.47 mg/100 g DW). The Zn content of catfish in this study was similar than those obtained by Beyza and Akif [31] in African catfish. Zn is usually taken to stimulate the immune system and human growth. It is a component of many metallo-enzymes, important for gene expression [41]. The Zn recommended dietary allowance for adult 8-11 mg and all these fish can provide 100% of the requirement in 100 g of fish, making them a good source of Zn. This species can be beneficial to population because help to combating malnutrition.

Table 4. Micro-nutrients contents (mg/100g DW) of four raw fish species collected in Maga Lake (Far North-Cameroon)

Fish species	Zn	Cu	Mn	Fe
Mullet	14.37±1.00^{a}	0.63±0.02^{c}	3.41±0.20^{c}	5.47±0.30^{b}
Carp	11.34±0.30^{c}	0.89±0.07^{a}	5.24±0.12^{b}	7.98±1.00^{a}
Catfish	12.28±0.28^{bc}	0.71±0.03^{b}	6.67±0.05^{a}	7.72±0.60^{a}
Kanga	12.75±0.80^{b}	0.71±0.01^{b}	1.64±0.02^{d}	5.72±0.40^{b}

Mean values in the same column with different superscript letters are significantly different (P< 0.05).

The Cu content of the four freshwater fish ranged from 0.63 to 0.89 mg/100 g DW. Carp has the highest content of Cu (0.89 mg/100 g DW) while mullet maintained the lowest content of Cu (0.63 mg/100 g DW). The Cu content obtained in this work is higher compared to those obtained by Mogabe *et al.* [40] in some common fish in Botswana (0.02-0.21 mg/100 g DW). Like Zn, Cu is also a part a many enzyme. It is involved in stimulating the body defence system and required for antibody development [42]. Cu occurs in very low level in food. According to Widman and Medeiros [43], the recommended daily requirement of copper in human nutrition is ranged to 1.5-2.5 mg. The Cu content of the different fish in this study is lower than the recommended daily intake, assuming a single serving of 100g of fish per day.

Manganese (Mn) concentration level in fish for this work ranged between 1.64-6.67 mg/100g. The highest content of Mg is found in cathfish (6.67 mg/100 g DW) while the lowest content is obtained to kanga (1.64 mg/100 g DW). These value are low compared to those found by Tenyang *et al.* [12] in some fish commercialized in Littoral region of Cameroon (102-466 mg/100g), but remain higher than those obtained by Mogobe *et al.* [40] for some local fish species in Botswana (0.06-1.08 mg/100g). Mn is a part of enzyme involved in urea formation. It activated several enzymes systems and in this capacity it can is required in for the synthesis of acid mucopolysaccharides to form the matrice of bones and egg shells. Consequently skeletal deformaties and defects in shell qualities occur when the Mn intakes is inadequate [44]. The recommended dietary allowance for Mn is 3.8 mg [45]. Mullet can contribute to 90% and Kanga to 43% of the daily Mn requirement if 100 g of these fresh fish is consumed. Consumption of 72.5 g of carp or 57 g of catfish can provide sufficient supply of Mn for the day.

Iron (Fe) value of the fish from this work ranged from 5.47-7.98 mg/100 g DW. Carp and catfish has the highest content of Fe (~8 mg/100 g DW) while mullet and kanga has the lower content (~5.5 mg/100 g DW). These results are in good agreement with those found by FAO [37] and Mogobe *et al.* [40] for some fish muscles. Fe has several vital functions in the body. It serves at carrier oxygen to the tissues from the lungs by red blood cell haemoglobin, as a transport medium for electrons within cells, and as an integrated part of important enzyme system in various tissues. Iron deficiencies cause anaemia, one of the commonest mineral deficiency diseases in Africa with 206 million people at risk [46]. The recommended nutrient intake of Fe for female adults between the ages to 19-55 years is 24 mg/day. Carp and catfish can provide 33% while mullet and kanga can contribute 23% of the daily Fe requirement for women if 100 g of fish is consumed.

The difference in mineral contents from those reported in the literature [12,37,40] possibly results from geographical place of harvest, seasonal, annual and environmental factors and species.

4. Conclusion

This study has demonstrated that the nutritional composition of raw Maga Lake fish used in this work change within the species. These fish species were found to be a fat fish. There were good sources of protein, fatty acids composition and essential mineral. DHA and EPA contained in these fish are good for health. Carp fish from this study was more beneficial for human health in developing country than others fish analysed due to the richer ω-3 fatty acid percentage composition and the ratio ω-3:ω-6. The data on the nutritional and nutraceutical composition of these fish analysed can be used as guideline for promoting the consumption of these fish and for further studies.

References

[1] Connor, W. E., Importance of n-3 fatty acids in health and disease. *American Journal of Clinical Nutrition*, 71, 1715-1755, 2000.

[2] Kris-Etherton, P.M., Harris, W.S. and Appel, L.J. Fish consumption, fish oil, Omega-3 fatty acids and Cardiovascular

disease. *Arteriosclerosis Thrombosis Vascular Biology*, 23, 20-31, 2003.

[3] Holub, D.J. and Holub, B.J. Omega-3 fatty acids from fish oils and cardiovascular disease. *Molecular Cell Biochemistry*, 263, 217-225, 2004.

[4] Gibson, R.A. Australian fish - an excellent source of both arachidonic acid and 3 polyunsaturated fatty acids. *Lipids*, 18, 743-752, 1983.

[5] Conner, W.E. The beneficial effects of omega-3 fatty acids: cardiovascular diseases and neurodevelopment. *Current Opinion in Lipidology*, 8, 1-3, 1997.

[6] Abd Rahman, S., The, S.H., Osman, H. and Daud, N.M. Fatty acid composition of some Malaysian fresh water fish. *Food Chemistry*, 54, 45-49, 1995.

[7] Burr, M.L. Fish and cardiovascular system. *Progress in Food and Nutrition Science*, 13, 291-316, 1989.

[8] Sargent, J.R. Fish oils and human diet. *British Journal of Nutrition*. 78 (Suppl.1), 5-13, 1997.

[9] Chyun, J.H. and Griminger, P. Improvement of nitrogen retention by arginine and glycine supplementation and its relation to collagen synthesis in traumatized mature and ageing rats. *Journal of Nutrition*, 114, 1705-1715, 1984.

[10] Field, C.J., Johnson, I.R. and Schley, P.D. Nutrients and their role in host resistance to infection. *Journal of Leukocyte Biology*, 71, 16-32, 2002.

[11] ASPN, Hallberyl, Iron nutrition in heath and desease. John libbey ed., London, 964, 1996.

[12] Tenyang, N. Womeni, H.M., Linder, M., Tiencheu, B., Villeneuve P. and Tchouanguep Mbiapo, F. Chemical composition, fatty acid, amino acid profiles and minerals content of six fish species commercialized in the Wouri river coast in Cameroon. *La Rivista Italiana Delle Sostanze Grasse*, Vol. XCI, 129-138, 2014.

[13] AOAC, Official Methods of Analysis, 16th ed. Washington, DC: Association of Official Analytical Chemists, 1990.

[14] Metcalfe, L.D., Schmitz, A.A. and Pelka, J.R. BF3-methanol procedure for rapid quantitative preparation of methyl esters from lipids. *Analytical Chemistry*, 38, 514, 1966.

[15] AOAC, Official Methods of Analysis of the Association of Analytical Chemists international, 18th ed. Gathersburg, MD U.S.A. Association of Official Analytical Chemists, 2005.

[16] Zuraini, A., Somchit, M.N., Solihah, M.H., Goh, Y.M., Arifah, A.K., Zakariad, M.S., Somchit, N., Rajion, M.A., Zakaria Z.A. and Mat, A.M. Jais. Fatty acid and amino acid composition of three local Malaysian Channa spp. Fish. *Food Chemistry* 97, 674-678, 2006.

[17] Yeannes, M.J. and Almandos, M.E. Estimation of fish proximate composition starting from water content. *Journal of Food Composition and Analysis*, 16, 81-92, 2003.

[18] Suriah, R.A., Huah, T.S., Hassan, O. and Daud, N.M. Fatty acid composition of some Malaysian fresh water fish. *Journal of Food Chemistry*, 54, 45-49, 1995.

[19] Visentainer, J.V., Souza, N.E., Makoto, M., Hayashi, C. and Franco, M.R.B. Influence of diets enriched with flaxseed oil on thealinolenic, eicosapentaenoic and docosahexaenoic fatty acid in Nile tilapia (*Oreochromis niloticus*). *Food Chemistry*, 90, 557-560, 2005.

[20] Souza, M.L.S., Baccarin, A.E., Viegas, E.M.M. and Kronka, S.N. Defumaçâo da Tilápia do Nilo (Oreochromis niloticus) Inteira Eviscerada e Filé: Aspectos Referentes às Características Organolépticas, Composicâo Centesimale Perdas Ocorridas no Processamento. *Revista Brasileira Zootecnia*, 33(1), 27-36, 2004.

[21] Piggot G.M. and Tucker, B.W. Effects of technology of nutrition. New York, USA: Marcel Dekker, 1990.

[22] Huss, Fresh fish quality and quality changes, Rome, Italy: F.A.O., 1998.

[23] Osman, H., Suriah, A.R and Law, E.C. Fatty composition and cholesterol content of selected marine fish in Malaysian waters. *Food chemistry*, 73, 55-60, 2001.

[24] Gutierrez, L.E. and du Silva, R.C.M. Fatty acid composition of commercially important fish from Brasil. Scientia Agricola Pivacicaba, 50(3), 478-483, 1993.

[25] Vlieg, P. and Body, D.B. Lipid contents and fatty acids composition of some New Zealand freshwater fish and marine fish, shellfish and roes. *New Zealand Journal of Marine Freshwater Research*, 22, 151, 1988.

[26] Fatima Aparecida Ferreira de Castro, Helena Maria Pinheira Sant'Ana, Flavia Millagres Campos, Neuza Maria Brunoro Costa, Marco Tulio Coeho Silva, Ana Lucia Saharo and Sylvia do carmo Castro Franceschini. Fatty acid composition of three freshwater fishes under different storage and cooking processes. *Food Chemistry*, 103, 1080-1090, 2007.

[27] Leaf, A. and Webber, P.C. Cardiovascular effect of n-3 fatty acids. *New England journal of Medicine*, 318, 549-555, 1998.

[28] San Giovanni, J.P. and Chew, E.Y. The role of omega-3 long chain polyunsaturated fatty acids in health and desease of the retina. *Progress in Retina and Eye Research*, 24(1), 87-138, 2005.

[29] Pompeia, C., Freitas, J.S., Kimj, S., Zyngier, S.B. and Curir. Arachidonic acid cytotoxicity in leukocytes: implications of oxidative stress and eicosanoid sintesis. *Biological Cellular*, 94(4), 251-265, 2002.

[30] Farhat, J. and Abdul, S.C. Chemical compositions and fatty acid profiles of three fresh water fish species. *Food chemistry*, 125, 991-996, 2011.

[31] Beyza E. and Akif O. The effect of cooking methods on mineral and vitamin contents of African catfish. *Food Chemistry*, 115, 419-422, 2009.

[32] Russel, E.W. Soil conditions and plant growth. Supergene Zone, M . Nedra, 19 (in Russian), 1973.

[33] Brody, T. Nutritional Biochemistry, San Diego, CA: Academic Press. 2nd Edn., 761-794, 1994.

[34] Fallon, S. and Enig, M.G. Nourishing Traditions. The Cook book that Challenges Policitally Correct Nutrition and the Diet Dictocrats. Revised 2nd Edn., 40-45, 2001.

[35] Wheaton, F.W. and Lawson, T.B. Processing aquatic food products. USA: John Wiley & Sons, 1985.

[36] Liu, Y.C. Food Chemistry. Chinese Light Industry Press, Beijing, 83, 1996.

[37] FAO/WHO, Human Vitamin and Mineral Requirements, Rome: FAO, 2001.

[38] Alas, A., Ozcan, M.M. and Harmankaya, M. Mineral contents of head, caudal, central fleshy part, and spinal columns of some fishes. *Environemental Monitoring and Assessment*, 186, 889-894, 2014

[39] Kawarazuka N. and Bene, C. The potential role of small fish species in improving micronutrient deficiencies in developing countries: Bulding evidence. *Public Health nutrition*, 14, 1927-1938, 2011.

[40] Magobe, O., Mosepele K. and Masamba, W.R.L. Essential mineral content of common fish species in Chanoga, Okavango delta, Botswana. *African Journal of Food Science*, 480-486, 2015.

[41] Wood, R. J. Assessment of marginal Zinc status in humans. *Journal of Nutrition*, 130, 1350-1354, 2000.

[42] Burke J.M. and Miller, J. E. Control of Haemonchus contortus in goats with a sustained-release multi-trace element/vitamin ruminal bolus containing copper, 2006.

[43] Wildman R. and Medeiros, D. Avanced human nutrition. Florida, USA: CRC Press LLC, Boca Raton, 2000.

[44] Gordon, R.F. Poultry Diseases. The English Language Book Society and Bailliere, Tindall, London, 1977.

[45] Pirestani, S., Sahari, A., Barzegar M. and Seyfabadi, S.J. Chemical compositions and minerals of some commercially important fish species from the south Caspian sea. *International Food Research Journal*, 16, 39-44, 2009.

[46] Lathan M.C. Human Nutrition in developing world. Rome: FAO Food and Nutrition Series №29widman R. And Medeiros D., Advanced human nutrition. Florida, USA: CRC Press LLC, Boca Raton, 1997.

32

Kithul Flour (*Caryota urens*) as a Potential Flour Source for Food Industry

J A A C Wijesinghe[1,*], I. Wicramasinghe[1], K.H Saranandha[2]

[1]Department of Food Science & Technology, University of Sri Jayewardenepura, Sri Lanka
[2]Food Research Unit, Gannoruwa
*Corresponding author: lucky.research@yahoo.com

Abstract This study search for underutilized flour source, obtained from Kithul (*Caryota urens*) trees in Sri Lanka which is potentially valuable as food ingredient for the industry. The aim of this paper is to discuss the Physico-chemical and functional properties of Kithul flour. The protein content of the flour was 1.0% while the total fat content was 0.33%. It contained considerable amount of Calcium, Potassium, Magnesium and Iron content as 70.1, 59.5, 66.6 and 14.0 mg/100g respectively. Total starch content was 66.82%. In the case of Amylose content, Kithul flour contained 28.42% while amylopectin presented 71.32%. The high moisture sorption value has presented by Kithul flour samples as 29.47%. And also being high density flour (as 0.69g/cm^3) this will be better thickener as well as a stabilizer in baking powders and as an emulsifier in the food industry. The measurement of gelatinization temperature which was obtained by Differential scanning colorimetric method was 76.74 °C, while enthalpy for gelatinization of Kithul flour was 11.12 J/g. The least gelation concentration (LGC) was 6. Length of the granules ranged from 26.5 to 64.7 μm and width of granules ranged from 7.0 to 56.9 μm. Three types of granular shapes as oval, spherical and irregular globular shape were exhibited. By considering all above Physico-chemical properties there is high possibility to use Kithul flour for empower the food industry, simultaneously strengthening the Kithul Industry with rural economy in Sri Lanka.

Keywords: *Kithul flour, Caryota urens, Physico-chemical composition, mineral content, DSC, FT-IR*

1. Introduction

Flour from many sources provides a better raw material for the production of starch and different sweeteners [1,2]. Starch has broad capability to move about wide variety of uses [3]. Due to functional properties affects the physical characteristics of many foods, starch is used to apportion those properties for processed food, in food industries. Starch is mainly applied as taste enhancer, thickener, binder, and filler [1,2,3,4] as well as stabilizer, edible film former in food industry [5], but also in non-food industries such as pharmaceutical, textile [6], Cosmetics, plastics, adhesives and paper industries [7] used to produce adhesive, agro-chemicals, cosmetics and toiletries, detergent, paper making additives, pharmaceuticals, paints, textiles, water purification agents, and biodegradable plastics [2,8].

Palms are proper source for starch production which belongs to oldest families of plants on earth [9]. Contemporary researchers pay their keen attention to discover novel sources of starch, which exist in the wild. Kithul (*Caryota urens*) is a better response for this requirement, which is still keep as semi-wild species [10]. This palm is native for India, Malaysia and Sri Lanka [11]. The bark yields around 100 kg -150 kg of pith per palm which used to preparation of flour [11]. The main usage of Kithul in current society is production of sap for toddy and jiggery [12]. So then common names for Kithul are Toddy palm and Jaggery palm [13]. The starch stored in trunk of the Kithul tree is the main edible food product of this palm. According to the reported values, palm generates 24 tons/hectare per year compared to rice which produce 6 tons / hectare per year while 5.5 tons/ hectare from corn and 2.5 tons/ hectare from potato [14]. According to the Rajalakshmi (2004) quality of flour from *Caryota urens* to be equal to the best sago of commerce extracted from *Metroxylon sagu rottb* [10].

The national survey conducted in 2009, has reported that the total number of Kithul trees found in the island amounts to 2,977,261 (vastly spread in Sabaragamuwa, Central and Uva provinces) [15]. Although matured or 'peedunu' trees was being 574,259, which sap was extracted amounts to only 89,855 (15.64%). Further evaluation proves that among the opportunities to develop Kithul industry there are 84% of the matured trees to be tapped [15]. This is the main point to deviate another expansion of Kithul industry as Kithul flour because non-tapping trees are more suitable for collecting. According to this statistic data 2,403,002 of new trees are available for maturation in 2014 with expanding the both tapping and flour industries [15].

Because of the high necessity due to number of industrial applications, a huge demand on few well-known starch sources. Therefore it is creating requirement on new sources for the continuous supply as commercially viable starch. The aim of the work reported herein was to undertake a comprehensive investigation on the physicochemical and functional properties of the flour obtained from Kithul (*Caryota urens)* tree.

2. Materials and Methods

2.1. Sample Collection

Five hundred grams of Kithul (*Caryota urens*) were purchased at random from both households and commercial market in Sri Lanka All chemical used was analytical grade.

2.2. Sample Preparation and Storage

Samples were sifted through a 355μm sieve and packed in air-tight containers, then stored in refrigerator (5°C) until further analysis.

2.3. Solubility Test

A 1g of Kithul flour was weighed and poured into a beaker containing 1ml, 2ml, 10ml, 1L and 10L distilled water at 30°C and was stirred, and the solubility was observed. Same procedure was repeated using 65% alcohol as a solvent. The procedures were repeated for all six samples [16].

2.4. Iodine Test

Using Musa et al, 2011 [16] starch identification test, 1g of flour was boiled with 15ml of water and allowed to cool. A few drops of 0.1N Iodine solution were added to 1ml of the mucilage and the colour changes recorded.

2.5. Moisture Content

One gram of each flour sample was weighted in a dish and reading was taken by placing in the Infra-red moisture content testing machine (ULTRA X) [6].

2.6. Proximate Analysis

Total fat, protein (N x 6.25), ash and crude fiber content were determined according to AOAC (1990) [17] methods. Carbohydrate contents were determined by difference sum of above categories. Mineral content was determined by dry ashing technique as described AOAC 999.11 [17]. The ash was dissolved in conc. HCl, filtered and dilute with distilled water. Prepared samples were analyzed with a standard series of solutions by Atomic Absorption Spectrophotometer (Thermo Scientific ICE3000 series).

2.7. Amylose and Amylopectin Determination

Amylose content of Kithul flour was determined in three steps using simple iodine colorimetric method as described by B.O. Juliano (1971) with slight modifications [18]. Amount of amylopectin was calculated by subtracting the amylose content from the starch content and expressed as g/100g of dry weight [19].

2.8. Total Starch Determination

Starch content was determined by the complete acid hydrolysis method. Flour sample of 2.5g was suspended in a mixture of 200 ml of water and 20ml of HCl acid (specific gravity 1.125).The mixture was heated in a flask provided with a reflux condenser for 2.5 hours. Contents were cooled, and neutralized with NaOH (5N).Volume was made to 250ml and sugar formed was determined as dextrose by Lane and Eynon reducing sugar estimation method. The dextrose multiplied by 0.9 was taken a starch [20].

2.9. Determination of Moisture Sorption Capacity

Moisture sorption capacity was determined by method described by shiihii et al (2011) [6]. Each dried flour samples were spread evenly in Petri dishes; the Petri dish was placed in a desiccator with 98% relative humidity at room temperature. The samples were periodically weighed until a constant weight was attained. The percentage increase in weight as calculated and taken as the moisture sorption capacity. The above experiments were repeated thrice and the mean of the readings was recorded.

2.10. Determination of Flour Density

Bulk density: Bulk density of the flour was determined according to the method of Musa et al.(2011) [16] with slight modifications.20g each, of individual flour respectively were poured through a short-stemmed glass funnel into a 250ml graduated glass cylinder and the volume occupied by the flour was read and the bulk density calculated in triplicate.

Tapped density: Graduated cylinder containing 20g of flour was dropped on a bench 50 times from a height of about 20mm and the respective volumes recorded. Same was done for all samples and the tapped densities were then calculated in g/ml [16].

Carr's Index: The difference between the tapped and bulk density divided by the tapped density was calculated and ratio expressed as a percentage [16].

Hausner ratio: Hausner ratio was calculated by taking the ratio of tapped density to bulk density for all the samples [16].

2.11. Comparison of Colour Variations of the Flour by Using Chromometer Minolta (CR 400) Colorimeter

The instrumental measurement of Kithul flour color was carried out with a Colorimeter Minolta CR-400 (Minolta, Spectrophotometer, Japan) and the results were expressed in accordance with the CIELAB system. It is based on the color perception of 92% of the population which has a good vision **[20, a]**. The samples were placed directly on the glass (6.4 mm diameter diaphragm with an optical glass), with the surface of the sample was manually made a flat. The color attributes were determined by colour coordinates of L^* (L^* = 0 [black] and L^* = 100 [white]), a^* ($-a^*$ = greenness and $+a^*$ = redness), and b^* ($-b^*$ = blueness and $+b^*$ = yellowness) [48].

2.12. Differential Scanning Calorimetry (DSC)

Thermograms for Kithul flour was taken by DSC (Model DSC TA instrument Q 200, USA).Flour was weighed on to the aluminum DSC pan and distilled water was added with micro syringe for 50% (w/w) mixture. Pan was sealed and allowed to stand for 1hr at room temperature. The scanning temperature range and heating rate were 30-140°C and 5°C/min, respectively, using empty pan as reference [20]. The onset temperature (To), peak temperature (Tp), conclusion temperature (Tc), and gelatinization enthalpy (ΔHG) were recorded.

2.13. FT – IR Spectroscopy

The transmission spectra were obtained between 4000 and 500 cm^{-1}, with 4 cm^{-1} resolution and 128 scans using Nicolet 6700, USA Fourier Transform Infrared Spectroscopy (FT-IR). Sample pellets were prepared for FT-IR by mixing vacuum dried 10 mg of samples with fused-110 mg of KBr. Spectral analyses were performed using OMNIC (version 8.0) software tools (Nicolet 6700, USA). [23].

2.14. Gelation Studies

Flour dispersions (2-18%w/v) were prepared in test tubes with distilled water (5ml). The flour suspensions were thoroughly mixed for 5 min. the test tubes were heated for 30min at 80°C in a water bath, followed by rapid cooling under running cold tap water for 2hr. Least gelation concentration was determined as lowest concentration when the sample from the inverted test tube did not fall down or slip [23].

2.15. Granular Morphological Variations of the Starch by Light Microscope (LM)

Morphology of Kithul starch granules was observed according to the method described by Jayakody et al, 2007 [24] with slight modifications. Starch solution was prepared by mixing with 1:1(w/v) mixture of distilled water to glycerin (1:1,v/v).The starch solution was stained with 0.05% iodine solution and a thin smear was prepared on a glass slide with a cover slip. Microscopic measurement of Kithul starch granules was done using light microscope (HITACHI SU 6600, Japan) under the magnification power of 400 x with Magnus live USB 2.0 viewer and UTHSCSA Image tool for windows software (Developed by University of Texas, 1995-2002).The length and diameter of the granules from each starch sample obtained from five different districts were measured and the shape of granules was observed.

2.16. Morphological Characteristics by Scanning Electron Microscope (SEM)

Morphology of native Kithul flour was studied by using scanning electron microscope (JSM 6380A, Jeol Japan). Flour samples placed in aluminum specimen holder were coated with gold and examined under different magnifications [20].

2.17. Statistical Analysis

Results were analyzed using one-way analysis of variance (ANOVA) at 0.05 probability level with MINITAB software package (version 17 for Windows).

3. Results and Discussion

3.1. Identification and Chemical Parameters

The Kithul flour treatments collected as per 2.1 were odourless. And these were insoluble in water and 65% ethanol with positive respond for Iodine test and acidity test (Table 1). Generally starch granules are insoluble in cold water. They undergo a reversible reaction in cold water, swelling slightly and becoming hydrated. Kithul flour was answered positively for Iodine test according to the Muazu,J *et al* the starches were all positive for this test [16].

Table 1. Results of identification and Chemical composition test of Kithul flour collected from different growing areas in Sri Lanka

Identification Test	**Result**
Odour	Odourless
Solubility Test	Insoluble
Iodine Test	Positive
Chemical composition Test	**Value**
Moisture Content %*	10.1± 2.1
Crude Protein (g/100g db)*	1.0 ± 0.2
Total Fat (g/100g db)*	0.33± 0.08
Crude Fiber (g/100g db)*	1.0± 0.5
Ash Content (g/100g db)*	0.8± 0.6
Carbohydrate (By Difference)*	87.5± 2.2
Calcium (mg/100g db)*	70.1 ± 26.9
Potassium (mg/100g db)*	59.5 ± 22.8
Sodium (mg/100g db)*	56.7 ± 35.5
Iron (mg/100g db)*	14.0 ± 4.9
Zinc (mg/100g db)*	3.3 ± 1.9
Magnesium (mg/100g db)*	66.6 ± 27.0

*Values were recorded in replicate of all six flour treatments.

According to the results moisture content ranged 8.0% to 12.2%, while protein content ranged from 0.8 to 1.2 g/100g (dry basis). Literature reveals that the crude protein content of *Caryota urens* flour has been reported as 1.13 [25].

According to the comparison with available data on economically important tuber and pith starch sources which are currently use as rich in starch, Kithul flour exhibited higher protein content (1.0%) than sweet potato (0.17%) [26] and sago pith flour was exhibited 0.1% [27] while potato (*Solanum tuberosum) starch* contained higher protein value than Kithul as 2.5 to 3.2% according to the Abbas *et al* [28]. Although mean value of the total fat content of Kithul flour (0.33± 0.08g/100g (dry basis) was presented little bit higher than reported value from Indian Scientist as 0.2 [25], it was comparably slightly lower than cassava starch (0.74 -1.49 g/100g) [29] as well as Sweet potatoes (1.1-1.7g/100g) [20]. This minimal amount of fat and protein of Kithul flour could be expected to have because of this palm is non-leguminous plant [30].

Reported crude fiber content of collected Kithul flour samples was ranged from 0.5 to 1.5% with aligned to literature, where the crude fiber content of Kithul flour has been reported as 0.2 to 3.5% [31].The fiber content can be different due to flour extraction method. Literature provides facts as in dry method removed portions of pith were dried, pounded and sieved through the household strainer, so it contains higher fiber content than wet method where pieces of pith were removed and pounded, mixed with water and sieved from a piece of cloth [12]. Crude fiber content showed comparable values with cassava root starch (1.4 to 3.2 g/100g) [30]. Depending on the extraction procedure ash content also can vary, by addition of sand particles could present by passing through the strainer or cloth. As per the results of this study total ash content of the flour samples were ranged of 0.2% to 1.4% while literature shows in Sri Lanka it was 0.65 % [24] and in Indian researchers have reported as the ash content is range to 1.5-2.5 % [31]. Carbohydrate content was calculated by difference of sum of above five components. Because of the moisture content of flour directly effect on this calculation there was slight variations of the carbohydrate content. Therefore total starch determination is the better way to identify the starch component in the flour.

Studied Kithul flour had a considerable amount of nutritionally important trace elements Iron (Fe) and Zinc (Zn), with mean values of 14.0 and 3.3 mg/100g (dry basis), respectively (Table 1). Mean concentrations of macro minerals, namely Sodium (Na), Potassium (K), Calcium (Ca), and Magnesium (Mg), were as 56.7, 59.5, 70.1, 66.6 mg/100g respectively. However Calcium content (70.1 mg/100g) of the Kithul flour was higher than corn starch18.6 mg/100g [32], while Iron content of Kithul flour (14.0 mg/100g) was exhibited same behaviour with corn (15.2 mg/100g) [32].

Usually, the functional properties are regulated by amylose and amylopectin content of the starch [33].As well as these parameters are key quality points of starches, which could impact characteristics of baked foods, and greatly influence important functional attributes like pasting and gelling properties [34]. During swelling amylose behaves as a restraint to gelatinization by distributing out of the granules. So it makes up the continuous phase by reticulating in outside the granules [35]. According to the reported previous data, waxy starches (High Amylose pectin) usually swell to a greater extent than non-waxy (High amylose) starches [36].

Based on the Rice classification of Lawal *et al* [37] Kithul flour was belonged to high amylose (25-33 %) category due to it has contained 28.42% Amylose content (Table 2). So it shows a non-waxy flour type as most native flours contain 20-30 % of amylose [31].

Table 2. Variations on Amylose, Amylopectin and total starch of native Kithul flour in Sri Lanka

Sample Name	Values
Amylose**	28.42±0.04
Amylopectin **	71.32±0.40
Total Starch	66.82±0.25

db = Dry basis.
* Data are the average of three repetitions ±standard deviation. The values in a column followed by the same letter are not
In base of 100% of starch.

Due to high amylose content water absorption index tend to be relatively low due to amylose is hardly converted into amorphous state compared to amylopectin [38]. Tester & Morrison, [39,40] reported a very similar ideas as amylopectin is efficient to swell due to the frailty of the intra- and inter-molecular bonds, while amylose inhibit the swelling of starch, [39,40].

Total starch content was determined using two steps as acid hydrolysis and determined the reducing sugar content. The Lane-Eynon method was used for determining the concentration of reducing sugars in the samples. A burette was used to add the flour solution being analyzed to a flask containing a known amount of boiling copper sulfate solution and a methylene blue indicator. The main disadvantage of this method is the results depend on the precise reaction times; temperatures and reagent concentrations used [41]. There for personnel errors must be carefully controlled. Kithul flour contained 66.82% of mean total starch content (Table 2) which slightly higher than value reported by Thanuja.P, [25] as 64% from India.

3.2. Functional Properties

Variation of the Moisture sorption value of native Kithul flour with the time is shown in the Figure 1 based on the data in Table 3. This behaving pattern is a reflection of the looseness between granules of Kithul flour and it confirms that the granules are not tightly associated together [27]. According to the literature, it is assumed that hydration of flour represents the water absorbed by the particles or particle surface [42].So high water sorption capacity means smaller particle size.

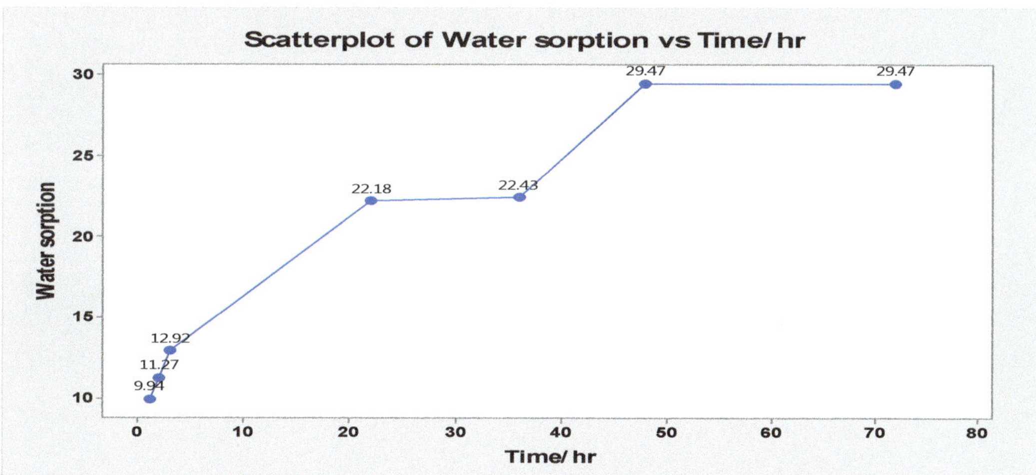

Figure 1. Scatterplot of water sorption behaviour of Native Kithul flour with the time

Table 3. Variations on water sorption value of Kithul flour with the time

Sample	1hr	2hr	3hr	22hr	36hrs	48hr	72hr
Raw Kithul Flour	9.94±1.05	11.27±0.64	12.92±1.06	22.18±0.52	22.43±0.46	29.47±0.4	29.47±0.4

Each value represents the means of three estimations of three samples of raw Kithul flour.

Bulk density is a measure of heaviness of a flour sample which is depended upon the particle size of the samples. It is important parameter for concluding industrial requirements as packaging, material handling and utilization in wet processes in the food industry [42]. Due to flours with high bulk densities are used as thickeners in food industry, Kithul flour could be used as a thickener [43], which has presented 0.69 of bulk density while bulk density of corn flour showed 0.51 [44]. The parameters called Carr's index and Hausner ratio predict the flow and compressibility of powders [42], which used in pharmaceutical industry. Based on the bulk and tapped densities, Kithul flour sample was obtained Carr's index as 22.58 (Table 4).

Table 4. Functional properties of native Kithul flour in Sri Lanka

Density	Value
Bulk Density	0.69±0.01
Tapped Density	0.89±0.005
Carrs Index= (tapped-bulk)*100/tapped	22.58±1.50
Hausner ratio=tapped/bulk	1.29±0.025
Colour attributes	
L* Mean	68.47±2.47
a* Mean	5.05±0.38
b* Mean	15.86±1.47
DSC Data	
Onset Temperature(°C)	72.02±1.08
Peak Temperature(°C)	76.74±0.81
Conclusion Temperature(°C)	87.46±0.54
Gelatinizing Enthalpy(J/g)	11.12±0.46

db = Dry basis.
Above all Data are the average of three repetitions ±standard deviation.

Hausner ratio values obtained in this study for Kithul flour samples were 1.29. As per the literature, Hausner ratio above 1.2 and Carr's index above 23% do not indicate good flow or good compressibility [26]. Kithul flour samples were indicated less suitable values as good flow and good compressibility as per the Muazu *et al* (2011) [42]. Starch modification would be answered easily for this situation. Colour analysis showed L*(lightness) values in the range of 60 to 70.94, a*(redness) values 4.67 to 5.43 and b* (yellowness) raked 14.39 to 17.33 (Table 4). The positive a* value for Kithul flour samples showed its propensity towards pinkish colour as positive a* values represent the redness of samples [22]. The colour of flour is depended on the presence of polyphenolic compounds, ascorbic acid and carotene has impact on its quality[45]. The crude Kithul flour contains a considerable amount of phenols, as 52.0 %, when compared to white coloured corn flour, which has no phenols [25]. However pigmentation in the flour is carried up to the final product. So it reduces the quality, as well as consumer's acceptability of starch product.

In the sense of food processing, the functionality of flour is mainly linked to its gelatinization and pasting characteristics [44]. In the presence of excess water, heating of starch cause to a phase transition from array to disarray known as gelatinization over a temperature range characteristic of the starch source. The Differential Scanning Calorimetry (DSC) provides an accurate and comprehensive method for monitoring the gelatinization characteristics of Kithul flour. One endothermic transition was raised during gelatinization of the flour (Figure 2). The data obtained for the DSC thermo grams for Kithul flour are given in Table 4. It can be seen that the, onset temperature, peak temperature and conclusion temperature values as 72.02°C, 76.74°C and 87.46 °C respectively, since gelatinization temperatures reflect the degree of orderliness of the molecules in the granules. Kithul flour has a comparatively slight lower gelatinization temperature than Corn starch (80°C) [35]. This is probably due to its higher amylose content. High amylose content has been found to influence gelatinization temperature [46]. FT-IR is one of the imperative instruments used to analyze food quality parameters [47]. Spatial distribution of starch, protein and lipid in flour-based products can be determined by FT-IR spectroscopy [48]. Figure 3 shows very strong absorption band around 3370-3340 cm^{-1} may be due to the presence of bonded N-H, C-H, O-H stretching in amines and amides [49].

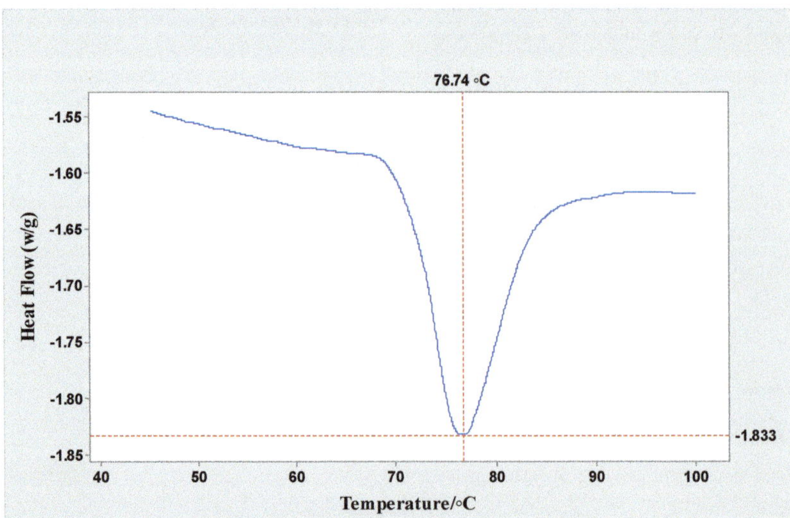

Figure 2. DSC thermo grams for Kithul flour

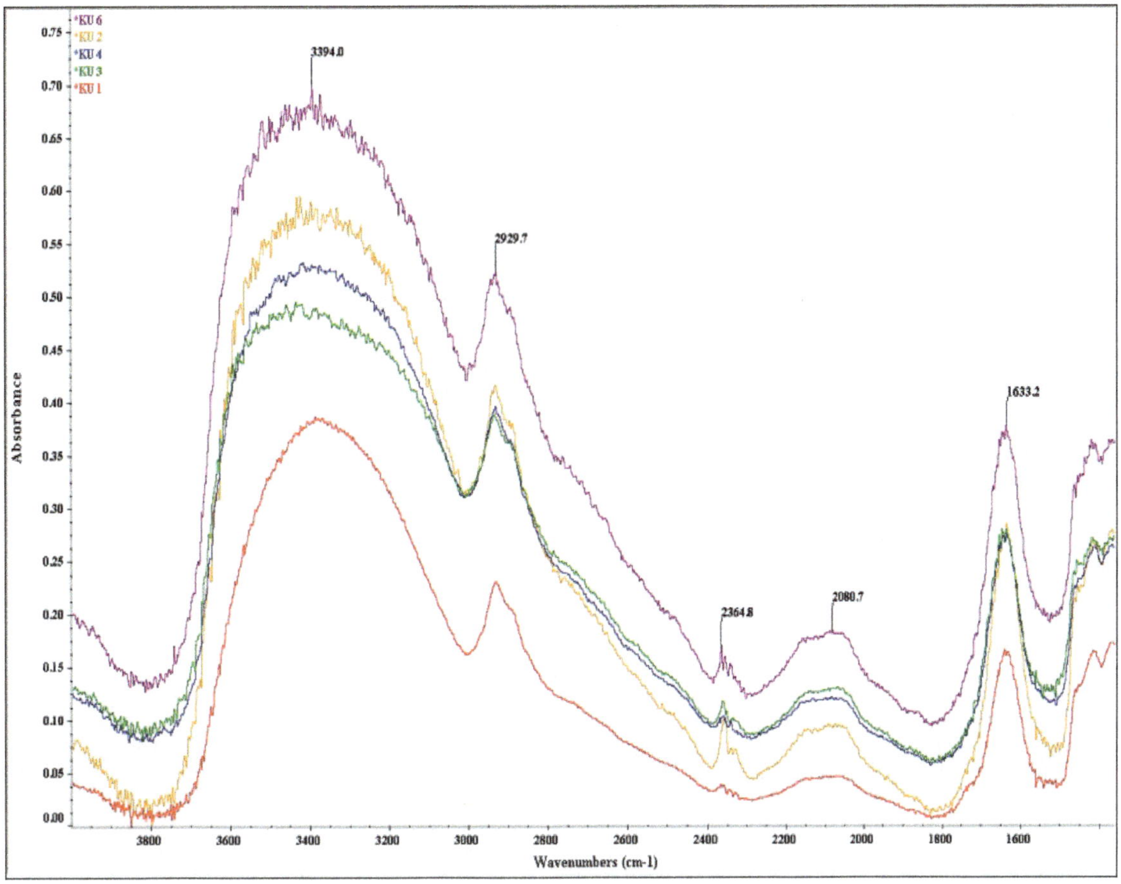

Figure 3. Transmission spectra obtained between 4000 and 1500 cm-1

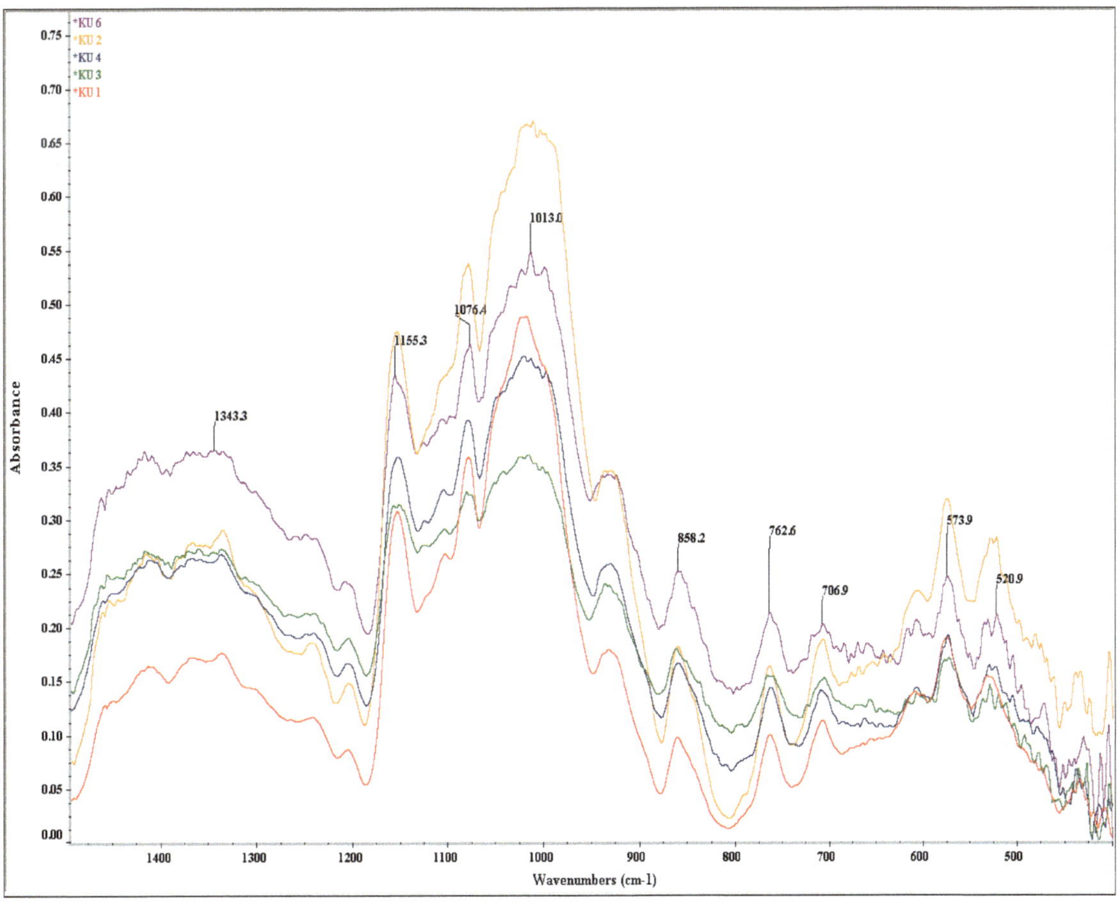

Figure 4. Transmission spectra obtained between 1500 and 4000 cm-1

The very strong absorption at 3394 cm^{-1} shows the presence of amino acids and the absorption band appearing the region 2932 cm^{-1} is as a result of N-H stretching [49]. In addition, a symmetrical stretching of NO_2 group results in strong absorption in the region 1660-1625 cm^{-1}. The observed absorption band at 1633 cm^{-1} indicates the presence of amines [49]. Figure 4 demonstrates the peak at around 1343 cm^{-1} as the C-H bending mode of CH_2 [49]. Absorption peak at 1155 cm^{-1} can be assign into C-O stretching or OH bending while CH bending can be assign to the peak at 1076 cm^{-1} [50]. The peak around 1013 cm^{-1} shows the C-O or C-C stretching modes [50]. Further, COH bending vibration is appeared at 855 cm^{-1} [50]. Peaks appeared at 573, 706 and 763 cm^{-1} can be assigned into the CH out-of-plane bending vibrations. The peak at around 520 cm^{-1} can be assign into the torsion and ring of phenyl [49].

3.3. Gelation Studies

In Table 5, gelation properties of native Kithul flour samples are presented. The least gelation concentration (LGC) was 6.This value is important for food processing activities which are based with the flour, to determine the correct water: flour ratio for desirable texture.

Values were recorded in replicate of all six flour treatments.

Table 5. Gelation properties of native

Concentration (% w/v)	Status
2	Viscous Liquid
4	Soft Gel
6	Soft gel+ Firm gel
8	Firm gel
10	Very firm gel
Least Gelation Concentration	6

3.4. Morphological Properties

Morphological properties of native Kithul flour were studied using both LM (Light Microscope) and SEM (Scanning Electron Microscope) at different magnifications ranging from 100x to 400x for LM (Figure 5) and 100x to 500x for SEM (Figure 6). Approximate size and shape of flour are presented in Table 6.

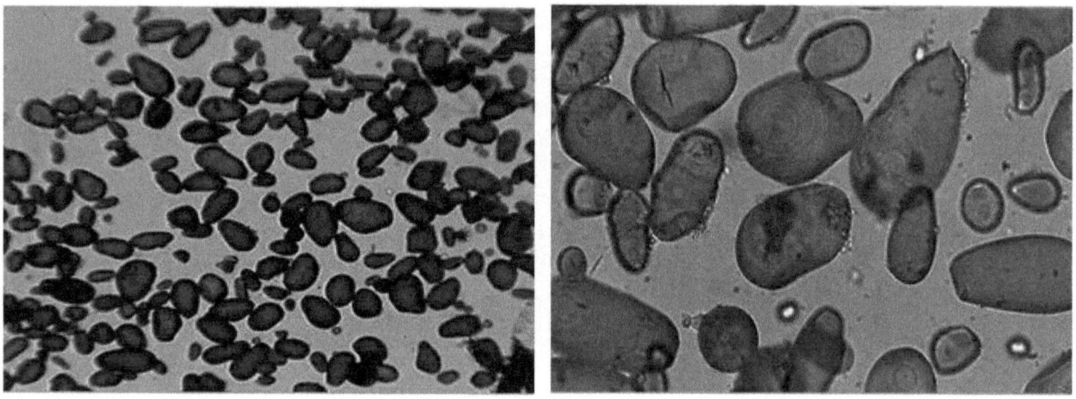

Figure 5. Light micrograph of Native Kithul (*Caryota urens*) flour under two magnifications (X100 and x400)

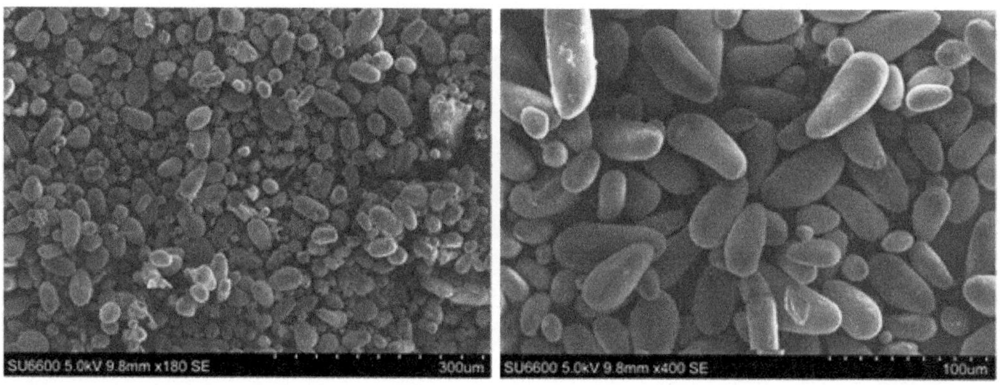

Figure 6. Scanning Electron Micrograph (SEM) of Native Kithul (*Caryota urens*) flour under two magnifications (300μm and 100 μm)

Table 6. Variations in Granular size and Granular shapes of Kithul flour collected from main growing areas in Sri Lanka (Based on LM and SME results)

Indices	Value
Length of the granule / μm	45.6 ± 19.1
Width of the granule / μm	26.4 ± 9.9
Spherical shape %	35.1 ± 17.3
Oval shape %	51.4 ± 14.6
Irregular globular shape %	13.5 ± 10.8
Values were recorded in means of three replicate of all flour treatments.	

Length of the granules ranged from 26.5 to 64.7 μm though mean value was 45.6 μm, which is compatible with previously reported value (44 μm) of *Caryota urens* pith flour [51], while the width of granules ranged from 7.0 to 56.9 μm with a mean value of 26.4 μm (Table 6). Three types of granular shapes as oval, spherical and irregular globular shape were exhibited. Mostly common in oval shape and with the walls that are generally smooth without edges, which represented 51.4 ± 14.6 % of granules. Spherical shape got the second representative percentage contributing 35.1 ± 17.3% of granular while irregular globular had lowest value as 13.5 ± 10.8 %.

4. Conclusion

Kithul flour has huge potentials of being used to food industry in order to provide better gelling property with other functional needs. Based on this study Kithul flour can be promoted because of its positive nutritional and Physico-chemical properties for the enhancement of different food products. However, the results from this study can be used by food technology researchers, agro-processors and investors as an input for their work. More specifically, the results can contribute to those who are interested in the production of starch based product developments which could be helped to empower the Kithul industry.

Acknowledgement

The authors thank the University of Sri Jayewardenepura, Sri Lanka for providing the financial assistance for this study (Grant No: ASP/06/RE/SCI/2012/04). And also authors are grateful to Dr. M.A.P.K Seneviratne, Department of Export Agriculture, District office, Peradeniya and special thanks to Dr.B.M.K.S Tilakaratne, Director of Institute of Post-Harvest Technology Research and Development Centre, Anuradhapura, Sri Lanka, Dr. Meththika Vithanage, Institute of Fundamental studies, Kandy, Sri Lanka for providing instrumental facility for carried out this study.

References

[1] Radosavljević M Cornstarch as an important renewable resource. Maize research Institute, Zemun Polje, Belgrade-Zemun, Yugoslavia. 2002: 23-31.

[2] Zehr BE, Eckhoff SR, Singh SK and PL Keeling Comparison of wet milling properties among maize inbred lines and their hybrids. Cereal Chemistry. 1995; 72: 491-497.

[3] Mengistu SL, AE Shimelis Evaluation Of The Physico-Chemical Properties Of Ethiopian Maize Variety (Bh-660) For Dextrose Production *Journal of Food,Agricultural Nutrition and Development,* Volume 12 No. 2012 ;6503-6516.

[4] Sprague GF Corn and corn improvement (2nd ed.). Madison, Wisconsin,USA. 1979: 721-732.

[5] Krisztina R.V,Starch bearing crops as food sources, Cultivated plants, primarily as food sources-vol.I (http://www.eolss.net/Eolss-sampleAllcharecter.aspx)visited on 2013.10.02

[6] Shiihii,S.U., Musa,H., Bhatia,P.G.,Martins, E.(2011) Evaluation of physicochemical properties of Eleusine corona starch, Nigerian Journals of Pharmaceutical Sciences, Vol.10, No.1, March, 2011, ISSN: 0189-823X.

[7] Kolawole S. A, Igwemmar N. C, Bello H. A. Comparison of the Physicochemical Properties of Starch from Ginger (*Zingiber officinale*) and Maize (*Zea mays*) International Journal of Science and Research (IJSR) ISSN (Online): 2319-7064 Volume 2 Issue 11, November 2013.

[8] Shimelis E, Meaza M and S Rakshit Physico-chemical properties, pasting behavior and functional characteristics of flours and starches from improved bean (*Phaseolus vulgaris* L.). *CIGR EJournal of Agricultural Engineering International.* 2006; 8: 1-19.

[9] Ishizuka, K., Hisujma, S., Darryl, RJ., Macer. 1995. "Sago palm, a promising renewable carbohydrate resource, a material for environmental conservation and sustainable development. In: Proceeding of the international seminar on traditional technology foe environmental conservation, development in the Asia- Pasif ic region, UNESCO", Tsukuba science city, Japan, pp. 75-76.

[10] Rajyalakshmi, P. 2004. "*Caryota* palm sago O A potential get underutilized natural resource for modern starch industry". Naturakl product redience vol 3(3), pp144-149.

[11] Council of Scientific and Industrial Research, New Delhi ;Wealth of India, 1992, "A dictionary of Indian raw materials and industrial product Vol.3", Publications and Information directorate., vol 3 ca –ci, pp 321-324

[12] Seneviratne, M.A.P.K. Unpublished phd thesis on "Utilization of indigenous knowledge for the uplistment of Kithul in Sri Lanka, 219, 217-222. (2011).

[13] Orwa,C., Mutua, A., Kindt, R., Jamnadass, R., Simons, A., Arecaceae L;. *Caryota urens.* Agroforestree Database: a tree reference and selection guide version 4.0. (2009). (http://www.worldagroforestry.org/af/treedb/) visited on 2013.10.02.

[14] Shoon Jong, Rain forest gifts; Sago & sago grubs, 2000. www.2000szgdocent org/ff/f-sago.htm. Visited last on 08.08.2013.

[15] Ministry of traditional industries and small enterprise development, 2009. "Statistic on Kithul palms in Sri Lanka; National programme for Kithul Development". pp 3-7.

[16] Musa, H., Gambo, A. And Bhatia., P.G.,(2011); Studies On Some Physicochemical Properties Of Native And Modified Starches From *Digitaria Iburua* And *Zea Mays.* International Journal Of Pharmacy And Pharmaceutical Sciences,Issn- 0975-1491 Vol 3, Issue 1, 2011.)

[17] Association of official Analytical chemistry (AOAC), 1990, Official methods of analysis of AOAC International, volume 1.

[18] Juliano, B.O. (1971). A simplified assay for milled rice amylose. *Cereal Science Today,* 16: 334-338.

[19] Patil U. H., And Gaikwad D. K., (2011), Seasonal Dynamics In The Nutritional And Antinutritional Status Of Stem Bark Of *Anogeissus Latifolia.*International Journal Of Applied Biology And Pharmaceutical Technology, *ISSN 0976-4550;* Vol.02 370 - 379.

[20] Senanayake, S., Gunarathne, A.,, Ranaweera, K.K.D.S., 2013. "Physico-chemical properties of five cultivars of Sweet potato (Ipomea batatas Lam(L))root tubers grown in Sri Lanka.", Trop.Agric.(Trinidad)Vol.90 No.2, pp 87-96.

[21] Hutchings J. R., Luo W., (2002); Color in Food, Improving quality, Chapter 14: Calibrated color imaging analysis of food: B.D. MacDougall, Woodhead Publishing in Food Science and Technology, Cambridge, England, 2002, pp. 352-366.

[22] CIE 15:2004. The Executive Director Office Of The Federal Register, Washington,.*CIE 15: Technical Report: Colorimetry,* 2004, 3rd edition.17, 16-27.

[23] Awokaya, K.N., Nwokocha, L.M., Moronkola, B.A., Moronkola, D.O., (2011). Studies on isolation,structural and functional properties of starch Succinate of Cococyam(*Colocasia antiquorum*) Prlagia research liabrary, Der chemical sinica, 2011, 2(60:228-244.

[24] Jayakody, S.W.P, Evaluation of Physiochemical properties and morphology of tarches of Kithul (*Caryota urens*), Kuru Kithul (*caryota plumose*) and their hybrid, Unpublished Bsc thesis,Dept. of Food Science & Technology, University of Peradeniya, pp 9, 4-27 (2009).

[25] Tanuja, P. and Rajyalakshmi, P. 2004. "Development of products with purified caryota palm (*caryota urens*) sago". J.Food Sci.Technol., 2004, vol, 41No.1, pp 80-82.

[26] Iheagwara, M. C. (2013), Isolation, Modification and Characterization of Sweet Potato (Ipomoea batatas L (Lam)) Starch; J Food Process Technol 2013, 4:1, ISSN:2157-7110 JFPT, an open access journal, Volume 4 • Issue 1 • 1000198.

[27] Swinkels, J. J. M. (1985). Sources of starch, its chemistry and physics. In G. M. A. VanBeynum & J. A. Roels (Eds.), Starch conversion technology (pp. 15-45). New York: Marcel Dekker, Inc.

[28] Abbas, G., K. Farooq, I.A. Hafiz, A. Husain, N.A. Abbasi and G. Shabbir. (2011). Assessment of processing and nutritional quality of potato genotypes in Pakistan. Pakistan J. Agr. Sci. 48(3): 169-175.

[29] Emmanuel, O.A., Clement, A., Agnes, S.B., Chiwona-Karlton, L. and Drinah, B.N, 2012. "Chemical composition and cyanogenic potential of traditional and high yielding CMD resistant cassava (*Manihot esculenta* Crantz) varieties", International Food Research Journal 19(1): pp 175-181.

[30] Omojola, M.O., Akinkunmi, Y.O., Olufunsho, K.O., Egharevba, H.O. and Martins, E.O. (2010); ISOLATION AND PHYSICO-

CHEMICAL CHARACTERIZATION OF COLA STARCH; Vol. 10, No. 07, pp. 2884-2900, 2009.

[31] Rajyalakshmi P and Greevani P. 1994. "Nutritive value of the foods cultivated and consumed by the tribals of South India", *Plant Food Human Nutr*, 1994, 46, pp 53-61

[32] Ihegwuagu, N.E., Omojola, O.O., Emeje,M.O., and Kunle, O.O., (2009), Isolation and evaluation of some physicochemical properties of *Parkia biglobosa* starch; *Pure Appl. Chem.*, Vol. 81, No. 1, pp. 97-104, 2009.

[33] Hermansson, M. and Svegmark, K. (1996). Developments in the understanding of starch functionality. Trends in Food Science and Technology, 7, 345-353.

[34] Lee I., Jin We G., Kim D E., ChoY S., Yoon M R., Shin M., Ko S., (2012) Classification of rice cultivars based on cluster analysis of hydration and pasting properties of their starches, LWT - Food Science and Technology 48 (2012) 164e168.

[35] Varavinit, S., Shobsngob, S., Varanyanond, W., Chinachoti, P. and Naivikul, O. (2003). Effect of amylose content on gelatinization, retrogradation and pasting properties of flours from different cultivars of Thai rice. Starch-Starke, 55(9), 410-415.

[36] Park, I.M., Ibanez, A.M., Zhong, F. and Shoemaker, C.F. (2007). Gelatinization and Pasting Properties of Waxy and Non-waxy Rice Starches. Starch-Starke, 59(8), 388-396.

[37] Lawal, O.S., Lapasin, R., Bellich, B., Olayiwola, T.O.,Cesaro, A., Yoshimura, M. and Nishinari, K. (2011).Rheology and functional properties of starches isolated from five improved rice varieties from West Africa. Food Hydrocolloids, 25 (7), 1785-1792.

[38] Lee, S. H., Han, O., Lee, H. Y., Kim, S. S., & Chung, D. H. (1989). Physicochemical properties of rice starch by amylose content. Korean Journal of Food Science and Technology, 21, 766e771.

[39] Morrison,W. R., Tester, R. F., Snape, C. E., Law, R., & Gidley, M. J. (1993). Swelling and gelatinization of cereal starches. IV. Some effects of lipid-complexed amylose and free amylose in waxy and normal barley starches. Cereal Chemistry, 70, 385e389.

[40] Tester, R. F., & Morrison, W. R. (1990). Swelling and gelatinization of cereal starches. I. Effects of amylopectin, amylose, and lipids. Cereal Chemistry, 67, 551e557.

[41] Patil, U. H. And Gaikwad, D. K. (2011), Seasonal Dynamics In The Nutritional And Antinutritional Status Of Stem Bark Of *Anogeissus Latifolia*.International Journal Of Applied Biology And Pharmaceutical Technology, *ISSN 0976-4550;* Vol.02 370-379.

[42] Muazu, J., Musa, H., Isah, A.B., Bhatia, P.G. and Tom, G.M.(2011) Extraction and characterization of Kaffir Potato Starch: A potential source of pharmaceutical raw material, Scholars Research Library,J. Nat. Prod. Plant Resour., 2011, 1 (2): 41-49.

[43] Ocloo, F.C.K., Bansa, D., Boatin, R., Adom, T., Agbemavor, W.S. (2010) Physico-chemical, functional and pasting characteristics of flour produced from Jackfruits (*Artocarpus heterophyllus*) seeds, AGRICULTURE AND BIOLOGY JOURNAL OF NORTH AMERICA, ISSN Print: 2151-7517, ISSN Online: 2151-7525.

[44] Saeleaw M. Schleining G., (2011): Composition, Physicochemical and Morphological Characterization of Pumpkin Flour. ICEF11-11th International Congress on Engineering and Food "FOOD PROCESS ENGINEERING IN A CHANGING WORLD", Athens, April 10-13, 2011.

[45] Gani A., Haq S.S, Masoodi, F.A., Broadway, A. A. and Gani, A., (2010)Physico-chemical, Morphological and Pasting Properties of Starches Extracted from Water Chestnuts *(Trapa natans)*from Three Lakes of Kashmir, India, Braz. Arch. Biol. Technol. v.53 n.3: pp. 731-740, May/June 2010.

[46] Ihegwuagu NE, Omojola MO, Emeje MO and OO Kunle Isolation and evaluation of some physicochemical properties of Parkia biglobosa starch. *J. Pure and Appl chem.* 2009; 81: 97-104.

[47] Amir, R.M., et al., *Application of Fourier Transform infrared (FTIR) Spectroscopy in Foods.* Internet journal of food safety, 2011. 13: p. 310-314.

[48] Kačuráková, M., et al., *Characterisation of xylan-type polysaccharides and associated cell wall components by FT-IR and FT-Raman spectroscopies.* Food Hydrocolloids, 1999. 13(1): p. 35-41.

[49] Kumar, J.K. and A.G.D. Prasad, *Identification and comparison of biomolecules in medicinal plants of Tephrosia Dentification and Comparison of Biomolecules in Medicinal Plants of Tephrosia Tinctoria and Atylosia Albicans by Using FTIR.* Romanian J. Biophys, 2011. 21(1): p. 63-71.

[50] Elizondo, N.J., P.J.A. Sobral, and F.C. Menegalli, *Development of films based on blends of Amaranthus cruentus flour and poly(vinyl alcohol).* Carbohydrate Polymers, 2009. 75(4): p. 592-598.

[51] Anilakumai, B., and Rajyalakshmi,P.(2000): Qualitative aspects, product suitability and in vitro starch digestibility of caryota palm sago-An uncommon food tribals.J.Food Sci.Technol., 2000, vol, 37 No.1, pp75-78. (2000).

Effect of Starch Type on the Physicochemical and Textural Properties of Beef Patties Formulated with Local Spices

P. D. Mbougueng[1,*], D. Tenin[1], C. Tchiégang[2], J. Scher[3]

[1]Department of Process Engineering, National School of Agro-Industrial Sciences, University of Ngaoundéré, Ngaoundéré, Cameroon
[2]Department of Food Engineering and Quality Control, University Institute of Technology, University of Ngaoundéré, Ngaoundéré, Cameroon
[3]Laboratory of Biomolecular Engineering, Lorraine University, Nancy, France
*Corresponding author: pides_mbougueng@yahoo.fr

Abstract This study was aimed at evaluating the effect of local potato and cassava starch on the proximate compositions, physicochemical and textural properties of beef patties. The control patty ($P_{control}$) was prepared with commercial Leader price™ potato starch incorporated at 50g/kg of ground meat, while other patties were formulated with starch from two local potatoes (*Sipiera* and *Tselefou*) and tree local cassava varieties (*2524, 4115* and *Seedling*) during which 20, 30, 40 and 50g starch /kg of ground meat was used.The results showed that moisture content varied according to starch type and starch quantity for both raw and cooked patties. Water contents of cooked patties were lower than their corresponding raw ones. The same trend was not observed for protein and fat contents of cooked patties. Starch content significantly affected the water holding capacity of raw patties ($P<0.05$) and for cooked patties, cooking yield was improved ($P<0.05$) at the lowest starch incorporation rate (20%) irrespective of the starch type. Patty prepared with Seedling starch at incorporation rate of 40g/kg of batter (*PS40*) is the most similar to the control one.

Keywords: *Beef patties, potato starch, cassava starch, physicochemical and textural properties*

1. Introduction

The use of extenders/binders/fillers in the elaboration of food products is on the rise because the latter presents the advantage of minimising product cost and improving or at least maintaining nutritional and sensory qualities of the final products [1]. The increasing demand for better quality and healthy meat products has also stimulated the use of new non-meat components. These non-meat components of natural or synthetic origin, known under the name of hydrocolloids or structuring additions, are introduced during processing and preservation of meat products [2]. Other non-meat additives tested as binders/extenders include: soya beans in meat products [3,4,5], faba beans, lentils, lupin and chickpeas in beef sausages [6], wheat flour in chicken nuggets [7], defatted sunflower meal in beef patties [8], cowpea and peanut flours in chicken nuggets [9,10]. Studies on functionality of various fillers including corn starch [11,12], rapeseed and mustard [13] as extenders in meat emulsions have also been reported. Starches are multifunctional food ingredients. They have many functional applications, including adhesion, binding, emulsion stabilisation, gelling and moisture retention [14]. Sunflower protein, corn germ flour and wild rice starches are used as binders and extenders in comminuted meat products to perform three basic functions: fat emulsification, water retention and to enhance the structure of meat products [5]. Readily available native starch could be a potentially cheap substitute for the higher priced modified starch and gums which are in common use [15,16]. Irish potatoes (*Solanum tuberosum*) and cassava (*Manihote sculanta*) tubers are abundant in Central Africa and other tropical areas and could serve as an economical source for starch. This is exemplified by the works of Mbougueng *et al.* [17] which revealed that some local native starches have physicochemical, functional and rheological properties similar to that of modified potato starch currently used as binders in meat products. In the present study, an investigation of the suitability of replacing commercial modified potato starch with native starch from local tubers in the production of beef patties was evaluated.

2. Materials and Methods

2.1. Materials

The *Sipiera* and *Tselefou* cultivars of Irish potato (*Solanum tuberosum*) were purchased at a local market while the *2425, 4115* and *Seedling* cultivars of Cassava (*Manihote sculanta* Crutz) were supplied by IRAD (Institute of Agricultural Research for Development) of Ngaoundere-Cameroon. These tubers were used at their commercial maturity, 6 months for potato and 12 for cassava. The control used in this study was the commercial Leader price™ potato starch.

Beef semi-membranous muscle (top round), udder fats and liver were obtained from an approved European abattoir (SOCOPA Mircourt slaughter-house: Nancy-France) using industrial slaughtering techniques. These samples were trimmed off all visible extra-muscular fat and connective tissue before storage at - 4°C for 72 hours. A formulation of local spices made up of fruit mixtures (*Hua gabonii, Xylopiaa ethiopica, Monodora myristica*) and the pulp of the fruit wings of *Tetrapleura tetraptera* [18] was used in the present study. For the exact composition of this spice mixtures contact the department of Food Science and Nutrition, University of Ngaoundere-Cameroon.

2.2. Methods

Starches used in this study were those extracted and characterised by Mbougueng *et al [17]*.

2.2.1. Native Starch Production

Starch extraction was carried out by the method of Alves *et al.* [19] with a slight modification. A total of 10 kg of tubers and roots were used in this study. All impurities and damaged tubers and roots were discarded. The remaining tubers and roots were first peeled, washed with distilled water, cut into small sizes and then chopped with a cutter (Manurhin, 03300 Cusset, n°426, France). The resulting product was mixed with distilled water. Fibres were separated by sieving through a 170-mesh screen. After washing several times, the starch obtained was oven dried at 45°C. Dried samples were then ground by using a Hobart mixer (Model 32BL79, New Hartford, CT 0657, USA).

2.2.2. Product Manufacture

Prior to processing, beef fats (udder) were boiled in water for 15min. and ground with liver and muscle meat through a 2mm plate. The ground meat, liver, beef fat, 2% sodium nitrate and locally formulated spices were thoroughly mixed for 5min. Six groups of patties were prepared and tested.

1) The control patty, containing 50g Leader price™ potato starch/kg ground meat ($P_{Control}$= PFPC).

2) The patties formulated with 20, 30, 40 and 50 g of *Sipiera* potato starch/kg of ground meat were designated, PSi_{20}, PSi_{30}, PSi_{40} and PSi_{50} respectively.

3) The patties formulated with 20, 30, 40 and 50 g of *Tselefou* potatoes starch/kg of ground meat were designated, PT20, PT_{30}, PT_{40} and PT_{50} respectively.

4) The patties formulated with 20, 30, 40 and 50 g of *2425* cassava starch/kg of ground meat were designated, PV_{20}, PV_{30}, PV_{40} and PV_{50} respectively.

5) The patties formulated with 20, 30, 40 and 50 g of *4115* cassava starch/kg of ground meat were designated, PQ_{20}, PQ_{30}, PQ_{40} and PQ_{50} respectively.

6) The patties formulated with 20, 30, 40 and 50 g of *Seedling* cassava starch/kg of ground meat were designated, PS_{20}, PS_{30}, PS_{40} and PS_{50} respectively.

Beef patties were cooked at 90°C in an oven (Memmert, UL 40, West Germany) to an internal temperature of 70°C. Cooked products were allowed to cool down at room temperature (22-25°C) for 30min. Cooled patties were cut into 25-30g portions, wrapped with aluminium foil and stored at 4°C prior to analyses.

2.2.3. Proximate Analysis of Patties

Moisture, protein, fat and ash contents were determined on raw and cooked products using AOAC methods [20]. Moisture was determined as weight loss of 3g sample after drying for 18h at 102°C. Crude protein was analysed by the micro Kjedahl method (Nx6.25). Fat was determined by weight loss after 16h extraction in a soxhlet apparatus with petroleum ether and ash by incineration of 3g sample at 550°C until a light grey ash result.

2.2.4. pH Determination

10 g of raw and cooked patties were homogenised with 90 ml of distilled water and the pH was determined with a pH-meter (Eutech Cybernetics, Cyberscan 1000, Singapore) [21].

2.2.5. Water Holding Capacity

The Tsai & Ockerman [22] press technique was used with some modification to measure the water holding capacity (WHC) of the raw patties. A sample (0.5g) was placed between 2 sheets of filter paper (Whatman n°1, stored over saturated KCL) which was placed between two Plexiglas sheets and pressed for 30minutes under 1kg load. The area of pressed meat and a spread juice was measured and the water holding capacity was calculated as follows:

$$\%FW = \frac{(Surface\ area - meat\ area)(mm^2)(6.11)}{(Total\ moisture\ in\ meat\ sample)(mg)} X100$$

%FW = % Free Water

WHC = 100 − %F W.

2.2.6. Cooking Loss

After formulating patties and placing them in mould of known weight, the filled moulds were weighed and then place in the oven for baking. At the end of the baking period, they were left to cool and then their weighed again.

$$\text{Cooking loss} = \frac{Raw\ weight - Cooked\ weight}{Raw\ weight}.$$

2.2.7. TBA Values

The degree of lipid oxidation of the raw and cooked beef patties was determined by the 2- thiobarbituric acid (TBA) cold extraction method, described by Wite *et al.* [23] was used. The results are expressed as mg malonaldehyde /kg of patty.

2.2.8. Color Measurement

Each patty sample was evaluated using a colorimeter (Lovibond RT Colour Measurement Kit V2.28) with a window of observation of 10° and one source of D65 light, the apparatus was gauged with a standard white plate (Lovibond RT100 N 319452) whose co-ordinates of color are: L* = 93,87, a* = 0,18 and b* = 2,71. L values range from 100 (white) to 0 (black), a values range from +a (green) to –a (red), and b values range from +b (yellow) to –b (blue). Average of the readings were computed and reported. Each result is the average of three determinations [24].

2.2.9. Texture Analysis

A texture profile analysis was applied to the cooked products based on a method described by Bourne [25]. Tree cores (diameter = 2.2 cm; height = 2cm) were cut from each cooked patty and were axially compressed to 50% of their original height in a two-cycle compression test using an Instron Universal Testing Machine (Model 4464, Instron Engineering Corp., Canton, MA). Determinations were performed at room temperature using tree replicates of cooked patties per treatment. Force-time deformation curves were obtained using a 5 kN load cell applied at a crosshead speed of 50mm/ min. The attributes reported are: hardness (N), cohesiveness (dimensionless), springiness (mm) and chewiness (Nxmm).

2.2.10. Statistical Analysis

The effect of each treatment was analyzed from the different preparations. Data were subjected to analysis of variance and the differences among means were obtained using Duncan's multiple range test (significance $p<0.05$) using the Statgraphics plus 5.0software.

3. Results and discussion

The starches used in this study were those extracted and characterised (Amylose, Phosphorus, Colour, Paste clarity, Particle size distribution, Scanning electron microscopy, Thermal properties, Swelling power, oil and water absorption capacity) by Mbougueng *et al* [17].

3.1. Proximate Composition of Raw and Cooked Patties

The results of proximate composition of raw and cooked patties are shown respectively on Table 1 and Table 2.

Table 1. Effect of starch types and starch rate on physicochemical properties of raw patties (%FM)

	Rates	$P_{Control}$	PSi	PT	PV	PQ	PS
Moisture	20		67.72±0.04^{kl}	67.55±0.07^{jkl}	67.42±0.48^{ijkl}	67.75±0.11^{kl}	67.30±0.08^{ijk}
	30	/	67.88±1.22^{l}	67.02±0.49^{ghi}	66.71±0.22^{fgh}	66.14±0.35^{bcde}	65.94±0.41^{abc}
	40		67.19±0.21^{hij}	66.75±0.46^{fgh}	66.59±0.15^{efg}	66.53±0.22^{defg}	65.68±0.19^{ab}
	50	66.03±0.60^{bc}	66.06±0.35^{bcd}	65.81±0.07^{ab}	66.43±0.29^{cdef}	65.54±0.61^{a}	66.62±0.37^{efg}
Protein (N x 6.25)	20		23.90±1.14^{cd}	24.95±0.15^{gh}	25.39±0.14^{hi}	23.21±0.38^{b}	22.12±0.42^{a}
	30	/	23.80±0.04^{cd}	25.55±0.14^{i}	25.51±0.10^{i}	23.16±0.05^{b}	23.06±0.87^{b}
	40		24.02±0.04^{cd}	23.73±0.43^{c}	24.49±0.49^{efg}	23.04±0.18^{b}	23.84±0.06^{cd}
	50	23.53±0.71^{bc}	24.03±0.18^{cde}	24.28±0.01^{def}	23.55±0.75^{bc}	23.80±0.16^{cd}	24.68±0.01^{fg}
Fat	20		5.77±0.09^{e}	6.42±0.07^{i}	6.16±0.30^{h}	6.18±0.11^{h}	5.93±0.03^{f}
	30	/	5.96±0.02^{fg}	5.97±0.00^{fg}	5.60±0.06^{cd}	6.04±0.11^{fg}	5.98±0.10^{fg}
	40		5.99±0.19^{fg}	5.92±0.00^{f}	5.46±0.02^{ab}	6.06±0.00^{gh}	5.63±0.04^{cd}
	50	5.69±0.01^{de}	5.55±0.01^{bc}	5.73±0.01^{de}	5.62±0.20^{cd}	5.38±0.16^{a}	5.67±0.02^{cde}
Ash	20		3.56±0.12^{defg}	3.46±0.07^{bcde}	3.55±0.20^{cdef}	3.47±0.11^{bcde}	3.71±0.26^{g}
	30	/	3.63±0.11^{fg}	3.31±0.01^{a}	3.47±0.09^{bcde}	3.57±0.05^{defg}	3.59±0.16^{efg}
	40		3.41±0.18^{abc}	3.51±0.15^{cdef}	3.34±0.10^{ab}	3.44±0.01^{abcd}	3.61±0.11^{efg}
	50	3.43±0.01^{abcd}	3.49±0.01^{bcdef}	3.47±0.17^{bdce}	3.60±0.19^{efg}	3.40±0.14^{abc}	3.52±0.08^{cdef}

Data represent averages of three independent trials± standard deviation.

[a-l] For the same parameter means within the same column and line inscribed with different superscript letters are significantly different ($p < 0.05$).

Table 2. Effect of starch types and starch incorporation level on physicochemical properties of cooked patties (%FM)

	Rates	$P_{Control}$	PSi	PT	PV	PQ	PS
Moisture	20		67.26±0.50^{hi}	67.15±0.21^{ghi}	66.43±0.30^{defgh}	67.56±0.21^{i}	66.01±1.82^{bcdef}
	30	/	66.83±0.16^{fghi}	66.86±0.60^{fghi}	66.42±0.26^{defgh}	66.48±0.41^{defgh}	65.97±0.67^{bcde}
	40		66.32±0.38^{cdef}	66.72±0.70^{efghi}	65.89±0.26^{abcde}	66.22±0.40^{bcdef}	65.79±0.61^{abcd}
	50	65.73±0.47^{abcd}	65.39±1.01^{ab}	65.84±0.25^{abcd}	65.65±0.36^{abcd}	65.53±0.45^{abc}	65.10±0.54^{a}
Proteins (N x 6.25)	20		24.16±0.90^{abc}	25.29±0.36^{def}	28.34±0.48^{g}	23.87±1.02^{ab}	25.05±0.43^{cdef}
	30	/	23.83±0.30^{ab}	27.74±0.71^{g}	25.57±0.44^{ef}	23.92±1.47^{ab}	27.32±2.33^{g}
	40		25.74±0.06^{f}	25.77±1.32^{f}	25.65±0.22^{ef}	23.14±0.10^{a}	23.97±1.37^{ab}
	50	23.80±0.16^{ab}	24.32±0.37^{bcd}	24.61±1.43^{bcde}	24.39±0.35^{bcd}	25.37±0.35^{def}	24.80±0.51^{bcdef}
Fat	20		6.42±0.91^{a}	6.15±0.68^{a}	6.41±0.38^{a}	6.36±0.43^{a}	6.05±0.54^{a}
	30	/	6.30±0.97^{a}	6.43±0.69^{a}	5.61±1.08^{a}	6.16±0.69^{a}	6.26±0.75^{a}
	40		6.44±0.70^{a}	5.94±0.59^{a}	6.62±0.60^{a}	5.86±0.94^{a}	6.18±0.90^{a}
	50	6.07±0.81^{a}	6.11±1.10^{a}	6.09±0.76^{a}	6.37±0.67^{a}	6.56±1.03^{a}	6.23±0.68^{a}
Ash	20		3.55±0.05^{bc}	3.51±0.15^{abc}	3.68±0.10^{c}	3.47±0.12^{abc}	3.54±0.38^{bc}
	30	/	3.54±0.07^{bc}	3.49±0.10^{abc}	3.50±0.23^{abc}	3.38±0.19^{ab}	3.54±0.26^{bc}
	40		3.43±0.05^{ab}	3.33±0.11^{ab}	3.47±0.20^{abc}	3.41±0.25^{ab}	3.42±0.17^{ab}
	50	3.42±0.06^{ab}	3.38±0.07^{ab}	3.31±0.16^{a}	3.47±0.20^{abc}	3.34±0.15^{ab}	3.41±0.13^{ab}

Data represent averages of three independent trials ± standard deviation.

[a-i] For the same parameter means within the same column and line inscribed with different superscript letters are significantly different ($p < 0.05$).

The moisture contents ranged between 65.54±0.61 and 67.88±1.22 % for raw patties and from 65.10±0.54 to 67.56±0.21 % for cooked ones. The results of variance analysis showed that moisture varied according to starch type and starch incorporation level used for both raw and cooked patties. This was expected and can be attributed to the differences in the physicochemical composition of the starches used and most especially to their moisture content.Water content of control ($P_{Control}$) raw patty was not significantly different (p>0.05) to those of the raw patties *Psi*, *PT* and *PV* at the same level of incorporation (50g/kg of batter), the same result was obtained in raw patties *PQ* and *PS* with only 30g of starch per kg of batter. After cooking the same trend was not observed. Moisture content of all the patties was not significantly different (p>0.05) from that of control and up till 20g/kg of batter, moistures content of *PSi*, *PT*, and PQ was significant (p<0.05) than that the control.Because of the moisture loss during cooking, water content of cooked patties was lower than that of their raw counter parts. For the same reason protein and fat percentage was higher in cooked patties than in raw ones. Results clearly showed that cooking increased the fat content on a percentage basis, in all formulations, these results are in good agreement with those obtained by Hoelscher *et al.* [26], Berry [27] and Troutt *et al.* [28]. Tornberg *et al.* [29] concluded that the dense meat protein matrix of low fat ground beef prevented fat migration. As for moisture content, proteins and fats content varied according to starch type and starch rate used for both raw and cooked patties.

Starch type and its rates of incorporation statistically influence (p<0.05) ash content of raw patties while after cooking, neither starch type nor it incorporation rate significantly influence ash content of patties. This observation is linked to low ash content of the starch used in this study (0,11±0,01 to 0,33±0,02%) [17] but also to the drainage of soluble minerals resulting from cooking losses and punching out with molds.

3.2. pH, Water Holding Capacity, Cooking Loss and TBA Index of Patties

pH values of cooked and uncooked patties were not significantly different (P>0.05) among treatments (Table 3).

Table 3. pH and Water Retention Capacity (WRC) of patties

	Rates	$P_{Control}$	PSi	PT	PV	PQ	PS
pH Raw	20		5.79±0.06^{a}	5.80±0.07^{a}	5.78±0.05^{a}	5.78±0.08^{a}	5.74±0.12^{a}
	30	/	5.79±0.10^{a}	5.80±0.07^{a}	5.81±0.08^{a}	5.76±0.12^{a}	5.79±0.10^{a}
	40		5.78±0.07^{a}	5.79±0.09^{a}	5.79±0.09^{a}	5.80±0.10^{a}	5.76±0.14^{a}
	50	5.72±0.09^{a}	5.80±0.08^{a}	5.76±0.12^{a}	5.74±0.15^{a}	5.79±0.08^{a}	5.77±0.08^{a}
pH Cooked	20		6.31±0.04^{a}	6.27±0.05^{a}	6.29±0.04^{a}	6.28±0.05^{a}	6.29±0.04^{a}
	30	/	6.30±0.05^{a}	6.30±0.05^{a}	6.29±0.04^{a}	6.29±0.04^{a}	6.31±0.04^{a}
	40		6.30±0.02^{a}	6.31±0.05^{a}	6.30±0.04^{a}	6.31±0.05^{a}	6.29±0.05^{a}
	50	6.29±0.06^{a}	6.28±0.06^{a}	6.27±0.03^{a}	6.31±0.04^{a}	6.30±0.04^{a}	6.28±0.04^{a}
WHC	20		77.72±0.38^{bcd}	76.47±0.98^{ab}	78.96±1.91^{cd}	77.90±0.60^{bcd}	78.95±0.09^{de}
	30	/	78.80±0.65^{de}	76.86±0.38^{bc}	77.93±0.90^{bcd}	79.59±1.86^{e}	79.05±1.91^{de}
	40		80.03±1.25^{ef}	77.04±0.76^{bc}	76.77±1.19^{bc}	78.90±0.90^{de}	76.98±1.88^{bc}
	50	76.78±1.95^{bc}	81.50±0.92^{f}	77.96±1.85^{cd}	75.20±0.58^{a}	78.11±1.91^{cd}	76.83±0.42^{bc}

Data represent averages of three independent repeats ± standard deviation.

$^{a-f}$ For the same parameter means within the same column and line effect with different superscript letters are significantly different (p < 0.05).

These results were similar to those obtained by Troutt *et al.* [28]. The pH values increased upon heating, similar results were observed on beef patties formulated with various animal fats and essential oils [30]. This increase of pH values could be related to the breaking of sulfur or imidazole linking of amino acids content in meat during cooking [31].

Starch rate significantly (P<0.05) affected the WHC of raw patties. The same trend was observed for some starch type at the same percentage of incorporation, this can be due the differences observed in functional properties of these starches [17] and their interaction with other constituent of the patties [32].

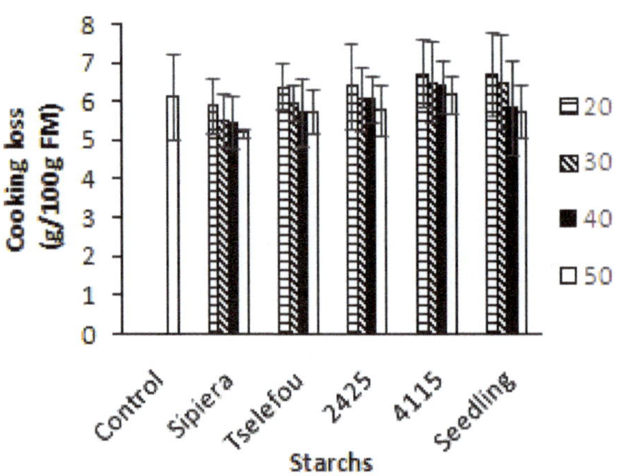

Figure 1. Influences of starch type and starch rate on Cooking loss of cooked patties

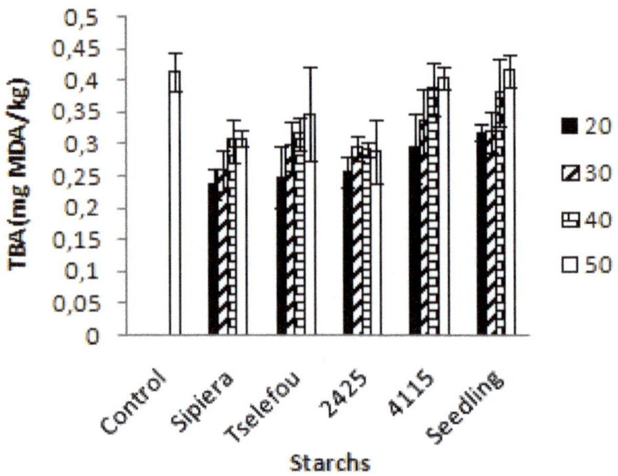

Figure 2. Influences of starch type and starch levels on Thiobartituric acid values of cooked patties

Cooking losses of patties (Figure 1) increased, but not significantly (P>0,05) with the percentage of starch incorporation. Cooking yield was improved (P<0.05) at the lowest starch incorporation (20%) irrespective of the starch type.

The results of the influence of starch type and incorporation rate on lipid oxidation of patties (Figure 2) are consistent with those of cooking losses Irrespective of the starch type the best yield is obtained with the lowest starch incorporation (20%), since TBA index increased (P<0,05) with starch incorporation except for starch *2425*. The Presence of minerals such as iron (pro-oxidant) in starch can explain the increase in lipids oxidation of patties [33].

3.3. Texture Attributes of Cooked Patties

Table 4 shows that the mean values for textural properties: hardness, cohesiveness, springiness and chewiness of patties were in the range 7.62±0.17 to 22.76±0.88 (N), 0.67±0.02 to 0.79±0.05, 0.57±0.00 to 0.97±0.02 (mm) and 3.43±0.26 to 16.68±0.16 (Nxmm), respectively. Starch incorporation rate increased (P<0.05) sample hardness and chewiness value. The increasing hardness might have been due to the reduction of moisture content of patties with increasing starch incorporation percentage (Table 1). These data were similar to the results of Ziegler *et al.* [34] who tested several types of dried and non dried sausages and reported that hardness decreased with moisture. Claus *et al.* [35] suggested that at higher water levels, the muscle proteins interact with the water rather than form cross-bridges that would increase hardness of beef/pork bologna. These results are also in good agreement with those reported by Carballo *et al.* [36] who indicated that the presence of starch had a significant increase in the hardness of bologna sausage. At the same percentage incorporation the starch types also influenced (p<0.05) hardness and chewiness. Cohesiveness and Springiness were significantly influenced (P<0.05) by starch type and the incorporated rate. For all texture attributes, values not significantly (P>0.05) different from control were obtained at some starch incorporation levels except for starch *2425* for Hardness, starches *Sipiera* and *Seedling* for Springiness and starches *Sipiera, Tselefou, 2425* and *4115* for Chewiness. It is evident from the Principal components analysis of texture attributes of cooked patties (Figure 3) that the patty prepared with Seedling starch at incorporation rate of 40g/kg of ground meat (*PS40*) is the most similar to the control patty.

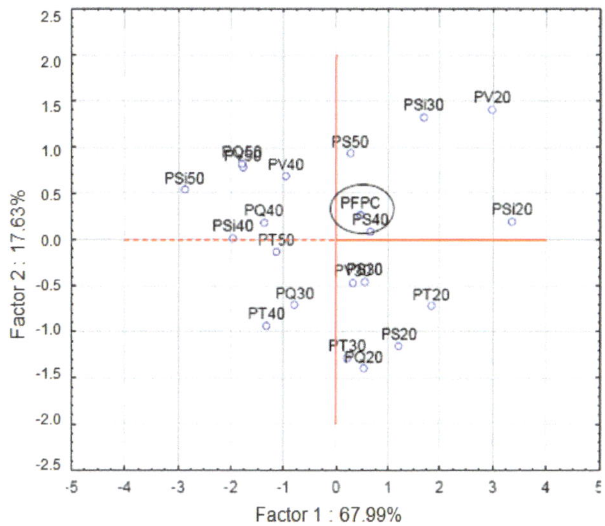

* $PFPC = P_{Control}$

Figure 3. Principal components analysis of texture attributes of cooked patties

Table 4. Texture attributes of cooked patties

	Rates	$P_{Control}$	PSi	PT	PV	PQ	PS
Hardness (N)	20		7.62 ± 0.17^{a}	9.89 ± 0.25^{b}	11.92 ± 0.20^{bcd}	10.94 ± 0.02^{bc}	11.48 ± 0.33^{bcd}
	30	/	14.90 ± 0.06^{e}	11.30 ± 0.40^{bc}	12.14 ± 0.14^{cd}	15.78 ± 0.51^{e}	13.62 ± 0.74^{de}
	40		19.02 ± 0.03^{fg}	15.44 ± 0.79^{e}	19.19 ± 1.12^{fg}	18.39 ± 0.35^{f}	15.15 ± 0.75^{e}
	50	15.54 ± 0.88^{e}	22.76 ± 0.88^{h}	18.19 ± 0.11^{f}	21.45 ± 0.16^{h}	20.71 ± 0.39^{gh}	18.03 ± 1.47^{f}
Cohesive-ness	20		0.69 ± 0.04^{ab}	0.70 ± 0.02^{abcd}	0.68 ± 0.01^{a}	0.74 ± 0.03^{cdefh}	0.67 ± 0.02^{a}
	30	/	0.71 ± 0.02^{abcde}	0.77 ± 0.01^{fghi}	0.79 ± 0.05^{i}	0.74 ± 0.02^{cdefg}	0.72 ± 0.01^{abcde}
	40		0.78 ± 0.03^{hi}	0.78 ± 0.00^{hi}	0.75 ± 0.01^{a}	0.77 ± 0.01^{ghi}	0.69 ± 0.06^{ab}
	50	0.71 ± 0.03^{abcde}	0.77 ± 0.04^{fghi}	0.73 ± 0.01^{bcdef}	0.75 ± 0.02^{efgh}	0.78 ± 0.01^{ghi}	0.70 ± 0.03^{abc}
Springi-ness (mm)	20		0.65 ± 0.00^{b}	0.84 ± 0.02^{de}	0.57 ± 0.00^{a}	0.94 ± 0.00^{i}	0.97 ± 0.02^{j}
	30	/	0.65 ± 0.01^{b}	0.92 ± 0.01^{h}	0.79 ± 0.06^{c}	0.97 ± 0.02^{j}	0.89 ± 0.03^{g}
	40		0.92 ± 0.00^{h}	0.97 ± 0.00^{i}	0.84 ± 0.01^{de}	0.88 ± 0.01^{fg}	0.86 ± 0.01^{ef}
	50	0.83 ± 0.03^{d}	0.95 ± 0.00^{ij}	0.96 ± 0.00^{ij}	0.89 ± 0.02^{g}	0.84 ± 0.02^{d}	0.80 ± 0.01^{c}
Chewi-ness (Nxmm)	20		3.43 ± 0.26^{a}	5.84 ± 0.43^{c}	4.68 ± 0.66^{b}	7.67 ± 0.34^{e}	7.49 ± 0.13^{de}
	30	/	6.83 ± 0.12^{d}	7.97 ± 0.33^{e}	7.64 ± 0.14^{e}	11.35 ± 0.89^{h}	8.68 ± 0.87^{f}
	40		13.72 ± 0.66^{kl}	11.73 ± 0.47^{hi}	12.05 ± 0.99^{ij}	12.52 ± 0.49^{j}	9.03 ± 0.23^{f}
	50	9.18 ± 0.57^{f}	16.68 ± 0.16^{m}	12.68 ± 0.17^{j}	14.23 ± 0.85^{l}	13.55 ± 0.48^{k}	10.10 ± 1.10^{g}

Data represent averages of three independent repeats ± standard deviation.

[a-f] For the same parameter means within the same column and line effect with different superscript letters are significantly different (p < 0.05).

3.4. CIE Lab Color Attributes of Cooked Patties

Table 5 shows mean color attributes values of patties. Lower starch incorporation (2%) led to significantly higher lightness (L*) for patties formulated with potato starches (*PSi* and *PT*) while the contrary is observed for patties prepared with cassava starches (*PQ* and *PS*), except for *PV*, a patty formulated with cassava starch that was not significantly influenced by starch incorporation. These differences in lightness of the cooked patties are probably due to differences of the lightness of starch as observed by Mbougueng *et al.* [17] and their interaction with the constituents of patties. At the same incorporation level, starch type significantly (P<0,05) influenced lightness of patties.

Redness (a*) and yellowness (b*) values were between 11.14±0.08 and 13.81±0.78 and between 14.35±0.06 and 17.67±0.14 respectively. Meat redness was due to the concentration of myoglobin which contributes to the darker color. The redness of patties decreases (p<0,05) with an increase in the incorporation of potato starches, while a reverse trend is observed for patties formulated with cassava starches (*PV, PQ* and *PS*).In the first case, the difference in colour of the patties can be attributed to the dilution of meat myoglobin up to some extent to the colour of the starches. Yellowness values of beef patties were also significantly affected (*P*<0.05) by starch type and rate of incorporation.

Table 5. The CIE Lab color attributes of patties

	Rates	$P_{Control}$	PSi	PT	PV	PQ	PS
L*	20		$51.40±1.12^{ghi}$	$53.30±0.61^{j}$	$50.60±1.52^{defgh}$	$49.65±1.26^{bcd}$	$49.57±0.34^{bcd}$
	30	/	$49.60±1.53^{bcd}$	$51.13±0.60^{fghi}$	$50.05±1.79^{bcdef}$	$49.14±0.15^{abc}$	$51.75±0.39^{i}$
	40		$49.94±0.37^{bcde}$	$50.68±0.48^{defghi}$	$50.35±0.43^{defg}$	$50.21±0.47^{cdef}$	$51.65±0.69^{hi}$
	50	$50.17±0.45^{cdef}$	$48.23±0.39^{a}$	$48.99±0.25^{ab}$	$50.86±0.37^{efghi}$	$51.36±1.87^{ghi}$	$50.44±0.17^{defg}$
a*	20		$13.24±0.56^{ij}$	$13.01±0.92^{fghi}$	$12.01±0.96^{bcde}$	$11.64±0.13^{abc}$	$11.55±0.50^{abc}$
	30	/	$13.15±0.19^{hi}$	$12.10±0.60^{cde}$	$11.89±0.54^{bcde}$	$12.18±0.02^{cde}$	$12.46±0.62^{defg}$
	40		$12.36±0.27^{def}$	$12.07±0.59^{cde}$	$11.40±0.24^{ab}$	$12.16±0.13^{cde}$	$12.51±0.03^{efgh}$
	50	$13.81±0.78^{j}$	$11.82±0.14^{bcd}$	$12.00±0.59^{bcde}$	$11.14±0.08^{a}$	$12.45±0.06^{defg}$	$13.05±0.09^{ghi}$
b*	20		$15.23±0.84^{b}$	$16.56±0.86^{cde}$	$17.32±0.38^{fgh}$	$16.89±0.45^{def}$	$16.07±0.25^{c}$
	30	/	$15.45±0.48^{b}$	$16.99±0.50^{efg}$	$16.32±0.16^{c}$	$17.29±0.11^{fgh}$	$15.47±0.31^{b}$
	40		$16.28±0.56^{c}$	$17.49±0.02^{gh}$	$16.59±0.07^{cde}$	$16.86±0.66^{def}$	$14.54±0.18^{a}$
	50	$15.40±0.29^{b}$	$17.67±0.14^{h}$	$17.47±0.03^{gh}$	$16.46±0.01^{cd}$	$16.93±0.32^{def}$	$14.35±0.06^{a}$

Data represent averages of three independent repeats ± standard deviation.
[a-f] For the same parameter means within the same column and line effect with different superscript letters are significantly different (p < 0.05).

4. Conclusion

The results of variance analysis showed that moisture content varied according to starch type and starch incorporation level for both raw and cooked patties. Cooking yield was improved (P<0.05) by the lowest rate of starch incorporation (20%) irrespective of starch type. Thus, as far as texture attributes of cooked patties are concerned cassava starch (Seedling) can be successfully used to control binding properties of beef patties. Local starches appear to have potential as an extender in finely ground meat products.

References

[1] McWatters, K.H., Functional characteristics of cowpea flours in feeds. *Journal of American Oil Chemists Society* 67: 272-276, 1990.

[2] Baranowska, H. M., Rezler, R., Poliszko, S., Dolata, W., Piotrowska E. & Piatek, M., Starch as a functional addition in meat batters. Starch. From starch containing sources to isolation of starches and their applications. New York: Nova Science Publishers, Inc., pp. 115-123, 2004.

[3] Ray, F. K., Parett, N. A., Van Stavern, B. D. & Ockerman, H. W., Effect of soya level and storage time on the quality characteristics of ground beef patties. J Food Sci 46: 1662-1664, 1981.

[4] Miles, C. W., Ziyad, J., Badwell, C. E. & Steele P. D., True and apparent retention of nutrients in hamburger patties made from beef or beef extended with different soy proteins. J Food Sci 49: 1167-1170, 1984.

[5] Tenin, D., Scher, J. & Hardy, J., Common bean flour as an extender in beef sausages. J Food Eng 52: 143-147, 2002.

[6] Abu Bakr, T. M., Shakib, L. A. & El Iragi, S. M., Mohammed MS, Upgrading and utilization of byproducts of slaughter houses. I. Fresh & canned sausage incorporating legume extenders in their meat emulsions. Alexandria Science Exchange, 7: 319-329, 1986.

[7] Rao, K. H., Singh, R. R. B., Anjaneyulu, A. S. R. & Rao, K. V. S. S. & Yadav, P. L., Effects of caseinases and refined wheat flour on the quality of chicken nuggets from spent hens. Indian J Anim Sci 67: 1004-1006, 1997.

[8] Rossi, M., Textured sunflower protein for use as meat extender. Lebensmittel Wissenschaft &Technologie 21: 267-270, 1988.

[9] Prinyawiwatkul, W., McWatters, K. H., Beuchat, L. R. & Phillips, R. D., Optimizing acceptability of chicken nuggets containing fermented cow pea and peanut flours. J Food Sci 62: 889-894, 1997a.

[10] Prinyawiwatkul, W., McWatters, K. H., Beuchat, L. R. & Phillips. R. D., Physicochemical and sensory properties of chicken nuggets extended with fermented cowpea and peanut flours. J Agri Food Chem 45: 1891-1899, 1997b.

[11] Bawa, A. S., Functional property measurement of mixtures of extender and filler sources of meat and beef. J Food Sci Tech. 22: 294-295, 1985.

[12] Correia, L. R. & Mittal, G. S., Kinetics of pH and color of meat emulsions containing various fillers during smoke housecooking. Meat Sci 29: 353-364, 1991.

[13] Mittal, G. S. & Usborne, W. R., Meat emulsion extenders. Food Technol 39: 121-130, 1985.

[14] Pietrasik, Z., Effect of content of protein, fat and modified starch on binding textural characteristics, and colour of comminuted scalded sausages. Meat Sci 51: 17-25, 1999.

[15] Hughes, E., Mullen, A. M. & Troy, D. J., Effects of fat level, tapioca starch and whey protein on frankfurters formulated with 5% and 12% fat. *Meat Sci* 48: 169-180, 1998.

[16] Wurzburg, O. B., Modified starches. In A. M. Stephen (Ed.), *Food polysaccharides and their applications,* New York: Marcel Dekker Inc. pp. 67-97, 1995.

[17] Mbougueng, P. D., Tenin, D., Tchiégang, C. & Scher, J., Physicochemical and Functional Properties of Cultivars of Irish Potato and Cassava Starches. Int J Food Eng 5:1-17, 2009.

[18] Mbougueng, P. D., Tenin, D., Tchiegang, C. & Scher, J., Effects of local spices on physico-chemical, microbiological and consumer acceptability of beef patties. J Food Technol 4: 16-21, 2006.

[19] Alves, R. M. L., Grossmann, M. V. E. & Silva, R. S. S. F., Gelling properties of extruded yam (*Dioscorea alata*) starch. Food Chem 67: 123-127, 1999.

[20] AOAC, Official Methods of Analysis of the Association of Official Analytical Chemists, In: Horwitz, W (Ed.) 15th ed. AOAC, Washington, DC, 1990.

[21] AOAC, Official Methods of Analysis of the Association of Official Analytical Chemists, (Ed) 13th ed. AOAC Washington, DC, 1980.

[22] Tsai TC, Ockerman HL (1981) Water-binding measurement of meat. J Food Sci *46*: 697-701, 707.

[23] Wite, V. C., Krause, G. F. & Bailey, M. E., A new extraction method for determining 2-thiobarbituric acid value of pork and beef during storage. J Food Sci 35: 582-585, 1970.

[24] Huda, N., Abdullah, A. & Babji, A. S., Chemical composition and quality of Malaysian fish balls. Proceedings of the Seminar Bersama ITB-UKM IV, April 12-13, UniversitiKebangsaan Malaysia, Malaysia, pp: 205-211, 2000.

[25] Boume, M. C., Texture profile analysis. Food Technol 37: 62-66, 1978.

[26] Hoelscher, L. M., Savell, J. W., Harris, J. M., Cross, H. R. & Rhee, K. S., Effect of initial fat level and cooking methods on cholesterol content and caloric value of ground beef patties. J Food Sci 52: 883-885, 1997.

[27] Berry, B. W., Low fat level effects on sensory, shear, cooking and chemical properties of ground beef patties. J Food Sci 57: 537-540, 1992.

[28] Troutt, E. S., Hunt, M. C., Johnson, D. E., Claus, J. R., Kastner, C. L. & Kropf, D. H., Characteristics of low-fat ground beef containing texture-modifying ingredients. J Food Sci 57: 19-24, 1992.

[29] Tornberg, E., Olsson, A. & Persson, K., A comparison in fat holding between hamburgers and emulsion sausages. Proceedings International Congress on Meat Science Technology 35: 753-757, 1989.

[30] Dzudie, T., Kouebou, C. P., Essia-ngang, J. J. & Mbofung, C. M. F., Lipid sources and oils effects on quality and stability of beef patties. J Food Eng 65: 67-72, 2004.

[31] Sherman, P., Food texture and rheology; *Academic press*, London, 1979.

[32] Bouton, P. E., Harris, P. V. & Carrol, F. D., The effect of cooking temperature and time on some mechanical properties of meat. *J. Food Sci.*, 37: 140-144, 1972.

[33] Zanardi, E., Ghidini, S., Battaglia, A. & Chizzolini, R., Lipolysis and lipid oxidation in fermented sausages depending on different processing conditionsand different antioxidants. Meat Sci 66: 415-423, 2004.

[34] Ziegler, G. R., Rizvi, S. S. H. & Acton, J. C., Relationship of water content to textural characteristics, water activity and thermal conductivity of some commercial sausages. J Food Sci 52: 901-905, 1987.

[35] Claus, J. R., Hunt, M. C. & Kastner, C. L., Effects of substituting added water for fat on the textural, sensory and processing characteristics of bologna. J Muscle Foods 1: 1-21, 1989.

[36] Carballo, J., Barreto, G. & JimeÂnez Colmenero F., Starch and egg white influence on properties of bologna sausage as related to fat content. J Food Sci 60: 673-677, 1995.

Permissions

All chapters in this book were first published in AJFST, by Science and Education Publishing; hereby published with permission under the Creative Commons Attribution License or equivalent. Every chapter published in this book has been scrutinized by our experts. Their significance has been extensively debated. The topics covered herein carry significant findings which will fuel the growth of the discipline. They may even be implemented as practical applications or may be referred to as a beginning point for another development.

The contributors of this book come from diverse backgrounds, making this book a truly international effort. This book will bring forth new frontiers with its revolutionizing research information and detailed analysis of the nascent developments around the world.

We would like to thank all the contributing authors for lending their expertise to make the book truly unique. They have played a crucial role in the development of this book. Without their invaluable contributions this book wouldn't have been possible. They have made vital efforts to compile up to date information on the varied aspects of this subject to make this book a valuable addition to the collection of many professionals and students.

This book was conceptualized with the vision of imparting up-to-date information and advanced data in this field. To ensure the same, a matchless editorial board was set up. Every individual on the board went through rigorous rounds of assessment to prove their worth. After which they invested a large part of their time researching and compiling the most relevant data for our readers.

The editorial board has been involved in producing this book since its inception. They have spent rigorous hours researching and exploring the diverse topics which have resulted in the successful publishing of this book. They have passed on their knowledge of decades through this book. To expedite this challenging task, the publisher supported the team at every step. A small team of assistant editors was also appointed to further simplify the editing procedure and attain best results for the readers.

Apart from the editorial board, the designing team has also invested a significant amount of their time in understanding the subject and creating the most relevant covers. They scrutinized every image to scout for the most suitable representation of the subject and create an appropriate cover for the book.

The publishing team has been an ardent support to the editorial, designing and production team. Their endless efforts to recruit the best for this project, has resulted in the accomplishment of this book. They are a veteran in the field of academics and their pool of knowledge is as vast as their experience in printing. Their expertise and guidance has proved useful at every step. Their uncompromising quality standards have made this book an exceptional effort. Their encouragement from time to time has been an inspiration for everyone.

The publisher and the editorial board hope that this book will prove to be a valuable piece of knowledge for researchers, students, practitioners and scholars across the globe.

List of Contributors

Abdeldaiem M. H.
Atomic Energy Authority, Nuclear Research Center, Egypt

Jehad Abbadi
Department of Biology, College of Science and Technology, Al-Quds University, Abu Dies, Jerusalem, Palestine

Ibrahim Afaneh and Ziad Ayyad
Department of Food Technology, College of Science and Technology, Al-Quds University, Abu Dies, Jerusalem, Palestine

Fuad Al-Rimawi, Wadie Sultan and Khalid Kanaan
Department of Chemistry and Chemical Technology, College of Science and Technology, Al-Quds University, Abu Dies, Jerusalem, Palestine

Fathima Hajara Aslam
Department of Food Science, Spectrum Institute of Science and Technology, Colombo, Sri-Lanka

Jagath Wansapala
Department of Food Science and Technology, Faculty of Applied Sciences, University of Sri-Jayewardenepura, Gangodawila, Nugegoda, Sri-Lanka

Manel Perera
MicroChem Laboratories, Nugegoda, Sri-Lanka

Abdelrahman R. Ahmed
Faculty of Specific Education, Home Economics Department, Ain Shams University, Cairo, Egypt
Institute for Food Technology and Food Chemistry, Department of Food Rheology, Technical University of Berlin, Sekr, KL-H1, Königin-Luise-Str, Berlin / Germany

Milene Teixeira Barcia, Vivian Caetano Bochi and Helena Teixeira Godoy
Departamento de Ciência dos Alimentos, Universidade Estadual de Campinas, Campinas, Brazil

Paula Becker Pertuzatti
Instituto de Ciências Exatas e da Terra, Engenharia de Alimentos, Universidade Federal de Mato Grosso, Barra do Garças, Brazil

Isidro Hermosín-Gutiérrez
Instituto Regional de Investigación Científica Aplicada, Universidad de Castilla-La Mancha, Ciudad Real, Spain

Paula Becker Pertuzatti and Mariana Bento da Cruz
Universidade Federal de Mato Grosso, Instituto de Ciências Exatas e da Terra, Curso de Engenharia de Alimentos, Barra do Garças, MT, Brazil

Denise Soares Dorneles de Oliveira, Kelly Cristina Campagnolo Port, Larissa Alves Vieira and Gabrielle Lázara Ribeiro Rodrigues de Barros Vinhal
Instituto Federal de Educação, Ciência e Tecnologia de Mato Grosso, Curso Técnico em Alimentos, Barra do Garças, MT, Brazil

Fana Haile
Arbamich University

Shimelis Admassu
Food, Beverage and Pharmaceutical Industry Development Institute, Ministry of Industry, Ethiopia

Abrehet Fisseha
Hawassa University, School of Nutrition, Food Science and Technology, Hawassa, Ethiopia

ChandanaVindika Kumara Lokunarangodage, Indira Wickramasinghe and Kamburawala Kankanamge Don Somathilaka Ranaweera
Department of Food Science and Technology, University of Sri Jayewardenepura, Nugegoda, Sri Lanka

Mustapha S., Mathew T. J. and Ibn-Aliyu A.
Department of Chemistry, Federal University of Technology, Minna, Nigeria

Mohammed U. M.
Department of Chemistry, Federal College of Education, Kontagora, Niger State, Nigeria

Adeosun N. O.
Department of Chemistry, Federal College of Education (Technical), Asaba, Nigeria

Muhammed S. S.
Department of Biochemistry, Federal University of Technology, Minna, Nigeria

Paula Becker Pertuzatti
Instituto de Ciências Exatas e da Terra, Engenharia de Alimentos, Universidade Federal de Mato Grosso, Barra do Garças, Brazil

Milene Teixeira Barcia
Escola de Química e Alimentos, Universidade Federal do Rio Grande, Rio Grande, Brazil

Md. Sultan Mahomud and Tajnuba Sharmin
Department of Food Engineering and Technology, Hajee Mohammad Danesh Science and Technology University, Dinajpur, Bangladesh

Md. Keramot Ali
Department of Food Technology and Rural Industries, Bangladesh Agricultural University, Mymensingh, Bangladesh

Md. Mostafijur Rahman and Md. Hafijur Rahman
Faculty of Agriculture, Hajee Mohammad Danesh Science and Technology University, Dinajpur, Bangladesh

Md. Jiaur Rahman
Department of Food Processing and Preservation, Hajee Mohammad Danesh Science and Technology University, Dinajpur, Bangladesh
Department of Biochemistry (MSc Student), Memorial University of Newfound, St. John's, NL, Canada

Armand Abdou Bouba and Joël Scher
Laboratory of Engineering and Biomolecules (LiBio), Nancy-University, Nancy, France

Roger Ponka and Goudoum Augustin
Department of Agriculture, Livestock and Derivate Products, University of Maroua, Maroua, Cameroon

Nicolas Njintang Yanou
Department of Biological Sciences, Faculty of Science, University of Ngaoundere, Ngaoundere, Cameroon

Mehanni Abul-Hamd El-Sayed
Department of Food Science and Nutrition, Faculty of Agriculture, Sohag University, Sohag, Egypt

Didier Montet
CIRAD, Montpellier, France

Carl Moses Mbofung
6Department of Food Science and Nutrition, University of Bamenda, Bamenda, Cameroon

Serah O. Alagbaoso, Justina N. Nwosu, Ijeoma .M. Agunwa, Njideka E. Njoku, Linda O. Akajiaku and Chioma N. Eluchie
Department of Food Science and Technology, Federal University of Technology, P.M.B 1526, Owerri, Imo State, Nigeria

Eluchie C. N. and Alagbaoso, S. O.
Department of food Science and Technology, Federal University of Technology P.M.B 1526, Owerri Imo State

Oranusi S. and Akujobi C. O.
Department of Industrial Microbiology, Federal University of Technology P.M.B 1526, Owerri Imo State

Anne I. Peter-Ikechukwu, Ngozi O. Kabuo, Serah O. Alagbaoso, Njideka E. Njoku, Chioma N.Eluchie. and Williams O. Momoh
Department of Food Science and Technology, Federal University of Technology, Owerri, Owerri, Imo State, Nigeria

Gouri Satpathy and Yogesh Kumar Tyagi
University School of Basic and Applied Sciences, GGS Indraprastha University, Delhi, India

Rajinder Kumar Gupta
University School of Biotechnology, GGS Indraprastha University, Delhi, India

Vishal. R. Parate, Mohammed. I. Talib
Department of Food Technology, UICT, North Maharashtra University, Jalgaon, 425001, India

Ngozi O. Kabuo, Moses Ojukwu, Gloria C Omeire and J. C Ibeabuchi
Department of Food Science and Technology, Federal University of Technology, Owerri, Imo State, Nigeria

Goudoum Augustin, Makambeu Ndomche Anne and Abdou Bouba Armand
The University of Maroua, The Higher Institute of Sahel, Department of Agriculture, Livestock and Derived Products, Maroua, Cameroon

Mbofung Carl Moses
The University of Bamenda, School of Technology, Department of Food and Bioresource Technology, Bambili, Cameroon

Srivastava A. K
Faculty, Mechanical Engineering Department, SRMGPC, Lucknow U.P, India

Shukla S. K
Centre For Energy Resource and Development (CERD), Mechanical Engineering Department, Indian Institute of Technology (Banaras Hindu University) Varanasi, India

Singh U. K
H.O.D Of Mechanical Engineering Department, K.N.I.T. Sultanpur

Radha C. and Hebina Babu K. T.
Department of Protein Chemistry & Technology, Central Food Technological Research Institute, Mysore, India

Ogunsina B. S.
Department of Agricultural & Environmental Engineering, Obafemi Awolowo University, Ile Ife, Nigeria

C. Martínez-Vera and M. G. Vizcarra-Mendoza
Departamento de Ingeniería de Procesos e Hidráulica, Universidad Autónoma Metropolitana-Iztapalapa. Apartado Postal 55-534, México D.F., 09340, México

I. Anaya-Sosa
Escuela Nacional de Ciencias Biológicas, Instituto Politécnico Nacional. Unidad Profesional Lázaro Cárdenas, Prolongación de Carpio y Plan de Ayala s/n, Col. Santo Tomas C.P. 11340. Del. Miguel Hidalgo. México, D.F

Marcio Schmiele, Georgia Ane Raquel Sehn, Valéria da Silva Santos, Yoon Kil Chang and Caroline Joy Steel
Department of Food Technology, School of Food Engineering, University of Campinas, Campinas, Brazil

Thaís de Souza Rocha
Department of Food Science and Technology, State University of Londrina, Londrina, Brazil

Eveline Lopes Almeida
Department of Biochemical Engineering, School of Chemistry, Federal University of Rio de Janeiro, Rio de Janeiro, Brazil

Elizabeth Harumi Nabeshima
Institute of Food Technology, Campinas, Brazil

Mohagir A. M.
Faculty of Pure and Applied Science, University of N'djamena, Chad

Bup N. D
Higher Institute of the Sahel, University of Maroua, Cameroon

Abi C. F.
Department of Chemistry, Higher Teachers' Training College, University of Yaounde1, Cameroon

Kamga R., Kapseu C.
National School of Agro-Industrial Science, University of Ngaoundere, Cameroon

M. T. Hasan, M. S. Rahman, M. N. Miah and K. U. Ahmed
Department of Biochemistry, Sher-e-Bangla Agricultural University, Dhaka, Bangladesh

M. H. A. Khatun
Department of Food Technology & Rural Industries, Bangladesh Agricultural University, Mymensingh, Bangladesh

M. A. M. Sajib
Department of Food Technology, Engineering and Nutrition, Lund University, SE Lund, Sweden

M. M. Rahman M. Roy
Department of Food Engineering and Tea Technology, Shahjalal University of Science and Technology, Sylhet, Bangladesh

Theophilus Maduabuchukwu Ikegwu
Department of Food science and Technology, Faculty of Applied Science and Technology, Ibrahim BadamasiBabangida University, Lapai, Niger State, Nigeria

Gabriel Ifeanyi Okafor
Department of Food Science and Technology, Faculty of Agriculture, University of Nigeria, Nsukka, Enugu State, Nigeria

Izuchukwu Shedrach Ochiogu
Department of Veterinary Obstetrics and Reproductive Diseases, Faculty of Veterinary Medicine, University of Nigeria, Nsukka, Enugu State, Nigeria

Etsuyankpa M. B.
Centre for Preliminary and Extra-mural Studies, Federal University of Technology, Minna, Niger State, Nigeria

Gimba C. E., Agbaji E. B. and Omoniyi K. I.
Chemistry Department, Ahmadu Bello University, Zaria, Kaduna State, Nigeria

M.M. Ndamitso and Mathew J. T.
Chemistry Department, Federal University of Technology, Minna, Niger State, Nigeria

Njoku N. E., Ubbaonu C. N., Alagbaoso S.O., Agunwa I.M and Eluchie C.N.
Department of Food Science and Technology, Federal University of Technology, Owerri, ImoState, Nigeria

Sanni L. A., Ogunsina B. S. and Ajayi M. O.
Department of Agricultural and Environmental Engineering, Obafemi Awolowo University, Nigeria

Adegbenjo A. O.
Department of Bioresource Engineering, McGill University Montréal, Quebec, Canada

Sandip I. Vidhani, Vijay G. Vyas, Heena J. Parmar, Viren M. Bhalani and Baljibhai A. Golakiya
Elemental Cell, Food Testing Laboratory, Department of Biotechnology, Junagadh Agricultural University, Junagadh. India

Mohammad M. Hassan and Ahmed Gaber
Scientific Research Center, Biotechnology and Genetic engineering Unit, Taif University, KSA

Noël Tenyang,
University of Maroua, Faculty of Science, Department of Biological Sciences, Maroua, Cameroon
University of Dschang, Faculty of Science, Department of Biochemistry, Dschang, Cameroon

Roger Ponka
University of Maroua, The Higher Institute of The Sahel, Department of Agriculture, Livestock and By-Products, Maroua, Cameroon

Bernard Tiencheu
University of Buea, Faculty of Science, Department of Biochemistry, Buea, Cameroon

Fabrice Tonfack Djikeng, and Hilaire Macaire Womeni
University of Dschang, Faculty of Science, Department of Biochemistry, Dschang, Cameroon

J A A C Wijesinghe and I. Wicramasinghe
Department of Food Science & Technology, University of Sri Jayewardenepura, Sri Lanka

K. H Saranandha
Food Research Unit, Gannoruwa

P. D. Mbougueng and D. Tenin
Department of Process Engineering, National School of Agro-Industrial Sciences, University of Ngaoundéré, Ngaoundéré, Cameroon

C. Tchiégang
Department of Food Engineering and Quality Control, University Institute of Technology, University of Ngaoundéré, Ngaoundéré, Cameroon

J. Scher
Laboratory of Biomolecular Engineering, Lorraine University, Nancy, France

Index

A

Acidity, 12-14, 17-19, 21, 47-49, 54, 80, 107, 109, 112, 133, 137, 141, 199-200, 203, 213, 226

Afnor, 139, 143, 168-169, 174

Agriculture Organization (FAO), 73

Alanine Amino Transferase (ALT), 184

Amino Acid, 34, 36-38, 40, 54, 85-90, 92-93, 205, 223

Autoclaving, 94-95, 97-99, 133

B

Biological Efficiency, 175, 179

Bulk Density, 51-52, 54-55, 66-67, 70, 125, 225, 228

C

Canavalia Plagiosperma Piper, 94-95, 97-99

Chemical Analysis, 3, 23-24, 33, 48, 80, 109, 139, 143, 193, 205

Chemical Evaluation, 84, 107, 205

Chicken Minced, 1

Citrobacter Freundi, 183

Cocoa Butter Substitute (CBS), 167

Colony Forming Units, 2

Compliance, 12, 58, 61, 113-114

Ctc, 59, 61, 63

D

Dehydrogenase Activity, 101-102, 105-106

Deterioration of Bleachability Index (DOBI), 139

Dhool, 58, 62-64

Diffusion Coefficient Estimation, 155

Dna Barcoding, 58, 64-65

E

Edible Coating, 1, 3-6, 8-9, 11

Emulsifying Capacity (EC), 67

Essential Amino Acid Content (EAA), 88

Ethanolic Extract, 1-6, 8-9, 106, 216

Evoo, 12, 14-18, 21-23

F

Fat Absorption Capacity (FAC), 67, 69

Fermentation, 28, 30-31, 41-44, 56, 61, 72, 74-77, 99-100, 102, 107-110, 112, 134, 138, 181, 191-198

Foam Stability (FS), 67

Foaming Capacity (FC), 67

Food Borne Pathogens, 5-6, 101, 106

Food Safety Management System (FSMS), 60

Free Fatty Acid (FFA), 139

Ft-ir, 224, 226, 228, 232

Functional Properties, 30, 40, 42, 45, 51-52, 54, 56-57, 66-67, 69-71, 92, 150, 153, 166, 206, 211, 224-225, 227-228, 236, 238

G

Gamma Irradiation, 1-6, 8-11

Gas Chromatography (GC), 113

Gc-ms, 113-115, 119-120, 212-214, 216

Gel Strength, 160, 162, 164

Gras, 75, 143, 173-174

I

Initial Adsorption Trial, 126

L

Lactobacillus Bulgaricus, 191, 193

M

Manufacturing, 58-61, 63-66, 122, 138

Method Validation, 113, 115, 121

Minerals, 32, 35, 42, 47, 51, 53, 66-67, 69-70, 78, 85, 150, 176, 182, 184, 192, 203-204, 212, 214, 218-219, 221, 223, 227, 236-237

Minimum Inhibitory Concentration (MIC), 101-102

Minitab, 27, 60, 226

Moisture Diffusivity, 144-149, 155, 158-159

Molecular Structure, 160, 166

Morphology, 102, 108, 122, 160, 162, 226, 231

Myddleton Shifter, 58

N

Nutritional Properties, 19, 32, 39, 66

O

Ocimum Sanctum, 212-217

Oil Quality Indicators, 14

Olea Europaea L, 12

Olive Oil, 12-21, 23-24, 91-92, 106, 112, 130, 202

Optimization, 77, 114-115, 119, 125-126, 128-129, 168, 174

Orthodox Black Tea, 58-64

P

Packaging Materials, 12-17, 19-20, 23-24, 78, 107, 110, 112, 132

Peroxide Value (PV), 9, 19, 141

Phenyl Isothiocyanate (PITC), 86

Physicochemical Parameters, 49, 123, 138-139, 141-142
Pleurotus Djamor, 175-176, 180-181
Polyunsaturated Fatty Acids (PUFA), 218
Proximate Analysis, 67, 182, 200, 219, 225, 234
Proximate Composition, 33, 51-53, 67, 92, 100, 175, 177, 179, 181, 200, 203, 205, 218-219, 235

Q
Quick Easy Cheap Effective Rugged And Safe (QuEChERS), 113

R
Radio Frequency Identification (RFID), 58
Randomized Stratified Sampling, 58-59
Rifd, 64
Rsm- Doehlert Experimental Design, 167

S
Sabouraud Dextrose Agar (SDA), 184
Saccharomyces Cerevisiae, 187, 189, 191
Sensory Evaluation, 3, 9, 11-14, 16-18, 21, 23-24, 34, 38, 47, 49-50, 56, 80-81, 131, 133, 137, 151
Shelf Life, 1-2, 10, 13, 18-19, 23-24, 32, 51-52, 55, 78, 80, 82-84, 131, 135, 140, 152, 183, 192
Stability Indicators, 12
Starch Extraction, 161, 234
Statistical Analysis, 3, 14, 27, 35, 48, 57, 87, 96, 100, 112, 131, 140, 162, 168, 177, 193, 209, 213, 226, 235
Storage Conditions, 12-18, 20, 23, 78, 80, 138
Sugar Assimilation Test, 102
Synsepalumdulcificum, 199-200, 203-205

T
Thermodynamic Parameters, 125
Thiobarbituric Acid Reactive Substances (TBARS), 1, 9
Total Phenolic Compounds, 16, 20
Total Volatile Nitrogen (TVN), 1, 3, 9
Traceability, 58-61, 63-65
Tripheny Tetrazolium Chloride (TTC), 101

U
Ultraviolet Extinction Coefficients, 13, 15, 19
Unsaturated Fatty Acids Usfa, 85

V
Viscosity, 28, 32, 51, 54-55, 66, 150, 152-153, 160, 162, 164-165

W
Water Absorption Capacity (WAC), 67, 69
Who, 3, 28, 34, 37-39, 48-49, 56, 58, 60-61, 64, 66, 85-86, 89-90, 92, 97-99, 143, 159, 188, 190-192, 197, 214, 216, 221, 223, 231, 237

Y
Yam Pounding Machine, 206-207, 210